Drug Transporters
Volume 1: Role and Importance in ADME and Drug Development

RSC Drug Discovery Series

Editor-in-chief
Professor David Thurston, *King's College, London, UK*

Series Editors:
Professor David Rotella, *Montclair State University, USA*
Professor Ana Martinez, *Centro de Investigaciones Biologicas-CSIC, Madrid, Spain*
Dr David Fox, *Vulpine Science and Learning, UK*

Advisor to the Board:
Professor Robin Ganellin, *University College London, UK*

Titles in the Series:
 1: Metabolism, Pharmacokinetics and Toxicity of Functional Groups
 2: Emerging Drugs and Targets for Alzheimer's Disease; Volume 1
 3: Emerging Drugs and Targets for Alzheimer's Disease; Volume 2
 4: Accounts in Drug Discovery
 5: New Frontiers in Chemical Biology
 6: Animal Models for Neurodegenerative Disease
 7: Neurodegeneration
 8: G Protein-Coupled Receptors
 9: Pharmaceutical Process Development
10: Extracellular and Intracellular Signaling
11: New Synthetic Technologies in Medicinal Chemistry
12: New Horizons in Predictive Toxicology
13: Drug Design Strategies: Quantitative Approaches
14: Neglected Diseases and Drug Discovery
15: Biomedical Imaging
16: Pharmaceutical Salts and Cocrystals
17: Polyamine Drug Discovery
18: Proteinases as Drug Targets
19: Kinase Drug Discovery
20: Drug Design Strategies: Computational Techniques and Applications
21: Designing Multi-Target Drugs
22: Nanostructured Biomaterials for Overcoming Biological Barriers
23: Physico-Chemical and Computational Approaches to Drug Discovery
24: Biomarkers for Traumatic Brain Injury
25: Drug Discovery from Natural Products
26: Anti-Inflammatory Drug Discovery
27: New Therapeutic Strategies for Type 2 Diabetes: Small Molecules
28: Drug Discovery for Psychiatric Disorders
29: Organic Chemistry of Drug Degradation
30: Computational Approaches to Nuclear Receptors

How to obtain future titles on publication:
A standing order plan is available for this series. A standing order will bring delivery of each new volume immediately on publication.

For further information please contact:
Book Sales Department, Royal Society of Chemistry, Thomas Graham House, Science Park, Milton Road, Cambridge, CB4 0WF, UK
Telephone: +44 (0)1223 420066, Fax: +44 (0)1223 420247,
Email: booksales@rsc.org
Visit our website at www.rsc.org/books

Drug Transporters
Volume 1: Role and Importance in ADME and Drug Development

Edited by

Glynis Nicholls
Independent Consultant, Wem, Shropshire, UK
Email: glynis.nicholls@gmail.com

Kuresh Youdim
F. Hoffman-La Roche AG, Basel, Switzerland
Email: kuresh.youdim@roche.com

RSC Drug Discovery Series No. 54

Print ISBN: 978-1-78262-069-3
Two volume set print ISBN: 978-1-78262-867-5
PDF eISBN: 978-1-78262-379-3
EPUB eISBN: 978-1-78262-868-2
ISSN: 2041-3203

A catalogue record for this book is available from the British Library

Published by The Royal Society of Chemistry,
Thomas Graham House, Science Park, Milton Road,
Cambridge CB4 0WF, UK

Registered Charity Number 207890

For further information see our web site at www.rsc.org

Printed in the United Kingdom by CPI Group (UK) Ltd, Croydon, CR0 4YY, UK

Preface

The original aim of this book was to produce a useable handbook on drug transporters, one that would be in the offices and laboratories of scientists from every discipline as a source of reference, be it on the fundamental aspects, or some of the newer, emerging disciplines of transporter science. To date, there are no books that specifically address this aspect, concentrating instead on in-depth reviews and discussions of particular topics. We hope that this book, divided into two volumes, will fill the current gap by encompassing a review of the available knowledge, techniques and tools within the transporter sciences as they relate to drug disposition and pharmacokinetics. It brings together the collective knowledge of over 200 years of expertise from a network of scientists from a variety of backgrounds and countries, with a particular focus being given to the role of membrane transporters in drug absorption, distribution, metabolism and elimination (ADME), and their impact on drug safety and drug efficacy. Since the book is intended for both newcomers and established scientists, coverage is given to almost all areas of the field, together with links to databases, references and reviews for the interested reader.

For ease of reference, the book as a whole is divided into four sections, with each section describing the current (2016) situation in the field of drug transporters. Given the extensive nature of the topics chosen, and our wish to include as many subject areas as possible, it has been necessary to divide the book into two separate volumes, with the first volume concentrating primarily on the theory and practice of those drug transporters currently used within the pharmaceutical industry and the second volume outlining some of the emerging areas within the field. However, it is intended that the two volumes should, where possible, be retained together for reference purposes and our use of the term 'book' will generally refer to both volumes. In the first volume, *Drug Transporters: Volume 1 Role and Importance in ADME*

RSC Drug Discovery Series No. 54
Drug Transporters: Volume 1: Role and Importance in ADME and Drug Development
Edited by Glynis Nicholls and Kuresh Youdim
© The Royal Society of Chemistry 2016
Published by the Royal Society of Chemistry, www.rsc.org

and Drug Development, some of the basic concepts are introduced in the initial chapter, with subsequent chapters focusing on the key ADME organs (liver, gastrointestinal tract, kidney, blood–brain barrier and lung) and how the differential expression of a multitude of characterized drug transporters can impact the fate of both endogenous and exogenous compounds within the human body. The following two sections cover preclinical models (*in silico*, *in vitro* and *in vivo*), and modelling approaches used within the pharmaceutical industry and how these can be valuable tools in determining the importance and clinical impact of transporter mediated drug–drug interactions. The current regulatory guidance is also discussed in the context of transporters and their potential clinical impact. The second volume *Drug Transporters: Volume 2: Recent Advances and Emerging Technologies*, is dedicated to our final section on emerging transporter science, introducing some of the newer areas and technologies where research is ongoing but is not necessarily part of routine investigations. Topics range from factors that may impact transporter form and function (regulation of expression, enzyme–transporter interplay and pharmacogenomics) through to more practical approaches to improve our understanding of transporter–mediated interactions (using microfluidics, proteomics, *in vivo* imaging and bioinformatics/cheminformatics). Transporters in other organs and tissues of the body, not covered in *Drug Transporters: Volume 1: Role and Importance in ADME and Drug Development*, are also briefly discussed. This section serves to illustrate not only how the transporter field is progressing in many different areas, but also how our knowledge is still incomplete, with much still to be done.

In completing our original aim, it was clear that a book such as this could only be achieved through collaboration—from the initial ideas and support of the Royal Society of Chemistry and the Drug Metabolism Discussion Group (DMDG), through to the cooperation and efforts of a total of 61 authors who are experts in their fields, drawn from industry, academia, commercial laboratories and regulatory authorities in Europe, the USA and Asia. The involvement and continued commitment to this book from all of our collaborators was outstanding, especially given their ongoing workloads, and we remain indebted to them. Mention should also be made of our steering team of scientists, for their invaluable input into formulating the overall book outline, and to our panel of external reviewers, whose input helped to ensure that the scientific content of the book was accurate and up to date.

Given the many months dedicated to this work by such an extensive network of expert scientists, we sincerely hope that this translates into a book that will be used by many people in the coming years as a valuable source of reference.

Glynis Nicholls
Kuresh Youdim

Acknowledgements

The editors would like to thank the following people for their input into reviewing the chapters of both volumes of this book:

Drug Transporters: Volume 1: Role and Importance in ADME and Drug Development
Members of the steering team and DMDG: Dr Peter Kilford (DMDG Committee/Covance Ltd), Dr Mohammed Ullah (Roche), John Keogh (JPK Consulting), Dr Pradeep Sharma (AstraZeneca), Dr Silke Simon (Roche), Susan Cole (MHRA) and Dr Mohammed Atari (Cyprotex), for their helpful input and independent review of several chapters.

Prof. Bruno Hagenbuch (University of Kansas) and Dr Bruno Stieger (University Hospital, Zurich) for their review of Chapter 7.

Dr Michael Gertz (Roche) for his review of Chapter 9.

Dr Agnès Poirier (Roche) for her review of Chapter 10.

Dr Terry Shepard and Dr David Wright (MHRA), and Dr Eva Gil Berglund (MPA) for their review of Chapter 11.

Drug Transporters: Volume 2: Recent Advances and Emerging Technologies
Prof. Per Artursson and Dr Fabienne Gaugaz (Uppsala University) for their review of Chapter 3.

Dr Pär Nordell and Dr Constanze Hilgendorf (AstraZeneca) for their review of Chapter 4.

Dr Jose Ulloa (Bioxydyn Ltd) for his review of Chapter 6.

Dr Marco Berrera (Roche) for his review of Chapter 7.

RSC Drug Discovery Series No. 54
Drug Transporters: Volume 1: Role and Importance in ADME and Drug Development
Edited by Glynis Nicholls and Kuresh Youdim
© The Royal Society of Chemistry 2016
Published by the Royal Society of Chemistry, www.rsc.org

Abbreviations

ABC	Adenosine triphosphate (ATP)-binding cassette transporter (gene superfamily)
ABCB	ATP-binding cassette gene subfamily B
ABCB1	Gene encoding MDR1 or P-gp
ABCB11	Gene encoding BSEP
ABCC	ATP-binding cassette gene subfamily C
ABCC2	Gene encoding MRP2
ABCG	ATP-binding cassette gene subfamily G
ABCG2	Gene encoding BCRP
ADE	Absorption, distribution and elimination
ADME	Absorption, distribution, metabolism and elimination
ADMET	Absorption, distribution, metabolism, elimination and toxicity
ADR	Adverse drug reaction
AhR	Aryl hydrocarbon receptor
AMP	Adenosine monophosphate
API	Active pharmaceutical ingredient
ARNT	Aryl hydrocarbon nuclear translocator
ASBT	Apical sodium dependent bile acid transporter
ATP	Adenosine triphosphate
AUC	Area under curve
AUMC	Area under the first order moment curve
BA	Bioavailability
BBB	Blood–brain barrier
BCRP	Breast cancer resistance protein (gene *ABCG2*)
BCS	Biopharmaceutics classification system
BCSFB	Blood–cerebrospinal fluid barrier
BDDCS	Biopharmaceutics drug disposition classification system

RSC Drug Discovery Series No. 54
Drug Transporters: Volume 1: Role and Importance in ADME and Drug Development
Edited by Glynis Nicholls and Kuresh Youdim
© The Royal Society of Chemistry 2016
Published by the Royal Society of Chemistry, www.rsc.org

BEI	Biliary excretion index
BLEC	Brain like endothelial cell
BRB	Blood Retinal Barrier
BSEP	Bile salt export pump (gene *ABCB11*)
Caco-2	human colon adenocarcinoma cell line
CAR	Constitutive androstane receptor (NR1I3)
CCK8	Cholecystokinin-8
CDF	5-(and-6)-carboxy-2′,7′-dichloro-fluorescein
CFTR	Cystic fibrosis transmembrane conductance regulator
CHMP	Committee for Medicinal Products for Human Use
CHO	Chinese hamster ovary (cell line)
ciPTEC	Conditionally immortalized proximal tubule epithelial cells
CIS-RR	Clash-detection guided iterative search with rotamer relaxation
CL or Cl	Clearance
CL_{ae}	Apical efflux clearance
CL_{bile}	Biliary clearance
CL_d	Diffusional clearance
CL_{efflux}	Active basolateral efflux clearance
CL_i/CL_o	Diffusional clearance in (i) or out (o) of the membrane
CL_{int}	Intrinsic clearance
$CL_{passive}$	Passive clearance
CL_{ren}	Renal clearance
CL_{uptake}	Active uptake clearance
C_{max}	Maximum observed blood concentration of drug
C_{media}	Substrate media concentration
CNS	Central nervous system
CNT	Concentrative nucleoside transporters (*SLC28A* subfamily)
CNV	Copy number variation
Co-med	Concurrent medication
CMVs	Canalicular membrane vesicles
CoMFA	Comparative molecular analysis
CoMSIA	Comparative molecular similarity indices analysis
COPD	Chronic obstructive pulmonary disease
CP	Choroid Plexus
CRISPR	Cluster regularly interspaced short palindromic repeats
CS	Candidate selection
CsA	Cyclosporine A
CSF	Cerebrospinal fluid
CT	Computed tomography
CTA	Clinical trial application
CTD	Common technical document
CYP(450)	Cytochrome P450
CYP3A4	Cytochrome P450 3A4
DCE-MRI	Dynamic contrast-enhanced magnetic resonance imaging
DDI	Drug–drug interaction

DIA	data independent acquisition
DILI	Drug induced liver injury
DME	Drug metabolizing enzyme
DMPK	Drug metabolism and pharmacokinetics
DNA	Deoxyribonucleic acid
DRM	Drug related material
ECM	Extracellular matrix
EMA	European Medicines Agency
ENT	Equilibrative nucleoside transporter (subfamily *SLC29A*)
E17βG	Estradiol-17β-glucuronide
ER	Endoplasmic reticulum
ESC	Embryonic stem cell
E1S	Estrone-1-sulfate
EGTA	ethylene glycol tetraacetic acid
ENT	Equilibrative nucleoside transporter (*SLC29A* subfamily)
F	Bioavailability
F_a	Fraction of the dose that is absorbed from the intestinal lumen to the intestinal enterocytes
F_g	Fraction of the dose that escapes pre-systemic intestinal first pass elimination
FAH	Fumarylacetoacetate hydrolase
FBS	Foetal bovine serum
FaSSIF	Fasted state simulated intestinal fluid
FeSSIF	Fed state simulated intestinal fluid
(US) FDA	(United States) Food and Drugs Administration
f_a	Fraction of dose absorbed
f_m	Fraction metabolized
f_t	Fraction transported
FTIH	First time in human
f_u	Fraction unbound
f_{ub}	Fraction unbound in blood
FXR	Farnesoid X receptor (NR1H4)
GFJ	Grape fruit juice
GFR	Glomerular filtration rate
GI tract/GIT	Gastro-intestinal tract
GLUT	Glucose transporter
GMO	Genetically modified organism
GR	glucocorticoid receptor (NR3C1)
GSH	Glutathione (reduced state)
GST	Glutathione *S*-transferase enzyme
HEK293	Human embryonic kidney cells
HIV	Human immunodeficiency virus
HGNC	Human Genome Organisation Gene Nomenclature Committee
HMG-CoA	3-Hydroxy-3-methyl-glutaryl-Coenzyme A reductase
HNF4α	Hepatic nuclear factor 4alpha (NR2A1)

HUGO	Human Genome Organisation
I	Inhibitor concentration
IC_{50}	The concentration of inhibitor required to inhibit transport by 50%
ID	injected dose
IHC	Immunohistochemistry
$I_{in.max}$	Maximum inhibitor concentration at the inlet to the liver
IND	Investigational new drug
iPSC	Induced pluripotent stem cell
IR	Immediate release
ISF	brain interstitial fluid
ITC	International Transporter Consortium
IVIVC	*In vitro–in vivo* correlation
IVIVE	*In vitro–in vivo* extrapolation
J_{max}	Maximum number of APIs translocated across area per time by a given transporter
k_a	Absorption rate constant of the inhibitor
K_d	dissociation constant
K_i	Inhibitory constant
$K_{m(app)}$	Michaelis constant in Michaelis–Menten kinetics (apparent)
K_p	Partition coefficient
LAT	L-type amino acid transporter
LC-MS/MS	Liquid chromatography tandem mass spectrometry
LD	Lethal dose
LLC-PK1	Lilly Laboratories Culture-Pig Kidney Type 1 (cell line)
logP	logarithm of the partition coefficient, a measure of lipophilicity
LOID	Lead optimization and identification
LTC4	Leukotriene C4
LXR	Liver X receptor (NR1H3)
m/z	Mass to charge ratio
MA	Marketing authorization
MALDI	Matrix assisted laser desorption ionization
MATE	Multidrug and toxin extrusion protein
MATE1	Multidrug and toxin extrusion protein (gene *SLC47A1*)
MATE2	Multidrug and toxin extrusion protein (gene *SLC47A2*)
MCT	Monocarboxylic acid transporter
MDCK	Madin Darby canine kidney (cell line)
MDCK II	Madin Darby canine kidney, type II (cell line)
MDR	Multidrug resistance protein
MDR1	Multidrug resistance protein 1 (also known as P-gp, gene *ABCB1*)
MHLW	Ministry of Health, Labour and Welfare
MMP	Matrix metalloproteinases
MPP+	1-methyl-4-phenylpyridinium

MQAPs	Model quality assessment programmes
mRNA	Messenger riboxynucleic acid
MRI	Magnetic resonance imaging
MRP	Multidrug resistance associated protein
MRP2	Multidrug resistance associated protein 2 (gene *ABCC2*)
MRP4	Multidrug resistance associated protein 4 (gene *ABCC4*)
MW	Molecular weight
NaDC3	Sodium dependent dicarboxylate transporter 3 (*SLC13A3*)
NAT	*N*-acetyltransferase
NBD	Nucleotide binding domain
NCE	New chemical entity
NDA	New drug application
NIR	Near infrared imaging
NME	New molecular entity
NMR	Nuclear magnetic resonance
NPT4	Sodium phosphate transporter 4
NR	Nuclear receptor
NSAID	Non-steroidal anti-inflammatory drug
NSF	Nephrogenic systemic fibrosis
NT	Nutrient transporter
NTBC	2-(2-nitro-4-fluoro-methylbenzoyl)-1,3-cyclohexanedione
NTCP	Na-taurocholate *co*-transporting polypeptide
NumHBA	Number of hydrogen bond acceptors
NumHBD	Number of hydrogen bond donors
OAT	Organic anion transporter
OAT1	Organic anion transporter 1 (gene *SLC22A6*)
OAT3	Organic anion transporter 3 (gene *SLC22A8*)
OATP	Organic anion transporting polypeptide
OATP1A2	Organic anion transporting polypeptide 1A2 (gene *SLCO1A2*)
OATP1B1	Organic anion transporting polypeptide 1B1 (gene *SLCO1B1*)
OATP1B3	Organic anion transporting polypeptide 1B3 (gene *SLCO1B3*)
OATP2B1	Organic anion transporting polypeptide 2B1 (gene *SLCO2B1*)
OCT	Organic cation transporter
OCT1	Organic cation transporter (gene *SLC22A1*)
OCT2	Organic cation transporter (gene *SLC22A2*)
OCTN	Organic cation/carnitine transporter
OCTN1	Organic cation/carnitine transporter 1 (gene *SLC22A4*)
OCTN2	Organic cation/carnitine transporter 2 (gene *SLC22A5*)
OSTα/β	Organic solute transporter α/β (gene *SLC51A/B*)
P_{app}	Apparent permeability studied *in vitro*
PAMPA	Parallel artificial membrane permeability assay
PBEC	Porcine brain endothelial cell
PBPK	Physiologically based pharmacokinetic
P_{car}	saturable carrier-mediated permeability
PCR	Polymerase chain reaction

PD	Pharmacodynamics
PDB	Protein Data Bank
Pdiff	passive diffusion
PDUFA	Prescription Drug User Fee Act
PDMS	Polydimethylsiloxane
$P_{\text{Eff,man}}$	Effective permeability in humans studied *in vivo*
PEPT	Peptide transporter
PEPT1	Peptide transporter 1 (gene *SLC15A1*)
PEPT2	Peptide transporter 2 (gene *SLC15A2*)
PET	Positron emission tomography
P-gp	P-glycoprotein (also known as MDR1, gene *ABCB1*)
PK	Pharmacokinetics
PK/PD	Pharmacokinetics/pharmacodynamics
PMDA	Pharmaceuticals and Medical Devices Agency (Japan)
POC or PoC	Proof of concept
P_{pas}	Passive diffusional driven permeability
PPAR	Peroxisome proliferator activated receptor
PS_{bile}	Intrinsic biliary clearance
PS_{eff}	Intrinsic sinusoidal efflux clearance
PS_{inf}	Intrinsic uptake clearance
PXR	Pregnane X receptor (NR1I2)
Q_{H}	Hepatic blood flow
QSAR	Quantitative structure–activity relationship
QWBA	Quantitative Whole Body Autoradiography
R&D	Research and development
RAF	relative activity factor
rag	Recombinant activating gene
RBE	Rat brain endothelial
rBSEP	Rat bile salt export pump (gene *ABCB11*)
RCSB	Research Collaboratory for Structural Bioinformatics
REF	Relative expression factor
ROC	Receiver operating characteristics
RPTEC	Renal proximal tubule epithelial cells
RT-PCR	Reverse transcriptase polymerase chain reaction
RXR	Retinoid X receptor
SCH	Sandwich-cultured hepatocytes
SCHH	Sandwich-cultured human hepatocytes
SCID	Severe combined immunodeficiency
SERT	Serotonin transporter
SF	Scaling factor
SGLT	Sodium glucose linked transporter
SLC	Solute carrier gene superfamily
SLCO	Solute carrier organic anion transporter gene subfamily
SLC22	Solute carrier subfamily 22
SLC47	Gene encoding MATE
SLCO1A2	Gene encoding OATP1A2

SLCO1B1	Gene encoding OATP1B1
SLCO1B3	Gene encoding OATP1B3
SLCO2B1	Gene encoding OATP2B1
SLC10A1	Gene encoding NTCP
SLC15A1	Gene encoding PEPT1
SLC15A2	Gene encoding PEPT2
SLC22A1	Gene encoding OCT1
SLC22A2	Gene encoding OCT2
SLC22A4	Gene encoding OCTN1
SLC22A5	Gene encoding OCTN2
SLC22A6	Gene encoding OAT1
SLC22A8	Gene encoding OAT3
SLC28A	Gene encoding CNT
SLC29A	Gene encoding ENT
SLC51A/B	Gene encoding OSTα/β
S log P	estimate of logP by summing the contribution of atom-weighted solvent accessible surface areas (SASA) and correction factors
SMR	Molar refractivity (including implicit hydrogens). This property is an atomic contribution model that assumes the correct protonation state (washed structures).
SNP(s)	Single nucleotide polymorphism(s)
SPECT	Single photon emission computed tomography
SRM	Selected reaction monitoring
SULT	Sulfotransferase enzyme
TALEN	Transcription activator-like effector nuclease
TCDB	Transporter classification database
TCDD	2,3,7,8-Tetrachlorodibenzo-*p*-dioxin
TEA	Tetraethylammonium
TEER	Transepithelial electrical resistance
TK	Toxicokinetics
$T_{\max}$	Time at which the $C_{\max}$ is observed
TMD	Trans-membrane domain
TPSA	Topological polar surface area
UDP	Uridine 5$'$-diphosphate
UGT	Uridine 5$'$-diphospho-glucuronosyltransferase or UDP-glucuronosyltransferase
URAT	Urate transporter
V_{D} or V_{d}	Volume of distribution
$V_{\max}$	Maximal velocity
ZFN	Zinc-finger nucleases

Contents

RSC Drug Discovery Series No. 54
Drug Transporters: Volume 1: Role and Importance in ADME and Drug Development
Edited by Glynis Nicholls and Kuresh Youdim
© The Royal Society of Chemistry 2016
Published by the Royal Society of Chemistry, www.rsc.org

Section II: Preclinical Models in Current Use within the Pharmaceutical Industry

Chapter 8 Knockout and Humanised Animal Models to Study Membrane Transporters in Drug Development 298
Nico Scheer, Xiaoyan Chu, Laurent Salphati and Maciej J. Zamek-Gliszczynski

Volume 2: Recent Advances and Emerging Technologies

Section I: The Role of Transporters in ADME

Membrane Transporters: Fundamentals, Function and Their Role in ADME

JOHN KEOGH,*[a,†] BRUNO HAGENBUCH,[b,†] CAROLINE RYNN,[c,†] BRUNO STIEGER[d,†] AND GLYNIS NICHOLLS[e,†]

[a] JPK Consulting, 26 Balmoral Road, Hitchin, Hertfordshire SG5 1XG, UK; [b] Department of Pharmacology, Toxicology and Therapeutics, The University of Kansas Medical Center, Kansas City, Kansas 66160, USA; [c] Metabolism and Pharmacokinetics, Novartis Institutes for Biomedical Research, Postfach CH-4002 Basel, Switzerland; [d] University Hospital, Department of Clinical Pharmacology and Toxicology, 8091 Zürich, Switzerland; [e] Independent Consultant, Wem, Shropshire, UK
*Email: JPKeogh@virginmedia.com

1.1   Introduction

In the middle of the 20th century, the absorption, distribution, metabolism and elimination (ADME) of pharmaceutical drugs was considered to be mediated primarily by simple diffusion and metabolism. The concept that membrane transporter proteins existed and could facilitate the flux of molecules across eukaryotic cell membranes was still in its infancy, but as knowledge and information expanded, it was recognised that transporter proteins could play an important role both in the movement of endogenous

---

[†] All authors contributed equally to this work.

RSC Drug Discovery Series No. 54
Drug Transporters: Volume 1: Role and Importance in ADME and Drug Development
Edited by Glynis Nicholls and Kuresh Youdim
© The Royal Society of Chemistry 2016
Published by the Royal Society of Chemistry, www.rsc.org

compounds within the body and in the ADME of drugs. Two superfamilies of transporters, the ATP binding cassette (ABC) and the solute carrier (SLC), comprising between them over 500 members, have now been identified in the human genome, although only a few transporters of specific interest to the pharmaceutical industry are described here.

This chapter summarises the discovery of transporters, their function in cellular processes, location and mechanism(s) of action, as well as outlining the key transporters currently considered to be clinically relevant. A description of how and why they are evaluated within drug discovery and development is included, outlining some of the key pharmacokinetic (PK) concepts useful to transporter scientists, and briefly discussing the methods and strategies used. While there are many different transporters within the body, this overview concentrates primarily on those transporters currently known to influence drug ADME and will not cover the transport of oligo-nucleotides or proteins. Links to current transporter databases and reviews are also included throughout the text for those wishing to pursue the area further.

1.2 The History of Transporter Science

1.2.1 The Discovery of Transport Processes

The basic functional unit of eukaryotic organisms is the cell, with each cell enclosed by a plasma membrane that forms an inherent physical barrier to the free transport of solutes. While small hydrophobic molecules are able to move freely across these phospholipid membranes by simple diffusion, the more water-soluble molecules require the presence of membrane proteins or channels embedded within the plasma membrane to gain access into and out of cells. This concept of transport *via* membrane proteins ('transporters' or 'drug transporters') and their involvement in the ADME of small drug molecules was first noted in the 1950's, although there were much earlier indications that transport processes may be present within the body.

Bile salts are highly amphipathic molecules that are synthesised and se-creted by the liver and considered as model compounds for enterohepatic circulation.[1,2] At the turn of the 18th century, Mauritius Reverhorst and Alfonso Borelli reported that the amount of bile recovered in faeces was much less than the amount of bile produced by the liver, and they coined the term *"motus circularis bili"* (or enterohepatic circulation, the cycling of compounds between the intestine and the liver).[3] This insight into the se-lective preservation and recycling of bile can be viewed as an early but major step towards the appreciation of transport processes for solutes. Focusing on specific mechanisms, Tappeiner noted in the 19th century that bile salts are preferentially absorbed in the ileum of dogs, but not in the upper part of the small intestine, suggestive of a transporter-mediated process.[4] In 1923, Eli Kennerly Marshal demonstrated the active secretion of phenolsulfonephthalein in the kidney, which may be considered the first demonstration of active (rather than passive) transport.[5] Prior to this, phenoltetrachlorophthalein was used as a liver function marker.[6] This work illustrated not only that both the liver and kidney are important drug clearance organs, but also that they can be selective, indicating the presence of specific, active mechanisms in these organs.[7] However, it was not until 1958 that Crane published his observations on active transport, demonstrating that 6-deoxyglucose can be transported against a concentration gradient[8] and subsequently that sugar absorption is dependent on the presence of sodium[9] and ultimately that the process is electrogenic.[10]

Applying the knowledge from these early findings, the very short half-life of penicillin in patients, concomitant with its rapid appearance in urine,[11] was believed to be caused by active tubular secretion of the drug in the kidney. It was therefore reasoned that *para*-aminohippurate, also known to be secreted by the kidney, may interfere with the secretion of penicillin. The subsequent finding that co-administration of *para*-aminohippurate with penicillin resulted in a marked retention of penicillin in the blood of dogs[12] may well be considered the first application of transporter research to ADME. These findings led to the development of probenecid, which inhibits penicillin secretion in the kidney and thereby prolongs its plasma half-life,[13,14] although the specific transporter proteins involved were not identified until many decades later. At the same time, it was discovered that probenecid enhanced renal urate elimination,[15,16] and probenecid is still used today as an uricosuric drug.[17]

In the 1960s, following the development of the first chemotherapeutic drugs for the treatment of cancers, it was noted that some tumours developed resistance against these drugs. This was followed by descriptions of cross-resistant cell lines[18] and the isolation and characterisation of several multidrug resistant cell lines, as reviewed by Gottesman and Ling.[19] For example, a Danish group reported on active drug export from Ehrlich ascites tumour cells[20] and Ling and Thompson, whilst isolating cells resistant to colchicine, observed that these cells were resistant to other chemotherapeutic drugs with different pharmacodynamic (PD) properties. The authors concluded that this cross-resistance was caused by a reduced permeability of the cell lines to drugs[21] and identified a glycoprotein of about 170 kDa that was only expressed in colchicine resistant and not in revertant colchicine sensitive cells. They named this protein P-glycoprotein (P-gp), whereby "P" designated permeability. Sequencing of a mouse multidrug resistance protein (MDR) sequence[22] demonstrated that P-gp is a member of the ABC superfamily of transporters, subfamily member B1 (*ABCB1*), which had previously been identified.[23] Purification and reconstitution of human multidrug resistance protein 1 (MDR1) finally demonstrated that MDR1 displayed drug-stimulated ATP hydrolysis[24] and drug transport.[25] Hence, P-gp is also known as MDR1 and ABCB1. These discoveries led to a series of programmes over many decades testing clinical *in vivo* inhibitors of P-gp to overcome multidrug resistance in chemotherapy, with limited success.[26]

                                                          Chapter 1

1.2.2 The Development of Transporter Science in Industry

Drug transporters first came to the attention of drug discovery and development specialists in the drug formulation and drug metabolism and PK (DMPK) fields, as these are disciplines that characterise how the body handles pharmaceutical compounds. Arguably, the earliest appreciation of differential drug–tissue distributions (and later of drug transport mechanisms) in the DMPK field arose with the development of semi-quantitative radiolabelled drug–tissue distribution studies and later quantitative whole body autoradiography in rats.[27,28] Distribution studies revealed that, for some drugs, the tissue concentrations of drug related material were much greater than blood concentrations, sometimes by many orders of magnitude, whereas in others, drug related material in tissues was virtually undetectable. These observations were presumed to be due to the physicochemical properties of the molecules (lipophilicity, charge at pH 7.4, solubility, *etc.*) and/or biological phenomena such as metabolism, and plasma and tissue protein binding.[29,30] However, with the discovery of P-gp, these assumptions began to be re-evaluated. Over a period of time, a body of evidence emerged, demonstrating that many marketed drugs were in fact both *in vitro* substrates and inhibitors of P-gp. This raised concerns that P-gp could influence drug absorption, distribution and elimination (ADE), and therefore be a potential mechanism driving both systemic and target organ exposure, as well as drug–drug interactions (DDIs). Given its abundant localisation at the blood–brain barrier, P-gp was most often evaluated as a modulator of drug penetration of the central nervous system (CNS),[31–35] becoming the first drug transporter routinely investigated during drug discovery and development in the pharmaceutical industry.

As the science, understanding and discovery of drug transporters and their function continued to develop, it became clear that a number of pharmaceutical products were substrates, inhibitors and inducers of multiple drug transporters of different types. These included widely prescribed drugs such as digoxin, transported by P-gp and the organic anion transporting polypeptide member 4C1 (OATP4C1);[36–38] metformin, transported by organic cation transporter member 1 (OCT1) and multidrug and toxin extrusion proteins (MATEs);[39–43] 3-hydroxy-3-methyl-glutaryl-CoA reductase (HMG-CoA) inhibitors, transported by OATP1B1;[44] fexofenadine, transported by OATPs and P-gp;[45,46] and acyclovir, transported by organic anion transporters (OATs),[47] to name just a few. One notable example is that of rosuvastatin, a HMG-CoA reductase inhibitor (also known as a 'statin') with a high distribution into the liver that was shown to be mediated *via* transporter dependent mechanisms.[48] Subsequent work showed that rosuvastatin and other statins were substrates of some of the OATPs, members of the SLC organic anion transporter (*SLCO*) gene family.[49] This and other work led to an explosion of interest in hepatic transporter systems, as it became clear that a number of DDIs between statins and other molecules could be readily explained by considering hepatic OATPs and other drug transporters, such as breast cancer resistance protein (BCRP; the protein product of the *ABCG2* gene), multidrug resistance associated protein 2 (MRP2; the protein product of the *ABCC2* gene), and the organic anion and cation transporters (OATs and OCTs; protein products of the *SLC22* gene family).[50–54] Even some molecules ultimately cleared by metabolism were discovered to be clinical substrates of drug transporters, which could result in the restriction or enhancement of systemic exposures and thereby influence the drugs' PK. For example, atorvastatin is a substrate of OATPs and P-gp, is primarily eliminated as metabolites, but is subject to DDIs when co-administered with transporter and/or metabolism inhibitors.[55]

Several findings regarding other transporter-mediated processes were also noted, and there are now many examples of the PK of marketed drugs being influenced by the action of drug transporters.[56] The clinical consequences of many of these interactions are still being investigated, but continue to result in changes and warnings on labels for drug use. This topic is discussed in more detail in Chapter 11. As knowledge of transporter tissue distribution and localisation grew, particularly for barrier tissues such as the gastro-intestinal tract (GIT) and in the major clearance organs (kidney and liver), there was a growing appreciation that drug transporters could significantly influence drug PK.[57–60] This was particularly true for molecules with low simple diffusion (see Figure 1.1a), as this limits their transfer across cellular membranes. In these cases, drug transporters were found to be an important mechanism for oral absorption,[61,62] systemic clearance,[63] and renal and faecal excretion.[40] Furthermore, evidence that drug molecules were substrates, inhibitors or inducers of multiple transporters and drug metabolising enzymes (DMEs), challenged commonly held assumptions that associated clinical DDIs were due to a single mechanism, and led to the re-evaluation of these DDIs.[64–67] However, identifying the precise contributions of multiple mechanisms to overall clinical PK remains a remarkably challenging goal. This is due in part to limitations of the current preclinical toolkit, as well as the reliance of PK on measurement of drug concentrations in the blood compartment rather than within organs or tissues.[57,58,63,68] While subtle changes in systemic drug concentrations can be measured, they may not always reflect much larger concentration changes in tissues and organs, which may result in unpredicted toxicology and/or pharmacology.[69]

These findings raised many questions for drug discovery and development scientists, as they alerted the scientific, medical and regulatory communities that they could have relevance to drug safety, efficacy and toxicity. A good example of the depth and breadth of a typical investigation now is that of bosentan DDIs and drug induced liver injury (DILI). Bosentan is a non-peptide dual endothelin receptor antagonist that is used to treat pulmonary arterial hypertension.[70] It is metabolised in the liver to three major oxidative metabolites,[71] undergoes biliary elimination[72] and was shown to be a substrate for OATPs expressed in human hepatocytes.[73] Studies in rats indicated that cyclosporine inhibited the uptake of bosentan into hepatocytes *via* members of the rat OATPs (rOATPs),[74] and *in vitro* studies identified

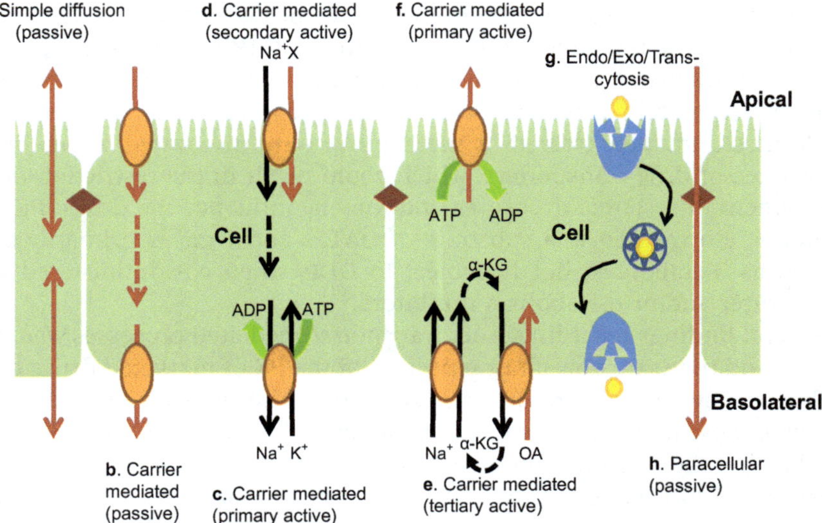

Figure 1.1   Mechanisms of transepithelial permeability. Ovals: transporter proteins; diamonds: tight junctions between cells; black arrows: direction of transport of solutes providing the driving force(s) for substrate transport; red arrows: direction of substrate (drug) transport; α-KG: alpha-ketoglutarate; OA: organic anion; X: solute/drug substrate. See text for further details.

rifampicin as an inhibitor of rat and human OATPs expressed in hepatocytes.[75,76] Several drugs were subsequently tested for PK interactions on co-administration with bosentan,[73,77] including cyclosporine[77] and rifampicin.[78] Both of these compounds increased the minimum blood concentrations of repeat-dose bosentan, suggesting inhibition of hepatic uptake and/or metabolism of bosentan by these drugs. Additionally, long term treatment of patients with rifampicin led to a reduction in serum concentrations of bosentan, which could be explained by the induction of cytochrome P450 (CYP450) 3A4 (CYP3A4),[78] which generated a metabolism-driven sink for bosentan in hepatocytes, stimulating its OATP mediated uptake.

In terms of DILI, elevated liver enzymes,[79,80] preceded by an increase in serum bile salts, were observed in patients receiving bosentan, which resolved on withdrawal of the drug. Interference with bile salt elimination, possibly through modulation of the bile salt export pump (BSEP; protein product of the *ABCB11* gene) was suspected. In addition, in a chronic heart failure trial, bosentan and glibenclamide were found to act synergistically on liver injury, elevating serum bile salt levels. This was also confirmed in rats.[80] *In vitro* investigations using isolated rat canalicular plasma membrane vesicles and vesicles from the insect cell line Sf9 over-expressing the rat bile salt export pump (rBSEP) demonstrated that bosentan and its three metabolites competitively inhibited rBSEP and human BSEP mediated

Table 1.1 Human SLC and ABC transporters currently considered of relevance to drug disposition.

Gene symbol	Protein name	Other/old protein name	Regulatory recommendations
SLCO1A2	OATP1A2	OATP-A, OATP	
SLCO1B1	OATP1B1	OATP-C, OATP2, LST-1	EMA/FDA/PMDA
SLCO1B3	OATP1B3	OATP8	EMA/FDA/PMDA
SLCO2B1	OATP2B1	OATP-B	
SLC10A1	NTCP	—	
SLC10A2	ASBT	ISBT	
SLC15A1	PEPT1[a]	Oligopeptide transporter 1	
SLC15A2	PEPT2	Oligopeptide transporter 2	
SLC22A1	OCT1	—	Consider for EMA, PMDA
SLC22A2	OCT2	—	EMA/FDA/PMDA
SLC22A4	OCNT1	ETT	
SLC22A5	OCTN2	CT1, CDSP	
SLC22A6	OAT1	PAHT, ROAT1, NKT	EMA/FDA/PMDA
SLC22A8	OAT3	—	EMA/FDA/PMDA
SLC22A12	URAT1	OAT4L, RST	
SLC29A1	ENT1[a]	—	
SLC29A2	ENT2[a]	—	
SLC47A1	MATE1[a]	—	PMDA, consider for EMA/FDA
SLC47A2	MATE2[a]	—	PMDA, consider for EMA/FDA
SLC51A	OSTα	OST alpha	
SLC51B	OSTβ	OST beta	
ABCB1	MDR1	P-gp	EMA/FDA/PMDA
ABCB4	MDR3	Phospholipid floppase	
ABCB11	BSEP[a]	SPGP	Consider for EMA, FDA, PMDA
ABCC1	MRP1	GS-X	
ABCC2	MRP2[a]	cMOAT	Consider for PMDA
ABCC3	MRP3[a]	MOAT-D	
ABCC4	MRP4[a]	MOAT-B	
ABCG2	BCRP	MXR	EMA/FDA/PMDA

[a]Emerging transporters, as referred to by the ITC.[103] Although not specifically recommended within the guidance for evaluation (apart from MATEs), many are now being 'considered'. Guidelines can be found on the following websites: EMA: www.ema.europa.eu/docs/en_GB/document_library/Scientific_guideline/2012/07/WC500129606.pdf; FDA: www.fda.gov/downloads/drugs/guidancecomplianceregulatoryinformation/guidances/ucm292362.pdf; PMDA: www.pmda.go.jp/english/. For a more complete list of transporters, readers are referred to the following websites: SLC transporters: http://www.genenames.org/cgi-bin/genefamilies/set/752 or www.bioparadigms.org/slc/intro.htm; ABC transporters: http://www.genenames.org/cgi-bin/genefamilies/set/417; PMDA: Japanese Pharmaceuticals and Medical Devices Agency.

transport of the bile salt taurocholate.[80,81] Glibenclamide also competitively inhibited rBSEP[82] and, as the serum markers for liver injury in patients spontaneously resolved after discontinuation of bosentan, the molecular mechanism of liver injury was linked to drug-induced cholestasis (*i.e.* the slowing or blockage of bile flow).[80,83] *In vivo* studies with bosentan in rats indicated no change in bile salt output, but there was stimulation of bile flow by bosentan mediated by rat MRP2 (rMRP2).[84] This was confirmed *in vitro* using human MRP2 and BSEP expressed in insect cells, indicating that bosentan inhibited BSEP but activated MRP2 transport.[85] Such a functional MRP2–BSEP interaction has also been observed for estradiol-17β-glucuronide[82] and gives an indication of the complexity of interactions that can occur for compounds that are substrates and/or inhibitors of membrane transporters.

Recently, it has been asserted that all solute transport across cell membranes is *via* transporter proteins, and that simple diffusion does not happen.[86,87] It should however be kept in mind that cells are unable to control simple diffusion and hence are vulnerable to the entry of potentially toxic substances, unless they possess protective mechanisms. Efflux transporters may limit the entry of such toxins, and indeed MDR1 has been dubbed a "vacuum cleaner", because of its ability to efflux substrates from the cell membrane to the external environment, thus protecting the cell.[88,89] Such considerations may support the concept that all solutes enter (or are expelled from) cells by transporter proteins. This is disputed by other workers in the field[90,91] and remains a matter of debate.

The interplay between metabolic enzymes and transporters is also considered to be of potential importance, as discussed in *Drug Transporters: Volume 2: Recent Advances and Emerging Technologies*, Chapter 4. It has been highlighted in regulatory guidance for DDIs[92–94] and is an active area of study for both academic[95,96] and industrial scientists.[63,97,98]

A list of drug transporters of current (2016) interest to the pharmaceutical industry is given in Table 1.1. The reader is advised that this list is not comprehensive and will almost certainly be updated as further advances and discoveries are published. It is recognised that as well as their direct impact on PK, drug transporters also influence other clearance mechanisms of ADME, such as the action of DMEs. Additionally, their expression in multiple barrier and clearance organs presents a major technical and logistical challenge to the pharmaceutical industry and to clinicians.

1.3 Transporter Form and Function

1.3.1 Transporter Families and Nomenclature

Transporters are large proteins (40–200 kDa) located in the plasma membrane of cells and organelles. They normally span the membrane many times and modulate the transfer of xenobiotics (including nutrients, micronutrients and pharmaceuticals), and endogenous substances such as neurotransmitters, hormones, signalling molecules, vitamins *etc.* across cellular membranes, tissues or organ barriers. There are two transporter superfamilies, the ABC transporters and the SLCs, numbering in excess of 500 members between them. Until the late 1990s, gene and protein nomenclature of transporters was haphazard and sometimes conflicting, resulting in multiple names for the same transporter, and occasionally the same name for different transporters. The Human Genome Organisation

(HUGO) Gene Nomenclature Committee (HGNC)[99] is now responsible for approving unique gene symbols and protein names; however, older publications still carry redundant names. Symbols for human and rodent proteins are given in all capitals (*e.g.* MDR1, OATP2B1, OCT1) while their corresponding gene symbols are always in italics and in all capitals for human genes (*e.g. ABCB1, SLCO2B1, SLC22A1*) and in lower case with an initial capital for rodents (*e.g. Abcb1, Slco2b1, Slc22a1*).

The transporters given in Table 1.1 include both the transporters quoted in the current regulatory guidance[93,94,100] and those considered to be of emerging importance in the pharmaceutical industry in 2016, which may be incorporated into future regulatory guidance. Additional information may be obtained by consulting white papers and the outputs from professional collaborations such as the International Transporter Consortium (ITC), which are regularly updated.[101–104]

1.3.1.1   ABC Superfamily

In mammals, there are seven families of ABC transporter genes (*ABCA* to *ABCG*), encoding 48 individual transporters (see the HGNC website for further details).[99] They are responsible for transporting a wide range of endogenous substrates, including conjugated bile salts, steroid hormones, cholesterol and unconjugated bilirubin. For drug transport, the *ABCB* family with MDR1 (*ABCB1*; also known as P-gp or ABCB1) and BSEP (*ABCB11*), *ABCC* family with the MRPs (*ABCC1* to *ABCC6*), and *ABCG* family with BCRP (*ABCG2*) are the most clinically relevant. All mammalian ABC transporters are efflux transporters and transport is driven by ATP hydrolysis. A functional ABC transporter consists of two transmembrane domains (TMD1 and TMD2), each fused at the C-terminus to a nucleotide binding domain (NBD1 and NBD2). These two units may be fused as a single protein or combined as homo- or hetero-dimers to form functional transporters. In addition, some members of the *ABCC* family contain an extra TMD (TMD0) at the *N*-terminus, spanning the membrane five times, which is connected by the cytoplasmic loop L0 to the TMD.[105]

1.3.1.2   SLC Superfamily

SLCs are classified into 52 different families based on their amino acid identities. Full listings of the families and their known or suggested function can be found in the HGNC and Bioparadigms websites.[106,107] They have recently been summarised in a special issue of *Molecular Aspects of Medicine*.[108] The different families are responsible for the transport of a hugely diverse range of biologically important Funt molecules, including sugars, amino acids, peptides, inorganic ions, organic anions and cations, metal ions, electrolytes and neurotransmitters, *etc.* They also transport, or are inhibited by, many drug molecules. Families that have been identified as important for drug transport include *SLCO, SLC22* and *SLC47*. The *SLCO* family members, including OATP1B1 (*SLCO1B1*), OATP1B3 (*SLCO1B3*), *SLC22* members OCT1 (*SLC22A1*), OCT2 (*SLC22A2*), OAT1 (*SLC22A6*) and OAT3 (*SLC22A8*) for drug uptake, and *SLC47* members MATE1 (*SLC47A1*) and MATE2K (*SLC47A2*) for drug efflux, are the most relevant in the study of clinical transporter-mediated DDIs, as they transport a number of important therapeutics. It should also be noted that other family members not listed here may be of clinical importance for specific compounds. All solute transporters are integral membrane proteins with 7–14 predicted membrane-spanning domains. Members of the *SLC5* and *SLC7* families are predicted to have 14 such TMDs,[109,110] while members of the *SLCO* and the *SLC22* family have 12 TMDs.[111] Members of the *SLC3* family span the membrane only once but they heterodimerise with a transporter of the *SLC7* family to form a functional transporter.[110] Most of these predictions are based on hydrophobicity analyses but, *e.g.* in the case of SGLT1, the topology has been deduced from the crystal structure of the related *Vibrio para-haemolyticus* sodium/galactose symporter.[112] With additional crystal structures to be solved in the future, the topology and number of TMDs of many more transporters will be elucidated.

1.3.2   Driving Forces for ABCs and SLCs

Transport can be defined as the movement of solutes from one aqueous compartment to a neighbouring compartment that is separated by a phospholipid bilayer containing transport proteins or by tight junctions between cells. Different mechanisms of transport are illustrated in Figure 1.1. It has been known for a long time that small molecules (*e.g.* NO and O_2) can move across cell membranes without a transport protein. Whether this is true for all small and lipophilic molecules is increasingly questioned.[87] This movement is always along the concentration gradient and is called "simple diffusion" (Figure 1.1a). It is worth noting that the phrases "passive permeability" or "passive membrane permeability" are very often used in the literature to describe simple diffusion processes. If solutes are hydrophilic or carry a net charge, they are less able to penetrate cell membranes without interacting with a membrane transporter protein (Figure 1.1b–f). With respect to energy requirements, protein-mediated transport can occur in two ways: either along the electrochemical gradient of the solute and thus be passive (Figures 1.1b and 1.2A); or against the electrochemical gradient and therefore be active (Figures 1.1c–f and 1.2B). Members of the SLC family of transporters can belong to either of these classes of passive or active transport. In contrast, the ABC transporters are all primary active transporters because they directly hydrolyse ATP to pump their substrates out of the cell (Figures 1.1f and 1.2B). The driving forces for some transporters have been demonstrated and, although the principles of primary, secondary or tertiary transport discussed below are valid, the precise mechanisms for many transporters remain unknown, particularly for SLCs. Passive protein-mediated transport, also called facilitated diffusion, is always along an electrochemical concentration gradient (Figures 1.1b and 1.2A), whereas active transport may work against an electrochemical gradient and can be primary (Figures 1.1c and 1.2B), secondary (Figures 1.1d and 1.2B) or even tertiary active (Figures 1.1e and 1.2B).

A primary active transporter is one that directly generates the energy required to move solutes against their electrochemical concentration gradient by hydrolysing ATP, such as ABC transporters (Figures 1.1f and 1.2B) or Na^+/K^+ ATPase (EC 3.6.3.9) (Figure 1.1c). Secondary active transporters often use the out-to-in sodium gradient, which is maintained by a primary active Na^+/K^+ ATPase, and couple sodium movement down its electrochemical gradient to provide substrate uptake against its electrochemical gradient (Figures 1.1c and 1.1d combined). Examples of such transporters include the different sodium dependent co-transporters responsible for cellular uptake of *e.g.* sugars,[109] amino acids,[113] dicarboxylates[114] or bile acids.[115] The organic anion transporters OAT1 and OAT3 are examples of tertiary active transport systems (Figure 1.1e). They mediate the uptake of organic anions against their electrochemical gradients into, for example, proximal tubule cells in exchange for α-ketoglutarate, which in turn is taken up into the cells *via* sodium dependent dicarboxylate transporter 3 (NaDC3; *SLC13A3*, a secondary active process).[116] The driving force is the sodium gradient, which is

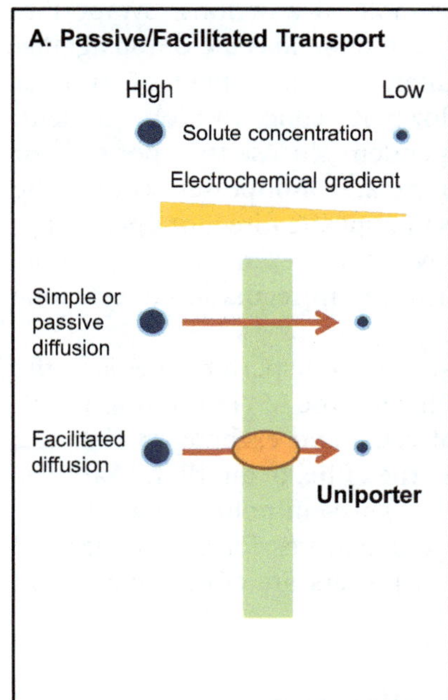

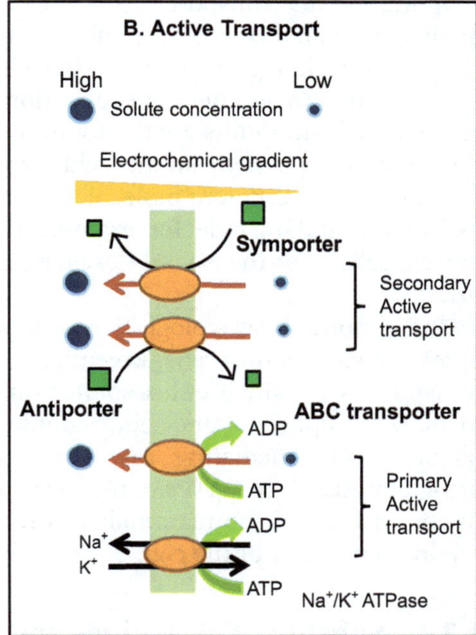

Figure 1.2　Energetics of transepithelial transport. Ovals and straight arrows: transporter proteins and the direction of solute transport; red arrows: solute/drug transport; curved arrows: co-transported solutes for symport and antiport transporters; large circles: high solute concentrations; small circles: low solute concentrations. See text for further details.

ultimately maintained by the Na^+/K^+ ATPase. All of these transporters are located in the same basolateral membrane of proximal tubule cells.[117] Although Na^+/K^+ ATPase is mentioned here, it is not in fact a member of the ABC superfamily and is not a recognised drug transporter, *per se*.

Endo- and exo-cytosis (Figure 1.1g) are primarily relevant to the transport of large biomolecules, such as polypeptides, and as such are not discussed further in this book. In addition, small ions such as Na^+ or Cl^- can move along their electrochemical gradient across tight junctions in a process called paracellular transport (Figure 1.1h). However, because this route is not an important route of drug transport under normal physiological conditions it will also not be discussed in this book.

1.3.3 Uptake, Efflux and Bi-directionality

Besides the classification based on driving forces outlined above, transporters are also classified according to their direction of transport, *i.e.* uniporters (Figure 1.2A), symporters and antiporters (Figure 1.2B), the latter two being co-transporters. Uniporters are found in most cells and mediate the uptake of sugars,[118] amino acids,[110] nucleosides[119] and other small molecules along their electrochemical concentration gradients without any coupling of this transport to that of other molecules or ions. Symporters mediate transport by coupling the movement of one molecule or ion against its concentration gradient with the simultaneous movement of another molecule or ion in the same direction down its concentration gradient. Examples of symporters are the sodium dependent glucose transporter[109] or the sodium dependent amino acid transporters.[113] Antiporters couple the transport of two different molecules or ions in opposite directions across the membrane and include for example the Na^+/Ca^{2+} exchanger in cardiac muscle cells[120] or the Na^+/H^+ exchanger involved in regulation of cytosolic pH.[121]

Under normal physiological conditions, SLC transporters mediate the uptake of solutes into cells. Exceptions to this are the MATE1 and MATE2K transporters,[122] which efflux solutes out of cells, and exchangers that can mediate the uptake of one solute coupled to the efflux of another. Examples include OAT1 mediating the exchange of *para*-aminohippurate for α-ketoglutarate,[117] or OATPs, which exchange substrates for bicarbonate.[123] In contrast, all of the mammalian ABC transporters are efflux pumps and export solutes out of the cell.

1.3.4 Substrate Specificities and Binding Sites

Membrane transporters are capable of transporting a wide range of structurally diverse substrates, often with overlapping substrate specificities, and are known as polyspecific drug transporters. Consequently, there are very few, if any, specific probes available for *in vitro* transporter studies, making the true substrate specificity of a transporter *in vivo* difficult to delineate. However, it is possible to assess the contributions of individual transporters in the disposition of a new molecular entity (NME) using methods such as relative activity factors[124] and PK modelling.

The overlap of substrate specificities may allow transporters to compensate for the loss of another that is either absent or non-functional (*e.g.* through disease or inhibition), so-called transporter redundancy. For example, the impaired biliary elimination of bilirubin in MRP2 deficient rats can be compensated for to some extent by its removal from the liver by MRP3, although the animals remain jaundiced.[125] However, this may not always be the case; the inherited mutation in *ABCB11*, which encodes the bile salt exporter BSEP in hepatocytes leads to severe liver disease in humans in childhood. This is despite the fact that bile salts can also be exported by other hepatic transporters, such as MRP4 and the organic solute transporter (OST; *SLC51*) α/β dimer (OSTα/OSTβ), as evidenced from *in vitro* data,[126] which might be expected to compensate for the loss of BSEP.

The issue of substrate specificity is further complicated by the fact that some transporters such as the OATPs have multiple substrate binding sites.[127] For example, OATP1B1 is reported to have two binding sites for estrone-3-sulfate, but only one binding site for fluvastatin.[66] Similarly, some MRPs transport certain substrates only in the presence of glutathione,[128] hence glutathione could be considered a modulator of the substrate specificity of these transporters. In addition, the kinetics of some transporters have been shown *in vitro* to be strongly influenced by membrane composition, particularly cholesterol content, with both MRP2 and P-gp showing altered transport kinetics in different lipid environments *in vitro*. MRP2 demonstrates allosteric kinetics with a low cholesterol content and Michaelis–Menten type transport with a high cholesterol content.[129,130] The reasons why this occurs, and the precise impact *in vivo,* are unclear, but appear to be related to the interaction of substrates with the lipids in the plasma membrane. What is clear is that the kinetics of substrate–transporter interactions are likely to be more complex than for DMEs.

To date, the structure of two mammalian ABC transporters (mouse P-gp[131] and human ABCB10, a mitochondrial transporter[132]) and one mammalian SLC transporter, human glucose transporter 1 (GLUT1; protein product of the *SLC2A1* gene[133]) have been crystallised. However, of these, only mouse P-gp is important for drug transport. In addition, numerous prokaryotic ABC transporters and several prokaryotic homologues of SLC transporters[134] have been crystallised. These bacterial crystal structures have been used to predict the structures of related mammalian transporters by comparative or homology modelling (*e.g.* for OATPs,[135] OAT1[136] and OCT1[137]). It has to be emphasised, however, that such models can have major limitations when homologues with less than 30% amino acid sequence identities are used, and therefore should only be used to formulate hypotheses, *e.g.* with respect to the location of charged amino acids that can then be tested experimentally using site-directed mutagenesis and functional assays.[138]

1.3.5 Transporter Localisation and Interplay

Membrane transporters are located in every cell in the body, although not necessarily just on the plasma membranes (Figure 1.3). They may also be found intracellularly (*e.g.* on mitochondrial membranes), but from an ADME perspective it is usually the plasma membrane transporters that are the main focus of interest, as reflected in the current regulatory guidance.[93,94,100] Some transporters can be exclusively expressed in one organ (*e.g.* OATP1B1 in the liver), some ubiquitously (*e.g.* OCT3) and others can have a restricted distribution (*e.g.* MATE1 in the liver and kidney) (Figure 1.3). Thus, to fully understand their role in drug PK, the location and expression of transporters at the organ, cell and sub-cellular levels needs to be determined. This in itself is challenging, for example: transporter expression levels are low relative to other proteins (*e.g.* DMEs), making quantitation difficult; specific antibodies are not routinely available for

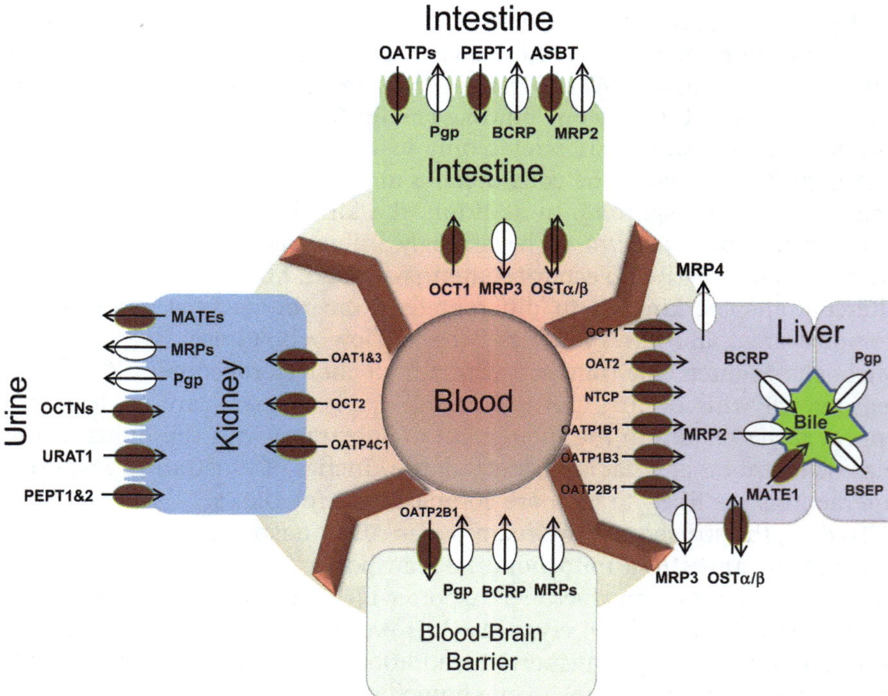

Figure 1.3 Schematic showing type, orientation, localisation and direction of transport of some major drug transporters in selected tissues. Transporters on the blood side of the cell concentrate substrates from the blood (SLC) or efflux substrates from the cell (ABC). Transporters on the non-blood side of the cell eliminate substrates from the cell (ABC and MATEs), or in the case of the GIT and kidney, facilitate uptake or re-uptake of substrates from the lumen (SLCs). Arrows: direction of transport of substrates; white ovals: ABC transporters; brown ovals: SLC transporters. See text for further details.

immunoquantification; and low yields of transporter protein in the membrane isolates used for quantitative techniques such as liquid chromatography tandem mass spectrometry (LC-MS/MS) can result in poor estimates of mass balance.[139]

The localisation of some key ADME transporters in human tissues are outlined schematically in Figure 1.3, which gives an indication of their distribution and orientation on specific membranes (apical or basolateral) of polarised cells such as hepatocytes or proximal tubule cells. This localisation of transporters is key to their biological function within those cells, tissues or organs.

For example, P-gp is expressed on the apical surface of cells and effluxes its substrates into the lumen of the organ where it is expressed. This means that in the enterocytes of the GIT, it effluxes substrates back into the GIT, thereby potentially limiting oral absorption. By contrast, in the liver, P-gp delivers substrates into the biliary tract from the hepatocytes, maintaining low intra-hepatic concentrations of substrates extracted from the blood. At the blood–brain barrier, P-gp effluxes substrates into the blood, thus protecting the brain from excessive drug exposure. Thus, a substrate of P-gp may have restricted oral absorption, enhanced biliary elimination and restricted CNS distribution. It is important therefore to consider transporter tissue distribution when evaluating the significance of a transporter–drug interaction.

The spatial distribution of transporters within or along the tissue or organ (*e.g.* small intestine to colon) is also relevant to their net impact. There are numerous instances of differential tissue distribution of transporters, for example P-gp, BCRP and OATP expression along the GIT,[140–142] and OATs, OCT1 and MATEs along the kidney proximal tubule.[143] This type of distribution along the length of luminal structures facilitates the selective uptake, efflux, re-uptake and recirculation of transported substrates, such as creatinine,[143,144] bile acids and salts,[145] and uric acid.[117,146] In addition, the balance of expression of different transporters acting in the same or opposing directions (*e.g.* MRPs and OATPs in the basolateral membrane of hepatocytes) may be relevant to their overall physiological impact.[46,147–149]

In the liver and GIT, transporters such as P-gp are co-expressed with DMEs, notably CYP3A4.[96,140,150] This has relevance in drug absorption, metabolism and elimination, and the DDI potential in those organs, as there is an apparent synergistic impact of these proteins on some processes. For example, the rate of appearance of a drug substrate in the enterocyte (and therefore presentation to the metabolising enzyme) will be decreased and/or delayed due to the influence of P-gp activity on drug permeability across the luminal membrane of the GIT; thus, intracellular drug concentrations may remain at sub-saturating levels for CYP3A4. A similar process is proposed for hepatic elimination, albeit that localisation of P-gp and CYP3A4 relative to the sequence of exposure to the drug is reversed.[96,97,150] Although the precise relationship and therefore the net impact on drug ADME is a matter of some debate, it has been demonstrated in genetically modified mice that the absence of either P-gp, Cyp3a or both results in substantial changes in the PK of drugs that are substrates for both mechanisms.[95]

1.3.6 Transporter Expression in Animal Species

In preclinical ADME studies, several different animal models and/or primary or immortalised cell lines from animal tissues can be used to assess drug–transporter interactions. Rodents, particularly mice, are a frequently used model.[39,101,151,152] Other species have also been used, although not routinely (*e.g.* dog[153] and monkey[154]). A typical objective of preclinical species studies is to aid in understanding human drug–transporter interactions, which may otherwise be difficult or impossible to study clinically. It is therefore important to be aware of differences between human and animal transporter expression, tissue distribution and function.

In vitro experiments suggest that transporters from different species have broadly similar substrate activities, and possibly affinities, although it has been reported that intrinsic transporter activity may differ between species (*e.g.* MRP2[155]). Genomic and *in vivo* data demonstrate some fundamental differences in transporter expression and distribution profiles across species, resulting in profoundly different PK between species for some molecules. This can have important implications for predicting the impact of human transporters on drug PK using data from preclinical species.

MDR1A and MDR1B in rodents are represented by MDR1 (P-gp) in humans, and both transporters must be eliminated (genetically engineered out, so called "knockout" mice, or chemically inhibited, so-called "chemical knockout") to reflect the impact of P-gp on human PK in mice.[31] The dual MDR knockout mouse model ($Mdr1a^{-/-}/Mdr1b^{-/-}$) has been used routinely and successfully to investigate, for example, P-gp-modulated CNS penetration of drug molecules, and has been extended to include BCRP and MRP2, creating triple knockout mouse models.[156–158] However, this relatively straightforward correlation is not applicable for all transporters. For instance, OCT1 and OCT2 are predominantly expressed in human liver and kidneys, respectively, whereas murine and rat orthologues are highly expressed in both tissues, resulting in substantially different excretion profiles in rodents *versus* humans for substrates of these transporters.[39]

A more complex scenario again is OATP and Na–taurocholate co-transporting polypeptide (NTCP; the gene product of *SLC10A1*) expression in hepatocytes. These transporters are responsible for the uptake of bile acids from the blood into the hepatocyte, and therefore important components in bile homeostasis, as well as being notable drug transporters. Although NTCP is expressed in both rodent and human hepatocytes, two members of the OATP1A family (OATP1A1 and OATP1A4) and OATP1B2 are expressed in rodents. Conversely, in humans, the OATP1B family predominates (OATP1B1 and OATP1B3) and OATP1As are absent.[159] Knockout of all of the rodent/murine *Oatp* genes is required to create a model that will give some indication of the net impact of contribution, loss or inhibition of these transporters in humans. However, ablation of the genes may not result in an otherwise "normal" animal, as these manipulations may lead to unpredictable alterations in the expression or function of other (transporter) proteins, as illustrated by van de Steeg *et al.*[160] and Slijepcevic *et al.*[161] Van de Steeg *et al.* showed that $Oatp1a^{-/-}$ and $Oatp1b^{-/-}$ knockout mice have elevated unconjugated bile acids compared with their wild-type littermates, indicating that NTCP is unable to compensate for the loss of the *Oatp* genes.[160] Similarly, characterisation of a NTCP knockout mouse strain showed changes in the expression and functional levels of a number of other bile acid transporters in the GIT, liver and kidney, including OATPs, impacting enterohepatic bile acid homeostasis *in vivo.*[161] This is also observed in humans, *e.g.* in a patient with non-functional NTCP, serum bile salt levels above 1 mM were found, demonstrating that OATPs cannot compensate for the loss of NTCP function despite their *in vitro* capacity to transport bile salts.[162]

Transgenic models, where the human gene has been added ("knock-in") or has replaced the rodent version(s) ("humanised") are increasingly available.[151] However, to demonstrate their utility in predicting human outcomes, they require careful validation. This is expanded upon in Chapter 8. Although preclinical species and genetically modified animals may have limited utility for clinical predictions or translations, and are highly dependent on the transporter(s) being tested, they are nonetheless useful tools to establish the *in vivo* impact of transporter interactions.[31,58,147]

1.3.7 Other Factors Affecting Transporter Form and Function

1.3.7.1 Regulatory Mechanisms

As for all membrane proteins, transporters are synthesised at the endoplasmic reticulum (ER), where post-translational modification begins. From the ER, transporters are shipped *via* the Golgi, where further post-translational modifications occur prior to delivery to their final destination, which is often the plasma membrane. Exit from the ER may be regulated for some transporters, hence affecting their expression levels at the plasma membrane, for example, the cystic fibrosis transmembrane conductance regulator.[163] Functional transporter activity may be conferred during trafficking to the plasma membrane, as has been demonstrated for Na^+/K^+ ATPase, which is not fully functional until it arrives at the cell surface.[164] Post-translational modifications, *e.g.* by phosphorylation or dephosphorylation of proteins, are important regulators of the activity of transporters. For example, the state of phosphorylation/dephosphorylation determines the protein levels of the bile acid transporter NTCP at the basolateral membrane of hepatocytes by cycling the transporter between an endosomal compartment and the plasma membrane.[165] The phosphorylation status of NTCP is regulated by several signalling pathways, including cyclic adenosine monophosphate, nitric oxide, Ca^{2+} and others.[165] Similarly, phosphorylation of OATPs may, depending on the transporter, lead to down-regulation of its activity or to its internalisation from the plasma membrane.[166] However, intracellular transporters may not always be functionally inactive; *e.g.* in the vinblastine-resistant human cervical carcinoma cell line KBV1, MDR1 is functionally expressed in lysosomes.[167] In addition, ABCA3 (involved in surfactant expression) is expressed in lysosome-like structures in alveolar type II cells,[168] while in BCR/ABL-positive leukemic cells, it is expressed in the lysosomal membrane, regulating imatinib sequestration.[169]

Cells can adapt protein expression levels for DMEs and transporters at the transcriptional level in response to extrinsic factors. Typically, this is achieved by nuclear receptors sensing intracellular ligand levels. These ligands can be drugs, environmental chemicals or endogenous ligands such as bile salts. The key nuclear receptors involved in regulating drug disposition are the pregnane X receptor (PXR), the constitutive androstane receptor (CAR)[170,171] and, for sensing environmental chemicals, the aryl hydrocarbon receptor (AhR).[172] The farnesoid X receptor (FXR) constitutes the bile salt sensor.[173] In order to activate transcription, these receptors need to bind a ligand and form, in the case of FXR, CAR and PXR, a heterodimer with the retinoid X receptor (RXR), and in the case of AhR, with the aryl hydrocarbon nuclear translocator.[172] This is discussed further in *Drug Transporters: Volume 2: Recent Advances and Emerging Technologies*, Chapter 2.

1.3.7.2 *Pharmacogenetics*

Several genetic polymorphisms (gene sequence variations that occur within a population) of transporter genes are known, some of which may cause disease, while others may have no obvious impact but are capable of influencing drug response (by increasing or decreasing the activity of the transporter protein, or by lack of expression). Those that cause disease are relatively rare (as they decrease evolutionary fitness), but do still occur, *e.g.* the genetically inherited Dubin–Johnson syndrome, where individuals have a non-functional MRP2 transporter and subsequent impairment of biliary secretion of both endogenous substrates and drugs.[174] Other genetic diseases associated with transporters include Rotor syndrome, associated with OATP mutations,[175] and respiratory distress syndrome, associated with ABCA3 mutations.[176] It should be noted that Dubin–Johnson and Rotor syndrome are benign human syndromes, meaning that such patients usually do not present with clinically relevant symptoms.

PK can be influenced by genetic variations in transporter genes (caused by single nucleotide polymorphisms (SNPs), insertions/deletions, or a gene or sequence copy number variation), but the field has not been as extensively evaluated as that of the metabolising enzymes. Recent studies have shown that some genetic variations in drug transporters can lead to changes in systemic exposure, as well as potentially affecting local (target) concentrations, although the latter is much more complex to monitor and its impact remains uncertain. Certain transporter polymorphisms appear to have only a minimal impact on drug ADME, including those of the P-gp transporter,[177] although some studies suggest that the SNP C3435T in the MDR1 gene may impact the expression of P-gp in some tissues and be a risk factor for certain diseases.[178,179] Other genetic variants, however, appear to have a substantial impact on drug PK, particularly those of the hepatic uptake transporter OATP1B1 (SNPs c.388 A > G and c.521 T > C)[44,180–182] and the efflux transporter BCRP (SNP c.421 C > A).[183] In some instances, this may cause an increased risk of adverse drug reactions (ADRs), as has been suggested for simvastatin induced myopathy.[184] Muscle toxicity (myopathy), and its extreme form rhabdomyolysis, can occur in association with statin therapy, especially when given at high doses and with certain other medications. The SEARCH genome-wide study identified a strong association between a common variant of OATP1B1 with reduced transport activity and the incidence of muscle myopathy, suggesting that genotyping for these variants could improve the safety of statin therapy. Several other studies have linked genetic variations in transporters to variations in drug exposure and clinical response, albeit on a much smaller scale. For example, genetic variants of the organic cation transporter OCT1 have been linked to changes in the PK and PD of metformin.[185–187] Recent work using PK/PD modelling of *in vitro* data for three genetic variants of OATP1B1 suggests a way forward for predicting the effect of polymorphisms for NMEs during pharmaceutical development.[188]

It has also been noted that the frequency of genetic variants can vary between different ethnic populations, suggesting that certain populations are more susceptible to variations in drug response for some xenobiotics. This has been particularly noted for variants of *SLCO1B1*,[189] and may be responsible at least in part for the observed differences in effect of statin drugs in different ethnic populations.[190,191]

As it has become apparent that some transporter polymorphisms can be a key determinant in inter-individual variability, the more recent pharmacogenetic guidelines from regulatory authorities specifically include references to drug transporter polymorphisms, as well as for metabolising enzymes. For example, the guideline from the European Medicines Agency (EMA) recommends that differences in exposure due to pharmacogenetic factors need to be evaluated during drug development, and provides a useful framework for when further investigations may be required or recommended for NMEs.[192] Papers have also been published on the industrial pharmaceutical perspective, with references to the transporter polymorphisms thought to be most relevant to drug development at present.[193,194]

The field of transporter pharmacogenetics is an emerging and expanding area, and there are ongoing areas of research to further the available knowledge base. One institution of note is the Pharmacogenomics of Membrane Transporters project at the University of California (USA), which has obtained DNA from many different sources for further investigation of functional variants, as well as forming a group of individuals willing to be called back for future pharmacogenetic studies (the SOPHIE cohort).[195]

1.3.7.3   Age, Gender and Disease

Knowledge of the effects of age, gender and disease on human transporter expression is limited in comparison to preclinical species. Many drug transporters have been observed to have age and/or gender specific expression in rodents, which if translatable to humans, could influence the PK and may explain age or gender related PK differences. For example, mouse OATP1A1 (*Slco1a1*) and OATP1A4 (*Slco1a4*) show both developmental and gender differences in expression in mice.[196]

Gender differences in expression have been observed for many more transporters and are summarised in a number of references.[197–199] It should, however, be kept in mind that transporter expression and function, as well as gender differences, are species dependent.[200,201] For example, liver expression of *Abcg2* messenger RNA (encoding BCRP) shows gender differences in mice but not in rats.[200] Expression of OATP1A1 in the apical membrane of the proximal tubule in the rat kidney is much higher in males than females, which may explain the much higher renal elimination of taurocholate and dibromosulfophthalein in females.[202] However, the same transporter is expressed at similar levels in rat liver, irrespective of gender. Similarly, expression of messenger RNA of *Slco1a1* and *Slco1a4* in mice shows gender differences in the liver and kidney, which is also transporter specific.[196] In humans, information on gender differences in expression of both transporters and DMEs is scarce.[203,204] Such differences in expression levels may well be the molecular basis for gender differences in PK.[205]

As DME levels change radically during growth and development from the newborn to adult, so do the expression levels of transporters during ontogenesis.[206,207] In humans, information on expression changes in drug transporters is very limited.[208,209] However, considerable knowledge has been obtained for species used in drug development, such as mice and rats. Interestingly, uptake transporters take longer to achieve adult expression levels than efflux transporters in both rats and mice.[199,210]

In disease states, transporter expression levels may be affected secondary to an underlying disease. Due to its central role in drug disposition, transporter expression in various forms of liver disease has been extensively studied in animal models.[211] In humans, various forms of liver disease lead to altered expression of drug uptake and efflux transporters.[209,212,213] Due to its devastating effect on patients with cancer, expression of MDR1 has been particularly well studied in clinical oncology.[214]

1.3.7.4   Dietary, Environmental and Lifestyle Factors

Dietary effects on drug PK are well documented in the literature. Mechanisms responsible for food–drug interactions may include physiological alterations in intestinal motility, gastric pH, gastric emptying or modulation of disposition pathways by dietary constituents. Clinical food–drug interactions have been attributed mainly to hepatic drug metabolism, with

CYP3A appearing to be particularly sensitive to dietary constituents, including grapefruit juice, garlic and St John's Wort.[215] However, the contribution of drug transporters in food–drug interactions is increasingly being recognised. For example, grapefruit juice can inhibit the activity of both P-gp[216] and some OATPs,[45] and has been reported to alter the PK of cyclosporine and fexofenadine.

Little is known about the impact of environmental and lifestyle factors on transporter form and function. A recently published article describes altered expression levels of DMEs and transporters in some alcoholics,[217] although the underlying mechanism has not been elucidated. Another recent review cites epigenetic and other factors that affect the expression and function of transporters in the placenta during gestation.[218] Finally van der Doelen *et al.* suggest that early life stress and the serotonin transporter (SERT or 5-HTT, *SLC6A4*) genotype are instrumental in mediating DNA methylation of corticotrophin releasing factor, leading to altered responses to stress in adult rats.[219] This area of research will undoubtedly develop, particularly regarding the impact of epigenetic factors on gene expression.

1.4 The Transporter Toolkit

1.4.1 *In Situ* and *In Vitro* Models: Basic Concepts, Limitations and Translation

The function of transporters can be investigated using several different experimental systems.[220] Many transporters have been sequenced,[221] a number have been identified and cloned,[134,222] and they have been characterised in different recombinant expression systems including *Xenopus laevis* oocytes, insect and mammalian cell lines, and even in humanised and knockout mice.[223] Consequently, because these systems are relatively specific for the transporter of interest, and good quality controls are available, the function of individual transporters can be carefully characterised.

Across academia and industry there is a heavy reliance on *in vitro* tools in a range of different formats to evaluate transporter interactions.[223] Immortalised human cell lines such as the colonic adenocarcinoma cell line (Caco-2),[224] and animal or insect cells stably or transiently transfected with one or more human genes [*e.g.* Madin–Darby canine kidney type II cells transfected with MDR1 (MDCKII-MDR1)] are common,[225–227] forming the workhorses for transporter investigations. They are also strongly advocated by regulatory agencies because of their relative ease of use, wide availability and, in the case of transfected cells, range of transporters expressed.[92,93] Notwithstanding the above advantages, these tools require careful and thorough characterisation and validation with suitable methodologies and controls to realise their full potential.[104,225,226,228,229] Isolated primary cells (*e.g.* hepatocytes[230] and kidney proximal tubule cells[231,232]) in a number of different formats can also yield useful and sometimes very elegant data, but are more variable, challenging and expensive than cell lines. Isolated plasma membranes (vesicles) containing the transporter of interest are a non-cell-based variant that can also be helpful, as the medium on both sides of the membrane can be accurately defined by the researcher, and concentration gradients can be established.[233,234] Some assay formats using end-points such as inorganic phosphate from the hydrolysis of ATP (ATPase assay) and a fluorescent assay, measuring intracellular production of calcein from calcein–acetoxymethyl (AM) following inhibition of P-gp efflux (Calcein–AM assay), are also available. The results are difficult to interpret, and often conflict with other more holistic assays.[235–238]

Experiments using whole organ perfusion, where transporters are working close to their native environment, or isolated cells grown on special supports or matrices, maintain many of the complex functions and interactions between proteins, but the contribution of individual transporters can be difficult to assess due to the influence of other mechanisms active in the system.[147]

Most routine *in vitro* assays measure the inhibitory potency of the NME against a well-characterised probe substrate. This approach can be very robust, and gives a good indication as to whether the NME will alter the PK of another drug *in vivo* (*i.e.* be a perpetrator of a DDI), but provides no information as to whether the NME itself is a substrate, and therefore a potential victim of a DDI. Establishing whether the NME is a substrate of a drug transporter can be challenging for a number of reasons, not least because of experimental and post-experimental variables that cannot easily be controlled. For example, the contribution of simple diffusion (Figure 1.1a) to overall transport is compound dependent, can be substantial and is difficult to correct for during data manipulation. Similarly, non-specific binding of the test drug to apparatus, cells or proteins may be difficult to monitor, making data unreliable. Post-experiment, the sample assay requires rigorous control, because analyte concentrations in donor and receiver samples of the relevant *in vitro* system may differ by many orders of magnitude, thus requiring both high sensitivity assays and rigorous sample isolation.[69]

One major problem with *in vitro* systems, and with the results obtained and published, is that not all experiments are performed under optimal conditions that allow determination of the substrate kinetics of transport.[101] For example, kinetic determinations have to be performed under so-called zero-*trans* conditions, where the function of the transporter is not limited by the external substrate concentration.[101] Thus, in instances where insufficient substrate is present, the apparent rate of transport may decrease, but only because the substrate has been depleted. Similarly, if the disappearance of the extracellular substrate concentration is measured, it is normally limiting, and the resulting data are not obtained under initial linear rate conditions. In addition, an uptake transporter may be inhibited by the increasing intracellular concentration of its substrate created by its own activity, reducing the transport rate compared with that observed under optimised conditions.[223,239] Furthermore, transporter functional expression levels should ideally be comparable across the different expression systems;

otherwise, a substrate for a given transporter that is not very efficiently transported might be classified as a non-substrate in a system with low functional transporter expression (*e.g.* estrone-3-sulfate for OATP1B3),[240] but classified as a substrate in one with much higher expression levels.[241]

It is also important to acknowledge that although many substrates and/or inhibitors can be identified using *in vitro* assays, they do not necessarily reflect what will happen *in vivo*. The presence of many other membrane-bound and soluble proteins, the (pharmaco)dynamic environment *in vivo*, and the nature of the extra- and intra-cellular matrices may be profoundly different *e.g.* hepatocytes bathed in albumin rich, circulating whole blood *versus* an albumin free uptake buffer. Therefore, it is necessary to establish and accurately record/report the general conditions under which the *in vitro* transporter assays are performed, as well as the data generated, in order to avoid misinformation due to suboptimal experimental design and/or misleading interpretation.[223]

Regardless of the technique used to generate data, interpretation is challenging, as there are few well-validated quantitative clinical translation methods available. However, physiologically based PK (PBPK) modelling tools (*e.g.* SimCyp and Gastroplus, among others) are rapidly improving and offer a realistic way forward to permit the integration of multiple-compartmental, multi-mechanistic data to model and predict clinical outcomes.[2,57,58,97,150,242,243] Nonetheless, scaling factors for transporters remain a major hurdle, as evidenced in a recent meta-analysis of OATP protein abundance across different hepatic *in vitro* models and whole liver.[244]

1.4.2   *In Vitro* Transporter Inhibition Studies

In order to determine whether a NME is an inhibitor of a particular transporter and to estimate its inhibitory potential (*e.g.* using the concentration of inhibitor required to inhibit transport by 50% (the IC_{50}) or the dissociation constant (K_i)), carefully designed experimental conditions must be maintained. For instance, probe substrate uptake should be measured under initial linear rate conditions in the absence and presence of a potential inhibitor.[223,225,233] These aspects are covered in more detail in Chapter 7. In order to characterise inhibition kinetics for a given transporter protein, substrate and inhibitor are co-applied to the *in vitro* or *ex vivo* system. This will give the most accurate estimation of inhibition kinetic constants. However, there are circumstances, *e.g.* high throughput assays, where it is not feasible to mix the inhibitors with the substrates before addition to the test system. In these cases, any subsequent secondary screens should be performed by co-administration of both inhibitor and substrate. Furthermore, it has become clear recently that several drug transporters (*e.g.* MRPs and OATPs) exhibit substrate dependent modulation; *i.e.* certain small molecules will inhibit the transport of a given substrate but they might not affect or might even stimulate the transport of other substrates.[127,129] The recent US Food and Drug Administration (FDA) guidance recommends testing the inhibitory potential of NMEs against at least one of several prototypical substrates, but given the practical and scientific challenges outlined above, it might be more appropriate to consider testing NMEs as inhibitors of several substrates to give the greatest chance of identifying an *in vitro* inhibitor.[245]

1.4.3   *In Vitro* Transporter Substrate Studies

Substrate studies rely on the same tools as inhibition studies, but are often more challenging to execute than inhibition studies, primarily because it is not possible to optimise experimental conditions in advance to guarantee robust experimental data for every NME. Solubility, non-specific and protein binding, cell toxicity and analytical sensitivity are all compound specific aspects and are generally unpredictable, requiring careful consideration and control in the chosen assay format. Substrates are identified and characterised by comparing drug transport in the presence and absence of a chemical inhibitor or against a non-expressing control, or both.[101,223]

Substrate assessment studies fall into two groups: the binary (positive or negative interaction) assay approach, where the molecule is defined as a substrate or a non-substrate (usually in one to three concentrations);[246,247] and the kinetic approach, where a kinetic constant such as the Michaelis constant (K_m) is determined using a wider range of drug concentrations.[223,248,249] Binary assays are traditionally used for P-gp and sometimes for other ABC transporters such as BCRP, where the transporter is expressed in a cell line. They are usually reasonably robust and sufficiently high throughput to permit a basic ranking of molecules. These assays are also often used to determine simple diffusion of molecules across the plasma membrane. Binary assays are also often used to make a preliminary assessment of substrate activity prior to embarking on a more comprehensive kinetic assessment. In practice, kinetic parameters for transporters are not often generated, as the experiments are time and resource consuming. As for inhibition experiments, substrate experiments require careful design to ensure that measurements are made during the linear phase of uptake and under optimal conditions.[223]

Alternatives to traditional Michaelis–Menten approaches for characterising transporter–substrate interactions have been published. These are generally more data-rich methods, and may provide more robust and translatable data.[226,228,250,251]

1.4.4   *In Vitro* Transporter Induction Studies

The mechanisms of induction of transporter expression appear to be closely linked with those of DME induction, as many of the same receptors (CAR, PXR, RXR, *etc.*) that are implicated in DME induction also induce transporter proteins.[127,252,253] Methodologies to investigate transporter induction therefore tend to be very similar to those employed for DMEs, *i.e.* polymerase chain reaction assays of liver samples, functional and expression assays in cultured hepatocytes, *etc.* Induction in humans has been observed clinically for some transporters.[98,254–257]

The transcriptional activation of an enzyme or transporter of interest can be studied by using the promoter of the gene of interest fused to a readout system such as luciferase in a transactivation reporter assay. Ideally, findings are confirmed using *in vivo* models, whereby the upregulation of the protein of interest is demonstrated. Transcriptional regulation can also occur indirectly by modulating the half-life of a messenger RNA.[258] Understanding the role of microRNAs in the regulation of key mechanisms of drug disposition is a very active and rapidly evolving research field.[259]

1.4.5 *In Vivo* Studies in Preclinical Species and Humans

The utility and limitations of preclinical *in vivo* models has been discussed previously (Section 1.3.6). When clinical investigations of transporter interactions are considered, they usually focus on DDI liability. However, other clinical PK data may yield additional information that can be used to drive and refine a transporter investigation strategy (*e.g.* non-linear PK with an increasing dose or on repeat dosing is indicative of saturation of a mechanism, which could be drug transporter related).

Selecting a clinical probe for transporter investigations is challenging, as there are currently few known probes (either as substrates or inhibitors) that are specific to a given transporter.[101,104,260] Most available clinical probes interact with multiple transporters, or transporters and DMEs. It may also be necessary to investigate the same transporter in different organs. Hence, multiple clinical investigations, and PBPK modelling, may be required in order to reach a point of clarity on the major DDI liability.[256,260] These investigations are necessary to support the safe clinical use of the NME and for inclusion in drug labels.

There is also an increasing trend to use PD or biomarkers such as creatinine clearance (OCTs and MATEs),[144] or levels of unconjugated bilirubin (OATPs)[261] or bile acids (BSEP)[262] as end-points in clinical investigations, acknowledging the fact that transporter interactions influence PD and toxicology as well as PK.[263–266]

1.4.6 Metabolite–Transporter Interactions

Most drug metabolites are more hydrophilic than the parent molecule and because of this are more likely to have a reduced ability to cross plasma membranes solely by simple diffusion. Therefore, drug transporters may be important in the cellular clearance of metabolites, for example, BSEP is inhibited by bosentan and its metabolites[80] or by troglitazone and its sulfated metabolite.[267] This is an area that has received relatively little investigation by academic or industrial transporter scientists to date. However, recent revisions of the regulatory DDI guidance have focused attention on the contribution of metabolites to drug safety and efficacy, obliging sponsors of new drug applications (NDAs) to include an evaluation of these in their submission data.[93,94,100]

1.5 Drug Transporters and PK

In order to understand the role of transporters in drug ADME, a basic understanding of the more common PK parameters can be useful to help determine the relevance, timing and suitability of transporter-related studies. For example, optimising drug clearance remains one of the most common and challenging DMPK activities in the discovery phase before selection of candidates for further investigation (often called candidate selection), as it needs to be low enough to enable an appropriate half-life and bioavailability of the drug candidate, yet not so low as to prolong exposure unnecessarily. Drug clearance is also an important consideration in the later stages of drug development, in order to understand how transporter interactions may affect overall drug disposition in health and disease. The following section discusses various aspects of drug permeability, absorption and clearance, and includes some brief descriptions of the PK parameters that are often considered when deciding upon an overall transporter strategy plan for NMEs.

1.5.1 Permeability

The plasma membrane is composed of a phospholipid bilayer containing a hydrophobic interior, which acts as a barrier to the free transport of solutes into and out of cells. Simple diffusion of drug molecules through the membrane is a key determinant of not only how they will enter the bloodstream, but also their susceptibility to clinically relevant transporter DDIs, and can be described by applying Fick's first law. This states that simple diffusion of a solute is the product of the concentration gradient of the solute across the cell membrane and the diffusivity of the uncharged form of the solute. The flux of a solute across a membrane can be expressed as:

$$J = D_m \frac{dC_m}{dx} \tag{1.1}$$

where J is the flux of the solute in mol cm^{-2} s^{-1}, D_m is the diffusivity of the solute within the membrane in cm^2 s^{-1} and the concentration gradient, dC_m/dx, is the difference in solute concentration inside and outside of the cell across a cell membrane of width dx.[268]

Simple diffusion is often measured early in the drug discovery process. The parallel artificial membrane permeability assay (also known as PAMPA) is suitable for primary screening in high throughput mode and measures the rate of simple diffusion (passive transmembrane) only, but it is possible to add a calculated passive paracellular component by using a biophysical approach[269] (see Figure 1.1). Primary screening may then be followed by lower throughput cell based models such as Caco-2 cells (a human colonic cell line possessing many intestinal transporters), wild-type MDCK cells or MDCKII-MDR1 cells. These cell line models afford an *in vitro* assessment of drug permeability and/or drug efflux potential[270–272] and can be considered to more closely reflect the *in vivo* situation. In order to categorise experimental outputs from these models, it is necessary to calibrate each system using molecules with known permeability characteristics, as the outputs can be variable from laboratory to laboratory and system to system.[271,272] Validation of the permeability method used is also important, in order to establish the rank order relationship between *in vitro* permeability and the human intestinal absorption values for a range of low- to well-absorbed compounds.[270–272]

Permeability is also a key factor that influences routes of drug absorption and elimination, and is used in the Biopharmaceutics Drug Disposition Classification System (BDDCS; covered in Sections 1.5.2 and 1.5.3).

1.5.2 Oral Absorption and Bioavailability

Absorption of a drug into the body will depend on its physicochemical properties, formulation and route of administration. Oral dosing in tablet or capsule form is the preferred route of dose administration unless another route is more appropriate (*e.g.* nasal/inhaled sprays for lung disorders, or creams or patches for dermal application). It requires an in-depth understanding of the NME (for example its solubility and dissolution profile) and the oral absorption process in order to achieve adequate levels of the drug in the bloodstream. As drugs need to be in solution to be absorbed, tablets or capsules must disintegrate and dissolve in the GIT in a reasonable timeframe and in a challenging environment.

Bioavailability describes the systemic availability of a drug. It is defined as the fraction of unchanged drug that reaches the systemic circulation following an extravascular (*e.g.* oral) dose. Bioavailability is the product of drug solubility, permeability, and GIT and hepatic transport/metabolism processes, and can be considered a key determinant of a successful oral drug product. Identifying the biological factors influencing bioavailability, essentially those associated with absorption and metabolism, is therefore important.[273] For example, efflux transporters present at the intestinal membrane may limit the passage of substrate NMEs at sub-saturating concentrations across the GIT, even if they have otherwise high simple diffusion. Conversely, soluble substrate NMEs at high concentrations in the GIT may saturate efflux transporters, thus minimising the impact of transporter interactions on bioavailability. Alternatively, absorptive transporters may enable the absorption of poorly permeable NMEs that are substrates. Compounds with low aqueous solubility, poor simple diffusion and/or high first pass hepatic extraction are usually associated with poor oral bioavailability.

Predicting oral bioavailability is challenging, however a number of different *in vitro*, *in vivo* and *in silico* approaches and systems have been reported over many years.[274–280] One system that has gained considerable acceptance in industry was developed by Amidon and co-workers in 1995.[281] They recognised that the fundamental parameters controlling the rate and extent of oral absorption of drugs were solubility and permeability, and proposed a Biopharmaceutics Classification System (BCS) that categorised drugs accordingly into four classes that could be used to predict the extent of oral drug absorption (Figure 1.4). In this system, class 1 molecules are considered most desirable, as they exhibit the highest levels of permeability and solubility, and are therefore likely to have good oral bioavailability. A modified version of this system, the BDDCS, was proposed some 10 years later to serve as a basis for predicting the importance of drug transporters and DMEs in determining disposition[282] (Figure 1.4). The premise of this system is that highly permeable, lipophilic drugs that are capable of crossing biological membranes are usually good substrates for CYP450 enzymes, and metabolism therefore is the major route of drug elimination. However, for compounds with low permeability that are less lipophilic, metabolism only plays a minor role in drug elimination, and these compounds are primarily eliminated unchanged in the bile and urine. Thus, class 1 compounds are predicted to be well absorbed at the GIT with minimal transporter effects, whereas for class 2 compounds, the GIT efflux transporter effects may predominate, affecting their rate of absorption and extent of oral bioavailability due to their lower solubility. For class 3 compounds, efflux transporter effects at the GIT will be minimal due to their good solubility, but absorptive transporter effects will be prominent due to their low permeability. Class 4 compounds are predicted to be substrates for both absorptive and efflux

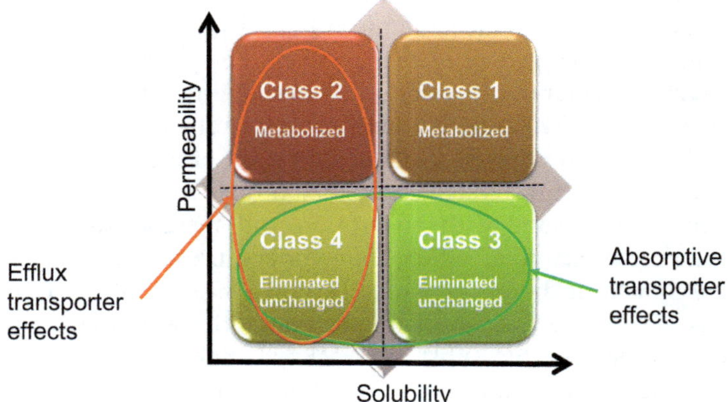

Figure 1.4   The combined BCS and BDDCS. The BCS categorises drugs into four classes according to their solubility and permeability to predict the extent of oral drug absorption as developability criteria. The BDDCS also considers routes of drug elimination and transporter effects on drug ADME. See text for further details.

transporters at the GIT due to their solubility and permeability deficiencies. It is necessary to validate the *in vitro* systems used to test these models with well-characterised molecules in order to establish optimal ranges for each class, otherwise classification of NMEs will be flawed.

In addition to the *in vitro* approaches described above, several experimental and computational models to simulate oral absorption have been developed, including commercial offerings from TNO Pharma[283] (the *in vitro* gastro-intestinal model; TIM) and PBPK modelling tools from Simulations Plus Inc.[284] (GastroPlus) and Certara[285] (SimCyp). These models combine drug physicochemical properties (*e.g.* aqueous solubility/dissolution, permeability, *etc.*) with GIT physiology (*e.g.* blood flow, GIT transit time, pH, *etc.*) and biological mechanisms (*e.g.* drug uptake and efflux transporters and DMEs), and can be useful tools in understanding absorption mechanisms.

1.5.3 Drug Clearance

In addition to oral dosing, drugs can be delivered by other routes such as intramuscular, intravenous, subcutaneous, intranasal, inhaled and dermal. Whatever delivery route is used, once the drug has reached the bloodstream and/or tissues, it will ultimately be removed (*i.e.* eliminated or cleared from the body) unless bound covalently to tissues. The description and calculation of clearance is covered extensively in the literature,[286–288] but in its simplest form it can be considered as the rate at which a drug is removed from the blood (or plasma) primarily (but not exclusively) *via* the liver or kidneys. The mechanism by which clearance is achieved in these organs can be *via* metabolic transformation to more hydrophilic metabolites, transporter-mediated uptake and/or efflux, or by passage of the unchanged drug into bile or urine (or a combination of these processes).[289] The total clearance (CL) is the sum of all these clearance routes:

$$\text{Total CL} = \text{hepatic CL} + \text{renal CL} + \text{other CL} \tag{1.2}$$

Drug clearance can be defined as the volume of blood that would contain the amount of drug eliminated in a given time interval (usually per minute). For first-order kinetics, the rate of elimination of a drug is proportional to its blood concentration, and clearance is described by:

$$\text{CL} = \text{elimination rate constant} \times V_D \tag{1.3}$$

Where volume of distribution, V_D, is a proportionality constant between the total amount of drug in the body and the concentration of the drug in the blood.[290] Each clearance route is described by an equation specific to that organ and is influenced by the extraction ratio of the drug (*i.e.* the relative efficiency of elimination of the drug after a single pass through the organ). The rate of clearance will be influenced not only by the physicochemical properties of the drug (lipophilicity, protein binding ability, ionisation, *etc.*) but also by intrinsic factors (*e.g.* blood flow through the organ) and thus variations can occur if organs are impaired (*e.g.* by disease, age or environmental factors) or if specific genetic polymorphisms of enzymes and/ or transporters are present.

In the kidney, renal clearance reflects the elimination of the drug through a combination of glomerular filtration, active tubular secretion and tubular reabsorption (passive or active).[291] For ionisable drugs, urine pH may also be a factor. Transporters are present on both the uptake (basolateral) and efflux (apical) membranes of the renal tubular cell (Figure 1.3) and, hence, for drugs that undergo active tubular secretion and/or reabsorption, drug clearance can be affected by inhibition or induction of these transporters.[288,292] For some drugs, this transport is the rate determining step in overall clearance.

For the liver, the main factors that determine hepatic clearance are hepatic blood flow (delivery of the drug to the liver, which in turn may be affected by the portal concentration of the drug), uptake of the unbound drug into hepatocytes, metabolic transformation by microsomal or other enzyme systems, and the rate of biliary secretion.[293,294] Thus, if a drug is a substrate of uptake and/or efflux transporters, inhibition or induction of these transporters may influence the overall clearance of the drug. For some drugs, transport either into or out of the hepatocyte is the rate determining step in the overall process and any change will have a direct impact on drug clearance, with potentially significant clinical effects. The statin drug class is the best documented in this respect, although there are several other examples in the literature.

Given the importance of clearance in drug disposition, attempts to predict human clearance from preclinical findings using *in vitro* techniques were developed more than 20 years ago following seminal publications on the well-stirred liver model.[295,296] This model describes the inter-relationship between hepatic clearance, intrinsic clearance (CL_{int}), blood flow and the unbound fraction of drug in the blood, with the term "intrinsic clearance" referring to the innate ability of an organ to excrete a drug when no limitations or barriers (blood flow restrictions, protein binding, *etc.*) exist. The measurement of CL_{int} in drug discovery and development has proved pivotal in the extrapolation of *in vitro* DMPK data to the *in vivo* situation. Prior to extrapolating metabolic hepatic clearance *in vivo*, the enzymatic CL_{int} from microsomes and hepatocytes is first "normalised" with scaling factors to account for the microsomal protein per gram of liver and/or hepatocellularity per gram of liver, as well as liver weight.[297] The relationship between the predicted (extrapolated) clearance *in vitro* and observed clearance *in vivo* is then determined using a mathematical model, referred to in the literature as either *in vitro–in vivo* clearance correlation (IVIVC) or *in vitro–in vivo* clearance extrapolation (IVIVE). Under-prediction of clearance has frequently been associated with the well-stirred model, and more recent extrapolation approaches have been developed to remove this systematic bias.[298–300] The widening gap between predicted and actual drug clearance in recent years has also been attributed to the changing molecular chemistry of pharmaceutical drugs, leading to a greater proportion of drugs with reduced permeability that are substrates of drug transporters.[301] As discussed in Section 1.5.2, the BDDCS system offers a framework to predict elimination routes, with more lipophilic class 1 and 2 compounds cleared by extensive metabolism and less permeable class 3 and 4 compounds primarily eliminated unchanged in bile and urine (Figure 1.4). The less permeable compound classes may therefore be good substrates for absorptive membrane transporters at major organs of elimination, raising free intracellular drug concentrations and facilitating the elimination processes (*e.g.* DMEs or efflux transporters at the hepatocyte sinusoidal or bile canalicular membranes, or renal tubules), resulting in the extensive elimination of unchanged drug in bile and/or urine.

Improvement in clearance predictions for substrates of hepatic transport requires accurate measurement of the intrinsic uptake and/or efflux processes in the hepatocyte using detailed kinetic study designs. Kinetic data may be incorporated into more complex equations to extrapolate and predict clearance, accounting for all five of the hepatobiliary pathways: simple diffusion, transporter-mediated uptake, sinusoidal efflux, metabolism and biliary efflux.[302–304] These can be combined with comprehensive mechanistic compartmental or PBPK models to predict *in vivo* clearance.[242,243,250] In most published cases, successful predictions have been obtained. These methods require detailed, time- and resource-intensive investigations of the mechanisms involved and are therefore impractical during drug discovery where multiple candidates may need to be assessed in a truncated time period. They are more readily applied during the later stages of drug development, where an in-depth understanding of the mechanisms of ADME is required, and a greater body of information is available. However, if there is reason to believe that hepatic uptake and/or efflux transporters may be the rate determining step in hepatic clearance for a class of NMEs, transporter assays with adequate throughput may sometimes be incorporated routinely into early drug discovery programmes,[104] so-called 'frontloading'.

1.6 Evaluating and Interpreting Drug Transporter Interactions in Drug Discovery and Development

Given the complexity of transporter expression, function, location and diversity in the body, it has often proved difficult to establish the precise impact of transporters on the ADE of drugs. This is further complicated by the potential changes in ADE due to interactions of transported drugs with DMEs, additional synergistic or opposing transport processes, or interaction with co-administered or endogenous molecules. Nonetheless, substantial progress has been made in recent years in understanding how transporters can affect the safety and efficacy of drugs and their importance in the clinical situation.[102,305–307] One consequence of this has been the routine incorporation of transporter assays into drug discovery and development strategies.[100,102,308] While such strategies vary between pharmaceutical companies, the overall objectives of the drug discovery and drug development roadmaps are generally the same (Figure 1.5). The initial aim of such strategies, once potential targets for drug action have been identified, is to test a library of compounds against the proposed targets using high throughput combinatorial chemistry approaches to find suitable "hits" (the "hit-to-lead" phase). This approach can test 10^3–10^6 different molecules, of which the "leads" (hits with undesirable characteristics filtered out) are then entered into the lead optimisation and identification (LOID) phase to further refine the process and rank their suitability for further development. By the end of this stage, which may test a few hundred molecules, a shortlist of NMEs is put forward for candidate selection, an important phase of the process where key decisions are made regarding which compounds will progress further. Only a few compounds from each drug class are chosen, with several tests being used to determine basic toxicokinetic (TK) and DMPK properties. Generally, only one NME is nominated for preclinical development, with others being kept as back-up compounds in case the lead compound fails to meet safety and efficacy criteria during the preclinical work. The procedures and methods used in preclinical development become more time consuming, concentrating mainly on *in vitro* methods and animal *in vivo* work, to try to predict ADME and any potential toxicity in humans. If suitable candidates are identified, they transition to the clinical phase and testing on humans can begin—a costly process taking many years as NMEs progress from first time in human (FTIH) in Phase 1 with healthy volunteers, through to Phase 2 and 3 clinical trials to assess their safety and efficacy in human patients. Proof of concept (PoC) studies in Phase 2 are a key phase of the process, to demonstrate clinical efficacy within a small group of patients and eliminate potential failures from the drug pipeline. The few compounds that do reach registration and launch continue to be monitored for unexpected side effects and DDIs; post-marketing (Phase 4) studies can be of benefit in obtaining further information on the risks/benefits of the drug and how to optimise its use.

Selection of drug candidates is a balance between adequate target potency/effect and optimised DMPK properties, to ensure the eventual elimination (clearance) from the body, with minimal DDI potential or ADRs. In this context, transporters are often considered in the later stages of drug discovery (late LOID and candidate selection), after other more critical considerations have been met, *e.g.* optimal molecular chemistry for target effect, minimal toxicity, *etc.* Investigations then generally take a stepwise approach, often with higher throughput assays in the earlier LOID discovery phase followed by more in-depth investigations as molecules progress towards clinical development. The methods used are outlined in more detail in Chapter 7.

1.6.1 Drug Discovery Approaches

The routine incorporation of permeability, oral absorption and bioavailability assessments of NMEs in the early phases of drug discovery have

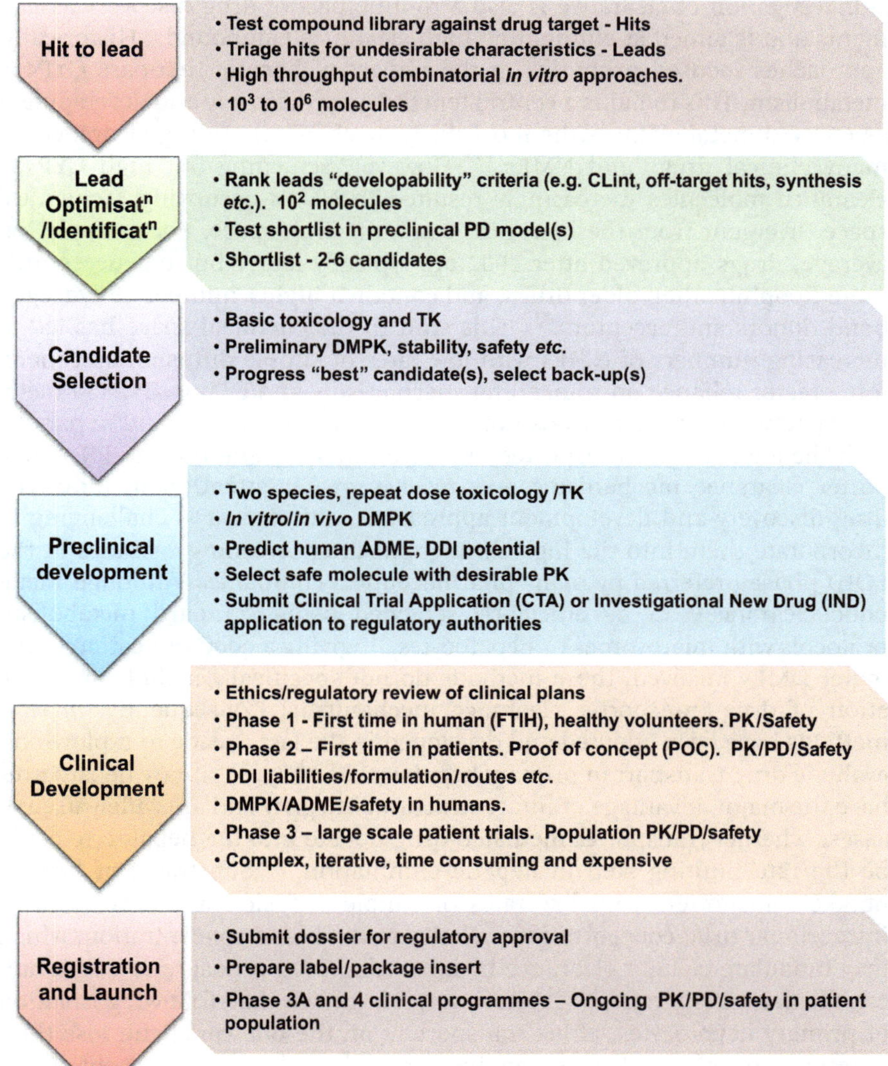

Figure 1.5 Overview of the drug discovery and development process.

already been mentioned in Section 1.5 and they are an important part of drug screening cascades. Thus, the solubility associated factors known to limit the oral absorption of drugs, which include high lipophilicity, high crystal lattice energy and low ionisation potential, can be optimised for some drug classes in early discovery to improve absorption. Optimisation of these factors can be productive, since simple diffusion is considered to be the most common mechanism by which compounds pass across the intestinal membrane, although this is disputed, as discussed elsewhere.[86,87]

Investigation of clearance is also a routine part of drug discovery assessments and is aimed at eliminating high clearance compounds. Historically, approaches focused primarily on the impact of hepatic first-pass CYP450 metabolism. This remains a central tenet of drug discovery and development as CYP450 metabolism is the most likely mechanism of drug clearance for many clinical drugs and NMEs.[309] However, screening out high CYP450 clearance molecules increasingly resulted in NMEs occupying a chemical space divergent from the historical common drug space. For example, on average, drugs approved after 2002 are typically larger but not necessarily more lipophilic than older drugs, and contain a higher number of hydrogen bond donors and acceptors.[310] This shift in the chemical space has led to increasing numbers of NMEs with low rates of simple diffusion and therefore greater reliance on transporter mechanisms and/or non-CYP450 metabolic pathways for their disposition.[294,305,311] Because of this, and in parallel with the advances in transporter science and its integration into PK, transporter clearance mechanisms now receive greater attention in industrial drug discovery and development approaches, although it is challenging to incorporate them into the high throughput drug screening cascades in the LOID phase preferred by many pharmaceutical companies. Although metabolic clearance can be efficiently screened using standard metabolism protocols with microsomes or hepatocytes,[312] giving a good indication of the major DMEs involved, these methods do not specifically include an evaluation of drug transporter clearance mechanisms. Consequently, *in vitro* methods have been adapted and developed in the last decade to explore and evaluate drug transport in primary hepatocytes.[66,250,313] Primary hepatocytes have the major advantage of intact structural integrity and are widely used to assess whether transporter-mediated drug uptake into the hepatocyte could be the rate limiting step in hepatic elimination.[294] Substrates of hepatic uptake transporters with low rates of simple diffusion will have elevated intracellular drug concentrations relative to the blood concentration, which may modulate cellular clearance by increasing drug exposure to DMEs and canalicular transporters.[304,314] However, it is known that, following isolation of primary hepatocytes, efflux transporters on the bile canaliculi lose their functionality, although this can be re-established to some extent if the cells are cultured in a collagen sandwich configuration or another three-dimensional cell culture system.[220,315–322] The sandwich-cultured hepatocyte model is useful for studying both uptake and biliary efflux transport of substrates *in vitro*.[323] Human hepatocytes in particular are subject to donor-to-donor and batch-to-batch variability. They are also more costly to obtain than cell lines, and this influences their routine use in industry.

Other clearance routes, *e.g.* renal elimination, usually receive little attention at these early stages of drug discovery unless there are known issues with a specific class of compounds. In these instances, specific screens or *in silico* approaches may be used to evaluate potential transporter issues during LOID (*e.g.* OCTs in the kidney for cisplatin, or OATs for non-steroidal anti-inflammatory drugs).

The investigation of specific transporter DDIs is often limited to projects where there is a historical precedent of transporter-mediated DDIs, which may be actively managed, or screened out. However, considering that certain widely prescribed drug classes are notable clinical substrates of drug transporters (*e.g.* statins and OATP1B1 and OATP1B3, digoxin and P-gp), there is an increasing trend towards early evaluation and screening out of these liabilities, and many pharmaceutical companies now include routine screening for P-gp and OATP1B1 inhibition in LOID.

1.6.2   Drug Development Approaches

Investigation of NMEs during development focuses mainly on understanding the properties and interactions of individual molecules, with a view to ensuring that drug candidates are both safe and efficacious in clinical use (Figure 1.5). There is a greater emphasis on evaluating the impact of transporter DDIs on drug PK, especially when there are known liabilities associated with a specific class of compounds, ensuring that NDAs are compliant with the current regulatory guidance.

1.6.2.1   *Regulatory Landscape*

Membrane transporter interactions were first included in regulatory guidance in 2006, as the importance of P-gp in clinically relevant DDIs became apparent.[324] As knowledge expanded, there was a growing awareness that the PK of a number of widely prescribed drugs (*e.g.* statins, metformin, digoxin and methotrexate) could be influenced by a range of drug transporters. In order to identify these transporters, and propose recommendations to guide preclinical and clinical studies, a consortium of industrial, regulatory and academic scientists with expertise in drug metabolism, transport and PK, named the ITC was formed. Its recommendations were subsequently incorporated into new regulatory guidance, which provides transporter investigation proposals and decision trees based on current knowledge.[93,94,100] Transporter-mediated DDIs now receive substantial attention in the drug development process[103,104] and regulatory guidance is often used to justify experimental strategies. The guidance relies heavily on a decision tree approach to develop an investigational strategy, making recommendations on appropriate selection and use of *in vitro* tools, and of clinical strategies. Although useful, it should be noted that decision trees can encourage a simplistic approach, which may not always be appropriate. In particular, the guidance recommends the use of cut-off values based upon a ratio of *in vitro* inhibitory potential (using IC_{50} or K_i values) *versus* circulating drug concentrations, assuming that *in vitro* inhibition values are wholly predictive of *in vivo* inhibition. In practice, this may not always be the case; as discussed in Section 1.4, *in vitro* components, study design and substrate/inhibitor selection can markedly influence inhibitory constant estimates. Similarly, the decision trees do not take into account specific drug classes; previous work may have indicated their susceptibility to interactions with additional transporters, some of which may not be included in the current guidance. However, the regulatory guidance acknowledges that transporter science is developing rapidly, and recommends that any experimental strategies should be devised with reference to the latest developments in the field.

1.6.2.2 Investigating DDI Liabilities and ADRs

Whereas transporter DDI investigations in early drug discovery focus primarily on screening out or actively managing potential liabilities, drug development approaches focus more on describing and understanding their impact on the safety and efficacy of the NME under investigation. A good drug product will have an acceptable safety, efficacy and toxicity profile, and ideally will have no ADRs, especially if the target population is one in which patients take several different medications (polypharmacy). Consequently, any transporter interaction will have the greatest relevance when it limits the clinical use of a NME because of a DDI with a co-medicated compound or endogenous compound, or if its transport through a membrane transporter leads to a response that is unpredictable or toxic. If oral bioavailability is altered, systemic exposure may be erratic, enhanced or reduced (depending on the nature and site of the interaction), which may increase the risk of unexpected DDIs and limit polypharmacy. Many drug transporters are inducible, or may have genetic polymorphisms that alter their function, which adds further complexity in predicting and understanding the impact of transporters on PK and patient safety.[43,127,253,325,326] Although in theory transporter-related ADRs could lead to the discontinuation of development of a NME, generally this only occurs as part of a wider consideration of its product profile, which can include its therapeutic target, pharmacological potency, patient population, expected co-medications and commercial considerations (Figure 1.5).

Given the complexity of possible interactions, the general approach for a NME entering the development phase is to create a transporter strategy tailored to that specific NME. While some studies are routinely executed in order to comply with the regulatory guidance, it is also necessary to take into account the physicochemical properties of the molecule, the existing preclinical data, the potential co-medications of the target population(s) and any data from previous candidates of the same compound class that exhibited specific transporter interactions. This can be done in several ways and some of the possible points that may be raised include, for example:

- What is the permeability or BDDCS class[282,327,328] of the NME? If poorly permeable, or categorised as a BDDCS class 3 or 4 molecule, the NME is more likely to interact with transporters, which will influence its ADME (see Figure 1.4).
- Do other physicochemical properties of the NME indicate the possibility of transporter involvement? For example, anion or cation

transporters may be implicated if the NME carries a net charge at physiological pH.

- Are there indications of non-linear or erratic PK that cannot be fully explained by other factors such as poor solubility/dissolution, poor simple diffusion/permeability or DME interactions, *etc.*?
- Is the NME systemically cleared predominantly by metabolism or eliminated unchanged? If unchanged, transporters may be an important clearance mechanism, particularly if non-linear PK is observed. Drug metabolites may also be substrates or inhibitors of transporters and may need to be considered if they have high circulating or tissue concentrations.
- What is the major route of clearance of the NME—is it eliminated hepatically or renally? For example, an anionic drug may interact with OATs in the kidney (*e.g.* non-steroidal anti-inflammatory drugs) or OATPs in the liver (*e.g.* rosuvastatin).

In addition to reviewing the properties of the NME, it may also be appropriate to consider other drug development aspects, for example:

- What are the common co-medications in the target population and are any transporters implicated in their PK or DDI profiles? What is the therapeutic dose level or concentration of the NME? The higher the dose level and/or systemic concentrations, the greater the possibility that the NME will cause a DDI.
- What is the therapeutic window of the NME (and any co-medications, if relevant)? If the margin between the effective drug concentration and the toxic concentration is small, then ADRs may occur more readily in the event of a DDI.
- What are the demographics of the target patient population (*e.g.* the young or elderly, ethnically biased)? Older patients often take many medications, leading to a greater risk of DDI liabilities; some drug transporter polymorphisms are more prevalent in certain ethnic populations, which can restrict the use of specific drugs in these populations.

Once all of the data have been reviewed, the individual plans for the NME are prepared and will include specific, in-depth assays of any transporters that may give rise to an ADR (as a substrate and/or inhibitor). Regulatory guidance is also consulted to ensure that the investigations will be compliant with current thinking, especially if the preclinical results indicate that clinical interaction studies are warranted. Many pharmaceutical companies now work with the regulatory authorities on these issues well before product launch. In practice, there are few if any hard and fast rules governing investigation of transporters in drug development. Discussions and plans should consider current transporter knowledge, including a risk–benefit analysis. Further discussion around this topic can be found in Chapter 10.

1.6.2.3 "Liability" Transporters

The list of drug transporters associated with DDI or ADRs (so-called liability transporters) continues to expand. Currently, regulatory authorities recommend that a minimum of seven transporters be considered as drug interaction targets (Table 1.1). These are P-gp, BCRP, OATP1B1, OATP1B3, OAT1, OAT3 and OCT2, with some regulatory bodies suggesting the inclusion of OCT1, BSEP, and possibly MATE1 and MATE2K (as detailed in Chapter 11). The guidance clearly advises the sponsors of NDAs that the list is not comprehensive, rather that investigations should be driven by the science and knowledge of the NME. Future published guidelines are likely to include other transporters such as MATE1, MATE2K and MRP2 in their recommended lists.

Liability transporters are assigned that status because there is a considerable body of evidence, usually a combination of *in vitro* mechanistic studies and clinical investigations, to implicate a given transporter in an ADR or DDI. For some transporters, such as P-gp and OATPs, the body of evidence overwhelmingly supports the designation. For others, particularly those more recently investigated, the body of evidence may be less conclusive. The common consideration for all liability transporters comes down to patient safety, which must take precedence for the pharmaceutical industry, for clinicians and for those responsible for approving medicines. However, it remains difficult to undertake definitive clinical investigations prior to the launch of the product due to the current lack of specific transporter probes for use in clinical trials.

1.7 Toxicity and Transporters

Toxicity due to transporter interactions is of concern and investigations are often a result of accumulated data/observations from drug discovery and development, and from drugs in clinical use. If toxicity is observed routinely for a given drug series, tests may be incorporated into the discovery programme to reduce or screen out the liability. For example, repeat dosing studies are routinely conducted in preclinical species to investigate PD, efficacy and tolerability in LOID or during candidate selection. These studies often include basic clinical chemistry assessments to evaluate and monitor potential organ toxicity and may give an indication of potential transporter mediated interactions. For example, bilirubin, a by-product of heme catabolism from red blood cells that is mainly eliminated in the liver, is a readily monitored circulating biomarker that can indicate hepatotoxicity. It enters hepatocytes *via* OATP1B1 and OATP1B3, and is conjugated by uridine 5′-diphospho-glucuronosyltransferases before being effluxed into bile *via* MRP2. Changes in plasma bilirubin (conjugated and unconjugated) levels may occur because of adaptive non-toxic changes or alternatively severe toxicity. Hyperbilirubinaemia induced by OATP and/or MRP2 is considered benign, so long as no other clinical chemistry findings or evidence of more severe hepatotoxicity are observed. However, these observations become critical for potential "Hy's cases" (hepatocellular DILI with jaundice), which would result in withdrawal of the drug from clinical use.

Once hyperbilirubinaemia has been observed *in vivo*, studies are performed during drug discovery to identify the mechanism, by investigating inhibition of hepatic transporters and/or enzymes, to gain a better understanding of the risk. Putting *in vitro* inhibition potency (IC_{50} or K_i) into context with *in vivo* exposure, using parameters such as the maximum observed drug concentration in blood (C_{max}) or the inhibitor concentration (I), may be useful to predict and understand the risk. The 2012 FDA guidance[92] is a useful reference to classify the magnitude of potential DDIs using this approach. If transporters are implicated, drug discovery activities will focus on reducing the inhibition liability in the chemical series through structure–activity relationship cycles.

BSEP, which constitutes the rate-limiting step in the biliary clearance of bile salts, is also considered to be of potential concern in terms of toxicity because its inhibition has been proposed to play a role in DILI.[329,330] However, DILI as a consequence of BSEP mechanisms may be complex and involve both direct and indirect inhibition of BSEP.[83] In recent years, *in vitro* BSEP inhibition assays[331] have been introduced into drug discovery screening strategies. However, *in vitro* BSEP inhibition data alone can be misleading, as they appear a poor predictor of DILI severity. Rather, it is suggested that consideration of the dosing regimen, route of administration and efficacious exposure (*e.g.* unbound plasma C_{max}) in the context of BSEP inhibition improves the assessment of DILI risk in humans.[332] The obvious challenges in making adequate risk assessments in discovery settings are related to the lack of knowledge on human efficacious exposure. There is also a need for assay panels that are physiologically relevant and can translate *in vitro* BSEP inhibition to meaningful predictions of the effects *in vivo* with respect to changes in bile acid profiles (*e.g.* in plasma, bile and urine) and increased plasma bile acids.[333]

1.8   Conclusions and Future Directions

Drug transporter science has progressed rapidly in the last few decades as the importance of transporters in human health and disease, as well as in pharmaceutical efficacy and safety, has become apparent. Given the extensive range, diversity of action and widespread tissue expression of transporters, it is likely to be some time before both the ABC and SLC superfamilies are fully characterised and their functions understood. Transporter interactions with both drugs and endogenous compounds are now widely acknowledged within the clinical and pharmaceutical sciences, and substantial efforts are being made to understand their impact on safe drug use. It is known that transporters can modulate the PK and PD of many drugs, and can influence, in negative and positive ways, all aspects of ADME, including DDIs and some toxicities. Hence, they are of specific interest to drug discovery and development scientists and clinicians. More recent work has shown that transporters are also subject to clinically relevant functional polymorphisms and mutations, an area that will require further study to fully understand its potential impact on drug use in populations with a higher incidence of specific polymorphisms.

As the action of transporters can influence other biological (*e.g.* metabolism and secretory) and physicochemical (*e.g.* simple diffusion) mechanisms, their impact ideally should be assessed by considering them as part of a more holistic system, *e.g.* in a mechanistic PBPK model, or in a clinical setting. This requires good preclinical experimental methods, which are continually improving. However, *in vitro* techniques generally require transporters to be expressed in a plasma membrane, in a specific orientation and in closed systems such as cells or vesicles, and their investigation and extrapolation to the human situation remains challenging. Similarly, *in vivo* studies in animals can be problematic due to species differences, although work is ongoing to try to overcome this by using transgenic and humanised animal models. Despite the use of all of these experimental approaches, many aspects of transporter biology, form, function, regulation and interplay are not fully elucidated or understood. Methods of transporter investigation and parameters derived from them have yet to be standardised and, until this happens, conflicting information will continue to be presented in the literature, as well as impacting on the predictive power of IVIVE tools. *In vivo* measurement of intracellular concentrations of drugs in tissues also requires improvements to allow a more realistic prediction of the impact of drug transporters in the clinic. As the science progresses and techniques evolve, demonstrations of ever more complex interactions such as the transporter-metabolism–transporter interplay can be expected, as well as integration of transporters into biochemical and metabolic schemes, acknowledging their pivotal roles in drug and solute disposition within the body.

References

1. A. F. Hofmann and L. R. Hagey, *J. Lipid Res.*, 2014, **55**, 1553.
2. A. F. Hofmann, G. Molino, M. Milanese and G. Belforte, *J. Clin. Invest.*, 1983, **71**, 1003.
3. U. Beuers and J. L. Boyer in *The Growth of Gastroentrologic Knowledge During the Twentieth Century*, ed. J. B. Kirsner, Lea & Febiger, Philadelphia, 1994, p. 267.
4. H. Tappeiner, *Sitzungsber. Kais. Akad. Wissensch. Wien*, 1878, 77, 281.
5. E. K. Marshall and J. L. Vickers, *Bull. Johns Hopkins Hosp.*, 1923, **34**, 1.
6. L. G. Rowntree, S. H. Hurwith and A. L. Bloomfield, *Johns Hopkins Hosp. Bull.*, 1913, **24**, 378.
7. J. J. Abel and L. G. Rowntree, *J. Pharmacol. Exp. Ther.*, 1909, **1**, 231.
8. R. K. Crane and T. H. Wilson, *J. Appl. Physiol.*, 1958, **12**, 145.
9. E. Riklis and J. H. Quastel, *Can. J. Biochem. Physiol*, 1958, **36**, 347.
10. H. Murer and U. Hopfer, *Proc. Natl. Acad. Sci. U. S. A.*, 1974, **71**, 484.

11. E. P. Abraham, A. D. Gardner, E. Chain, N. G. Heatley, C. M. Fletcher, M. A. Jennings and H. W. Florey, *Lancet*, 1941, **6155**, 177.
12. K. H. Beyer, R. Woodward, L. Peters, W. F. Verwey and P. A. Mattis, *Science*, 1944, **100**, 107.
13. K. H. Beyer, H. F. Russo, E. K. Tillson, A. K. Miller, W. F. Verwey and S. R. Gass, *Am. J. Physiol.*, 1951, **166**, 625.
14. W. P. Boger, J. O. Beatty, F. W. Pitts and H. F. Flippin, *Ann. Intern. Med.*, 1950, **33**, 18.
15. A. B. Gutman, *Bull. N. Y. Acad. Med.*, 1951, **27**, 144.
16. A. B. Gutman and T. F. Yu, *Trans. Assoc. Am. Physicians*, 1951, **64**, 279.
17. N. Robbins, S. E. Koch, M. Tranter and J. Rubinstein, *Cardiovasc. Toxicol.*, 2012, **12**, 1.
18. J. A. Sheps and V. Ling, *Pflugers Arch.*, 2007, **453**, 545.
19. M. M. Gottesman and V. Ling, *FEBS Lett.*, 2006, **580**, 998.
20. K. Dano, *Biochim. Biophys. Acta*, 1973, **323**, 466.
21. V. Ling and L. H. Thompson, *J. Cell. Physiol.*, 1974, **83**, 103.
22. P. Gros, J. Croop and D. Housman, *Cell*, 1986, **47**, 371.
23. C. F. Higgins, I. D. Hiles, K. Whalley and D. J. Jamieson, *EMBO J.*, 1985, **4**, 1033.
24. S. V. Ambudkar, I. H. Lelong, J. Zhang, C. O. Cardarelli, M. M. Gottesman and I. Pastan, *Proc. Natl. Acad. Sci.*, 1992, **89**, 8472.
25. A. B. Shapiro and V. Ling, *J. Biol. Chem.*, 1995, **270**, 16167.
26. A. Palmeira, E. Sousa, M. H. Vasconcelos and M. M. Pinto, *Curr. Med. Chem.*, 2012, **19**, 1946.
27. C. H. Morris, J. E. Christian, T. S. Miya and W. G. Hansen, *J. Pharm. Sci.*, 1970, **59**, 325.
28. P. G. Waser and J. Reller, *Agents Actions*, 1972, **2**, 170.
29. H. Suzuki, H. Furukawa, H. Miyazaki and M. Hashimoto, *Radioisotopes*, 1977, **26**, 152.
30. R. C. Thomas, R. S. Hsi, H. Harpootlian and R. W. Judy, *J. Pharm. Sci.*, 1975, **64**, 1360.
31. A. H. Schinkel, U. Mayer, E. Wagenaar, C. A. Mol, L. van Deemter, J. J. Smit, M. A. van der Valk, A. C. Voordouw, H. Spits, O. van Tellingen, J. M. Zijlmans, W. E. Fibbe and P. Borst, *Proc. Natl. Acad. Sci. U. S. A.*, 1997, **94**, 4028.
32. A. H. Schinkel, E. Wagenaar, C. A. Mol and L. van Deemter, *J. Clin. Invest.*, 1996, **97**, 2517.
33. W. Löscher and H. Potschka, *J. Pharmacol. Exp. Ther.*, 2002, **301**, 7.
34. M. Chishty, A. Reichel, J. Siva, N. J. Abbott and D. J. Begley, *J. Drug Target.*, 2001, **9**, 223.
35. V. V. Rao, J. L. Dahlheimer, M. E. Bardgett, A. Z. Snyder, R. A. Finch, A. C. Sartorelli and D. Piwnica-Worms, *Proc. Natl. Acad. Sci. U. S. A.*, 1999, **96**, 3900.
36. Y. Tanigawara, N. Okamura, M. Hirai, M. Yasuhara, K. Ueda, N. Kioka, T. Komano and R. Hori, *J. Pharmacol. Exp. Ther.*, 1992, **263**, 840.

37. I. A. de Lannoy and M. Silverman, *Biochem. Biophys. Res. Commun.*, 1992, **189**, 551.
38. T. Mikkaichi, T. Suzuki, T. Onogawa, M. Tanemoto, H. Mizutamari, M. Okada, T. Chaki, S. Masuda, T. Tokui, N. Eto, M. Abe, F. Satoh, M. Unno, T. Hishinuma, K.-I. Inui, S. Ito, J. Goto and T. Abe, *Proc. Natl. Acad. Sci. U. S. A.*, 2004, **101**, 3569.
39. J. W. Higgins, D. W. Bedwell and M. J. Zamek-Gliszczynski, *Drug Metab. Dispos.*, 2012, **40**, 1170.
40. X. Li, X. Sun, J. Chen, Y. Lu, Y. Zhang, C. Wang, J. Li, Q. Zhang, D. Zhao and X. Chen, *Xenobiotica*, 2014, 1.
41. N. Kimura, S. Masuda, Y. Tanihara, H. Ueo, M. Okuda, T. Katsura and K.-I. Inui, *Drug Metab. Pharmacokinet.*, 2005, **20**, 379.
42. D.-S. Wang, J. W. Jonker, Y. Kato, H. Kusuhara, A. H. Schinkel and Y. Sugiyama, *J. Pharmacol. Exp. Ther.*, 2002, **302**, 510.
43. L. Badolo, B. Jensen, C. Säll, U. Norinder, P. Kallunki and D. Montanari, *Xenobiotica*, 2014, 1.
44. M. Niemi, M. K. Pasanen and P. J. Neuvonen, *Pharmacol. Rev.*, 2011, **63**, 157.
45. H. Glaeser, D. G. Bailey, G. K. Dresser, J. C. Gregor, U. I. Schwarz, J. S. McGrath, E. Jolicoeur, W. Lee, B. F. Leake, R. G. Tirona and R. B. Kim, *Clin. Pharmacol. Ther.*, 2007, **81**, 362.
46. S. Matsushima, K. Maeda, H. Hayashi, Y. Debori, A. H. Schinkel, J. D. Schuetz, H. Kusuhara and Y. Sugiyama, *Mol. Pharmacol.*, 2008, **73**, 1474.
47. H. Izzedine, V. Launay-Vacher and G. Deray, *AIDS*, 2005, **19**, 455.
48. K. Nezasa, K. Higaki, M. Takeuchi, M. Nakano and M. Koike, *Xenobiotica*, 2003, **33**, 379.
49. R. H. Ho, R. G. Tirona, B. F. Leake, H. Glaeser, W. Lee, C. J. Lemke, Y. Wang and R. B. Kim, *Gastroenterology*, 2006, **130**, 1793.
50. D. W. Schneck, B. K. Birmingham, J. A. Zalikowski, P. D. Mitchell, Y. Wang, P. D. Martin, K. C. Lasseter, C. D. A. Brown, A. S. Windass and A. Raza, *Clin. Pharmacol. Ther.*, 2004, **75**, 455.
51. P. J. Neuvonen, M. Niemi and J. T. Backman, *Clin. Pharmacol. Ther.*, 2006, **80**, 565.
52. P. J. Neuvonen, *Curr. Opin. Invest. Drugs*, 2010, **11**, 323.
53. A. J. Allred, C. J. Bowen, J. W. Park, B. Peng, D. D. Williams, M. B. Wire and E. Lee, *Br. J. Clin. Pharmacol.*, 2011, **72**, 321.
54. R. G. Tirona, *Clin. Pharmacol. Ther.*, 2005, **78**, 311.
55. H. Lennernäs, *Clin. Pharmacokinet.*, 2003, **42**, 1141.
56. K. M. Morrissey, C. C. Wen, S. J. Johns, L. Zhang, S.-M. Huang and K. M. Giacomini, *Clin. Pharmacol. Ther.*, 2012, **92**, 545.
57. R. H. Rose, S. Neuhoff, K. Abduljalil, M. Chetty, A. Rostami-Hodjegan and M. Jamei, *CPT: pharmacometrics Syst. Pharmacol.*, 2014, **3**, e124.
58. S. Bosgra, E. van de Steeg, M. L. Vlaming, K. C. Verhoeckx, M. T. Huisman, M. Verwei and H. M. Wortelboer, *Eur. J. Pharm. Sci.*, 2014, **65C**, 156.

59. L. Wang and D. H. Sweet, *AAPS J.*, 2013, **15**, 53.
60. Y. Lai, M. Varma, B. Feng, J. C. Stephens, E. Kimoto, A. El-Kattan, K. Ichikawa, H. Kikkawa, C. Ono, A. Suzuki, M. Suzuki, Y. Yamamoto and L. Tremaine, *Expert Opin. Drug Metab. Toxicol.*, 2012, **8**, 723.
61. J. Imanaga, T. Kotegawa, H. Imai, K. Tsutsumi, T. Yoshizato, T. Ohyama, Y. Shirasaka, I. Tamai, T. Tateishi and K. Ohashi, *Pharmacogenet. Genomics*, 2011, **21**, 84.
62. I. Ieiri, Y. Doi, K. Maeda, T. Sasaki, M. Kimura, T. Hirota, T. Chiyoda, M. Miyagawa, S. Irie, K. Iwasaki and Y. Sugiyama, *J. Clin. Pharmacol.*, 2012, **52**, 1078.
63. P. Nordell, S. Winiwarter and C. Hilgendorf, *Mol. Pharm.*, 2013, **10**, 4443.
64. J. Kindla, M. F. Fromm and J. König, *Expert Opin. Drug Metab. Toxicol.*, 2009, **5**, 489.
65. I. Bachmakov, H. Glaeser, M. F. Fromm and J. König, *Diabetes*, 2008, **57**, 1463.
66. J. Noé, R. Portmann, M.-E. Brun and C. Funk, *Drug Metab. Dispos.*, 2007, **35**, 1308.
67. Y. Shitara, T. Itoh, H. Sato, A. P. Li and Y. Sugiyama, *J. Pharmacol. Exp. Ther.*, 2003, **304**, 610.
68. H. van de Waterbeemd and E. Gifford, *Nat. Rev. Drug Discovery*, 2003, **2**, 192.
69. X. Chu, K. Korzekwa, R. Elsby, K. Fenner, A. Galetin, Y. Lai, P. Matsson, A. Moss, S. Nagar, G. R. Rosania, J. P. F. Bai, J. W. Polli, Y. Sugiyama and K. L. R. Brouwer, *Clin. Pharmacol. Ther.*, 2013, **94**, 126.
70. A. Benigni and G. Remuzzi, *Lancet*, 1999, **353**, 133.
71. P. L. van Giersbergen, A. Treiber, M. Clozel, F. Bodin and J. Dingemanse, *Clin. Pharmacol. Ther.*, 2002, **71**, 253.
72. C. Weber, R. Gasser and G. Hopfgartner, *Drug Metab. Dispos.*, 1999, **27**, 810.
73. A. Treiber, R. Schneiter, S. Hausler and B. Stieger, *Drug Metab. Dispos.*, 2007, **35**, 1400.
74. A. Treiber, R. Schneiter, S. Delahaye and M. Clozel, *J. Pharmacol. Exp. Ther.*, 2004, **308**, 1121.
75. K. Fattinger, V. Cattori, B. Hagenbuch, P. J. Meier and B. Stieger, *Hepatology*, 2000, **32**, 82.
76. S. R. Vavricka, J. Van Montfoort, H. R. Ha, P. J. Meier and K. Fattinger, *Hepatology*, 2002, **36**, 164.
77. J. Dingemanse and P. L. van Giersbergen, *Clin. Pharmacokinet.*, 2004, **43**, 1089.
78. P. L. van Giersbergen, A. Treiber, R. Schneiter, H. Dietrich and J. Dingemanse, *Clin. Pharmacol. Ther.*, 2007, **81**, 414.
79. H. Krum, R. J. Viskoper, Y. Lacourciere, M. Budde and V. Charlon, *N. Engl. J. Med.*, 1998, **338**, 784.
80. K. Fattinger, C. Funk, M. Pantze, C. Weber, J. Reichen, B. Stieger and P. J. Meier, *Clin. Pharmacol. Ther.*, 2001, **69**, 223.

81. J. Noe, B. Stieger and P. J. Meier, *Gastroenterology*, 2002, **123**, 1659.
82. B. Stieger, K. Fattinger, J. Madon, G. A. Kullak Ublick and P. J. Meier, *Gastroenterology*, 2000, **118**, 422.
83. B. Stieger, *Drug Metab. Rev.*, 2010, **42**, 437.
84. L. Fouassier, N. Kinnman, G. Lefevre, E. Lasnier, C. Rey, R. Poupon, R. P. Elferink and C. Housset, *J. Hepatol.*, 2002, **37**, 184.
85. Y. Mano, T. Usui and H. Kamimura, *Biopharm. Drug Dispos.*, 2007, **28**, 13.
86. D. B. Kell and S. G. Oliver, *Front. Pharmacol.*, 2014, **5**, 1.
87. D. B. Kell, *Trends Pharmacol. Sci.*, 2015, **36**, 15.
88. Y. Raviv, H. B. Pollard, E. P. Bruggemann, I. Pastan and M. M. Gottesman, *J. Biol. Chem.*, 1990, **265**, 3975.
89. C. F. Higgins and M. M. Gottesman, *Trends Biochem. Sci.*, 1992, **17**, 18.
90. S. Balaz, *Drug Discov. Today*, 2012, **17**, 1079.
91. K. Sugano, M. Kansy, P. Artursson, A. Avdeef, S. Bendels, L. Di, G. F. Ecker, B. Faller, H. Fischer, G. Gerebtzoff, H. Lennernaes and F. Senner, *Nat. Rev. Drug Discovery*, 2010, **9**, 597.
92. Food and Drugs Administration, http://www.fda.gov/downloads/drugs/GuidancecomplianceRegulatoryInformation/Guidances/UCM292362.pdf.
93. European Medicines Agency, http://www.ema.europa.eu/docs/en_GB/document_library/Scientific_guideline/2012/07/WC500129606.pdf.
94. N. Nagai, *Drug Metab. Pharmacokinet.*, 2010, **25**, 3.
95. R. A. B. van Waterschoot and A. H. Schinkel, *Pharmacol. Rev.*, 2011, **63**, 390.
96. L. Z. Benet, *Mol. Pharm.*, 2009, **6**, 1631.
97. A. S. Darwich, S. Neuhoff, M. Jamei and A. Rostami-Hodjegan, *Curr. Drug Metab.*, 2010, **11**, 716.
98. H. X. Zheng, Y. Huang, L. A. Frassetto and L. Z. Benet, *Clin. Pharmacol. Ther.*, 2009, **85**, 78.
99. http://www.genenames.org/genefamilies/ABC.
100. U.S. Food and Drugs Administration, http://www.fda.gov/Drugs/DevelopmentApprovalProcess/DevelopmentResources/DrugInteractionsLabeling/ucm080499.htm.
101. M. J. Zamek-Gliszczynski, C. a. Lee, A. Poirier, J. Bentz, X. Chu, H. Ellens, T. Ishikawa, M. Jamei, J. C. Kalvass, S. Nagar, K. S. Pang, K. Korzekwa, P. W. Swaan, M. E. Taub, P. Zhao, A. Galetin and C. International Transporter, *Clin. Pharmacol. Ther.*, 2013, **94**, 64.
102. Y. Lai and P. Hsiao, *Curr. Pharm. Des.*, 2013, **20**, 1577.
103. K. M. Hillgren, D. Keppler, A. A. Zur, K. M. Giacomini, B. Stieger, C. E. Cass and L. Zhang, *Clin. Pharmacol. Ther.*, 2013, **94**, 52.
104. D. Tweedie, J. W. Polli, E. G. Berglund, S. M. Huang, L. Zhang, A Poirier, X. Chu and B. Feng, *Clin. Pharmacol. Ther.*, 2013, **94**, 113.
105. I. Klein, B. Sarkadi and A. Váradi, *Biochim. Biophys. Acta*, 1999, **1461**, 237.
106. http://www.genenames.org/genefamilies/SLC.

107. M. A. Hediger, T. Hegedus and T. Mehdi, http://www.bioparadigms.org/slc/intro.htm.
108. B. Hagenbuch and B. Stieger, *Mol. Aspects Med.*, 2013, **34**, 396.
109. E. M. Wright, *Mol. Aspects Med.*, 2013, **34**, 183.
110. D. Fotiadis, Y. Kanai and M. Palacín, *Mol. Aspects Med.*, 2013, **34**, 139.
111. M. Roth, A. Obaidat and B. Hagenbuch, *Br. J. Pharmacol.*, 2012, **165**, 1260.
112. S. Faham, A. Watanabe, G. M. Besserer, D. Cascio, A. Specht, B. A. Hirayama, E. M. Wright and J. Abramson, *Science*, 2008, **321**, 810.
113. S. Broer, *Pflugers Arch.*, 2014, **466**, 155.
114. A. M. Pajor, *Pflugers Arch.*, 2014, **466**, 119.
115. T. Claro da Silva, J. E. Polli and P. W. Swaan, *Mol. Aspects Med.*, 2013, **34**, 252.
116. H. Shimada, B. Moewes and G. Burckhardt, *Am. J. Physiol.*, 1987, **253**, F795.
117. H. Koepsell, *Mol. Aspects Med.*, 2013, **34**, 413.
118. M. Mueckler and B. Thorens, *Mol. Aspects Med.*, 2013, **34**, 121.
119. J. D. Young, S. Y. Yao, J. M. Baldwin, C. E. Cass and S. A. Baldwin, *Mol. Aspects Med.*, 2013, **34**, 529.
120. D. Khananshvili, *Mol. Aspects Med.*, 2013, **34**, 220.
121. M. Donowitz, C. Ming Tse and D. Fuster, *Mol. Aspects Med.*, 2013, **34**, 236.
122. H. Motohashi and K. Inui, *Mol. Aspects Med.*, 2013, **34**, 661.
123. S. Leuthold, B. Hagenbuch, N. Mohebbi, C. A. Wagner, P. J. Meier and B. Stieger, *Am. J. Physiol. Cell Physiol.*, 2009, **296**, C570.
124. A. Yamada, K. Maeda, E. Kamiyama, D. Sugiyama, T. Kondo, Y. Shiroyanagi, H. Nakazawa, T. Okano, M. Adachi, J. D. Schuetz, Y. Adachi, Z. Hu, H. Kusuhara and Y. Sugiyama, *Drug Metab. Dispos.*, 2007, **35**, 2166.
125. M. Kuroda, Y. Kobayashi, Y. Tanaka, T. Itani, R. Mifuji, J. Araki, M. Kaito and Y. Adachi, *J. Gastroenterol. Hepatol.*, 2004, **19**, 146.
126. D. Keppler, *Z. Gastroenterol.*, 2011, **49**, 1553.
127. B. Hagenbuch and B. Stieger, *Mol. Aspects Med.*, 2014, **34**, 396.
128. D. Keppler, *Handb. Exp. Pharmacol.*, 2011, 299.
129. C. Guyot, L. Hofstetter and B. Stieger, *Mol. Pharmacol.*, 2014, **85**, 909.
130. F. J. Sharom, *Front. Oncol.*, 2014, **4**, 41.
131. J. Li, K. F. Jaimes and S. G. Aller, *Protein Sci.*, 2014, **23**, 34.
132. C. A. Shintre, A. C. W. Pike, Q. Li, J.-I. Kim, A. J. Barr, S. Goubin, L. Shrestha, J. Yang, G. Berridge, J. Ross, P. J. Stansfeld, M. S. P. Sansom, A. M. Edwards, C. Bountra, B. D. Marsden, F. von Delft, A. N. Bullock, O. Gileadi, N. A. Burgess-Brown and E. P. Carpenter, *Proc. Natl. Acad. Sci. U. S. A.*, 2013, **110**, 9710.
133. D. Deng, C. Xu, P. Sun, J. Wu, C. Yan, M. Hu and N. Yan, *Nature*, 2014, **510**, 121.
134. M. A. Hediger, B. Clémençon, R. E. Burrier and E. A. Bruford, *Mol. Aspects Med.*, 2013, **34**, 95.

135. F. Meier-Abt, Y. Mokrab and K. Mizuguchi, *J. Membr. Biol.*, 2005, **208**, 213.
136. J. L. Perry, N. Dembla-Rajpal, L. A. Hall and J. B. Pritchard, *J. Biol. Chem.*, 2006, **281**, 38071.
137. C. Popp, V. Gorboulev, T. D. Müller, D. Gorbunov, N. Shatskaya and H. Koepsell, *Mol. Pharmacol.*, 2005, **67**, 1600.
138. Y. M. Weaver and B. Hagenbuch, *J. Membr. Biol.*, 2010, **236**, 279.
139. V. Kumar, B. Prasad, G. Patilea, A. Gupta, L. Salphati, R. Evers, C. E. C. A. Hop and J. D. Unadkat, *Drug Metab. Dispos.*, 2015, **43**, 284.
140. S. Berggren, C. Gall, N. Wollnitz, M. Ekelund, U. Karlbom, J. Hoogstraate, D. Schrenk and H. Lennernäs, *Mol. Pharm.*, 2007, **4**, 252.
141. Q. Wang, R. K. Bhardwaj, D. Herrera-Ruiz, N. N. Hanna, I. T. Hanna, O. S. Gudmundsson, T. Buranachokpaisan, I. J. Hidalgo and G. T. Knipp, *Mol. Pharm.*, 2004, **1**, 447.
142. T. G. Tucker, A. M. Milne, S. Fournel-Gigleux, K. S. Fenner and M. W. H. Coughtrie, *Biochem. Pharmacol.*, 2012, **83**, 279.
143. H. Motohashi, Y. Nakao, S. Masuda, T. Katsura, T. Kamba, O. Ogawa and K.-I. Inui, *J. Pharm. Sci.*, 2013, **102**, 3302.
144. H. Motohashi and K. Inui, *AAPS J.*, 2013, **15**, 581.
145. A. F. Hofmann and L. R. Hagey, *J. Lipid Res.*, 2014, **55**, 1553.
146. O. Sperling, *Mol. Genet. Metab.*, 2006, **89**, 14.
147. M. Hobbs, C. Parker, H. Birch and K. Kenworthy, *Xenobiotica*, 2012, **42**, 327.
148. H. Sun, Y.-Y. Zeng and K. S. Pang, *Drug Metab. Dispos.*, 2010, **38**, 769.
149. K. Kubo, S. Sekine and M. Saito, *Biosci. Biotechnol. Biochem.*, 2009, **73**, 2432.
150. B. S. Abuasal, M. B. Bolger, D. K. Walker and A. Kaddoumi, *Mol. Pharm.*, 2012, **9**, 492.
151. L. Salphati, X. Chu, L. Chen, B. Prasad, S. Dallas, R. Evers, D. Mamaril-Fishman, E. G. Geier, J. Kehler, J. Kunta, M. Mezler, L. Laplanche, J. Pang, A. Rode, M. G. Soars, J. D. Unadkat, R. A. B. van Waterschoot, J. Yabut, A. H. Schinkel and N. Scheer, *Drug Metab. Dispos.*, 2014, **42**, 1301.
152. A. H. Schinkel, J. J. Smit, O. van Tellingen, J. H. Beijnen, E. Wagenaar, L. van Deemter, C. A. Mol, M. A. van der Valk, E. C. Robanus-Maandag and H. P. te Riele, *Cell*, 1994, **77**, 491.
153. A. Menter, B. Thrash, C. Cherian, L. H. Matherly, L. Wang, A. Gangjee, J. R. Morgan, D. Y. Maeda, A. D. Schuler, S. J. Kahn and J. A. Zebala, *J. Pharmacol. Exp. Ther.*, 2012, **342**, 696.
154. Y. Lai, K. E. Sampson, L. M. Balogh, T. G. Brayman, S. R. Cox, W. J. Adams, V. Kumar and J. C. Stevens, *J. Pharmacol. Exp. Ther.*, 2010, **334**, 936.
155. M. Ninomiya, K. Ito, R. Hiramatsu and T. Horie, *Drug Metab. Dispos.*, 2006, **34**, 2056.
156. I. Y. Gong, S. E. Mansell and R. B. Kim, *Basic Clin. Pharmacol. Toxicol.*, 2013, **112**, 164.

157. S. Marchetti, N. A. de Vries, T. Buckle, M. J. Bolijn, M. A. J. van Eijndhoven, J. H. Beijnen, R. Mazzanti, O. van Tellingen and J. H. M. Schellens, *Mol. Cancer Ther.*, 2008, **7**, 2280.

158. C. R. Hill, D. Jamieson, H. D. Thomas, C. D. A. Brown, A. V Boddy and G. J. Veal, *Biochem. Pharmacol.*, 2013, **85**, 29.

159. B. Hagenbuch and P. J. Meier, *Pflugers Arch.*, 2004, **447**, 653.

160. E. van de Steeg, E. Wagenaar, C. M. M. van der Kruijssen, J. E. C. Burggraaff, D. R. de Waart, R. P. J. O. Elferink, K. E. Kenworthy and A. H. Schinkel, *J. Clin. Invest.*, 2010, **120**, 2942.

161. D. Slijepcevic, C. Kaufman, C. G. K. Wichers, E. H. Gilglioni, F. A. Lempp, S. Duijst, D. R. de Waart, R. P. J. Oude Elferink, W. Mier, B. Stieger, U. Beuers, S. Urban and S. F. J. van de Graaf, *Hepatology*, 2015, **62**, 207.

162. F. M. Vaz, C. C. Paulusma, H. Huidekoper, M. de Ru, C. Lim, J. Koster, K. Ho-Mok, A. H. Bootsma, A. K. Groen, F. G. Schaap, R. P. J. Oude Elferink, H. R. Waterham and R. J. A. Wanders, *Hepatology*, 2015, **61**, 260.

163. J. R. Riordan, *Annu. Rev. Biochem.*, 2008, **77**, 701.

164. M. J. Caplan, B. Forbush, G. E. Palade and J. D. Jamieson, *J. Biol. Chem.*, 1990, **265**, 3528.

165. M. S. Anwer and B. Stieger, *Pflugers Arch.*, 2014, **466**, 77.

166. B. Stieger and B. Hagenbuch, *Curr. Top. Membr.*, 2014, **73**, 205.

167. T. Yamagishi, S. Sahni, D. M. Sharp, A. Arvind, P. J. Jansson and D. R. Richardson, *J. Biol. Chem.*, 2013, **288**, 31761.

168. C. Albrecht and E. Viturro, *Pflugers Arch.*, 2007, **453**, 581.

169. B. Chapuy, M. Panse, U. Radunski, R. Koch, D. Wenzel, N. Inagaki, D. Haase, L. Truemper and G. G. Wulf, *Haematologica*, 2009, **94**, 1528.

170. B. L. Urquhart, R. G. Tirona and R. B. Kim, *J. Clin. Pharmacol.*, 2007, **47**, 566.

171. A. di Masi, E. De Marinis, P. Ascenzi and M. Marino, *Mol. Aspects Med.*, 2009, **30**, 297.

172. H. Li and H. Wang, *Expert Opin. Drug Metab. Toxicol*, 2010, **6**, 409.

173. S. Modica, E. Bellafante and A. Moschetta, *Front. Biosci. Landmark Ed.*, 2009, **14**, 4719.

174. J. Kartenbeck, U. Leuschner, R. Mayer and D. Keppler, *Hepatology*, 1996, **23**, 1061.

175. E. van de Steeg, V. Stránecký, H. Hartmannová, L. Nosková, M. Hřebíček, E. Wagenaar, A. van Esch, D. R. de Waart, R. P. J. Oude Elferink, K. E. Kenworthy, E. Sticová, M. al-Edreesi, A. S. Knisely, S. Kmoch, M. Jirsa and A. H. Schinkel, *J. Clin. Invest.*, 2012, **122**, 519.

176. J. A. Wambach, D. J. Wegner, K. Depass, H. Heins, T. E. Druley, R. D. Mitra, P. An, Q. Zhang, L. M. Nogee, F. S. Cole and A. Hamvas, *Pediatrics*, 2012, **130**, e1575.

177. T. Sakaeda, *Drug Metab. Pharmacokinet.*, 2005, **20**, 391.

178. K. Jamroziak, W. Młynarski, E. Balcerczak, M. Mistygacz, J. Trelinska, M. Mirowski, J. Bodalski and T. Robak, *Eur. J. Haematol.*, 2004, **72**, 314.

179. Y.-H. Li, Y.-H. Wang, Y. Li and L. Yang, *Yi Chuan Xue Bao*, 2006, **33**, 93.

180. K. Maeda, I. Ieiri, K. Yasuda, A. Fujino, H. Fujiwara, K. Otsubo, M. Hirano, T. Watanabe, Y. Kitamura, H. Kusuhara and Y. Sugiyama, *Clin. Pharmacol. Ther.*, 2006, **79**, 427.

181. D. A. Katz, R. Carr, D. R. Grimm, H. Xiong, R. Holley-Shanks, T. Mueller, B. Leake, Q. Wang, L. Han, P. G. Wang, T. Edeki, L. Sahelijo, T. Doan, A. Allen, B. B. Spear and R. B. Kim, *Clin. Pharmacol. Ther.*, 2006, **79**, 186.

182. L. T. de Lima, C. T. Bueno, D. Vivona, R. D. C. Hirata, M. H. Hirata, V. T. de, M. Hungria, C. S. Chiattone, M. A. Zanichelli, M. de, L. L. F. Chauffaille and E. M. Guerra-Shinohara, *Hematology*, 2015, **20**, 137.

183. K. M. Giacomini, P. V Balimane, S. K. Cho, M. Eadon, T. Edeki, K. M. Hillgren, S.-M. Huang, Y. Sugiyama, D. Weitz, Y. Wen, C. Q. Xia, S. W. Yee, H. Zimdahl and M. Niemi, *Clin. Pharmacol. Ther.*, 2013, **94**, 23.

184. E. Link, S. Parish, J. Armitage, L. Bowman, S. Heath, F. Matsuda, I. Gut, M. Lathrop and R. Collins, *N. Engl. J. Med.*, 2008, **359**, 789.

185. M. M. H. Christensen, C. Brasch-Andersen, H. Green, F. Nielsen, P. Damkier, H. Beck-Nielsen and K. Brosen, *Pharmacogenet. Genomics*, 2011, **21**, 837.

186. M. V. Tzvetkov, S. V. Vormfelde, D. Balen, I. Meineke, T. Schmidt, D. Sehrt, I. Sabolić, H. Koepsell and J. Brockmöller, *Clin. Pharmacol. Ther.*, 2009, **86**, 299.

187. Y. Shu, C. Brown, R. A. Castro, R. J. Shi, E. T. Lin, R. P. Owen, S. A. Sheardown, L. Yue, E. G. Burchard, C. M. Brett and K. M. Giacomini, *Clin. Pharmacol. Ther.*, 2008, **83**, 273.

188. R. Li, H. A. Barton and T. S. Maurer, *CPT: pharmacometrics Syst. Pharmacol.*, 2014, **3**, e151.

189. M. K. Pasanen, P. J. Neuvonen and M. Niemi, *Pharmacogenomics*, 2008, **9**, 19.

190. R. H. Ho, L. Choi, W. Lee, G. Mayo, U. I. Schwarz, R. G. Tirona, D. G. Bailey, C. Michael Stein and R. B. Kim, *Pharmacogenet. Genomics*, 2007, **17**, 647.

191. Y. Tomita, K. Maeda and Y. Sugiyama, *Clin. Pharmacol. Ther.*, 2013, **94**, 37.

192. European Medicines Agency, http://www.ema.europa.eu/docs/en_GB/document_library/Scientific_guideline/2012/02/WC500121954.pdf.

193. J. A. Williams, T. Andersson, T. B. Andersson, R. Blanchard, M. O. Behm, N. Cohen, T. Edeki, M. Franc, K. M. Hillgren, K. J. Johnson, D. A. Katz, M. N. Milton, B. P. Murray, J. W. Polli, D. Ricci, L. A. Shipley, S. Vangala and S. A. Wrighton, *J. Clin. Pharmacol.*, 2008, **48**, 849.

194. Y. Lai, M. Varma, B. Feng, J. C. Stephens, E. Kimoto, A. El-Kattan, K. Ichikawa, H. Kikkawa, C. Ono, A. Suzuki, M. Suzuki, Y. Yamamoto and L. Tremaine, *Expert Opin. Drug Metab. Toxicol.*, 2012, **8**, 723.

195. D. L. Kroetz, S. W. Yee and K. M. Giacomini, *Clin. Pharmacol. Ther.*, 2010, **87**, 109.

196. X. Cheng, J. Maher, C. Chen and C. D. Klaassen, *Drug Metab. Dispos.*, 2005, **33**, 1062.
197. M. E. Morris, H.-J. Lee and L. M. Predko, *Pharmacol. Rev.*, 2003, **55**, 229.
198. C. D. Klaassen and L. M. Aleksunes, *Pharmacol. Rev.*, 2010, **62**, 1.
199. A. Emami Riedmaier, A. T. Nies, E. Schaeffeler and M. Schwab, *Pharmacol. Rev.*, 2012, **64**, 421.
200. Y. Tanaka, A. L. Slitt, T. M. Leazer, J. M. Maher and C. D. Klaassen, *Biochem. Biophys. Res. Commun.*, 2005, **326**, 181.
201. X. Chu, K. Bleasby and R. Evers, *Expert Opin. Drug Metab. Toxicol.*, 2013, **9**, 237.
202. Y. Kato, K. Kuge, H. Kusuhara, P. J. Meier and Y. Sugiyama, *J. Pharmacol. Exp. Ther.*, 2002, **302**, 483.
203. I. Sabolić, A. R. Asif, W. E. Budach, C. Wanke, A. Bahn and G. Burckhardt, *Pflugers Arch.*, 2007, **455**, 397.
204. M. L. Ruiz, A. D. Mottino, V. A. Catania and M. Vore, *Compr. Physiol.*, 2013, **3**, 1721.
205. O. P. Soldin, S. H. Chung and D. R. Mattison, *J. Biomed. Biotechnol.*, 2011, **2011**, 187103.
206. G. L. Kearns, S. M. Abdel-Rahman, S. W. Alander, D. L. Blowey, J. S. Leeder and R. E. Kauffman, *N. Engl. J. Med.*, 2003, **349**, 1157.
207. S. N. de Wildt, *Expert Opin. Drug Metab. Toxicol.*, 2011, **7**, 935.
208. P. Myllynen, E. Immonen, M. Kummu and K. Vähäkangas, *Expert Opin. Drug Metab. Toxicol.*, 2009, **5**, 1483.
209. K. Wlcek and B. Stieger, *Biofactors*, 2014, **40**, 188.
210. B. Gao, M. V. St Pierre, B. Stieger and P. J. Meier, *J. Hepatol.*, 2004, **41**, 201.
211. A. Geier, M. Wagner, C. G. Dietrich and M. Trauner, *Biochim. Biophys. Acta*, 2007, **1773**, 283.
212. B. Stieger, M. Heger, W. de Graaf, G. Paumgartner and T. van Gulik, *Eur. J. Pharmacol.*, 2012, **675**, 1.
213. M. J. Pollheimer, P. Fickert and B. Stieger, *Mol. Aspects Med.*, 2014, **37**, 35.
214. A. Tamaki, C. Ierano, G. Szakacs, R. W. Robey and S. E. Bates, *Essays Biochem.*, 2011, **50**, 209.
215. R. Z. Harris, G. R. Jang and S. Tsunoda, *Clin. Pharmacokinet.*, 2003, **42**, 1071.
216. D. J. Edwards, M. E. Fitzsimmons, E. G. Schuetz, K. Yasuda, M. P. Ducharme, L. H. Warbasse, P. M. Woster, J. D. Schuetz and P. Watkins, *Clin. Pharmacol. Ther.*, 1999, **65**, 237.
217. D. Theile, T. T. Schmidt, W. E. Haefeli and J. Weiss, *J. Pharm. Pharmacol.*, 2013, **65**, 1518.
218. F. Staud and M. Ceckova, *Expert Opin. Drug Metab. Toxicol.*, 2015, 1.
219. R. H. A. van der Doelen, I. A. Arnoldussen, H. Ghareh, L. van Och, J. R. Homberg and T. Kozicz, *Dev. Psychopathol.*, 2015, **27**, 123.

220. K. Brouwer, D. Keppler, K. Hoffmaster, D. A. Bow, Y. Cheng, Y. Lai, J. Palm, B. Stieger, R. Evers and International Transporter Consortium, *Clin. Pharmacol. Ther.*, 2013, **94**, 95.

221. M. A. Hediger, T. Hegedus and T. Mehdi, http://www.bioparadigms.org/slc/intro.htm.

222. B. Stieger and C. F. Higgins, *Pflugers Arch.*, 2007, **453**, 543.

223. K. L. R. Brouwer, D. Keppler, K. A. Hoffmaster, D. A. Bow, Y. Cheng, Y. Lai, J. E. Palm, B. Stieger, R. Evers and International Transporter Consortium, *Clin. Pharmacol. Ther.*, 2013, **94**(1), 95.

224. R. Elsby, D. D. Surry, V. N. Smith and A. J. Gray, *Xenobiotica*, 2008, **38**, 1140.

225. J. P. Keogh and J. R. Kunta, *Eur. J. Pharm. Sci.*, 2006, **27**, 543.

226. J. Bentz, M. P. O'Connor, D. Bednarczyk, J. Coleman, C. Lee, J. Palm, Y. A. Pak, E. S. Perloff, E. Reyner, P. Balimane, M. Brännström, X. Chu, C. Funk, A. Guo, I. Hanna, K. Herédi-Szabó, K. Hillgren, L. Li, E. Hollnack-Pusch, M. Jamei, X. Lin, A. K. Mason, S. Neuhoff, A. Patel, L. Podila, E. Plise, G. Rajaraman, L. Salphati, E. Sands, M. E. Taub, J.-S. Taur, D. Weitz, H. M. Wortelboer, C. Q. Xia, G. Xiao, J. Yabut, T. Yamagata, L. Zhang and H. Ellens, *Drug Metab. Dispos.*, 2013, **41**, 1347.

227. J. Weiss, J. Rose, C. H. Storch, N. Ketabi-Kiyanvash, A. Sauer, W. E. Haefeli and T. Efferth, *J. Antimicrob. Chemother.*, 2007, **59**, 238.

228. T. T. Tran, A. Mittal, T. Gales, B. Maleeff, T. Aldinger, J. W. Polli, A. Ayrton, H. Ellens and J. Bentz, *J. Pharm. Sci.*, 2004, **93**, 2108.

229. D. Gartzke and G. Fricker, *J. Pharm. Sci.*, 2014, **103**, 1298.

230. N. D. Pfeifer, K. Yang and K. L. R. Brouwer, *J. Pharmacol. Exp. Ther.*, 2013, **347**, 727.

231. A. Verhulst, R. Sayer, M. E. De Broe, P. C. D'Haese and C. D. A. Brown, *Mol. Pharmacol.*, 2008, **74**, 1084.

232. C. D. A. Brown, R. Sayer, A. S. Windass, I. S. Haslam, M. E. De Broe, P. C. D'Haese and A. Verhulst, *Toxicol. Appl. Pharmacol.*, 2008, **233**, 428.

233. R. Elsby, V. Smith, L. Fox, D. Stresser, C. Butters, P. Sharma and D. D. Surry, *Xenobiotica*, 2011, **41**, 764.

234. H. Tang, D. R. Shen, Y.-H. Han, Y. Kong, P. Balimane, A. Marino, M. Gao, S. Wu, D. Xie, M. G. Soars, J. C. O'Connell, A. D. Rodrigues, L. Zhang and M. E. Cvijic, *J. Biomol. Screen.*, 2013, **18**, 1072.

235. S. V Ambudkar, *Methods Enzymol.*, 1998, **292**, 504.

236. Y. Shirasaka, Y. Onishi, A. Sakurai, H. Nakagawa, T. Ishikawa and S. Yamashita, *Biol. Pharm. Bull.*, 2006, **29**, 2465.

237. P. Szerémy, A. Pál, D. Méhn, B. Tóth, F. Fülöp, P. Krajcsi and K. Herédi-Szabó, *J. Biomol. Screen.*, 2011, **16**, 112.

238. M. R. Ansbro, S. Shukla, S. V Ambudkar, S. H. Yuspa and L. Li, *PLoS One*, 2013, **8**, e60334.

239. T. Tachibana, S. Kitamura, M. Kato, T. Mitsui, Y. Shirasaka, S. Yamashita and Y. Sugiyama, *Pharm. Res.*, 2010, **27**, 442.

240. M. Hirano, K. Maeda, Y. Shitara and Y. Sugiyama, *J. Pharmacol. Exp. Ther.*, 2004, **311**, 139.

241. C. Gui, Y. Miao, L. Thompson, B. Wahlgren, M. Mock, B. Stieger and B. Hagenbuch, *Eur. J. Pharmacol.*, 2008, **584**, 57.

242. S. W. Paine, A. J. Parker, P. Gardiner, P. J. Webborn and R. J. Riley, *Drug Metab. Dispos.*, 2008, **36**, 1365.

243. A. Poirier, T. Lave, R. Portmann, M. E. Brun, F. Senner, M. Kansy, H. P. Grimm and C. Funk, *Drug Metab. Dispos.*, 2008, **36**, 2434.

244. J. Badee, B. Achour, A. Rostami-Hodjegan and A. Galetin, *Drug Metab. Dispos.*, 2015.

245. K. M. Giacomini, S.-M. Huang, D. J. Tweedie, L. Z. Benet, K. L. R. Brouwer, X. Chu, A. Dahlin, R. Evers, V. Fischer, K. M. Hillgren, K. A. Hoffmaster, T. Ishikawa, D. Keppler, R. B. Kim, C. A. Lee, M. Niemi, J. W. Polli, Y. Sugiyama, P. W. Swaan, J. A. Ware, S. H. Wright, S. W. Yee, M. J. Zamek-Gliszczynski and L. Zhang, *Nat. Rev. Drug Discov.*, 2010, **9**, 215.

246. K. M. Mahar Doan, J. E. Humphreys, L. O. Webster, S. A. Wring, L. J. Shampine, C. J. Serabjit-Singh, K. K. Adkison and J. W. Polli, *J. Pharmacol. Exp. Ther.*, 2002, **303**, 1029.

247. J. W. Polli, S. A. Wring, J. E. Humphreys, L. Huang, J. B. Morgan, L. O. Webster and C. S. Serabjit-Singh, *J. Pharmacol. Exp. Ther.*, 2001, **299**, 620.

248. X. Li, J. Hu, B. Wang, L. Sheng, Z. Liu, S. Yang and Y. Li, *Toxicol. Appl. Pharmacol.*, 2014, **275**, 163.

249. F. Colombo, H. Poirier, N. Rioux, M. A. Montecillo, J. Duan and M. D. Ribadeneira, *Xenobiotica*, 2013, **43**, 915.

250. K. Menochet, K. E. Kenworthy, J. B. Houston and A. Galetin, *J. Pharmacol. Exp. Ther.*, 2012, **341**, 2.

251. H. Ellens, S. Deng, J. Coleman, J. Bentz, M. E. Taub, I. Ragueneau-Majlessi, S. P. Chung, K. Herédi-Szabó, S. Neuhoff, J. Palm, P. Balimane, L. Zhang, M. Jamei, I. Hanna, M. O'Connor, D. Bednarczyk, M. Forsgard, X. Chu, C. Funk, A. Guo, K. M. Hillgren, L. Li, A. Y. Pak, E. S. Perloff, G. Rajaraman, L. Salphati, J.-S. Taur, D. Weitz, H. M. Wortelboer, C. Q. Xia, G. Xiao, T. Yamagata and C. A. Lee, *Drug Metab. Dispos.*, 2013, **41**, 1367.

252. R. Hayeshi, C. Hilgendorf, P. Artursson, P. Augustijns, B. Brodin, P. Dehertogh, K. Fisher, L. Fossati, E. Hovenkamp, T. Korjamo, C. Masungi, N. Maubon, R. Mols, A. Müllertz, J. Mönkkönen, C. O'Driscoll, H. M. Oppers-Tiemissen, E. G. E. Ragnarsson, M. Rooseboom and A.-L. Ungell, *Eur. J. Pharm. Sci.*, 2008, **35**, 383.

253. X.-Q. Yu, C. C. Xue, G. Wang and S.-F. Zhou, *Curr. Drug Metab.*, 2007, **8**, 787.

254. B. J. Kirby, A. C. Collier, E. D. Kharasch, D. Whittington, K. E. Thummel and J. D. Unadkat, *Drug Metab. Dispos.*, 2011, **39**, 1070.

255. H. Kusuhara, M. Miura, N. Yasui-Furukori, K. Yoshida, Y. Akamine, M. Yokochi, S. Fukizawa, K. Ikejiri, K. Kanamitsu, T. Uno and Y. Sugiyama, *Drug Metab. Dispos.*, 2013, **41**, 206.

256. B. J. Kirby, A. C. Collier, E. D. Kharasch, D. Whittington, K. E. Thummel and J. D. Unadkat, *Drug Metab. Dispos.*, 2012, **40**, 610.

257. S. Misaka, F. Müller and M. F. Fromm, *Curr. Opin. Pharmacol.*, 2013, **13**, 847.

258. G. A. Cipolla, *Front. Genet.*, 2014, **5**, 337.

259. I. P. Pogribny and F. A. Beland, *Expert Opin. Drug Metab. Toxicol.*, 2013, **9**, 713.

260. C. A. Lee, M. A. O'Connor, T. K. Ritchie, A. Galetin, J. A. Cook, I. Ragueneau-Majlessi, H. Ellens, B. Feng, M. E. Taub, M. F. Paine, J. W. Polli, J. A. Ware and M. J. Zamek-Gliszczynski, *Drug Metab. Dispos.*, 2015, **43**, 490.

261. D. Keppler, *Drug Metab. Dispos.*, 2014, **42**, 561.

262. E. Jacquemin, *Clin. Res. Hepatol. Gastroenterol.*, 2012, **36**(Suppl. 1), S26.

263. T. Dujic, K. Zhou, L. A. Donnelly, R. Tavendale, C. N. Palmer and E. R. Pearson, *Diabetes*, 2015, **64**, 1786.

264. M. Kwon, Y. A. Choi, M.-K. Choi and I.-S. Song, *Arch. Pharm. Res.*, 2015, **38**, 849.

265. A. Mahrooz, H. Parsanasab, M. B. Hashemi-Soteh, Z. Kashi, A. Bahar, A. Alizadeh and M. Mozayeni, *Clin. Exp. Med.*, 2015, **15**, 159.

266. S. K. Cho, C. O. Kim, E. S. Park and J.-Y. Chung, *Br. J. Clin. Pharmacol.*, 2014, **78**, 1426.

267. C. Funk, C. Ponelle, G. Scheuermann and M. Pantze, *Mol. Pharmacol.*, 2001, **59**, 627.

268. A. Avdeef, *Curr. Top. Med. Chem.*, 2001, **1**, 277.

269. B. Faller, *Curr. Drug Metab.*, 2008, **9**, 886.

270. J. D. Irvine, L. Takahashi, K. Lockhart, J. Cheong, J. W. Tolan, H. E. Selick and J. R. Grove, *J. Pharm. Sci.*, 1999, **88**, 28.

271. P. V Balimane, Y. H. Han and S. Chong, *AAPS J*, 2006, **8**, E1.

272. V. E. Thiel-Demby, J. E. Humphreys, L. A. St John Williams, H. M. Ellens, N. Shah, A. D. Ayrton and J. W. Polli, *Mol. Pharm.*, **6**, 11.

273. V. H. Thomas, S. Bhattachar, L. Hitchingham, P. Zocharski, M. Naath, N. Surendran, C. L. Stoner and A. El-Kattan, *Expert Opin. Drug Metab. Toxicol.*, 2006, **2**, 591.

274. A. Somogyi, M. Eichelbaum and R. Gugler, *Eur. J. Clin. Pharmacol.*, 1982, **22**, 85.

275. T. Past, S. Tapsonyi and T. Jávor, *Pol. J. Pharmacol. Pharm.*, 1984, **36**, 401.

276. M. J. Ginski and J. E. Polli, *Int. J. Pharm.*, 1999, **177**, 117.

277. T. Sawamoto, S. Haruta, Y. Kurosaki, K. Higaki and T. Kimura, *J. Pharm. Pharmacol.*, 1997, **49**, 450.

278. G. Sánchez-Castaño, A. Ruíz-García, N. Bañón, M. Bermejo, V. Merino, J. Freixas, T. M. Garriguesx and J. M. Plá-Delfina, *J. Pharm. Sci.*, 2000, **89**, 1395.

279. P. R. N. Wolohan and R. D. Clark, *J. Comput. Aided. Mol. Des.*, 2003, **17**, 65.

280. A. R. Hilgers, D. P. Smith, J. J. Biermacher, J. S. Day, J. L. Jensen, S. M. Sims, W. J. Adams, J. M. Friis, J. Palandra, J. D. Hosley, E. M. Shobe and P. S. Burton, *Pharm. Res.*, 2003, **20**, 1149.

281. G. L. Amidon, H. Lennernas, V. P. Shah and J. R. Crison, *Pharm. Res.*, 1995, **12**, 413.

282. C. Y. Wu and L. Z. Benet, *Pharm. Res.*, 2005, **22**, 11.

283. https://www.tno.nl/en/focus-area/healthy-living/predictive-health-technologies/pharma/.

284. http://www.simulations-plus.com/Products.aspx?pID=11.

285. http://www.simcyp.com/.

286. M. Rowland, L. Z. Benet and G. G. Graham, *J. Pharmacokinet. Biopharm.*, 1973, **1**, 123.

287. K. S. Pang and M. Rowland, *J. Pharmacokinet. Biopharm*, 1977, **5**, 625.

288. D. W. Seldin, *J. Nephrol.*, 2004, **17**, 166.

289. Y. Shitara, T. Horie and Y. Sugiyama, *Eur. J. Pharm. Sci.*, 2006, **27**, 425.

290. P. L. Toutain and A. Bousquet-Mélou, *J. Vet. Pharmacol. Ther.*, 2004, **27**, 441.

291. S. H. Wright and W. H. Dantzler, *Physiol. Rev.*, 2004, **84**, 987.

292. T. Watanabe, H. Kusuhara, T. Watanabe, Y. Debori, K. Maeda, T. Kondo, H. Nakayama, S. Horita, B. W. Ogilvie, A. Parkinson, Z. Hu and Y. Sugiyama, *Drug Metab. Disp.*, 2011, **39**, 1031.

293. K. S. Pang and M. Rowland, *J. Pharmacokinet. Biopharm.*, 1977, **5**, 625.

294. A. J. Parker and J. B. Houston, *Drug Metab. Dispos.*, 2008, **36**, 1375.

295. J. B. Houston, *Biochem. Pharmacol.*, 1994, **47**, 1469.

296. G. R. Wilkinson and D. G. Shand, *Clin. Pharmacol. Ther.*, 1975, **18**, 377.

297. O. Pelkonen and M. Turpeinen, *Xenobiotica*, 2007, **37**, 1066.

298. L. M. Berezhkovskiy, *J. Pharm. Sci.*, 2011, **100**, 1167.

299. P. Poulin, J. R. Kenny, C. E. Hop and S. Haddad, *J. Pharm. Sci.*, 2012, **101**, 838.

300. A. K. Sohlenius-Sternbeck, C. Jones, D. Ferguson, B. J. Middleton, D. Projean, E. Floby, J. Bylund and L. Afzelius, *Xenobiotica*, 2012, **42**, 841.

301. L. Huang, L. Berry, S. Ganga, B. Janosky, A. Chen, J. Roberts, A. E. Colletti and M. H. Lin, *Drug Metab. Dispos.*, 2010, **38**, 223.

302. L. Liu and K. S. Pang, *Drug Metab. Dispos.*, 2005, **33**, 1.

303. K. Umehara and G. Camenisch, *Pharm. Res.*, 2012, **29**, 603.

304. P. J. Webborn, A. J. Parker, R. L. Denton and R. J. Riley, *Xenobiotica*, 2007, **37**, 1090.

305. M. G. Soars, P. J. Webborn and R. J. Riley, *Mol. Pharmacol.*, 2009, **6**, 1662.

306. W. J. Hua, W. X. Hua and H. J. Fang, *Cardiovasc. Ther.*, 2012, **30**, e234.

307. L. Sanchez-Covarrubias, L. M. Slosky, B. J. Thompson, T. P. Davis and P. T. Ronaldson, *Curr. Pharm. Des.*, 2014, **20**, 1422.

308. J. Yu, T. K. Ritchie, A. Mulgaonkar and I. Ragueneau-Majlessi, *Drug Metab. Dispos.*, 2014, **42**, 1991.

309. L. C. Wienkers and T. G. Heath, *Nat. Rev. Drug Discovery*, 2005, **4**, 825.

310. B. Faller, G. Ottaviani, P. Ertl, G. Berellini and A. Collis, *Drug Discovery Today*, 2011, **16**, 976.

311. C. Funk, *Expert Opin. Drug Metab. Toxicol.*, 2008, **4**, 363.

312. R. Stringer, P. L. Nicklin and J. B. Houston, *Xenobiotica*, 2008, **38**, 1313.

313. S. Shimada, H. Fujino, T. Morikawa, M. Moriyasu and J. Kojima, *Drug Metab. Pharmacokinet.*, 2003, **18**, 245.

314. S. Shugarts and L. Z. Benet, *Pharm. Res.*, 2009, **26**, 2039.

315. R. Chang, K. Emami, H. Wu and W. Sun, *Biofabrication*, 2010, **2**, 045004.

316. F. Consolo, G. B. Fiore, S. Truscello, M. Caronna, U. Morbiducci, F. M. Montevecchi and A. Redaelli, *Tissue Eng., Part C. Methods*, 2009, **15**, 41.

317. L. Di, P. Trapa, R. S. Obach, K. Atkinson, Y.-A. Bi, A. C. Wolford, B. Tan, T. S. McDonald, Y. Lai and L. M. Tremaine, *Drug Metab. Dispos.*, 2012, **40**, 1860.

318. K. Domansky, W. Inman, J. Serdy, A. Dash, M. H. M. Lim and L. G. Griffith, *Lab Chip*, 2010, **10**, 51.

319. N. T. Elliott and F. Yuan, *J. Pharm. Sci.*, 2011, **100**, 59.

320. V. N. Goral, Y.-C. Hsieh, O. N. Petzold, J. S. Clark, P. K. Yuen and R. A. Faris, *Lab Chip*, 2010, **10**, 3380.

321. M. A. Lancaster and J. A. Knoblich, *Science*, 2014, **345**, 1247125.

322. J. Kasuya and K. Tanishita, *Biomatter*, 2012, **2**, 290.

323. X. Liu, J. P. Chism, E. L. Lecluyse, K. R. Brouwer and K. L. Brouwer, *Drug Metab. Dispos.*, 1999, **27**, 637.

324. http://www.fda.gov/OHRMS/DOCKETS/98fr/06d-0344-gdl0001.pdf.

325. I. Cascorbi and S. Haenisch, *Methods Mol. Biol.*, 2010, **596**, 95.

326. M. M. H. Christensen, R. S. Pedersen, T. B. Stage, C. Brasch-Andersen, F. Nielsen, P. Damkier, H. Beck-Nielsen and K. Brøsen, *Pharmacogenet. Genomics*, 2013, **23**, 526.

327. L. Z. Benet, *J. Pharm. Sci.*, 2013, **102**, 34.

328. C. A. Larregieu and L. Z. Benet, *Mol. Pharm.*, 2014, **11**, 1335.

329. R. Kubitz, C. Dröge, J. Stindt, K. Weissenberger and D. Häussinger, *Clin. Res. Hepatol. Gastroenterol.*, 2012, **36**, 536.

330. M. S. Padda, M. Sanchez, A. J. Akhtar and J. L. Boyer, *Hepatology*, 2011, **53**, 1377.

331. R. E. Morgan, M. Trauner, C. J. van Staden, P. H. Lee, B. Ramachandran, M. Eschenberg, C. A. Afshari, C. W. Qualls, R. Lightfoot-Dunn and H. K. Hamadeh, *Toxicol. Sci.*, 2010, **118**, 485.

332. S. Dawson, S. Stahl, N. Paul, J. Barber and J. G. Kenna, *Drug Metab. Dispos.*, 2012, **40**, 130.

333. S. Brown, *J. Invest. Dermatol.*, 2013, **133**, 1709.

CHAPTER 2

Drug Transporters in the Liver: Their Involvement in the Uptake and Export of Endo- and Xeno-biotics

BRUNO STIEGER*[a] AND BRUNO HAGENBUCH[b]

[a] University Hospital, Department of Clinical Pharmacology and Toxicology, 8091 Zürich, Switzerland; [b] Department of Pharmacology, Toxicology and Therapeutics, The University of Kansas Medical Center, Kansas City, Kansas 66160, USA, Email: bhagenbuch@kumc.edu
*Email: bruno.stieger@uzh.ch

2.1 Introduction

The liver is situated in the upper right of the abdomen and is, by weight, one of the largest organs of the body. Based on its external appearance, four lobes can be distinguished. However, for clinical practice, a system of eight schematic portal segments introduced in the early 1950s by the French surgeon Couinaud was considered more appropriate.[1] Today, a more complex system is discussed that, based on portal branching, consists of 20 different liver territories.[2] This has to be used in particular in the context of very sophisticated surgical procedures.

The liver receives 25–30% of the cardiac output, amounting to about 100 ml of blood per 100 g of liver tissue per minute. However, the blood supply to the liver is unique in that only about 20–30% is supplied as oxygen rich blood

RSC Drug Discovery Series No. 54
Drug Transporters: Volume 1: Role and Importance in ADME and Drug Development
Edited by Glynis Nicholls and Kuresh Youdim
© The Royal Society of Chemistry 2016
Published by the Royal Society of Chemistry, www.rsc.org

by the hepatic artery, while the remaining 70–80% is supplied as oxygen poor blood by the portal vein. The venous and arterial blood are mixed in the sinusoids to various extents as the arterial and the venous microvasculature are contractile-independent from each other. In addition, the liver contains a rich network of "vessels" necessary for bile formation and drainage of the bile, as well as a lymphatic system. Consequently, the liver is composed of several cell types with very different structures and functions, among them hepatocytes, biliary epithelial cells, vascular endothelial cells, lymphatic endothelial cells, Ito cells and Kupffer cells.[3] Of these different cell types, the vast majority are hepatocytes. Unlike the kidney, which has nephrons as functional units, the liver is organised differently, being divided anatomically into liver lobules or functionally into liver acini (Figure 2.1).[3] The liver lobule has a more or less hexagonal shape. It is delineated by six so-called portal triads and has, in the centre of the hexagons, a central vein. The portal triad is formed by a small branch of the hepatic artery, a portal vein and a bile duct, and is surrounded by connective tissue. Starting from the central vein, where blood leaves the liver lobule, hepatocytes appear arranged as radiating string-like structures, the so-called hepatocyte plates. The liver acinus has a more or less elliptic shape with the short axis of the ellipse formed by a line connecting to neighbouring portal triads. This is the site of blood and nutrient entry into the acinus and is named zone 1. The long axis of the ellipse is formed by connecting two central veins,

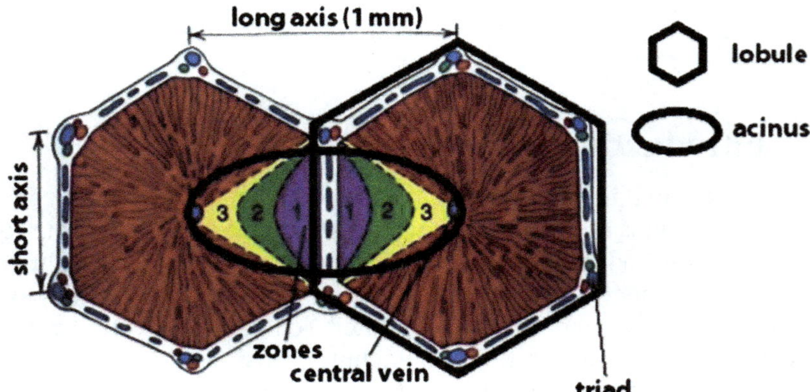

Figure 2.1   Organisation of the liver lobule and acinus. The hexagonal liver lobule is formed by six so-called portal triads, named because of the close proximity of an artery, a vein and a bile duct. In the centre of the hexagon is a central vein. The acinus, which spans from central vein to central vein, can be divided into three zones. (1) The periportal zone close to the portal triads is the zone with the most oxygenated blood, while (3) the perivenous zone is close to the central vein and has the lowest content of oxygen. (2) In between these two zones is the so-called transition zone or zone 2.
Reproduced with permission from Godoy *et al.* (2013),[3] published under the Springer Science + Business Media CC-BY license.

whereby the area close to the central vein is delineated as zone 3 of the acinus. In between zones 1 and 3, zone 2 is defined. From zone 1 to zone 3, there is a diminishing oxygen and nutrient supply. It is important to realise that the metabolic compartmentalisation is reflected by a specific protein expression pattern. For example, the enzymes of the urea cycle are highly expressed around the portal triad in zone 1, whereas glutamine synthetase as an ammonia scavenger is only expressed in the innermost area around the central vein.[3–5] Glycogen synthesis occurs around the portal triad, while xenobiotic metabolism occurs around the central vein.

The fate of a drug in hepatocytes can be divided into four different steps or phases,[6,7] two of which require the presence of transport proteins. Phase 0 is defined as the transporter-mediated uptake of endo- and xeno-biotics into the hepatocytes. This occurs predominantly from the portal blood plasma across the sinusoidal (or basolateral) membrane of hepatocytes. The main transporters known to be involved in drug uptake are three members of the organic anion transporting polypeptide (OATP) family (*SLCO* gene family), namely OATP1B1, OATP1B3 and OATP2B1,[8] two members of the organic anion transporter (OAT) family (*SLC22* gene family), OAT2 and OAT7,[9] two members of the organic cation transporter (OCT) family (*SLC22* gene family), OCT1 and OCT3,[9] and to a lesser extent the Na^+/taurocholate cotransporting polypeptide (NTCP; *SLC10* gene family).[10,11] The main function of NTCP is believed to be the uptake of conjugated bile salts, an important step in the enterohepatic circulation of bile salts. Based on more recent research, NTCP has also been recognised as a receptor for the hepatitis B and D virus.[12]

Currently, to the best of our knowledge, no evidence is available suggesting that xenobiotics may (re)enter hepatocytes at the canalicular membrane. There is however experimental evidence for endocytosis at the canalicular membrane of hepatocytes.[13] Such a process could in principle lead to a slow reuptake of solutes from the canaliculus. Whether uptake of drugs could also occur in a carrier-mediated process across the canalicular membrane from the canalicular bile compartment remains open for debate. It is however conceivable that canalicular-expressed multidrug and toxin extrusion 1 (MATE1; *SLC47* gene family) may mediate reuptake of substrates assuming an alkaline pH in the canaliculus.[14]

Once in the hepatocytes, drugs and xenobiotics may undergo metabolism, a process requiring chemical activation of drugs followed by conjugation. Phase I metabolism involves activation of drugs and xenobiotics, which often occurs by cytochrome P450-mediated reactions, whereby the vast majority of drugs and xenobiotics are metabolised by cytochrome P450 3A4, which is also most abundant in human liver.[15] Conjugation reactions constitute phase II of drug metabolism. This is performed by transferases such as glucuronosyl-, glutathione- or sulfo-transferases, and renders the formed drug metabolite more water soluble.[16]

In order to be eliminated from the body, the resulting metabolites, or parent drugs that are hydrophilic or charged, have to be exported from hepatocytes by transport systems residing in the plasma membrane.

This step constitutes phase III and can occur across the basolateral membrane back into the portal blood or across the canalicular membrane into bile. These export systems are often, if not always, members of the ATP binding cassette (ABC) transporter superfamily. ABC transporters expressed in the basolateral membrane of hepatocytes involved in this process are multidrug resistance associated protein (MRP, *ABCC* gene)3, MRP4 and MRP5. They transport their substrates back into the sinusoidal blood plasma, from where they travel to the kidney *via* the systemic circulation for renal excretion. In addition, the organic solute transporters α and β (OSTα/OSTβ; *SLC51* gene family) form a heterodimeric transporter expressed as an efflux transporter in the basolateral membrane of hepatocytes.[17] ABC transporters expressed in the canalicular (or apical) plasma membrane of hepatocytes, namely multidrug resistance protein (MDR, *ABCB* gene)1, MRP2 and ABCG2 (breast cancer resistance protein (BCRP, *ABCG*)), export their substrates, sometimes against a steep concentration gradient, into the canaliculus, from where they are moved *via* the bile into the duodenum for faecal excretion or, after reabsorption in the intestine, are transported back to the liver in a process called enterohepatic circulation. The following sections briefly summarise the current key knowledge about the individual drug transporters expressed in hepatocytes and will cover function, clinical roles and importance, polymorphisms and involvement in certain diseases. The use of *in vitro* and *in vivo* tools for studying molecular mechanisms of drug transporters and the impact of drug transporters on drug disposition will be covered in Chapters 7 and 8. In the future, the number of relevant drug transporters may well increase due to the development of novel drugs that might be transported by or even targeted to additional transporter proteins.[18]

2.2 Solute Carrier Superfamily Members Expressed in Hepatocytes

Several solute carrier (SLC) superfamily members are known to be expressed at the basolateral membrane of hepatocytes, and are mainly involved in the uptake of endo- and xeno-biotics for eventual elimination by the liver (Figure 2.2 and Table 2.1).[19] As outlined in Chapter 1, SLCs are currently classified into 52 different families based on their amino acid identities (http://www.genenames.org/cgi-bin/genefamilies/set/752 or http://slc.bioparadigms.org/) and have recently been summarised in a special issue of *Molecular Aspects of Medicine*.[20] We will focus in the following section on the families that are known to be involved in drug transport in human hepatocytes.

2.2.1 The *SLCO* Family of OATPs

The *SLCO* family contains 11 genes and one pseudogene. For most of these genes, one functional protein has been characterised, but for OATP1B3 and OATP3A1, functional splice variants have also been reported.[21,22] Substrates

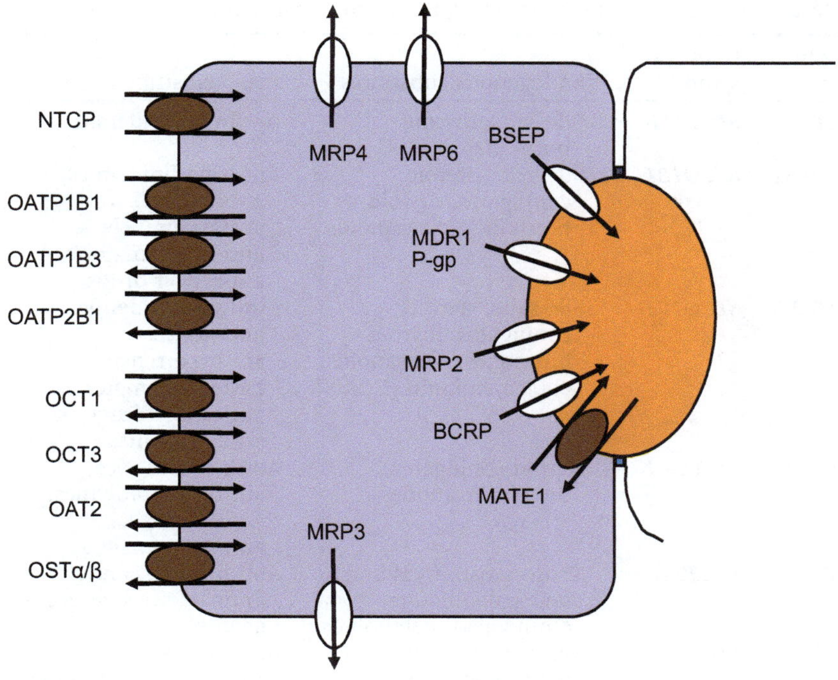

Figure 2.2 Schematic representation of transporters expressed in the basolateral (sinusoidal) and canalicular membranes of human hepatocytes. Members of the SLC superfamily are indicated by brown ovals while members of the ABC superfamily are indicated by white ovals.

of OATPs are in general large (molecular weight >350 Da) and amphipathic, which typically applies for compounds that are cleared by the liver.[19] Several of the OATPs are considered to play a role in the transport of drugs in the liver; in particular OATP1B1, OATP1B3 and OATP2B1, which are all expressed at the basolateral membrane of human hepatocytes, have been shown to be able to transport a wide variety of structurally diverse drugs.[23] Certain adverse drug–drug interactions (*e.g.* statins with fibrates or cyclosporine) can also be explained at least in part by inhibition of OATP-mediated drug uptake.[24] Although it is generally accepted that OATPs are sodium-independent and electroneutral uptake transporters, and thus indirectly take part in drug metabolism by mediating drug uptake into hepatocytes, one study has suggested that they may not always be uni-directional, since efflux of thyroxine from preloaded HEK293 cells expressing rat OATP1C1 has been demonstrated.[25]

Several studies have been published that investigated aspects of the transport mechanism of OATPs, but the exact mechanism by which they are able to mediate uptake or efflux of such a wide variety of structurally un-related compounds has not yet been elucidated. OATPs are electroneutral exchangers; for example, taurocholate uptake into *Xenopus laevis* oocytes can be stimulated by preloading the oocytes with taurocholate.[26] This so-called

Table 2.1 Human SLC transporters expressed in hepatocytes.

Protein symbol	Gene symbol	Endogenous substrates	Drug substrates
NTCP	*SLC10A1*	Bile salts, thyroid hormones	Antihyperlipidemics
OATP1B1	*SLCO1B1*	Bile salts, steroid conjugates, thyroid hormones, eicosanoids	Antihyperlipidemics, antibiotics, antihypertensives, anti-inflammatories, anticancer drugs
OATP1B3	*SLCO1B3*	Bile salts, steroid conjugates, thyroid hormones, eicosanoids, cholecystokinin (CCK-8)	Antihyperlipidemics, antibiotics, antihypertensives, anti-inflammatories, anticancer drugs, antihistamines
OATP2B1	*SLCO2B1*	Steroid conjugates, thyroid hormones	Antihypertensives, antihyperlipidemics, antidiabetics, antihypertensives
OCT1	*SLC22A1*	Neurotransmitters and neuromodulators, amino acid metabolites	Antidiabetics, antivirals, antiparasitics, anticancer agents
OCT3	*SLC22A3*	Neurotransmitters and neuromodulators	Anesthetics, antiarrhythmics, antivirals, anticancer agents
OAT2	*SLC22A7*	Dicarboxylates, cyclic nucleotides, prostaglandins, neurotransmitters, sulfated steroid hormones	Diuretics, antibiotics, antivirals, anticancer agents, antihistamines
MATE1	*SLC47A1*	Neurotransmitters and neuromodulators	Anticancer agents, antivirals, antibiotics, antidiabetics
OSTα/β	*SLC51A1/A2*	Bile salts	Digoxin

trans-stimulation indicates counter-transport and demonstrates that the transport protein works bi-directionally and could mediate either the uptake or release of its substrate. Furthermore, bicarbonate,[27,28] glutathione[29] or glutathione conjugates[26] can be exchanged for substrate uptake. However, not every OATP seems to work the same way and the final proof of such an exchange mechanism is still missing. Moreover, it is currently unclear if OATPs are able to work against a substrate concentration gradient.

2.2.1.1   *OATP1B1 (*SLCO1B1*)*

OATP1B1 is considered to be a "liver specific" OATP, because to date it has only been shown to be functionally expressed at the basolateral membrane of hepatocytes.[8] Its endogenous substrates include bile acids, bilirubin and its conjugates, eicosanoids, thyroid hormones, and steroids and their conjugates.[8] Already in 1999 it was recognised that OATP1B1 could transport the HMG-CoA reductase inhibitor (statin) pravastatin[30] and during the following years many drugs including other statins, anti-hypertensives, antibiotics and anticancer drugs, were shown to either inhibit OATP1B1-mediated substrate uptake or to be substrates of OATP1B1.[23] Today, OATP1B1 is known as an important drug uptake system in human hepatocytes, and functional inactivation of OATP1B1—either by inhibitors such as fibrates or cyclosporine,[24] or due to the presence of polymorphisms[31]—has been recognised as the likely mechanism of statin-induced myopathy. In hepatocellular carcinoma, expression of OATP1B1 is reduced in tumour tissues,[32] potentially leading to elevated plasma exposures for compounds that are exclusively cleared by OATP1B1. Besides statin-induced myopathy, OATP1B1 is also known to be involved in Rotor syndrome, a rare genetic disease presenting with hyperbilirubinaemia caused by the lack of function of both OATP1B1 and OATP1B3.[33]

2.2.1.2   OATP1B3 (SLCO1B3)

OATP1B3 is the second "liver specific" OATP. Its endogenous substrates are very similar to those for OATP1B1,[8] but in addition, OATP1B3 transports the octapeptide hormone cholecystokinin (CCK-8).[34] Besides human OATP1B3, only rat OATP1B2 has been shown to mediate the uptake of CCK-8[34] and therefore CCK-8 is considered to be an OATP1B3 specific substrate in human hepatocytes. Again, similar to OATP1B1, OATP1B3 can transport numerous drug substrates including several anticancer drugs such as methotrexate, paclitaxel, docetaxel, imatinib, rapamycin and SN-38.[32] Unlike OATP1B1, OATP1B3 expression has been detected in numerous cancer cell lines and in several cancers, including colon, lung, prostate, breast and pancreatic.[32,35] Thus, this increased expression in cancer, together with the fact that many anticancer drugs are substrates of OATP1B3, led to the suggestion that targeting of OATP1B3 as a potential drug uptake transporter could be used to improve anticancer therapy.[32,36,37] However, more recently, it was realised that the OATP1B3 protein expressed in cancer cells was the result of alternative splicing[38] and represented a protein also called cancer-type OATP1B3, which is missing the first 28 amino acids and as a consequence has a greatly reduced transport function.[22,39] Because the liver-expressed full-length OATP1B3 will very efficiently remove anticancer drugs from the circulation, it is very unlikely that drugs can be targeted specifically to cancer cells expressing the cancer type OATP1B3. It should also be noted that, as mentioned above, non-functional hepatic OATP1B3 in combination with non-functional OATP1B1 results in Rotor syndrome.[33] In addition, polymorphisms in the *SLCO1B3* gene have been associated with unconjugated hyperbilirubinaemia.[40,41]

2.2.1.3   OATP2B1 (SLCO2B1)

OATP2B1 is the third member of the *SLCO* family expressed at the basolateral membrane of hepatocytes. OATP2B1 protein has a more ubiquitous expression than the other hepatic OATPs, being present in multiple organs including the small intestine, placenta, mammary gland, blood–brain barrier, heart, skin and skeletal muscle.[8] Its substrate specificity appears to be more restricted than that of the OATP1B family and includes endogenous substrates such as steroid conjugates and thyroid hormones. Transport mediated by OATP2B1 is pH-dependent, being enhanced at pH values as low as 5.0,[42,43] and it is known that several drugs including statins are transported by OATP2B1.[23] Enhanced transport at a low pH might be explained, at least in part, by the presence of an extracellular histidine residue close to extracellular loop 2, which seems to be important for stimulation of substrate transport for several of the OATPs.[28] Given that OATP2B1 is expressed in the small intestine,[42] and that the intestinal pH is slightly acidic,[44] this pH-dependent stimulation of transport could be a necessary prerequisite for OATP2B1-mediated drug uptake from the small intestine.[45]

Since all three OATPs are expressed in the basolateral membrane of human hepatocytes, it is difficult to assess the contribution of each OATP to drug uptake using isolated hepatocytes unless each OATP can be selectively inhibited. So far, no selective inhibitors have been reported. However, some substrates are available that are preferentially transported by either OATP1B1 or OATP1B3. OATP1B1 transports estrone-3-sulfate *via* two binding sites/translocation pathways with different apparent Michaelis constant (K_m) values, with the high affinity component having a K_m value well below 1 μM.[46–48] This is in contrast to OATP1B3, which transports estrone-3-sulfate with a K_m value of about 60 μM.[49] Thus, at very low substrate concentrations (nanomolar range), estrone-3-sulfate is predominantly transported by OATP1B1 and can therefore be considered as an OATP1B1 selective substrate at these concentrations.[50] The octapeptide CCK-8[34] and the angiotensin receptor blocker telmisartan[51] have both been shown to be selective substrates for OATP1B3 and thus can be used to selectively assess OATP1B3 function. However, to date, there are no known specific inhibitors for OATP2B1 and more research is needed to identify selective inhibitors for each OATP.

2.2.2   The *SLC22* Family of OCTs and OATs

The *SLC22* family contains 23 genes encoding proteins that mediate the transport of organic cations, organic anions and zwitterions.[9] However, for around ten of the proteins in this family a transport function has not yet been identified. With respect to drug transport in hepatocytes, OCT1, OCT3 and OAT2 (Figure 2.2) are involved in drug uptake into hepatocytes across the basolateral membrane. Although OAT7 is also expressed at the basolateral membrane of hepatocytes, it so far has only been demonstrated to transport endogenous compounds such as steroid conjugates and butyrate.[9] In the kidneys, OCT2, OAT1 and OAT3 are important for the secretion of numerous drugs.[9] In general, smaller (<350 Da) and more water soluble compounds that are normally cleared by the kidneys are handled by members of the *SLC22* family.[19]

2.2.2.1 *OCT1* (SLC22A1)

OCT1 was the first member of the *SLC22* family to be identified and cloned from the rat kidney.[52] The human orthologue, however, is mainly expressed in the liver, where it plays an important role in the uptake of cationic drugs such as the antidiabetic metformin, antivirals and several anticancer agents.[9] In addition, it seems to be involved in the clearance of neurotransmitters such as serotonin, dopamine and norepinephrine.[53] Its function can be studied *in vitro* by using the cationic model substrates tetraethylammonium (TEA), *N*-methylquinine, and 1-methyl-4-phenylpyridinium (MPP+).[19]

Besides OCT1, OCT3 (*SLC22A3*) is also expressed in hepatocytes. It can transport very similar substrates to OCT1 and may be involved as a "backup" transporter for the hepatic elimination of cationic compounds, when OCT1 is either inhibited or otherwise non-functional.[9] OCT1 and OCT3 can transport organic cations in both directions across the membrane, the driving force being the electrochemical gradient.[9] Thus, in the case of hepatocytes, which have an inside-negative membrane potential, organic cations will be taken up in an electrogenic manner and transport can thus be monitored either using radiolabelled compounds or *via* electrophysiology.[54]

For both OCT1 and OCT3, polymorphisms have been described that potentially can affect drug disposition.[55] Metformin pharmacokinetics in healthy volunteers has been shown to be affected by the OCT1 genotype[56] although, because additional genes and transporters are involved in metformin disposition, the situation is more complex and not entirely dependent on OCT1 expression.[57]

2.2.2.2 *OAT2* (SLC22A7)

OAT2 is expressed at the basolateral membrane of human hepatocytes where it mediates the efflux of glutamate into the sinusoidal space.[58] With respect to drugs, OAT2 has been shown to transport nonsteroidal anti-inflammatory drugs, diuretics, antibiotics, antivirals and some anticancer agents.[59] Only a few polymorphisms have been identified for *SLC22A7* encoding OAT2 so far and, based on a study investigating OAT2 protein expression in the liver of Korean subjects, the difference in protein expression levels obtained for the genetic variants did not correlate with the *SLC22A7* genotype.[60] This suggests that polymorphisms in the *SLC22A7* gene might not be important for drugs that are transported by OAT2.

In contrast to the liver specific OATPs, where the regulatory agencies recommend that investigational drugs be tested for interactions with transporters (see Section 2.4), such requirements so far only exist for OCT2, OAT1 and OAT3 (the transporters expressed in the kidney[61]), and not for the hepatocellular OCT3 and OAT2; a recommendation to consider OCT1 does, however, currently exist within some regulatory agencies.

2.2.3 *SLC10*: the Sodium Bile Salt Cotransporter Family

The *SLC10* family consists of seven genes including *SLC10A1* (NTCP), which is expressed at the basolateral membrane of hepatocytes.[10] Originally cloned and characterised as a transporter for the uptake of conjugated bile salts into hepatocytes,[62] NTCP was also shown to be able to transport several xenobiotics when expressed in oocytes or mammalian cells.[10] Of particular interest was the finding that, *in vitro*, NTCP is able to transport several statins[11,63–65] and it was speculated that NTCP could be involved in statin disposition, at least under conditions where OATPs are inhibited. However, NTCP does not seem to play an important role in statin handling under normal physiological conditions.[66] More recent studies have demonstrated that NTCP functions as the receptor for the hepatitis B and D viruses,[67,68] suggesting that inhibition of NTCP could be a possible new therapeutic approach to treat these important liver diseases.[69] Also of note is the recent development of the first NTCP knockout mouse, which could be an important tool in the elucidation of NTCP function beyond bile salt transport.[70]

2.2.4 Multidrug and Toxin Extrusion (MATE) Family (*SLC47*)

MATE1 (*SLC47A1*) is a polyspecific drug transporter capable of transporting several anticancer drugs, as well as certain antivirals, antibiotics and antidiabetics.[71,72] It is expressed in the canalicular membrane of hepatocytes and works electroneutrally *via* proton exchange, suggesting that a transmembrane proton gradient may act as a driving force. However, even though the exact pH of canalicular bile is not known, bile at its earliest point of collection has an alkaline pH. Hence, it is not clear whether MATE1 does indeed mediate the canalicular export of drugs. To the best of our knowledge, no data so far exist with respect to the genetic regulation of MATE1 expression in the liver or its altered expression in liver disease. Recent work has caused the authorities to consider the inclusion of MATE1 in future guidance.[73]

2.2.5 The Heterodimeric OSTα/OSTβ in the *SLC51* Family

OSTα/OSTβ (*SLC51A1/SLC51A2*) was identified in 2001 in skate livers by an expression cloning approach.[74,75] In order to be functional it needs to form a heterodimer, hence its name OSTα/OSTβ. This transport system is present in the basolateral membrane of hepatocytes and was found to transport taurocholate, estrone-3-sulfate, digoxin and prostaglandin E_2.[74] It later became clear that OSTα/OSTβ is responsible for the basolateral efflux of bile salts from enterocytes in the terminal ileum.[17] OSTα/OSTβ also transports dehydroepiandrosterone-3-sulfate and is inhibited by bromosulfophthalein, indomethacin, probenecid and spironolactone.[17] In the cholestatic liver, OSTα/OSTβ is induced at the mRNA and protein level, but with a discordant upregulation of the two subunits.[76] Nevertheless, due to its ability to efflux bile acids, it might play a protective role in liver diseases where canalicular bile secretion is impaired.[77] However, firm evidence supporting this concept is still missing. It should also be noted that the role of OSTα/OSTβ as an efflux system for drug metabolites has not been investigated thoroughly.[78]

2.3 ABC Transporters in Hepatocytes

Several ABC transporters are expressed at the plasma membrane of hepatocytes (Figure 2.2 and Table 2.2).[79] All ABC transporters have as minimal structural requirements two nucleotide binding domains (NBD) and two *trans*-membrane domains (TMD), which are fused in the order: TMD1–NBD1–TMD2–NBD2 to a variable degree.[80] MRPs and MDR1 are so-called full transporters with the following building blocks fused: TMD1–NBD1–TMD2–NBD2, and MRP2, MRP3 and MRP6 have an additional trans-membrane domain TMD0 at the N-terminus.[81] ABCG2 is a so-called half transporter, forming a homodimer $(TMD1-NBD1)_2$.[81] ABC transporters hydrolyse ATP and use the energy to transport their substrates out of cells, sometimes against steep concentration gradients. MRP3, MRP4 and MRP6 are expressed in the basolateral membrane of hepatocytes in adult livers.[7,82] At the canalicular membrane, MDR1 (or P-gp), MRP2 and BCRP (also called ABCG2) are involved in the export of many drugs and their metabolites. The bile salt export pump (BSEP) is also expressed at the canalicular membrane and is the only export system for primary bile salts. Although it is not considered to be a drug transporter, it is known to be susceptible to inhibition by some drugs. Since BSEP is the key transporter for secreting bile salts from

Table 2.2 Human ABC transporters expressed in hepatocytes.

Protein symbol	Gene symbol	Endogenous substrates	Drug substrates
MDR1, P-gp	*ABCB1*	Bulky organic cations	Wide variety of rather bulky, neutral or cationic drugs including many anticancer agents
BSEP	*ABCB11*	Bile salts	
MRP2	*ABCC2*	Conjugated bilirubin and glutathione	Anticancer drugs, antibiotics, statins, antihypertensives, numerous glucuronidated drug metabolites
MRP3	*ABCC3*	Conjugated bilirubin, bile salts, leukotriene C_3	Anticancer agents, glucuronidated and sulfated drug metabolites
MRP4	*ABCC4*	Cyclic monophosphate nucleotides, eicosanoids and bile salts	Antivirals, antibiotics, cardiovascular compounds
MRP6	*ABCC6*	ATP, leukotriene C_4, *N*-ethylmaleimide *S*-glutathione	
ABCG2, BCRP	*ABCG2*	Estrone-3-sulfate, urate and other endogenous organic anions	Antivirals, antibiotics and anticancer agents, sulfated drug metabolites

hepatocytes, impairment of its activity, *e.g.* by drugs or their metabolites, may result in the development of acquired cholestatic liver disease (see Section 2.3.7).[83] A brief description of these transporters is given below.

2.3.1 MRP3 (*ABCC3*)

Endogenous substrates of MRP3 include conjugated bilirubin, bile salts with low affinity and a K_m larger than 200 μM for glycocholate, and the inflammatory mediator leukotriene C_3, as well as several others.[84–86] In addition, MRP3 transports a multitude of drug metabolites, whereby it seems to prefer glucuronidated metabolites over sulfated ones.[84] The anticancer drugs etoposide, leucovorin and methotrexate are also substrates of this ABC transporter.[86] Importantly, MRP3 also mediates the transport of environmental compounds, such as glucuronides of phytoestrogens, leaving the possibility that metabolites from dietary components modulate the activity of MRP3.[87] MRP3 substrates are excreted from hepatocytes into the portal blood plasma, from where they can reach the kidney for elimination from the body.

Under normal physiologic conditions MRP3 is expressed at relatively low levels (compared with canalicular MRP2) and is subject to considerable inter-individual variability (greater than 50-fold).[88] In conditions of liver disease, such as cholestasis, MRP3 is upregulated as a potential salvage pathway for bile salts that accumulate in hepatocytes.[85,86] In patients with Dubin–Johnson syndrome, who do not have a functional MRP2 and therefore cannot export bilirubin conjugates into bile, MRP3 is also upregulated in the basolateral hepatocyte membranes.[89] *In vitro*, MRP3 was shown to be inducible in the human hepatoma cell line HepG2 by phenobarbital[90] and in HepG2 cells transfected with the estrogen receptor α by 17α-ethinylestradiol.[91] The cardiac glycoside digoxin has also been shown to lead to induction of MRP3 in the colon cancer cell line T84.[92]

2.3.2 MRP4 (*ABCC4*)

MRP4 has broad substrate specificity and transports a wide variety of endogenous substrates, such as cyclic monophosphate nucleotides, eicosanoids and bile salts.[84,85] Notably, for transport of some eicosanoids and for bile salts, MRP4 requires the presence of cellular glutathione (GSH) in the millimolar range.[85] MRP4 also exports a wide variety of drugs, including some antivirals, antibiotics and cardiovascular compounds.[85,93]

As for MRP3, MRP4 is also expressed at low levels (compared with canalicular MRP2) in normal human liver and is subject to considerable inter-individual variability.[94] In cholestatic liver disease, human MRP4 is upregulated,[94,95] albeit much less than in rodent liver, and hence might provide a salvage pathway for bile salts.[85] Similarly, in end-stage liver disease of variable etiology, expression of both the MRP4 mRNA and protein has been found to be highly upregulated.[96] Notably, in a study analysing a small number of human liver samples, MRP4 was found to be upregulated both at the mRNA and protein level after acetaminophen induced liver injury.[97] Using *in vitro* promoter transactivation assays, it has been demonstrated that the MRP4 promoter is under the control of the aryl-hydrocarbon receptor,[98] which is an important sensor of environmental toxins. In human bronchoalveolar H358 cells, MRP4 expression attenuated the formation of DNA adducts after treatment with the environmental carcinogen 2,3,7,8-tetrachlorodibenzo-*p*-dioxin (TCDD), indicating a cyto-protective role of MRP4.[99]

2.3.3 MRP6 (*ABCC6*)

The substrate specificity of MRP6 had until recently remained unknown. Early studies identified leukotriene C_4 and *N*-ethylmaleimide *S*-glutathione as substrates of MRP6, suggesting that MRP6 may play a role in the export of some drug metabolites from hepatocytes.[100] It is known that mutations in the gene encoding MRP6 lead to pseudoxanthoma elasticum, a disease that presents with ectopic calcification that leads to progressive mineralisation of connective tissues, with clinical manifestations in the skin, eyes and cardiovascular system.[101] This clinical phenotype led to the formulation of a "metabolic hypothesis", *i.e.* that a circulating substance exported by MRP6 may in its absence cause a pathogenetic mechanism for pseudoxanthoma elasticum,[102] and hence supported the necessity for continuing the search for endogenous MRP6 substrates. Very recently, it was demonstrated that MRP6 induces, by a yet unknown mechanism, the cellular release of nucleoside triphosphates.[103] This remarkable finding makes it rather unlikely, but does not completely rule out the possibility, that MRP6 is involved in the export of drug metabolites from hepatocytes.

2.3.4 MDR1 (*ABCB1*)

MDR1 is a very important drug export pump that transports a wide variety of drugs, including many anticancer drugs, which are generally rather bulky, neutral or cationic compounds,[59,104–108] thus acting to protect hepatocytes from the potential cytotoxicity of certain drugs.

MDR1 expression in humans is subject to considerable inter-individual variability, depending on both inherited and acquired factors.[109] MDR1 expression is regulated by the pregnane X receptor (PXR)[110] and both herbal components and drugs have been shown to induce its expression in the liver as well as in other organs such as the gut.[111–115] In addition, in several forms of liver disease, MDR1 expression is upregulated.[42,116] An upregulation of MDR1 has also been observed in livers from patients with acetaminophen toxicity.[97] Besides its well-documented function as an anticancer drug transporter, MDR1 is an important drug export system in hepatocytes, mediating, for example, export of statins such as cerivastatin and pravastatin[117] into bile.

2.3.5   ABCG2 (BCRP, *ABCG2*)

ABCG2 is also called BCRP because it was initially cloned from a multidrug-resistant breast cancer cell line. It transports various endogenous compounds such as estrone-3-sulfate and urate, a wide variety of drugs including antivirals, antibiotics and anticancer drugs (*e.g.* lamivudine, ciprofloxacin and mitoxantrone), and also sulfated metabolites.[118,119] ABCG2 may be a cancer stem cell marker and as such be regulated by miR-21.[120]

It is also subject to considerable inter-individual variability at the protein level, which is pharmacokinetically relevant.[121,122] The regulation of ABCG2 expression has not been well characterised, but its promoter contains estrogen and progesterone response elements as well as a peroxisome proliferator activated receptor γ (PPARγ) response element.[123] Again, in a limited number of liver samples from patients suffering from acetaminophen toxicity, ABCG2 expression was found to be elevated[97] and *in vitro* transactivation of its promoter by the arylhydrocarbon receptor has been demonstrated.[59,124] In addition, in primary human hepatocytes, transfection with the nuclear receptor constitutive androstane receptor (CAR) leads to activation of ABCG2 transcription.[125] In nonalcoholic fatty liver disease, ABCG2 expression was found to increase with disease progression.[126] Also, in human liver samples from patients with alcohol-induced cirrhosis and with hepatitis C infections, expression of ABCG2 was found to be increased.[127,128]

2.3.6   MRP2 (*ABCC2*)

The physiologic function of MRP2 is the export of conjugated bilirubin and glutathione into the canaliculus, a process that is defective in Dubin–Johnson syndrome.[85,129] Having a broad substrate specificity, MRP2 is able to transport a wide variety of drugs, including anticancer drugs, antibiotics, statins and antihypertensives, as well as environmental chemicals. In addition, it exports predominantly glucuronidated drug metabolites, some of them in the presence of glutathione.[85,86]

MRP2 expression is under the control of PXR and CAR, which may result in induction by a considerable number of drugs and environmental chemicals.[113–115] For example, treatment of patients with rifampicin (600 mg day^{-1} for 1 week) leads to an induction of MRP2 in human liver.[130] Against the background of induction of MRP2 by drugs, it should be kept in mind that expression of MRP2 in human liver is also subject to large inter-individual variability.[109,121] In liver disease, MRP2 was found to be reduced in patients with cholestatic liver diseases[95,131–133] as well as in patients with hepatitis C virus infection[134] and with primary sclerosing cholangitis.[135]

2.3.7   BSEP (*ABCC11*)

BSEP is not currently considered a major transporter for drugs, but it does show a high specificity for primary bile salts and it is indispensable for the export of bile salts from hepatocytes.[11,136] It has been reported that BSEP can transport statins with low affinity *in vitro.*[137] As mentioned above, statins are substrates of multiple ABC transporters in the canalicular membrane.[138] Hence, it is challenging to predict *in vivo* the specific ABC transporter(s) involved in, for example, pravastatin transport. However, inhibition of BSEP by drugs and drug metabolites, if sustained, can lead to acquired cholestasis in susceptible patients.[83,139] Inhibition of BSEP by drugs and drug metabolites may be a cause of drug induced liver injury (DILI), which, because of the risks involved, requires careful consideration during drug development, despite the fact that the regulatory guidelines differ in this aspect.

2.4 Implications for Drug Development

In addition to identifying and characterising the disposition of novel drugs, the drug development process includes a thorough assessment of potential liabilities of drugs such as their toxicity and drug–drug interaction risk. This is particularly important in terms of a major excretory organ such as the liver, which expresses, as outlined above, a wide range of transporters capable of transporting drugs from several major drug classes. Such interactions can occur at the uptake site (basolateral membrane) of hepatocytes or at the canalicular membrane. For example, inhibition of OATPs by sildenafil, rifampicin or cyclosporine A leads to a marked systemic exposure to bosentan;[140] in the case of coadministration of cerivastatin and gemfibrozil, OATP inhibition and other factors led to rhabdomyolysis and withdrawal of cerivastatin from the market.[141]

Following the published recommendations of the International Transporter Consortium[61] (ITC, with experts from industry, academia and the regulatory agencies), the Food and Drug Administration (FDA), the European Medicines Agency (EMA) and latterly the Japanese Pharmaceuticals and Medical Devices Agency (PMDA) have published draft guidelines that recommend that all new investigational drugs with significant hepatic elimination, *i.e.* drugs that are estimated to have $\geq$25% hepatic elimination, should be evaluated using *in vitro* systems to determine whether they are substrates of OATP1B1 and/or OATP1B3.[142,143] Similarly, the regulatory authorities also recommend that investigational drugs be evaluated as inhibitors of these two OATPs. According to the recommendations of the ITC,[61] OATP interactions can be determined by measuring the inhibition of a single, well-established model substrate, such as estradiol-17β-glucuronide, a statin or bromosulfophthalein, in a recombinant cell line expressing the respective OATP. However, it has recently become clear that inhibition of OATP-mediated transport can be substrate dependent.[47,144,145] This means that while a compound might inhibit the OATP-mediated transport of one model substrate (*e.g.* estradiol-17β-glucuronide) it might not inhibit the transport of another model substrate (*e.g.* estrone-3-sulfate).[47,144,145] Similar findings have also been reported for OCT2.[146] Therefore, the fact that a compound inhibits OATP- or OCT-mediated uptake of one model substrate does not necessarily mean that this compound will inhibit uptake of all substrates transported by the respective transporter.[147] Again, more research is needed before it is possible to fully predict inhibitors and substrates for the different OATPs.

With respect to ABC transporters, both the FDA and the EMA recommend the investigation of the interaction of a new molecular entity (NME) with MDR1 and ABCG2, both of which are present in several tissues, including liver. However, proposals for investigating BSEP interactions are less well defined by the regulatory agencies. The FDA recommends studying the interaction of a NME with BSEP "when appropriate", whereas the EMA suggests that BSEP "should also preferably be investigated". However, if elevations of serum bile salts and/or alkaline phosphatase are observed during drug development, it is certainly warranted to perform BSEP inhibition studies, as any potential interaction with BSEP should be considered when designing studies in humans to test for a potential cholestatic liver injury. Currently, other hepatocellular ABC transporters are not listed by the regulatory agencies, but the guidance may be updated to include additional hepatocellular transporters including ABC transporters.[73]

Today, rather sophisticated software tools are available, allowing not only modelling of the pharmacokinetic properties of drugs prior to human application, but also modelling of the potential need for individualisation of doses for specific patient groups, *e.g.* taking into account genetic variants of systems involved in the disposition of the NME, underlying diseases, race, age and so on.[148,149] Testing physiologically-based pharmacokinetic modelling used for simulating drug disposition in patients revealed the need for empirically derived scaling factors for the liver.[150] The determination of such scaling factors remains challenging, because for example for hepatic OATPs, the scaling factors are not transporter but drug dependent.[150] An additional limitation is the fact that all models require input of data generated *in vitro* or *in vivo* and therefore the quality of the modelling results critically depends on the quality of such data.

With respect to transporter research, so far no accepted standardised and validated systems are available for generating parameters for drug transport.[151] Hence, even for the same transporter tested with the same drugs, vastly different parameters can be obtained at different laboratories.[152] It is clear that transport experiments, regardless of the system used, need to be performed under carefully controlled conditions.[147] In addition, so far only very limited data on the inter-individual variability of transporter expression in humans are available,[153] and these data often show about a ten-fold or higher range of expression values. This makes accurate predictions of potential drug–drug interactions very challenging. It should also be kept in mind that the unbound intracellular drug concentration remains unknown or can only be extrapolated at best,[154] making it practically impossible to predict the role of hepatocellular efflux systems in drug clearance. Finally, drug sanctuaries, *e.g.* the brain[155] or neonates,[156] will at best require adaptations of the currently used models for the successful modelling of drug disposition from *in vitro* data.

2.5 Summary and Conclusions

The human hepatocyte contains a set of both uptake and export transporters that work together to ensure that the liver can fulfil its important role of metabolising and excreting endo- and xeno-biotics. Although the original function of hepatic transporters was probably to assist in the removal of potentially toxic endogenous compounds and food components from the body, their function in drug clearance is now well-documented, in particular for the multi-specific OATPs and OCTs expressed at the basolateral membrane, which currently are considered important hepatic drug transporters. They work in concert with the drug export transporters of the ABC family, which are mainly expressed in the canalicular membrane, to maintain a vectorial movement of cholephilic substrates such as bile salts, bilirubin, and drugs and their metabolites from the blood to bile.

Although there is a large amount of information available about the different drug substrates for hepatic transporters, additional research is needed to elucidate the roles these transporters play under normal physiological conditions. We are only starting to obtain information on the extent of inter-individual variations of transporter expression in human liver as well as the adaptation of expression levels under various physiologic and pathophysiologic situations. Knockout mice may be helpful tools in this regard, although it needs to be kept in mind that species differences exist that may complicate drawing conclusions from these animal models to the human situation.

References

1. J. Rodes, J. P. Benhamou, A. Blei, J. Reichen and M. Rizzetto, *Textbook of Hepatology*, Blackwell Publishing, 3rd edn, 2007.
2. P. Majno, G. Mentha, C. Toso, P. Morel, H. O. Peitgen and J. H. Fasel, *J. Hepatol.*, 2014, **60**, 654–662.
3. P. Godoy, N. J. Hewitt, U. Albrecht, M. E. Andersen, N. Ansari, S. Bhattacharya, J. G. Bode, J. Bolleyn, C. Borner, J. Bottger *et al.*, *Arch. Toxicol.*, 2013, **87**, 1315–1530.
4. K. Jungermann and N. Katz, *Physiol. Rev.*, 1989, **69**, 708–764.
5. R. Gebhardt, *Pharmacol. Ther.*, 1992, **53**, 275–354.
6. S. R. Vavricka, J. Van Montfoort, H. R. Ha, P. J. Meier and K. Fattinger, *Hepatology*, 2002, **36**, 164–172.
7. B. Doring and E. Petzinger, *Drug Metab. Rev.*, 2014, **46**, 261–282.
8. B. Hagenbuch and B. Stieger, *Mol. Aspects Med.*, 2013, **34**, 396–412.
9. H. Koepsell, *Mol. Aspects Med.*, 2013, **34**, 413–435.
10. T. Claro da Silva, J. E. Polli and P. W. Swaan, *Mol. Aspects Med.*, 2013, **34**, 252–269.
11. B. Stieger, *Handb. Exp. Pharmacol.*, 2011, 205–259.
12. H. Yan, G. Zhong, G. Xu, W. He, Z. Jing, Z. Gao, Y. Huang, Y. Qi, B. Peng, H. Wang, L. Fu, M. Song, P. Chen, W. Gao, B. Ren, Y. Sun, T. Cai, X. Feng, J. Sui and W. Li, *eLife*, 2012, **1**, e00049.

13. C. Rahner, B. Stieger and L. Landmann, *Gastroenterology*, 2000, **119**, 1692–1707.
14. H. Motohashi and K. Inui, *Mol. Aspects Med.*, 2013, **34**, 661–668.
15. M. Ingelman-Sundberg, *Naunyn Schmiedebergs Arch. Pharmacol.*, 2004, **369**, 89–104.
16. W. E. Evans and M. V. Relling, *Science*, 1999, **286**, 487–491.
17. P. A. Dawson, M. L. Hubbert and A. Rao, *Biochim. Biophys. Acta*, 2010, **1801**, 994–1004.
18. A. Cesar-Razquin, B. Snijder, T. Frappier-Brinton, R. Isserlin, G. Gyimesi, X. Bai, R. A. Reithmeier, D. Hepworth, M. A. Hediger, A. M. Edwards and G. Superti-Furga, *Cell*, 2015, **162**, 478–487.
19. B. Hagenbuch, *Clin. Pharmacol. Ther.*, 2010, **87**, 39–47.
20. M. A. Hediger, B. Clemencon, R. E. Burrier and E. A. Bruford, *Mol. Aspects Med.*, 2013, **34**, 95–107.
21. R. D. Huber, B. Gao, M. A. Sidler Pfandler, W. Zhang-Fu, S. Leuthold, B. Hagenbuch, G. Folkers, P. J. Meier and B. Stieger, *Am. J. Physiol.*, 2007, **292**, C795–C806.
22. N. Thakkar, K. Kim, E. R. Jang, S. Han, K. Kim, D. Kim, N. Merchant, A. C. Lockhart and W. Lee, *Mol. Pharmaceutics*, 2013, **10**, 406–416.
23. M. Roth, A. Obaidat and B. Hagenbuch, *Br. J. Pharmacol.*, 2012, **165**, 1260–1287.
24. P. J. Neuvonen, M. Niemi and J. T. Backman, *Clin. Pharmacol. Ther.*, 2006, **80**, 565–581.
25. D. Sugiyama, H. Kusuhara, H. Taniguchi, S. Ishikawa, Y. Nozaki, H. Aburatani and Y. Sugiyama, *J. Biol. Chem.*, 2003, **278**, 43489–43495.
26. L. Li, P. J. Meier and N. Ballatori, *Mol. Pharmacol.*, 2000, **58**, 335–340.
27. L. M. Satlin, V. Amin and A. W. Wolkoff, *J. Biol. Chem.*, 1997, **272**, 26340–26345.
28. S. Leuthold, B. Hagenbuch, N. Mohebbi, C. A. Wagner, P. J. Meier and B. Stieger, *Am. J. Physiol.*, 2009, **296**, C570–C582.
29. L. Q. Li, T. K. Lee, P. J. Meier and N. Ballatori, *J. Biol. Chem.*, 1998, **273**, 16184–16191.
30. B. Hsiang, Y. Zhu, Z. Wang, Y. Wu, V. Sasseville, W.-P. Yang and T. G. Kirchgessner, *J. Biol. Chem.*, 1999, **274**, 37161–37168.
31. S. C. Group, E. Link, S. Parish, J. Armitage, L. Bowman, S. Heath, F. Matsuda, I. Gut, M. Lathrop and R. Collins, *N. Engl. J. Med.*, 2008, **359**, 789–799.
32. A. Obaidat, M. Roth and B. Hagenbuch, *Annu. Rev. Pharmacol. Toxicol.*, 2012, **52**, 135–151.
33. E. van de Steeg, V. Stranecky, H. Hartmannova, L. Noskova, M. Hrebicek, E. Wagenaar, A. van Esch, D. R. de Waart, R. P. Oude Elferink, K. E. Kenworthy, E. Sticova, M. al-Edreesi, A. S. Knisely, S. Kmoch, M. Jirsa and A. H. Schinkel, *J. Clin. Invest.*, 2012, **122**, 519–528.
34. M. G. Ismair, B. Stieger, V. Cattori, B. Hagenbuch, M. Fried, P. J. Meier and G. A. Kullak-Ublick, *Gastroenterology*, 2001, **121**, 1185–1190.

35. A. Hays, U. Apte and B. Hagenbuch, *Pharm. Res.*, 2013, **30**, 2260–2269.
36. M. Svoboda, J. Riha, K. Wlcek, W. Jaeger and T. Thalhammer, *Curr. Drug Metab.*, 2011, **12**, 139–153.
37. T. Liu and Q. Li, *J. Drug Targeting*, 2014, **22**, 14–22.
38. M. Nagai, T. Furihata, S. Matsumoto, S. Ishii, S. Motohashi, I. Yoshino, M. Ugajin, A. Miyajima, S. Matsumoto and K. Chiba, *Biochem. Biophys. Res. Commun.*, 2012, **418**, 818–823.
39. S. Imai, R. Kikuchi, Y. Tsuruya, S. Naoi, S. Nishida, H. Kusuhara and Y. Sugiyama, *Pharm. Res.*, 2013, **30**, 2880–2890.
40. S. Sanna, F. Busonero, A. Maschio, P. F. McArdle, G. Usala, M. Dei, S. Lai, A. Mulas, M. G. Piras, L. Perseu, M. Masala, M. Marongiu, L. Crisponi, S. Naitza, R. Galanello, G. R. Abecasis, A. R. Shuldiner, D. Schlessinger, A. Cao and M. Uda, *Hum. Mol. Genet.*, 2009, **18**, 2711–2718.
41. L. Alencastro de Azevedo, T. Reverbel da Silveira, C. G. Carvalho, S. Martins de Castro, R. Giugliani and U. Matte, *Pediatr. Res.*, 2012, **72**, 169–173.
42. D. Kobayashi, T. Nozawa, K. Imai, J. Nezu, A. Tsuji and I. Tamai, *J. Pharmacol. Exp. Ther.*, 2003, **306**, 703–708.
43. T. Nozawa, K. Imai, J. Nezu, A. Tsuji and I. Tamai, *J. Pharmacol. Exp. Ther.*, 2004, **308**, 438–445.
44. H. Daniel, C. Fett and A. Kratz, *Am. J. Physiol.*, 1989, **257**, G489–G495.
45. I. Tamai and T. Nakanishi, *Curr. Opin. Pharmacol.*, 2013, **13**, 859–863.
46. I. Tamai, T. Nozawa, M. Koshida, J. Nezu, Y. Sai and A. Tsuji, *Pharm. Res.*, 2001, **18**, 1262–1269.
47. J. Noe, R. Portmann, M. E. Brun and C. Funk, *Drug Metab. Dispos.*, 2007, **35**, 1308–1314.
48. C. Gui and B. Hagenbuch, *Protein Sci.*, 2009, **18**, 2298–2306.
49. C. Gui, Y. Miao, L. Thompson, B. Wahlgren, M. Mock, B. Stieger and B. Hagenbuch, *Eur. J. Pharmacol.*, 2008, **584**, 57–65.
50. M. Hirano, K. Maeda, Y. Shitara and Y. Sugiyama, *J. Pharmacol. Exp. Ther.*, 2004, **311**, 139–146.
51. N. Ishiguro, K. Maeda, W. Kishimoto, A. Saito, A. Harada, T. Ebner, W. Roth, T. Igarashi and Y. Sugiyama, *Drug Metab. Dispos.*, 2006, **34**, 1109–1115.
52. D. Gründemann, V. Gorboulev, S. Gambaryan, M. Veyhl and H. Koepsell, *Nature*, 1994, **372**, 549–552.
53. K. H. Boxberger, B. Hagenbuch and J. N. Lampe, *Drug Metab. Dispos.*, 2014, **42**, 990–995.
54. A. Sturm, V. Gorboulev, D. Gorbunov, T. Keller, C. Volk, B. M. Schmitt, P. Schlachtbauer, G. Ciarimboli and H. Koepsell, *Am. J. Physiol.*, 2007, **293**, F767–F779.
55. B. Stieger and P. J. Meier, *Pharmacogenomics*, 2011, **12**, 611–631.
56. Y. Shu, C. Brown, R. A. Castro, R. J. Shi, E. T. Lin, R. P. Owen, S. A. Sheardown, L. Yue, E. G. Burchard, C. M. Brett and K. M. Giacomini, *Clin. Pharmacol. Ther.*, 2008, **83**, 273–280.

57. M. L. Becker, E. R. Pearson and I. Tkac, *Int. J. Endocrinol.*, 2013, **2013**, 686315.
58. C. Fork, T. Bauer, S. Golz, A. Geerts, J. Weiland, D. Del Turco, E. Schomig and D. Grundemann, *Biochem. J.*, 2011, **436**, 305–312.
59. G. Burckhardt and B. C. Burckhardt, *Handb. Exp. Pharmacol.*, 2011, 29–104.
60. H. J. Shin, C. H. Lee, S. S. Lee, I. S. Song and J. G. Shin, *Clin. Chim. Acta*, 2010, **411**, 99–105.
61. K. M. Giacomini, S. M. Huang, D. J. Tweedie, L. Z. Benet, K. L. Brouwer, X. Chu, A. Dahlin, R. Evers, V. Fischer, K. M. Hillgren, K. A. Hoffmaster, T. Ishikawa, D. Keppler, R. B. Kim, C. A. Lee, M. Niemi, J. W. Polli, Y. Sugiyama, P. W. Swaan, J. A. Ware, S. H. Wright, S. W. Yee, M. J. Zamek-Gliszczynski and L. Zhang, *Nat. Rev. Drug Discovery*, 2010, **9**, 215–236.
62. B. Hagenbuch and P. J. Meier, *J. Clin. Invest.*, 1994, **93**, 1326–1331.
63. H. Fujino, T. Saito, S. Ogawa and J. Kojima, *J. Pharm. Pharmacol.*, 2005, **57**, 1305–1311.
64. R. Greupink, L. Dillen, M. Monshouwer, M. T. Huisman and F. G. Russel, *Eur. J. Pharm. Sci.*, 2011, **44**, 487–496.
65. R. H. Ho, R. G. Tirona, B. F. Leake, H. Glaeser, W. Lee, C. J. Lemke, Y. Wang and R. B. Kim, *Gastroenterology*, 2006, **130**, 1793–1806.
66. A. Vildhede, M. Karlgren, E. K. Svedberg, J. R. Wisniewski, Y. Lai, A. Noren and P. Artursson, *Drug Metab. Dispos.*, 2014, **42**, 1210–1218.
67. H. Yan, G. Zhong, G. Xu, W. He, Z. Jing, Z. Gao, Y. Huang, Y. Qi, B. Peng, H. Wang, L. Fu, M. Song, P. Chen, W. Gao, B. Ren, Y. Sun, T. Cai, X. Feng, J. Sui and W. Li, *eLife*, 2012, **1**, e00049.
68. Y. Ni, F. A. Lempp, S. Mehrle, S. Nkongolo, C. Kaufman, M. Falth, J. Stindt, C. Koniger, M. Nassal, R. Kubitz, H. Sultmann and S. Urban, *Gastroenterology*, 2014, **146**, 1070–1083.
69. F. A. Lempp and S. Urban, *Intervirology*, 2014, **57**, 151–157.
70. D. Slijepcevic, C. Kaufman, C. G. K. Wichers, E. H. Gilglioni, F. A. Lempp, S. Duijst, D. R. de Waart, R. P. Oude Elferink, W. Mier, B. Stieger, U. Beuers, S. Urban and S. F. J. van de Graaf, *Hepatology*, 2015, **62**, 207–219.
71. K. Damme, A. T. Nies, E. Schaeffeler and M. Schwab, *Drug Metab. Rev.*, 2011, **43**, 499–523.
72. A. Yonezawa and K. Inui, *Br. J. Pharmacol.*, 2011, **164**, 1817–1825.
73. K. Maeda and Y. Sugiyama, *Mol. Aspects Med.*, 2013, **34**, 711–718.
74. W. Wang, D. J. Seward, L. Li, J. L. Boyer and N. Ballatori, *Proc. Natl. Acad. Sci. U. S. A.*, 2001, **98**, 9431–9436.
75. N. Ballatori, W. V. Christian, S. G. Wheeler and C. L. Hammond, *Mol. Aspects Med.*, 2013, **34**, 683–692.
76. J. L. Boyer, M. Trauner, A. Mennone, C. J. Soroka, S. Y. Cai, T. Moustafa, G. Zollner, J. Y. Lee and N. Ballatori, *Am. J. Physiol.*, 2006, **290**, G1124–G1130.

77. C. J. Soroka, N. Ballatori and J. L. Boyer, *Semin. Liver Dis.*, 2010, **30**, 178–185.
78. A. S. Grandvuinet, H. T. Vestergaard, N. Rapin and B. Steffansen, *J. Pharm. Pharmacol.*, 2012, **64**, 1523–1548.
79. K. Wlcek and B. Stieger, *Biofactors*, 2014, **40**, 188–198.
80. C. F. Higgins, *Annu. Rev. Cell Biol.*, 1992, **8**, 67–113.
81. I. Klein, B. Sarkadi and A. Varadi, *Biochim. Biophys. Acta*, 1999, **1461**, 237–262.
82. N. D. Pfeifer, R. N. Hardwick and K. L. Brouwer, *Annu. Rev. Pharmacol. Toxicol.*, 2014, **54**, 509–535.
83. B. Stieger, *Drug Metab. Rev.*, 2010, **42**, 437–445.
84. P. Borst, C. de Wolf and K. van de Wetering, *Pflugers Arch.*, 2007, **453**, 661–673.
85. D. Keppler, *Handb. Exp. Pharmacol.*, 2011, 299–323.
86. L. W. van der Schoor, H. J. Verkade, F. Kuipers and J. W. Jonker, *Expert Opin. Drug Metab. Toxicol.*, 2015, **11**, 273–293.
87. K. van de Wetering, W. Feddema, J. B. Helms, J. F. Brouwers and P. Borst, *Gastroenterology*, 2009, **137**, 1725–1735.
88. T. Lang, M. Hitzl, O. Burk, E. Mornhinweg, A. Keil, R. Kerb, K. Klein, U. M. Zanger, M. Eichelbaum and M. F. Fromm, *Pharmacogenetics*, 2004, **14**, 155–164.
89. J. Konig, D. Rost, Y. Cui and D. Keppler, *Hepatology*, 1999, **29**, 1156–1163.
90. Y. Kiuchi, H. Suzuki, T. Hirohashi, C. A. Tyson and Y. Sugiyama, *FEBS Lett.*, 1998, **433**, 149–152.
91. M. L. Ruiz, J. P. Rigalli, A. Arias, S. S. Villanueva, C. Banchio, M. Vore, A. D. Mottino and V. A. Catania, *Biochem. Pharmacol.*, 2013, **86**, 401–409.
92. K. Naruhashi, Y. Kurahashi, Y. Fujita, E. Kawakita, Y. Yamasaki, K. Hattori, A. Nishimura and N. Shibata, *Drug Metab. Pharmacokinet.*, 2011, **26**, 145–153.
93. F. G. Russel, J. B. Koenderink and R. Masereeuw, *Trends Pharmacol. Sci.*, 2008, **29**, 200–207.
94. U. Gradhand, T. Lang, E. Schaeffeler, H. Glaeser, H. Tegude, K. Klein, P. Fritz, G. Jedlitschky, H. K. Kroemer, I. Bachmakov, B. Anwald, R. Kerb, U. M. Zanger, M. Eichelbaum, M. Schwab and M. F. Fromm, *Pharmacogenomics J.*, 2008, **8**, 42–52.
95. V. Keitel, M. Burdelski, U. Warskulat, T. Kuhlkamp, D. Keppler, D. Haussinger and R. Kubitz, *Hepatology*, 2005, **41**, 1160–1172.
96. M. Kurzawski, V. Dziedziejko, M. Post, M. Wojcicki, E. Urasinska, J. Mietkiewski and M. Drozdzik, *Pharmacol. Rep.*, 2012, **64**, 927–939.
97. S. N. Barnes, L. M. Aleksunes, L. Augustine, G. L. Scheffer, M. J. Goedken, A. B. Jakowski, I. M. Pruimboom-Brees, N. J. Cherrington and J. E. Manautou, *Drug Metab. Dispos.*, 2007, **35**, 1963–1969.
98. Y. Li, H. Yuan, K. Yang, W. Xu, W. Tang and X. Li, *Curr. Med. Chem.*, 2010, **17**, 786–800.

99. S. L. Gelhaus, O. Gilad, W. T. Hwang, T. M. Penning and I. A. Blair, *Toxicol. Lett.*, 2012, **209**, 58–66.
100. A. Ilias, Z. Urban, T. L. Seidl, O. Le Saux, E. Sinko, C. D. Boyd, B. Sarkadi and A. Varadi, *J. Biol. Chem.*, 2002, **277**, 16860–16867.
101. J. Uitto, L. Bercovitch, S. F. Terry and P. F. Terry, *Am. J. Med. Genet. A*, 2011, **155A**, 1517–1526.
102. N. Chassaing, L. Martin, P. Calvas, M. Le Bert and A. Hovnanian, *J. Med. Genet.*, 2005, **42**, 881–892.
103. R. S. Jansen, A. Kucukosmanoglu, M. de Haas, S. Sapthu, J. A. Otero, I. E. Hegman, A. A. Bergen, T. G. Gorgels, P. Borst and K. van de Wetering, *Proc. Natl. Acad. Sci. U. S. A.*, 2013, **110**, 20206–20211.
104. C. Marzolini, E. Paus, T. Buclin and R. B. Kim, *Clin. Pharmacol. Ther.*, 2004, **75**, 13–33.
105. J. H. Lin and M. Yamazaki, *Drug Metab. Rev.*, 2003, **35**, 417–454.
106. S. F. Zhou, *Xenobiotica*, 2008, **38**, 802–832.
107. I. Cascorbi, *Handb. Exp. Pharmacol.*, 2011, 261–283.
108. R. Callaghan, F. Luk and M. Bebawy, *Drug Metab. Dispos.*, 2014, **42**, 623–631.
109. Y. Meier, C. Pauli-Magnus, U. M. Zanger, K. Klein, E. Schaeffeler, A. K. Nussler, N. Nussler, M. Eichelbaum, P. J. Meier and B. Stieger, *Hepatology*, 2006, **44**, 62–74.
110. H. Wang and E. L. LeCluyse, *Clin. Pharmacokinet.*, 2003, **42**, 1331–1357.
111. S. Zhou, L. Y. Lim and B. Chowbay, *Drug Metab. Rev.*, 2004, **36**, 57–104.
112. S. Marchetti, R. Mazzanti, J. H. Beijnen and J. H. Schellens, *Oncologist*, 2007, **12**, 927–941.
113. B. L. Urquhart, R. G. Tirona and R. B. Kim, *J. Clin. Pharmacol.*, 2007, **47**, 566–578.
114. A. D. Mottino and V. A. Catania, *World J. Gastroenterol.*, 2008, **14**, 7068–7074.
115. L. Badolo, B. Jensen, C. Sall, U. Norinder, P. Kallunki and D. Montanari, *Xenobiotica*, 2015, **45**, 177–187.
116. J. E. Ros, L. Libbrecht, M. Geuken, P. L. Jansen and T. A. Roskams, *J. Pathol.*, 2003, **200**, 553–560.
117. S. Matsushima, K. Maeda, C. Kondo, M. Hirano, M. Sasaki, H. Suzuki and Y. Sugiyama, *J. Pharmacol. Exp. Ther.*, 2005, **314**, 1059–1067.
118. H. E. Meyer zu Schwabedissen and H. K. Kroemer, *Handb. Exp. Pharmacol.*, 2011, 325–371.
119. Q. Mao and J. D. Unadkat, *AAPS J.*, 2014, **17**, 65–82.
120. V. Haghpanah, P. Fallah, R. Tavakoli, M. Naderi, H. Samimi, M. Soleimani and B. Larijani, *Tumour Biol.*, 37(1), 1299–1308.
121. T. G. Tucker, A. M. Milne, S. Fournel-Gigleux, K. S. Fenner and M. W. Coughtrie, *Biochem. Pharmacol.*, 2012, **83**, 279–285.
122. B. Prasad, Y. Lai, Y. Lin and J. D. Unadkat, *J. Pharm. Sci.*, 2013, **102**, 787–793.
123. R. W. Robey, K. K. To, O. Polgar, M. Dohse, P. Fetsch, M. Dean and S. E. Bates, *Adv. Drug Delivery Rev.*, 2009, **61**, 3–13.

124. K. P. Tan, B. Wang, M. Yang, P. C. Boutros, J. Macaulay, H. Xu, A. I. Chuang, K. Kosuge, M. Yamamoto, S. Takahashi, A. M. Wu, D. D. Ross, P. A. Harper and S. Ito, *Mol. Pharmacol.*, 2010, **78**, 175–185.

125. S. Benoki, K. Yoshinari, T. Chikada, J. Imai and Y. Yamazoe, *Arch. Biochem. Biophys.*, 2012, **517**, 123–130.

126. R. N. Hardwick, C. D. Fisher, M. J. Canet, G. L. Scheffer and N. J. Cherrington, *Drug Metab. Dispos.*, 2011, **39**, 2395–2402.

127. V. R. More, Q. Cheng, A. C. Donepudi, D. B. Buckley, Z. J. Lu, N. J. Cherrington and A. L. Slitt, *Drug Metab. Dispos.*, 2013, **41**, 1148–1155.

128. A. R. Zekri, Z. K. Hassan, A. A. Bahnassy, G. M. Sherif, E. L. D. Eldahshan, M. Abouelhoda, A. Ali and M. M. Hafez, *Asian Pac. J. Cancer Prev.*, 2012, **13**, 5433–5438.

129. D. Keppler, *Drug Metab. Dispos.*, 2014, **42**, 561–565.

130. H. U. Marschall, M. Wagner, G. Zollner, P. Fickert, U. Diczfalusy, J. Gumhold, D. Silbert, A. Fuchsbichler, L. Benthin, R. Grundstrom, U. Gustafsson, S. Sahlin, C. Einarsson and M. Trauner, *Gastroenterology*, 2005, **129**, 476–485.

131. J. Shoda, M. Kano, K. Oda, J. Kamiya, Y. Nimura, H. Suzuki, Y. Sugiyama, H. Miyazaki, T. Todoroki, S. Stengelin, W. Kramer, Y. Matsuzaki and N. Tanaka, *Am. J. Gastroenterol.*, 2001, **96**, 3368–3378.

132. G. Zollner, P. Fickert, R. Zenz, A. Fuchsbichler, C. Stumptner, L. Kenner, P. Ferenci, R. E. Stauber, G. J. Krejs, H. Denk, K. Zatloukal and M. Trauner, *Hepatology*, 2001, **33**, 633–646.

133. H. L. Chen, Y. J. Liu, H. L. Chen, S. H. Wu, Y. H. Ni, M. C. Ho, H. S. Lai, W. M. Hsu, H. Y. Hsu, H. C. Tseng, Y. M. Jeng and M. H. Chang, *Pediatr. Res.*, 2008, **63**, 667–673.

134. E. Hinoshita, K. Taguchi, A. Inokuchi, T. Uchiumi, N. Kinukawa, M. Shimada, M. Tsuneyoshi, K. Sugimachi and M. Kuwano, *J. Hepatol.*, 2001, **35**, 765–773.

135. M. Oswald, G. A. Kullak-Ublick, G. Paumgartner and U. Beuers, *Liver*, 2001, **21**, 247–253.

136. R. Kubitz, C. Droge, J. Stindt, K. Weissenberger and D. Haussinger, *Clin. Res. Hepatol. Gastroenterol.*, 2012, **36**, 536–553.

137. M. Hirano, K. Maeda, H. Hayashi, H. Kusuhara and Y. Sugiyama, *J. Pharmacol. Exp. Ther.*, 2005, **314**, 876–882.

138. A. C. Rodrigues, *Expert Opin. Drug Metab. Toxicol.*, 2010, **6**, 621–632.

139. C. G. Dietrich and A. Geier, *Expert Opin. Drug Metab. Toxicol.*, 2014, **10**, 1533–1551.

140. A. Treiber, R. Schneiter, S. Hausler and B. Stieger, *Drug Metab. Dispos.*, 2007, **35**, 1400–1407.

141. Y. Shitara, M. Hirano, H. Sato and Y. Sugiyama, *J. Pharmacol. Exp. Ther.*, 2004, **311**, 228–236.

142. Food and Drugs Administration, http://www.fda.gov/downloads/drugs/GuidancecomplianceRegulatoryInformation/Guidances/UCM292362.pdf.

143. European Medicines Agency, http://www.ema.europa.eu/docs/en_GB/document_library/Scientific_guideline/2012/07/WC500129606.pdf.

144. M. Roth, J. J. Araya, B. N. Timmermann and B. Hagenbuch, *J. Pharmacol. Exp. Ther.*, 2011.

145. M. Roth, B. N. Timmermann and B. Hagenbuch, *Drug Metab. Dispos.*, 2011, **39**, 920–996.

146. M. Belzer, M. Morales, B. Jagadish, E. A. Mash and S. H. Wright, *J. Pharmacol. Exp. Ther.*, 2013, **346**, 300–310.

147. K. L. Brouwer, D. Keppler, K. A. Hoffmaster, D. A. Bow, Y. Cheng, Y. Lai, J. E. Palm, B. Stieger, R. Evers and C. International Transporter, *Clin. Pharmacol. Ther.*, 2013, **94**, 95–112.

148. M. Rowland, C. Peck and G. Tucker, *Annu. Rev. Pharmacol. Toxicol.*, 2011, **51**, 45–73.

149. S. M. Huang, D. R. Abernethy, Y. Wang, P. Zhao and I. Zineh, *J. Pharm. Sci.*, 2013, **102**, 2912–2923.

150. R. Li, H. A. Barton, P. D. Yates, A. Ghosh, A. C. Wolford, K. A. Riccardi and T. S. Maurer, *J. Pharmacokinet. Pharmacodyn.*, 2014, **41**, 197–209.

151. R. Li, H. A. Barton and M. V. Varma, *Clin. Pharmacokinet.*, 2014, **53**, 659–678.

152. J. Bentz, M. P. O'Connor, D. Bednarczyk, J. Coleman, C. Lee, J. Palm, Y. A. Pak, E. S. Perloff, E. Reyner, P. Balimane, M. Brannstrom, X. Chu, C. Funk, A. Guo, I. Hanna, K. Heredi-Szabo, K. Hillgren, L. Li, E. Hollnack-Pusch, M. Jamei, X. Lin, A. K. Mason, S. Neuhoff, A. Patel, L. Podila, E. Plise, G. Rajaraman, L. Salphati, E. Sands, M. E. Taub, J. S. Taur, D. Weitz, H. M. Wortelboer, C. Q. Xia, G. Xiao, J. Yabut, T. Yamagata, L. Zhang and H. Ellens, *Drug Metab. Dispos.*, 2013, **41**, 1347–1366.

153. X. Qiu, H. Zhang and Y. Lai, *AAPS J.*, 2014, **16**, 714–726.

154. X. Chu, K. Korzekwa, R. Elsby, K. Fenner, A. Galetin, Y. Lai, P. Matsson, A. Moss, S. Nagar, G. R. Rosania, J. P. Bai, J. W. Polli, Y. Sugiyama, K. L. Brouwer and C. International Transporter, *Clin. Pharmacol. Ther.*, 2013, **94**, 126–141.

155. R. Gomeni, *Curr. Opin. Pharmacol.*, 2014, **14**, 23–29.

156. K. Allegaert and J. N. van den Anker, *Neonatology*, 2014, **105**, 344–349.

Drug Transporters in the Intestine

BENTE STEFFANSEN

Department of Physics, Chemistry and Pharmacy, Faculty of Sciences, University of Southern Denmark, 55 Campusvej, DK-5230 Odense, Denmark
Email: steffansen@sdu.dk

3.1 The Intestinal Tract and Drug Absorption

Most drugs are administered as oral drug formulations, usually as immediate release (IR) tablet formulations that disintegrate in the stomach. To understand the role of intestinal drug transporters (DTs) in apparent drug absorption, a brief review of the relevant physiological parameters of the gastro-intestinal (GI) tract as well as drug and nutrient absorption is presented here. The GI tract can be divided into several compartments, *i.e.* the stomach, duodenum, jejunum, ileum and colon. The primary function of the GI tract is to store, propel and decompose nourishment into a semifluid mass of partly digested food, which is first expelled by the stomach *via* the pyloric sphincter, as chyme, into the duodenal compartment. The chyme residence time in the stomach of healthy humans depends mainly on the fed or fasted state. In the fasted state, transit time ranges from 7 to 202 min and the pH of gastric juice varies from 1.7 to 4.7.[1] The transit time is generally longer and pH higher in the fed state stomach compared with the fasted.[1] In the duodenal compartment, chyme and gastric juice are mixed with pancreatic juice at pH 7.9–8.4, pancreatic enzymes and bile, including bile acids, resulting in bulk duodenal fluid at pH 4–5.5.[2,3] Intestinal transit time

RSC Drug Discovery Series No. 54
Drug Transporters: Volume 1: Role and Importance in ADME and Drug Development
Edited by Glynis Nicholls and Kuresh Youdim

Published by the Royal Society of Chemistry, www.rsc.org

ranges, in the duodenum, jejunum, ileum and colon are, respectively, 0.5–0.75, 1.5–2.0, 5–7 and 1–60 hours, with the intestinal fluid pH increasing along these compartments from pH 5.5 in the jejunum to 7.0–7.5 in ileum and rectum.[1–3] In the intestinal bulk fluids, the nutritional peptides/ proteins, carbohydrates, glycerides and vitamins are released from the chyme and dissolve or form micelles with bile acids. The dissolved nutrients are then partly hydrolysed, mainly by pancreatic enzymes in the intestinal fluid, to the respective amino acids, saccharides and charged fatty acids. Proteins and peptides are to a large extent metabolised in the intestinal fluid by pancreatic proteases and further during intestinal permeability by peptidases in the enterocytes. Proteases in the GI fluid and enterocytes have been reviewed elsewhere.[4] Absorption of these hydrophilic or charged molecular nutrients across the lipophilic intestinal membrane relies mainly on intestinal carriers/transporters, *i.e.* di-/tri-peptides rely on the solute carrier (*SLC*)*15A1*, amino acids on a wide range of amino acid transporters from the *SLC7A, 6A* and *36A* families, glucose on the glucose transporters from the *SLC5A* and *2A* families, and fatty acids on fatty acid transporter 4 (FATP4)/*SLC27A*4 (Table 3.1).

Intrinsic drug absorption to the mesenteric blood network, which confluences to portal blood, is illustrated in Figure 3.1. Intrinsic drug absorption depends primarily on the three following parameters: (1) the active pharmaceutical ingredient (API) passive diffusional-driven intestinal permeability (P_{pas}); (2) its solubility; and (3) the probability of metabolism in the GI fluid. The first two parameters that may define the intrinsic drug absorption depend, apart from the intestinal physiological dimensions and the mesenteric blood flow that is discussed in Section 3.2, on the physico-chemical characteristics of the API and thus on its ionisation state within the physiological pH range of the GI fluid. Consequently, for drugs that are acids, the increase of pH in the intestinal fluid from the duodenum to ileum results in a progressively greater fraction of API present as an ionised species along with its intestinal transit. Whereas the P_{pas} is generally inversely proportional to the fraction of ionised species of APIs that are acids, solubility is generally directly proportional. This coherence is generally *vice versa* for APIs that are bases. For these reasons, it is complicated to predict the influence of intestinal fluid pH on intrinsic drug absorption of APIs that are acids and bases. Nevertheless, studying how pH influences drug absorption is important in order to optimise the pharmaceutical formulation of new drug candidates as well as in lifecycle strategies. An example is the investigation of the reason for the observed variability in human desvenlafaxine bioavailability (BA), a serotonin–norepinephrine reuptake inhibitor. The intrinsic absorption of the amphoteric API desvenlafaxine was based on *in vitro* solubility and permeability studies, predicted by a physiologically based pharmacokinetic (PBPK) model and compared with absorption from clinical data. The fraction of deionised/neutral species of desvenlafaxine, which refers to its basic amine functionality (pK_a 9.18), increases with the pH in the intestinal fluid during transit. One may therefore speculate how

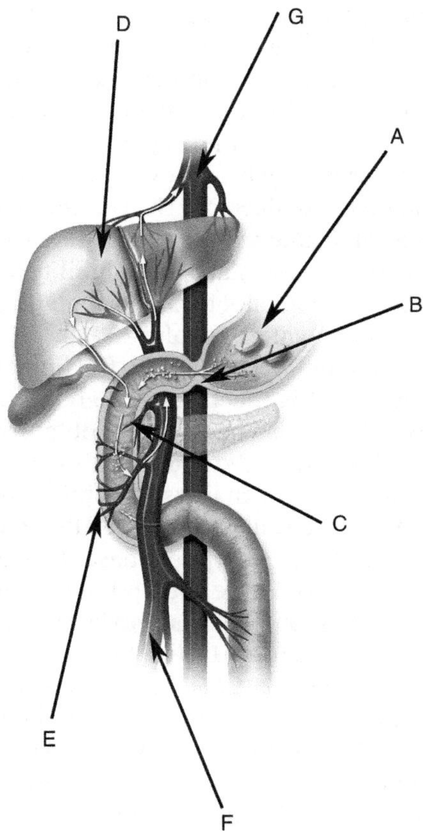

Figure 3.1 Schematic illustration of the GI tract. (A) Disintegration of a tablet formulation in the gastric lumen/stomach; (B) the pyloric sphincter; (C) the duodenum with bile and pancreatic ducts emptying the respective bile and pancreatic juice into the lumen; (D) the liver; (E) the mesenteric blood network surrounding the intestine; (F) portal blood by which mesenteric blood is distributed to the liver; and (G) central blood from which samples for plasma concentration profiles are taken. Figure 3.1 is printed with permission from illustrator Henning Dalhoff.

increases in neutral species during transit may impact on the intestinal absorption of desvenlafaxine, since its P_{pas} across the intestine is expected to increase with pH and its solubility in the intestinal fluid is expected to decrease. However, based on the PBPK predictions, it was suggested that absorption of desvenlafaxine, administered as an immediate release (IR) formulation, is restricted by pH-dependent permeability across the intestine rather than pH-dependent solubility in GI fluid.[5] It is likely that pH-partitioning may assist in the intrinsic absorption of desvenlafaxine. Conceding the fact that only a limited fraction of desvenlafaxine is present as the neutral species, *i.e.* 6.6×10^{-6} to 2×10^{-2} out of 1.00 at the respective intestinal fluid pH of 4–7.5, this efficacious species concentration nevertheless results in a concentration gradient across the intestinal membrane that seems to facilitate the complete absorption of desvenlafaxine. The BA of desvenlafaxine may therefore be very sensitive towards pH variability in intestinal fluid. One strategy to compensate for such a variable BA of desvenlafaxine may be to develop an extended release formulation, which restricts desvenlafaxine release in the upper small intestinal fluid where the pH variability is large. Such an extended release formulation of desvenlafaxine was predicted to decrease the variability in its absorption.[5] If the absorption of desvenlafaxine was solubility restricted, the formulation strategy might be to increase desvenlafaxine solubility in the intestinal fluid, *e.g.* by adding solubilising excipients to the formulation.

For APIs that are acids and with pK_a values <7.4, pH partitioning may be even more efficacious than for bases, since the fraction of neutral API species in mesenteric blood at pH 7.4 is much lower than the fraction in the upper intestinal fluid at pH 4–5.5. This continuously obtained concentration difference of neutral species across the intestinal membrane, caused by the inward-directed pH gradient between intestinal fluids and mesenteric blood, assists in particular the intrinsic absorption of APIs that are acids.[6,7]

Although the intrinsic drug absorption described above is important, transporters do play a large role in the absorption of many drugs.[8] Thus, the absorption of APIs may rely on the absorptive DTs/nutrient transporters (NTs) such as peptide transporter 1 (PEPT1)/*SLC15A1* or organic anion transporting polypeptide (OATP) 2B1/*SLCO2B1* (Table 3.1). Designing APIs in which absorption is improved by administering them as (pro)drugs for intestinal absorptive DTs/NTs is increasingly being considered as a strategy during drug development.[9–12] Such a strategy is especially important for those APIs that have permeability-limited absorption and thus belong to Biopharmaceutics Classification System (BCS) class 3 or 4. Other APIs are victims of exsorptive DTs that limit their BA because these APIs are re-shuffled directly to the intestinal tract by luminal/apical exsorptive DTs such as breast cancer resistance protein (BCRP)/*ABCG2*, P-glycoprotein (P-gp)/*ABCB1* and/or multidrug resistance associated protein (MRP) 2/*ABCC2* (Table 3.1). The BA of such APIs, especially those APIs that have solubility limited absorption, *i.e.* belonging to BCS class 2 or 4, may be further compromised by cytochrome P450 3A4 (CYP3A4), which is a metabolising enzyme expressed in enterocytes (Table 3.2) that limits the BA of many APIs such as cyclosporine and midazolam.[13,14] APIs may also inhibit or induce the absorption or exsorption of APIs that are substrates for the same DTs and thus perpetrate drug–drug interactions (DDIs). It is generally a challenge to elucidate the separate impact of intestinal transporters or enzymes on the apparent plasma concentration–time profile of the investigated APIs. The reasons for this are that intestinal and hepatic transport/metabolism processes of the APIs occur in tandem prior to their appearance in central blood, and that the APIs may interact with multiple transporters and/or enzymes that, furthermore, may be expressed in both enterocytes and hepatocytes. Examples of transporters expressed in both intestinal and hepatic tissue include BCRP, P-gp, MRP2 and OATP2B1. The CYP3A4 enzyme is also expressed in both enterocytes and hepatocytes, and Bailey *et al.* have shown that, in clinical situations, grapefruit juice (GFJ) consumed together with the APIs felodipine or nifedipine increased the area under the plasma concentration time curve (AUC) of these APIs.[15] It was later shown that the furanocoumarins in GFJ inactivate intestinal CYP3A.[16] Nevertheless, it is a challenge to elucidate whether an observed interaction with transporters and/or enzymes takes place in the intestine and/or liver. It is important to elucidate and predict the site(s) of interactions in order to avoid the subsequent risk of drug-induced toxicity due to such interactions. The intestinal availability (F_G), which is defined as the fraction of drug transferred from the intestinal enterocytes into the liver escaping from intestinal metabolism, can be predicted for substrates of CYP3A from *in vitro* studies on human intestinal microsomes.[17] In these studies, GFJ can be used to inactivate the CYP3A-mediated metabolism and F_G can be used to further predict the separate impact of intestinal CYP3A metabolism on API absorption.[17] Another *in vivo* method estimated inhibition constants (K_i) from a clinical study and used them to predict, using a PBPK model, the impact of co-administration of the known CYP3A4 inhibitor ketoconazole or other inhibitors on the AUCs of the CYP3A4 API substrates vardenafil and aprepitant. The PBPK model gave a more accurate prediction when both intestinal and hepatic CYP3A4 were taken into account.[18,19] Interactions between drugs and beverages on enzymes/transporters have recently been reviewed by An *et al.*[20]

3.2   The Enterocyte Monolayer

Irrespective of the intestinal segment, the intestinal membrane is composed of mucosa, sub-mucosa, circular smooth muscle layer, longitudinal smooth muscle layer and serosa, as illustrated in Figure 3.2. The mucosa comprises an innermost monolayer of enterocytes and is shaped in plicae or folds that expose villi and further brush border microvilli at each enterocyte. In an adult man, the intestinal length is altogether approximately 8 m, but the surface area of the monolayer of enterocytes amounts to approximately 450 m^2 due to the extensively folded structure of the mucosa.[1,2,21] The large monolayer dimensions are richly vascularised with the mesenteric blood flow, at approximately 19 l h^{-1}.[22] Most drugs, nutrients and endogens (*i.e.* endogenous compounds such as bile acids) are absorbed into the mesenteric blood, but exceptions to this are very lipophilic compounds with log $P > 5$, which to some extent are absorbed into lymph lacteals.[23,24] The monolayer of the enterocytes accounts for more than 90% of the duodenal, jejunal and ileal cells that generally make up the intestinal barrier.[25,26] Most dissolved nutrients, endogens and APIs are directed to the liver, *via* the mesenteric/portal blood, with the likelihood of being first-pass metabolised, as illustrated in Figure 3.1, whereas the highly lipophilic substances may, to some extent, undergo lymphatic absorption and thus bypass the portal vein and first-pass metabolism in the liver.[23,24,27]

The enterocytes are closely connected by tight junctions, which enable cell polarisation and differentiation into apical and basolateral membranes. Apart from the overall physiological parameters of the GI tract described in Section 3.1, the transporter-related parameters of the enterocytes of interest in drug development are discussed in this section. Thus, the transporter-related parameters and API physicochemical characteristics might be combined in *e.g.* PBPK models that include the physiological parameters of importance for the prediction and risk evaluation of DDIs between APIs that, based on *in vitro* studies, are shown to be substrates, inhibitors and/or inducers of the intestinal DTs. The transporter-related parameters include: (1) intestinal localisation and abundance/expression levels of NTs/DTs in enterocytes within the different intestinal segments; (2) the surface area of the enterocyte monolayer in the different segments; (3) luminal/apical *versus* serosal/basolateral localisation of NTs/DTs in the enterocytes, as well as their functionality as described by kinetic parameters and the structural requirements for substrates, inhibitors and inducers; and (4) the predicted effective intestinal permeability of APIs in humans ($P_{\text{eff, man}}$).

The NTs/DTs that are expressed in enterocytes are shown in Table 3.1 and, where possible, described by their localisation in the enterocyte membrane (column 1), *i.e.* apical and/or basolateral. NT/DT abundance is generally described either by the expression level of the corresponding mRNA sequence or by the expression levels of the transport protein itself, where known. Since the expression level of the transport protein is believed to be more closely related to actual transport functionality than mRNA expression, the protein expression levels are of most interest; however, they are also more difficult to determine. A study on the abundance of a number of intestinal NT/DT proteins in healthy epithelium from six organ donors found expression levels of between 0.2 and 1.6 pmol mg^{-1} with the exception of organic cation transporter (OCT)3 (<0.1 pmol mg^{-1}) and PEPT1 (2.6–4.9 pmol mg^{-1}). Protein abundances, given in picomoles of protein per milligram of intestinal tissue, of some important DTs as well as of CYP3A4 are shown in Table 3.2.[28,29] From Table 3.2 it can be seen that BCRP is expressed at higher levels in enterocytes belonging to the small intestine than the colon. The protein expression of P-gp seems to increase in enterocytes along the GI tract, in the order duodenum<jejunum<ileum<colon. The mRNA and protein levels of clinically-relevant DTs in the different segments of the intestine were also studied, and increasing P-gp protein expression as well as ASBT protein expression along the intestine was observed.[28] The MRP2 protein was equally expressed in enterocytes from the different compartments of the small intestine, but was expressed to a larger extent in colon enterocytes. In contrast to P-gp expression levels, another study found that the CYP3A4 protein was expressed at the highest level in the enterocytes from the duodenum followed by the jejunum, whereas the expression level was below the level of quantification (BLQ; 0.02 pmol mg^{-1}) in the ileum and

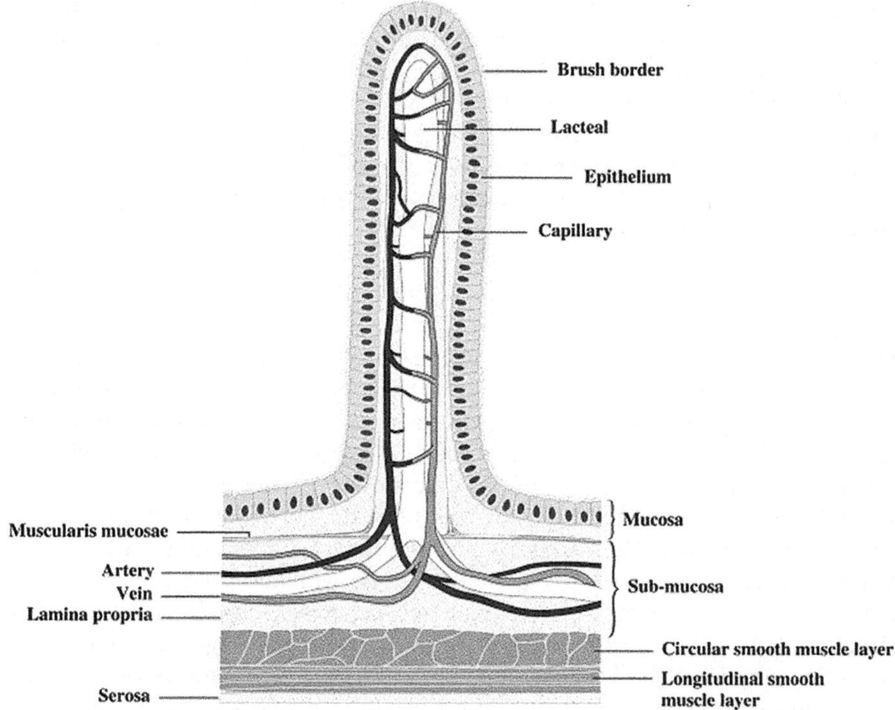

Figure 3.2 Structure of an intestinal villus. From the luminal perspective: mucosa (lined by enterocytes), sub-mucosa, circular smooth muscle layer, longitudinal smooth muscle layer and serosa.
Reprinted from ref. 3, copyright 2001 with permission from Elsevier.

colon.[29] Although it is not shown here, OCT3 and MRP3 were also detected, which is in accordance with the study by Englund *et al.*, who also found OCT3 and MRP3 mRNA expression in human intestinal tissues, although they studied diseased tissue.[30]

The kinetic parameters of substrates for specific transporters include K_m and J_{max}; K_m is the Michaelis–Menten constant for a specific API of a specific NT/DT, given in concentration units such as millimolarity, and J_{max} is the maximal number of APIs that are translocated by the given amount of NT/DT across an area per time, and is given in flux units, such as millimoles per second $\times$ centimeters-squared. Thus, J_{max} depends on the transporter protein abundance, which is one of the reasons why the abundance of transporter proteins is important to determine. The kinetic parameter of inhibition is generally determined as the inhibitory constant (K_i), which is the concentration of API that gives rise to 50% inhibition of a specific substrate/probe flux, mediated by a specific NT/DT. More detailed descriptions of how these kinetic parameters are determined *in vitro* are described elsewhere.[31–33]

$P_{\text{eff, man}}$ was studied in humans by Lennernäs *et al.*, who determined $P_{\text{eff, man}}$ across the human intestine *in vivo* by studying the disappearance of APIs by the Loc-I-Gut® method. From $P_{\text{eff, man}}$, they predicted the fraction of API absorbed (f_{a}), that is, the fraction of API fluxed across the intestine.[34–36] These studies formed the basis for the permeability parameter that, together with the solubility parameter, defines the four API classes in the BCS.[36–38] $P_{\text{eff, man}}$ of an API relates, according to Fick's first law, to the f_{a} by the following boundary condition:

$$J = P_{\text{eff, man}} \times C_{\text{s}} \tag{3.1}$$

where J is the API flux in units, such as nanomoles per second × centimetres-squared, and C_{s} is the API solubility (in molarity units) at the luminal fluid interface.[38] Dependent on the API interaction with NTs/DTs in the enterocytic membrane, $P_{\text{eff, man}}$ may encompass some saturable carrier-mediated permeability (P_{car}). $P_{\text{eff, man}}$ must be determined *in vivo* and thus is a very limited and expensive method. For this reason, there is a great amount of interest in correlating the apparent permeability (P_{app}), which is determined *in vitro*, to $P_{\text{eff, man}}$. There are principally two well-accepted and available *in vitro* methods that are applied for such *in vitro–in vivo* correlations: bidirectional API P_{app} studies across human intestinal segments mounted in Ussing chambers[39,40] or studies across filter grown monolayers of differentiated cells, such as filter grown Caco-2 cells or transfected cells.[33,41] The kinetic parameters of P_{car} should also be determined, and are generally determined *in vitro* either from bidirectional studies of flux or P_{app} using the above described methods.[31] This may include (stable) transfected cell lines that specifically express the transporter of interest, such as the BCRP–Madin–Darby canine kidney II (MDCKII) or MRP4–human embryonic kidney (HEK) 293 cells.[41,42] Prediction of human $P_{\text{eff, man}}$ or f_{a} from *in vitro* P_{app} values obtained by the Ussing chamber method was recently described by Sjöberg *et al.*[40] In this study, the P_{app} ranged from $0.09 \pm 0.06 \times 10^{-6}$ cm s^{-1} for sulfasalazine to $140 \pm 40.8 \times 10^{-6}$ cm^{-1} for L-leucine; such P_{app} values generally correlate well with $P_{\text{eff, man}}$ by a dimensionless factor of approximately 100. The P_{app} in this study also correlated well with the corresponding literature data on f_{a}.[40] Similarly, P_{app} values across monolayers of differentiated filter grown Caco-2 cells generally correlate well with $P_{\text{eff, man}}$ by a dimensionless factor of approximately 100 for APIs that primarily permeate by P_{pas}. However, if NTs/DTs are involved in drug transport, then the P_{app} correlation to $P_{\text{eff, man}}$ is less reliable.[43] The reason for this may be that the protein expression levels of NTs/DTs in Caco-2 cells vary, which may give rise to the poor *in vitro–in vivo* correlation between P_{app} determined in Caco-2 cells and $P_{\text{eff, man}}$ when carriers are determining for API absorption. P_{app} and the kinetic parameters for NT/DT can be applied for the prediction of $P_{\text{eff, man}}$ and thus f_{a} in the advanced dissolution, absorption and metabolism (ADAM) model, which is part of the SimCyp/Certara® PBPK model,[44–46] or in other similar prediction models.

3.3 Drug Transporters in Absorption

APIs that enter the mesenteric blood are further distributed by NTs/DTs to the hepatocytes in the liver, from where they are either redistributed across the sinusoidal membrane to the central blood as the parent API or metabolites, or they are excreted by canalicular efflux DTs to bile. From bile, the compounds generally re-enter the intestinal tract for further reabsorption (enterohepatic recycling) or for faecal excretion, together with insoluble drugs and indigestible chyme components. Enterohepatic recycling is reviewed by Roberts *et al.*[47]

A large number of APIs are substrates and/or inhibitors of DTs. In the present chapter the focus is on the NTs/DTs expressed in enterocytes and their role in API, nutrient and endogen absorption to mesenteric blood. These transporters belong to the SLC or ABC transporter families. The SLC family is nicely systemised in the Bioparadigms database (www.bioparadigms.org), which is based on data from the special issue in *Pflügers Archives* on SLC transporters[48] and later updated with data from the SLC special issue in *Molecular Aspects of Medicine*.[49] The ABC family was reviewed in 2012[50] and DTs from both the SLC and ABC families are included in the database (The UCSF-FDA TransPortal: A Public Drug Transporter Database).[51]

The main enterocyte transporters are presented in Table 3.1 and are, apart from cellular localisation, described by their protein name, gene name and identification number (ID) from the National Center for Biotechnology Information (NCBI). The transporters are, where possible, also described by transport type, *i.e.* SLC exchanger (Ex), SLC co-transporter (Co), SLC facilitative transporter (F) or ABC. The primary co-transported and exchanged substrates are mentioned in the same column as the transport type. Table 3.1 also includes, to my best knowledge, the most common substrates, *i.e.* APIs, endogens and nutrients, together with, for some selected transporters, the reported inhibitors known to date (2016).

From Table 3.1 it can be seen that many APIs have been suggested as substrates and/or inhibitors for the following apical absorptive transporters, in particular: proton-coupled amino acid transporter 1 (PAT1)/*SLC36A1*;[52,53] apical sodium-dependent bile acid transporter (ASBT)/*SLC10A2*;[54] sodium glucose transporter (SGLT)1/*SLC5A1*;[55,56] mono-carboxylate transporter (MCT)1/*SLC16A1*;[57] concentrative nucleoside transporter (CNT) 1/*SLC28A1*, CNT2/*SLC28A2* and CNT3/*SLC28A3*;[58–60] OATP1A1/*SLCO1A2*[61] and OATP2B1/ *SLCO2B1*;[62,63] OCTN1/*SLC22A4*;[64] as well as PEPT1/*SLC15A1*.[11,65] For a comprehensive but not exhaustive list of substrates/inhibitors for these transporters see Table 3.1. There are several examples of API interactions with these transporters in the GI tract. For example, the oral absorption of the tri-peptidomimetic antibiotic cefadroxil in mice relies on PEPT1/ *SLC15A1*, with the systemic exposure of cefadroxil being reduced by 90% and maximal plasma concentration (C_{max}) by 15–20-fold in PEPT1 knockout mice compared with wild type.[66] A similar study using PEPT1 knockout mice demonstrated the importance of PEPT1 in the rate and extent of oral absorption of the antiviral API valaciclovir.[67] For APIs where their absorption relies on the presence of membrane transporters, there is a risk that the API will restrict its own absorption if the applied dose is sufficient to saturate the transporter.[68] However, PEPT1/*SLC15A1* is expressed at high levels in the intestine and is a high capacity transporter with K_m values in the millimolar range[69] and thus the risk of saturation for this transporter is believed to be minimal. To the best of my knowledge, there is no example of a clinical study in which nonlinear drug absorption can be ascribed to the API being a victim drug for PEPT1. One example of an API that is a perpetrator for PEPT1 has been observed in rats, following administration of bestatin and cefixime. After oral co-administration of bestatin and cefixime, the C_{max} of bestatin decreased and the time taken to reach C_{max} (T_{max}) increased when co-administered with cefixime and *vice versa*, which was attributable to interactions at PEPT1.[70]

Many APIs have also been found to be substrates and/or inhibitors or inducers of the following apical efflux/exsorptive DTs: BCRP/*ABCG2*, MRP2/*ABCC2* and P-gp (MDR1/*ABCB1*);[51] for further references see Table 3.1. The clinical relevance of these DTs in intestinal drug exsorption, using selected examples from clinical studies and including the potential impact of polymorphisms, is covered in a recent review by Misaka *et al.*[71] Whereas inhibition by concurrently administered APIs that interact with exsorptive DTs may lead to increased plasma concentrations, induction reduces the absorption of the substrate API and thereby decreases its plasma concentration. The exsorptive DTs generally have lower capacities (K_m in micromolar ranges) than NTs, and for this reason the risks of DDIs and drug–food interactions for these DTs are believed to be more widespread.[72,73] An example is the observed clinical interaction between sulfasalazine and curcumin at intestinal BCRP/*ABCG2*.[72] When curcumin (2 g) was given orally to healthy subjects 30 min before oral sulfasalazine (100 µg or 2 g), the C_{max} and AUC of sulfasalazine increased.[74] Therefore, sulfasalazine's AUC can be altered by some co-administered compounds. Based on this study, it has been suggested that sulfasalazine could be used as a probe to investigate the possibility of clinical DDIs of new molecular entities (NMEs) on intestinal BCRP.[75] It should also be noted that, *in vitro*, sulfasalazine was found to be a substrate for the absorptive DT OATP2B1, expressed in both the jejunum and ileum, which may have some impact on its intestinal absorption.[30,74]

Basolateral transporters have also been suggested as sites for possible drug interactions, altering oral absorption, *i.e.* MRP1/*ABCC1* and MRP3/*ABCC3*, OSTα/β/*SLC51A/B*, and OCT1/*SLC22A1*; for references see Table 3.1. For example, an interaction between digoxin and bile acids at OSTα/β has been suggested.[76,77] The International Transporter Consortium has also highlighted the potential risk of DDIs at these basolateral transporters.[73]

Absorption of nutrients from intestinal fluid to mesenteric blood, as well as distribution of endogens such as amino acids,[78] bile acids and estrone-1-sulfate (E1S), may depend on several NTs/DTs, as outlined in Figure 3.3. NTs/DTs involved in absorption of neutral amino acids (AA0) are outlined in

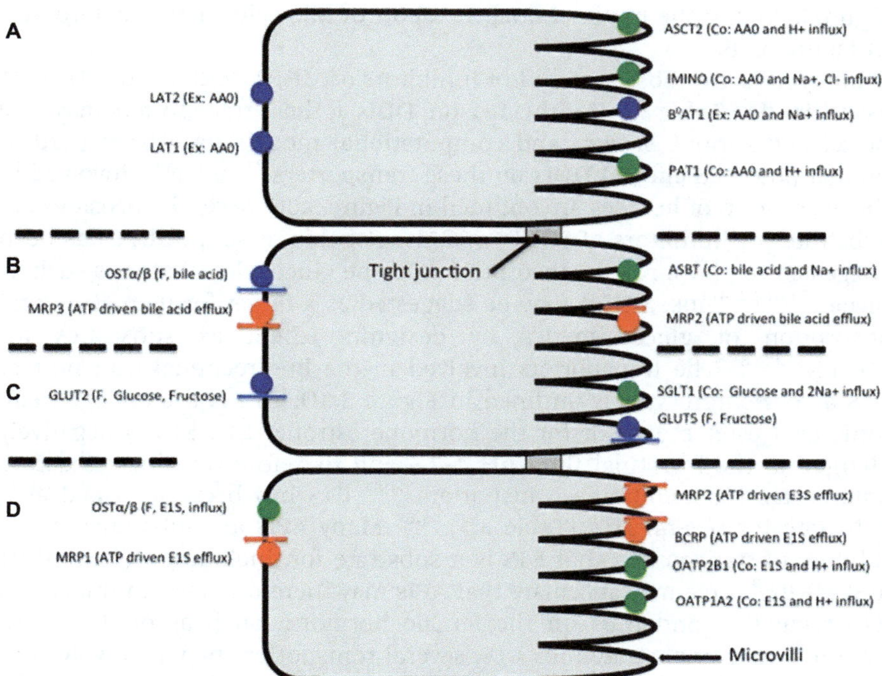

Figure 3.3 NTs and transporters for endogens involved in (A) neutral amino acid (AA0) translocation across the apical membrane of enterocytes by PAT1/*SLC36A1*, ASCT2/*SLC1A5*, IMINO/*SLC6A20* and B⁰AT1/*SLC6A19*, and across the basolateral membrane by the LAT1/*SLC7A5* and LAT2/*SLC7A8* exchangers. The function of these basolateral NTs is dependent on the co-expression of the 4F2hc/*SLC3A2* protein, and has been reviewed in more detail elsewhere.[78] (B) Bile acid transport. In the absorptive direction three different transporters are involved, ASBT/*SLC10A2*, basolateral bidirectional OSTα/β/*SLC51A/B* and MRP3/*ABCC3*.[79] In the exsorptive direction, two transporters, the basolateral OSTα/β/*SLC51A/B* and apical MRP2/*ABCC3* are involved.[79,80] (C) Absorption of hexoses. At the apical membrane SGLT1/*SLC5A1* is responsible for glucose and galactose influx, whereas GLUT5/*SLC2A5* facilitates fructose influx. In addition apical GLUT7 facilitates flux of both glucose and fructose (not shown). At the basolateral membrane GLUT2/*SLC2A2* facilitates translocation of fructose and glucose as reviewed in more detail elsewhere.[78] (D) E1S transport. In the absorptive direction, E1S relies on OATP1A2/*SLCO1A2*, OATP2B1/*SLCO2B1* in the apical direction and MRP1/*ABCC1* in the basolateral membrane. In the exsorptive direction, basolateral OSTα/β/*SLC51A/B*, and apical BCRP/*ABCG2* and MRP2/*ABCC2* are suggested to be involved.[86,87] Facilitative transporters are shown as blue globes with indicative bi-directional arrows. Exchange and co-transporters are shown in green, while ABC transporters are shown in red with indicative efflux arrows. The apical membrane is illustrated with microvilli facing the intestinal lumen and the basolateral membrane, facing the mesenteric blood. Enterocytes are interlinked by tight junctions, which also define the border between apical and basolateral membranes and thereby enable the cells to polarise their expression of membrane transporters. Co: co-transporter; E3S: estrone-3-sulfate; Ex: exchanger; F: facilitative transporter.

Figure 3.3A and the bi-directional transport of bile acids across enterocytes in Figure 3.3B.

Many APIs are substrates and/or inhibitors of MRP2, MRP3 and OSTα/β,[80] as outlined in Table 3.1. For this reason, DDIs at these transporters may take place, and various *in vitro* and computational models have been used to predict potential clinical DDIs on these transporters.[31,79,80] NTs involved in the absorption of hexoses are outlined in Figure 3.3C. Several glucosides are substrates or inhibitors of SGLT1/*SLC5A1*. For this reason, SGLT1 has been suggested to play a role in absorption of some glucoside substrates such as piceid.[55,81,82] Thus, SGLT1 may be suggested as a target for increasing oral absorption of glucomimetics by designing them as substrates for SGLT1.[55,81–83] The transporters involved in the bi-directional transport of E1S across enterocytes is outlined in Figure 3.3D. E1S is the endogen sulfonic acid ester precursor for the hormone estrone (E1). E1S is negatively charged at the intestinal fluid pH 4–7.5 and therefore translocates across enterocytes *via* membrane transporters.[84,85] E1S has been described as a substrate for several DTs (Table 3.1).[86,87] Many APIs are substrates or inhibitors of transporters that E1S is a substrate for, including BCRP, MRP1 and MRP2.[31] One may speculate that APIs may therefore have an impact on E1S regulation and thus on the female hormone, such as on E1 distribution.[88] As demonstrated for E1S, several transporters may be involved in the absorption and exsorption of a single compound. An API example is methotrexate, which is highlighted in bold in Table 3.1 and is a substrate of seven transporters that are expressed in enterocytes, *i.e. ABCC1, ABCC2, ABCC3, ABCC4, ABCC5, ABCG2* and *SLCO1A2*. Interactions between methotrexate and bilirubin at intestinal transporters were observed in rats and this has been suggested to be one of the reasons why methotrexate induces intestinal toxicity in some patients.[89]

3.4 Conclusions

The aim of the present chapter was to review transporters in the intestine and to exemplify their roles in nutrient, endogen and API absorption and exsorption. A total of 51 intestinal transporters from both the SLC and ABC families have been identified to play a role in the absorption of nutrients, and/or the absorption/exsorption of APIs and/or endogens. To date (2016), 14 of these transporters have been found to interact with APIs. Many APIs are described to be substrates and/or inhibitors of the efflux/exsorptive apical transporters BCRP, P-gp and MRP2. This may give rise to DDIs, food–drug interactions or endogen–drug interactions at intestinal transporters, with a potential impact on API absorption and drug safety/efficacy. This is the case for sulfasalazine absorption, which is increased when administered with high concentrations of curcumin. In addition, some endogens such as E1S and bile acids are substrates of several intestinal transporters that also transport APIs. One may therefore speculate that these APIs could also have an impact on the regulation of endogens and thus have implications for health.

Table 3.1 Intestinal transporters from the SLC and ABC superfamilies represented by cellular localisation, protein name, gene name and identification number (ID) in the National Center for Biotechnology Information (NCBI).[a]

Cellular localisation	Protein name	Gene name and ID	Transport type[b]	Substrates	Inhibitors
Apical	LAT1	*SLC7A5*[c] 8140	Ex: AA0	Large AA0, baclofen, gabapentin, methylmercury-L-cysteine, melphalan, L-DOPA, L-methyl-DOPA, 3-*O*-methyl-dopatriiodothyronine, thyroxine[103,e]	–
Basolateral	LAT2	*SLC7A8*[c] 23428	Ex: AA0	AA0, cysteine[103,e]	–
Apical	b[0,+] AT	*SLC7A9*[d] 11136	Ex: AA0	AA[+], large AA0, cysteine, gabapentin[103,e]	–
Not specified	B[0]AT1	*SLC6A19* 340024	Co: Na[+]	AA0[103,e]	–
Basolateral	Y[+]LAT1	*SLC7A7*[c] 9056	Ex: Na[+]/large AA0	Large AA0 for Na[+] and AA[+] for AA0[103,e]	–
Basolateral	Y[+]LAT2	*SLC7A6*[c] 9057	Ex: Na[+]/large AA0	Large AA0 for Na[+] and AA[+] for AA0[103,e]	–
Not specified	ASC1	*SLC7A10*[c] 56301	Ex: AA0	Small D- and L-AA0, α-aminobutyric acid, β-alanine, D-serine[103,e]	–
Basolateral	CAT1	*SLC7A1* 6541	Ex	AA[+103,e]	–
Apical	IMINO	*SLC6A20* 54716	Na[+]/Cl[−]	AA0, imino acids[103,e]	–
Apical	ATB[0,+] (colon)	*SLC6A14* 11254	Na[+]/Cl[−]	AA[+], AA0, pregabalin[103,e]	–
Not specified	TAUT	*SLC6A6* 6533	Na[+]/Cl[−]	Taurine, β-alanine, GABA[90,103]	–
Basolateral	TAT1	*SLC16A10* 117247	F	AAA, L-DOPA, thyroxine, triiodothyronine	–
Not specified	EAAC1	*SLC1A1* 6505	Co: K[+]/Na[+]/Cl[−]	AA-, Glu-1a, Asp-3[103,e]	–

Table 3.1 (*Continued*)

Cellular localisation	Protein name	*Gene name* and ID	Transport type[b]	Substrates	Inhibitors
Apical	PAT1	*SLC36A1* 206358	Co: H^+	AA0, *trans*-4-aminocrotonic acid, δ-aminopentanoic acid, δ-aminolevulinic acid, arecaidine, L-azetidine-2-carboxylic acid, betaine, *cis*-4-hydroxy-proline, D-cycloserine, 3,4-dehydro-D,L-proline, glycylglycine, GABA, gaboxadol, β-guanidinopropionic acid, guvacine, isoguvacine, isonipecotic acid, muscimol, nipecotic acid, sarcosine, D-serine, taurine, 4,5,6,7-tetraisoxazolo[4,5-*c*]pyridine-3-ol, vigabatrin[91]	–
Apical	SVCT1	*SLC23A1* 9962	Co: Na^+	Ascorbic acid[31,103]	–
Apical	ASBT	*SLC10A2* 6555	Co: Na^+	Cholesterol, cholate, glycochenodeoxycholate, glycoursodeoxycholate, glycodeoxycholate, taurocholate[26,31]	A3309, cyclosporin, EGCG, PR835, S8921, SC435, 264W94 and many more. For a comprehensive list see ref. 80
Not specified	SMVT	*SLC5A6* 8884	Co: Na^+	Biotin, lipoate, pantothenate[103,e]	–
Apical	BCRP	*ABCG2* 9429	ABC	Albendazole sulfoxide, atorvastatin, azidopine, bisantrene, cerivastatin, cimetidine, ciprofloxacin, daunorubicin, $E_2$17βG, DHEAS, diclofenac, dipyridamole, doxorubicin, enrofloxacin, epirubicin, E1S, fluvastatin, folic acid,	Abacavir, amprenavir, atazanavir, atorvastatin, beclomethasone, betamethasone, cerivastatin, corticosterone, daunorubicin, delavirdine, dexamethasone, DHEAS, diclofenac, digoxin, dipyridamole, doxorubicin, efavirenz, elacridar (GF120918),

				glibenclamide, grepafloxacin, hematoporphyrin, imatinib, irinotecan, lamivudine, **methotrexate**, 4-methylbelliferone sulfate, mitoxantrone, nitrendipine, norfloxacin, ofloxacin, oxfendazole, pitavastatin, pravastatin, prazosin, rosuvastatin, riboflavin, simvastatin, SN-38, SN.38 glucuronide, sulfasalazine, topotecan, uric acid, ulifloxacin, vitamin K, zidovudine[31,51]	erlotinib, estradiol sulfate, E1S, fluvastatin, fumitremorgin C, glibenclamide, Ko143, leflunomide, lopinavir, **methotrexate**, methylprednisolone, mitoxantrone, nelfinavir, nicardipine, nifedipine, nimodipine, nilotinib, nitrendipine, novobiocin, pitavastatin, prednisolone, ritonavir, rosuvastatin, saquinavir, simvastatin, SN38, sulfasalazine, taurolithocholate sulfate, triamcinolone, vitamin K3[31,51]
Not specified	FATP4	*SLC27A4* 376497	NR	Palmitate, oleate, linolate, butyrate, myristate, oleate, palmitate, stearate, arachidonate, linoleate[103,e]	–
Apical	SGLT1	*SLC5A1* 6523	Co: $Na^+(H^+)$	D-glucose, galactose, arbutin, benzyl-β-glucoside, helicon, indican, iodide, α-methyl-D-glycopyranoside, β-methyl-D-glycopyranoside, 3-O-methyl-glycoside myoinositol, 2-naphthyl-β-D-glucoside, pantothenate, piceid, proline, urea[81,83,93]	Esculin, 4-β-D-glucopyranosyl-aminobenzenesulfonamide, 8-hydroxyquinoline-β-D-glucoside, 4-methylumbelliferyl-β-glucopyranoside, 4-nitrophenyl β-glucoside, phlorizin, quercetin-3-β-glucoside, quercetin-4′-β-glucoside, sergliflozin-A[83,93]
NR	SGLT3	*SLC5A4* 6527	Channel	Glucose, Na^+, H^+	–
NR	SGLT4	*SLC5A9* 200010	Co: Na^+	Mannose, fructose, glucose	–

Table 3.1 (*Continued*)

Cellular localisation	Protein name	*Gene name* and ID	Transport type[b]	Substrates	Inhibitors
Basolateral	GLUT2	*SLC2A2* 6514	F	D-glucose, D-fructose, glucosamine, mannose, streptozotocin[103]	–
Not specified	GLUT5	*SLC2A5* 6518	F	Fructose	–
Apical	MCT1	*SLC16A1* 6566	Co: H^+, Ex: monocarboxylate	Benzoic acid, butyrate, L-lactate, nicotinic acid propionate, pyruvate, pravastatin, salicylic acid[51,103]	Υ-hydroxybutyrate, D-lactate, L-lactate, neteglinide[51,103]
NR	MCT8	*SLC16A2* 6567	F	Thyroid hormones, diiodothyronine, thyroxine, reverse triiodothyronine	–
Basolateral	MRP1	*ABCC1* 4363	ABC	p-Aminohippuric acid, bilirubin diglucuronide, bilirubin monoglucuronide, citalopram, daunorubicin, DHEAS, DNP-GS, doxorubicin $E_2 17\beta G$, E1S, epirubicin, etoposide folic acid, glutathione, idarubicin, irinotecan leucovorin, leukotriene C_4, leukotriene D_4, leukotriene E_4, **methotrexate**, Mitoxantrone, Saquinavir, SN-38, teniposide, topotecan, trivalent antimony, vinblastine, vincristine[50,94–98]	Atazanavir, benzbromarone, benzylpenicillin, bilirubin diglucuronide, bilirubin monoglucuronide, curcumin, doxorubicin, $E_2 17\beta G$, estradiol sulfate, $E_2 3SO_4 17\beta G$, $E_3 16\beta G$, $E_3 17\beta G$, E1S, daunorubicin, flurbiprofen, furosemide, genistein, glibenclamide, glycholate, glycolithocholate-3-sulfate, indinavir, indomethacin, indoprofen, kaempferol, ketoprofen, leukotriene C_4, lopinavir, MK-571, myricetin, nelfinavir, probenecid, quercetin, ritonavir, saquinavir, sulfinpyrazone, taurocholate, valspodar (PSC-833), verapamil, vinblastine, vincristine[51,96–98]

Apical	MRP2	*ABCC2* 1244	ABC	*p*-Aminohippuric acid, bilirubin diglucuronide, bilirubin monoglucuronide, DHEAS, DNP-GS, $E_2 17\beta G$, etoposide, E1S, fexofenadine, glutathione, indinavir, irinotecan, leucovorin, leukotriene C_4, **Methotrexate**, olmesartan, pitavastatin, pravastatin, rifampicin, ritonavir, rosuvastatin, saquinavir, SN-38, sulfinpyrazone[31,51]	Bilirubin diglucuronide, bilirubin monoglucuronide, benzbromarone, curcumin, cyclosporine, daunorubicin, delavirdine, efavirenz, emtricitabine, etoposide, furosemide, gemifloxacin, glibenclamide, indomethacin, ketoprofen, MK-571, leukotriene C_4, PAK-104P, probenecid, reserpine, sulfinpyrazone, valspodar (OSC-833), vinblastine, vincristine[31,51]
Basolateral	MRP3	*ABCC3* 4363	ABC	DNP-GS, $E_2 17\beta G$, etoposide, glucuronide, fexofenadine, folic acid, glibenclamide, glycocholate, hyocholate-glucuronide, leucovorin, leukotriene C_4, **methotrexate**, taurocholate, taurolithocholate-3-sulfate[31,51]	Benzbromarone, cyclosporin, DNP-GS, delavirdine, doxorubicin, efavirenz, emtricitabine, etoposide, glycocholate, glycolithocholate sulfate, indomethacin, lithocholate sulfate, **methotrexate**, probenecid, sulfinpyrazone, taurochenodeoxycholate, taurocholate, taurodeoxycholate, taurolithocholate sulfate[31,51]
Basolateral	MRP4	*ABCC4* 10257	ABC	Adefovir, chenodeoxycholylglycine, chenodeoxycholytaurine, cholate, cholyltaurine, cAMP, cGMP, dehydroepiandrosterone sulfate, $E_2 17\beta G$, folic acid, **methotrexate**, olmesartan, *p*-aminohippurate, prostaglandin E_1, prostaglandin E_2, tenofovir, topotecan[31,51]	Benzbromarone, candesartan, celecoxib, diclofenac, dipyridamole, indomethacin, ketoprofen, losartan, MK-571, nitrobenzylmercaptopurine riboside, probenecid, sildenafil, sulfinpyrazone, sulindac, trequinsin, zaprinast[31,51]

Table 3.1 (*Continued*)

Cellular localisation	Protein name	*Gene name* and ID	Transport type[b]	Substrates	Inhibitors
NR	MRP5	*ABCC5* 10057	ABC	cAMP, cGMP, DNP-GS, folic acid, leucovorin, 6-mecaptorine, **methotrexate**[25,31]	cAMP, folic acid, leucovorin, probenecid, sildenafil, sulfinpyrazone[25,31]
NR	MRP6	*ABCC6* 368	ABC	DNP-GS, Etoposide, Leutrokiene C_4,[95,99]	Benzbromarone, DNP-GS, indometacin, probenecid[31,51]
Basolateral	ENT1	*SLC29A1* 2030	F	Purine, pyrimidine nucleosides, cladribine, cytarabine, fludarabine, gemcitabine, zalcitabine, didanoside (ddI)[103]	Dipyridamole, dilazep
Apical	CNT1	*SLC28A1* 9154	Concentrative: Na^+	Pyrimidine nucleosides, 3′-azido-3′-deoxythymidine (AZT), cladribine, cytarabine 2′,3′-dideoxycytidine (ddC), 5′deoxy-5-fluorouridine, gemcitabine, urdine, zaltidabine[103]	–
Apical	CNT2	*SLC28A2* 9153	Concentrative: Na^+	Purine nucleosides, cladribine, ddI[103]	–
Apical	CNT3	*SLC28A3* 64078	Concentrative: Na^+	Purine and pyrimidine nucleosides, AZT, 3-deazauridine 5-flurouridine, floxuridine, gemcitabine, ribavirin zebularine, zalcitabine[103]	–
Apical	OATP1A2	*SLCO1A2* 6579	NR	Bilirubin, cholate, chlorambucil-taurocholate, cholate, ciprofloxacin, DHEAS, $E_2$17βG, enoxacin, e3s, fexofenadine, gatifloxacin, glycholate, imatinib, levofloxacin, lomefloxacin, **methotrexate**, norfloxacin,	Chenodeoxycholate, cholate, dexamethasone, $E_2$17βG, erythromycin, E3S grepafloxacin, indinavir, 3-iodothyronamine, ketoconazole, lovastatin, moxifloxacin, nelfinavir, quinidine, saquinavir,

				pitavastatin, prostaglandin E$_2$, reverse triiodothyronine, rocuronium, rosuvastatin, saquinavir, taurocholate, taurochenodeoxycholate, tauroursodeoxycholate, tebipenem pivoxil, thyroxine, triiodothyronine[31,51]	taurocholate, tauroursodeoxycholate, ursodeoxycholate, verapamil[31,51]
Apical	OATPB2B1	*SLCO2B1* 11309	Co: H$^+$	Aliskiren, atorvastatin, benzylpenicillin, bosentan, DHEAS, E3S, fexofenadine, fluvastatin, glibenclamide, pitavastatin, pravastatin, pregnenolone sulfate, progesterone, prostaglandin E$_2$, rosuvastatin, taurocholate, tebipenem pivoxil[31,51]	Aliskiren, *p*-aminohippuric acid, atazanavir, atorvastatin, cerivastatin, cimetidine, cyclosporin, darunavir, DHEAS, digoxin, efavirenz, β-estradiol, estriol, estrone, gemfibrozil, glibenclamide, indinavir, indomethacin, lopinavir, lovastatin, memantine, mifepristone, nelfinavir, nicotinic acid, pregnenolone sulfate, probenecid, repaglinide, rifampicin, rifamycin SV, ritonavir, rofecoxib, rosiglitazone, saquinavir, simvastatin, taurocholate, testosterone, tipranavir[31,51]
Not specified	OATP3A1	*SLCO3A1* 28232	NR	E3S, benzylpenicillin prostaglandin E$_1$, prostaglandin E$_2$, prostaglandin F$_{2\alpha}$, thyroxine, vasopressin[31,51]	–
Not specified	OATP4A1	*SLCO4A1* 28231	NR	Benzylpenicillin, E$_2$17βG, E3S, prostaglandin E$_2$, reverse triiodothyronine, taurocholate, thyroxine, triiodothyronine[31,51]	–

Table 3.1 (*Continued*)

Cellular localisation	Protein name	Gene name and ID	Transport type[b]	Substrates	Inhibitors
Not specified	OAT10	*SLC22A13* 9390	Ex: organic anions	–	–
Basolateral	OSTα/β	*SLC51A/B*	F	DHEAS, digoxin, E3S, glycochenodeoxycholate, glycodeoxycholate, glycoursodeoxycholate, prostaglandin E_2, taurochenodeoxycholate, taurocholate, taurodeoxycholate, tauroursodeoxycholate[31,51]	Digoxin, E3S, glycholithocholic acid, glycholithocholic acid sulfate, indomethacin, lithocholic acid sulfate, probenecid, spironolactone, taurolithocholate, taurolithocholate sulfate[31,51]
Basolateral	OCT1	*SLC22A1* 6580	F	Organic cations such as acetyl-L-carnitine, aciclovir, desipramine, dopamine, ganciclovir, histidine, noradrenaline, quinidine, serotonin, spermine, spermidine, tetraethylammonium (TEA), thiamine, tyramine[103]	–
Basolateral	OCT2	*SLC22A2* 6582	F	Acriflavine, agmatine, amantadine, amiloride, choline, cimetidine, cisplatin creatinine, 4-[4-(dimethylamino)-styryl]-*N*-methylpyridinium (ASP), dopamine, famotidine, histamine, memantine, metformin, *N*-1-methylnicotinamide (NMN), 1-methyl-4-phenylpyridinium (MPP+), noradrenaline (norepinephrine), oxaliplatin ranitidine, serotonin, TEA[103]	–

Apical	OCTN1	*SLC22A4* 6583	F Ex: H$^+$	Pyrilamine, quinidine, verapamil, carnitine derivatives, betaine, cephaloridine, choline, emetine, quinidine, TEA, valproate[103]	–
Not specified	OCTN2	*SLC22A5* 6584	Co: Na$^+$, L-carnitine F		–
Apical	PEPT1	*SLC15A1* 6564	Co: H$^+$	L,L-dipeptides, L,L-tripeptides alafosfalin, δ-amino levulinic acid, AT264, L-α-methyl-DOPA-pro, ampicillin, bestatin, captopril, cefaclor, cefadroxil, cefamandole, cefdinir, cefixime, ceftibuten, cefuroximaxetil, cephalexin, cephalothin, cephradine, ceronapril, cyclacillin, enalapril, fosinopril, glycyl-sarcosine, isoleucine-thiazolidide, lisinopril, loracarbef, phenylalanine, gabapentin, ramipril, L-valaciclovir, L-val-L-valacyclovir, L-valgancyclocir[100–102]	Glibenclamide, ibuprofen, nateglinide quinapril, spirapril[29]
NR	PHT1	*SLC15A4* 121260	Co: H$^+$	L,L-dipeptides, L,L-tripeptides histidine[103]	–
NR	PHT2	*SLC15A3* 51296	Co: H$^+$	L,L-dipeptides, L,L-tripeptides, histidine[103]	–
P-gp	MDR1	*ABCB1* 5243	ABC	Abamectin, actinomycin D, amebicides, amprenavir, berberine, biotin, bisantrene, cimetidine, colchicine, corticosterone cyclosporine A daunorubicin, dexamethasone, digoxin, docetaxel, domperidone,	Amiodaron, astemizole, azelastine, azithromycin, clarithromycin, cyclosporine, desethylamiodarone, dipyridamole, elacridar (GF120918) erlotinib, erythromycin, itraconazole,

Table 3.1 (*Continued*)

Cellular localisation	Protein name	*Gene name* and ID	Transport type[b]	Substrates	Inhibitors
				doxorubicin, emetine, epirubicin, erythromycin etoposide, fexofenadine, FK506, gramicidin D, hydrocortisone indinavir, lopinavir, irinotecan, ivermectin, loperamide, **methotrexate**, mitoxantrone, morphine, nelfinavir, nicardipine, ondansetron, paclitaxel, ritonavir, rhodamine 123, saquinavir, asimadoline, teniposide, topotecan, triamcinolone, verapamil, vinblastine, vincristine, valinomycin[51]	ketoconazole, OC144-093, paclitaxel, quinidine, reserpine, ritonavir, roxithromycin, tamoxifen, telithromycin, valspodar (PSC 833), verapamil, vinblastine, XR9576, zosuquidar (LY335979)[51]
NR	NaPiIIb	*SLC34A4* 10568	Co: Na	HPO_4^{2-}	–

[a]AA0: Neutral amino acids; AA−: anionic amino acids; AA+: cationic amino acids; AAA: aromatic amino acids; Co: co-transporter; Ex: exchanger; F: facilitative transporter; GABA: gamma-amino butyric acid; NR: To the best of my knowledge, not reported.
[b]The transporter type refers to driving force. For some selected transporters, inhibitors are also mentioned. Primary references can be found in the following reviews or databases: ref. 31, 50, 51, 80, 81, 83, 90–98, 100–103.
[c]The transporters are related to their driving force and substrates. Transport function is dependent on the co-expression of 4F2hc/*SLC3A2* protein or;
[d]rBAT/*SLC3A1* and has been reviewed in more detail elsewhere.[78]
[e]http://www.bioparadigms.org/slc/intro.htm.

Table 3.2 Protein expression of DTs and the CYP3A4 protein.

pmol mg^{-1}	Small intestine			Colon
	Duodenum[a] (1.9 m^2 and 0.35 m)	Jejunum[a] (184.0 m^2 and 2.8 m)	Ileum[a] (276.0 m^2 and 4.2 m)	Including rectum[a] (1.3 m^2 and 1.5 m)
ABCG2/BCRP protein expression	0.19–0.41[28]			0.04–0.16[28]
ABCB1/MDR1/P-gp protein expression	0.37–0.98[29] 0.29–1.06[28]	0.19–1.22[29]	0.43–1.00[29]	0.14–0.61[29] 0.15–1.77[28]
MRP2 protein expression	0.05–0.24[29] 0.76–1.03[28]	0.53–0.87[29]	0.79–1.33[29]	1.80–2.13[29] 0.95–1.77[28]
CYP3A4 protein expression	0.64–2.11[29]	0.12–1.07[29]	BDQ[b,28]	BDQ[b,28]
ASBT protein expression	0.01–1.58[28]			BDQ[b,28]
PEPT1 protein expression	2.63–4.89[28]			0.19–0.31[28]
OCT1 protein expression	0.57–0.84[28]			0.47–0.73[28]
OATP2B1 protein expression	0.43–0.56[28]			0.48–0.73[28]

[a]Surface area and length, respectively, of the various intestinal sections without considering the fold expansion of duodenum and jejunum.[2]
[b]BDQ: below quantification.

References

1. M. Koziolek, M. Grimm, D. Becker, V. Iordanov, H. Zou, J. Shimizu, C. Wanke, G. Garbacz and W. Weitschies, *J. Pharm. Sci.*, 2015, **104**, 2855.
2. A. L. Daugherty and R. J. Mrsny, *Pharm. Sci. Technol. Today*, 1999, **4**, 144.
3. J. M. DeSesso and C. F. Jacobsen, *Food Chem. Toxicol.*, 2001, **39**, 209.
4. B. Steffansen, C. U. Nielsen and S. Frokjaer, *Eur. J. Biopharm.*, 2005, **60**, 241.
5. F. Franek, A. Jarlfos, F. Larsen, P. Holm and B. Steffansen, *Eur. J. Pharm. Sci.*, 2015, **77**, 303.
6. S. Neuhoff, A.-L. Ungell, I. Samora and P. Artursson, *Eur. J. Pharm. Sci.*, 2005, **25**, 211.
7. P. A. Shore, B. B. Brodie and C. A. Hogben, *J. Pharmacol. Exp. Ther.*, 1957, **119**, 361.
8. K. Sugano, M. Kansy, P. Artursson, A. Avdeef, S. Bendels, L. Di, G. F. Ecker, B. Faller, H. Fischer, G. Gerebtzoff, H. Lennernäs and F. Senner, *Nat. Rev. Drug Discovery*, 2010, **9**, 597.
9. A. E. Thomsen, M. S. Christensen, M. A. Bagger and B. Steffansen, *Eur. J. Pharm. Sci.*, 2004, **23**, 319.
10. A. H. Eriksson, M. V. Varma, E. J. Perkins and C. L. Zimmerman, *J. Pharm. Sci.*, 2010, **99**, 1574.
11. B. Steffansen, S. Frøkjær, M. E. Taub, 1998 Novo Nordisk A/S, *WO9802451A1*, Patent application.

12. S. K. Nigam, *Nat. Rev. Drug Discovery*, 2015, **14**, 29.

13. J. C. Kolars, W. M. Awni, R. M. Merion and P. B. Watkins, *Lancet*, 1991, **338**, 1488.

14. M. F. Paine, D. D. Shen, K. L. Kunze, J. D. Perkins, C. L. Marsh, J. P. McVicar, D. M. Barr, B. S. Gillies and K. E. Tummel, *Clin. Pharmacol. Ther.*, 1996, **60**, 14.

15. D. G. Bailey, J. D. Spence, C. Munoz and J. M. Arnold, *Lancet*, 1991, **337**, 268.

16. P. Schmiedlin-Ren, D. J. Edwards and M. E. Fitzsimmons, *Drug Metab. Dispos.*, 1997, **25**, 1228.

17. M. Gertz, A. Harrison, J. B. Houston and A. Galetin, *Drug Metab. Dispos.*, 2010, **38**, 1147.

18. M. Kato, Y. Shitara, H. Sato, K. Yoshisue, M. Hirano, T. Ikeda and Y. Sugiyama, *Pharm. Res.*, 2008, **25**, 1891.

19. Y. Mano, Y. Sugiyama and K. Ito, *J. Pharm. Sci.*, 2015, **104**, 3183.

20. G. An, J. K. Mukker, H. Derendorf and R. F. Frye, *J. Clin. Pharmacol.*, 2015, **55**, 1313.

21. J. Sobotta and H. Becher, *Atlas of Human Anatomy*, ed. H. Ferner and J. Staubesand, 9th edn, Urban and Schwarzenberg, München, 1975, pp. 101–143.

22. D. N. Granger, P. D. Richardson, P. R. Kvietys and N. A. Mortillaro, *Gastroenterology*, 1980, **78**, 837.

23. S. M. Caliph, F. W. Faassen and C. J. Porter, *J. Pharm. Pharmacol.*, 2014, **66**, 1377.

24. N. L. Trevaskis, W. N. Charman and C. J. Porter, *Adv. Drug Delivery Rev.*, 2008, **17**, 702.

25. L. Shen, *Curr. Top. Microbiol. Immunol.*, 2009, **337**, 1.

26. J. R. Turner, *Nat. Rev. Immunol.*, 2009, **9**, 799.

27. J. Iqbal and M. M. Hussain, *Am. J. Endocrinol. Metabol.*, 2009, **296**, E1183.

28. M. Drozdzik, C. Gröer, J. Penski, J. Lapczuk, M. Ostrowski, Y. Lai, B. Prasad, J. D. Unadkat, W. Siegmund and S. Oswald, *Mol. Pharm.*, 2014, **11**, 3547.

29. A. Bruyére, Decléves, F. Bouzom, K. Ball, C. Marques, X. Treton, M. Pocard, P. Valleur, Y. Bouhnik, Y. Panis, J.-M. Scherrmann and S. Mouly, *Mol. Pharm.*, 2010, **7**, 1596.

30. G. Englund, F. Rorsman, A. Rönnblom, U. Karlbom, L. Lazorova, J. Gråsjö, A. Kindmark and P. Artursson, *Eur. J. Pharm. Sci.*, 2006, **29**, 269.

31. A. S. Grandvuinet, H. T. Vestergaard, N. Rapin and B. Steffansen, *J. Pharm. Pharmacol.*, 2012, **64**, 1523.

32. C. U. Nielsen, B. Brodin and B. Steffansen, *Molecular Biopharmaceutics. Aspects of Drug Characterization, Drug Delivery and Dosage Form Evaluation*, ed. B. Steffansen, B. Brodin and C. U. Nielsen, Pharmaceutical Press, London, p. 153.

33. P. Artursson, P. Matsson and M. Karlgren, *Transporters in Drug Development*, ed. Y. Sugiyama and B. Steffansen, AAPS Press, Springer, New York, 2013, pp. 37–65.

34. H. Lennernäs, Ö. Ahrenstedt, R. Hallgren, L. Knutson, M. Ryde and L. K. Paalzow, *Pharm. Res.*, 1992, **9**, 1243.

35. H. Lennernäs, D. Nilsson, S.-M. Aquilonius, Ö. Ahrenstedt, L. Knutson and L. K. Paalzow, *Br. J. Clin. Pharmacol.*, 1993, **35**, 243.

36. G. L. Amidon, H. Lennernäs, V. P. Shah and J. R. Crison, *Pharm. Res.*, 1995, **12**, 413.

37. CDER/FDA, U.S. Department of Health and Human Services Food and Drug Administration Center for Evaluation and Research (CDER), In: Guidance for industry: waiver of in vivo bioavailability and bioequivalence studies for immediate release solid dosage forms based on Biopharmaceutics Classification System, 2000.

38. V. P. Shah and G. L. Amidon, *AAPS J.*, 2014, **16**, 894.

39. H. H. Ussing and K. Zerahn, *Acta Physiol. Scand.*, 1951, **23**, 110.

40. Å. Sjöberg, M. Lutz, C. Tannergren, C. Wingolf, A. Borde and A. L. Ungell, *Eur. J. Pharm. Sci.*, 2013, **48**, 166.

41. I. Hubatsch, E. G. E. Ragnarsson and P. Artursson, *Nat. Protoc.*, 2007, **2**, 2111.

42. I. S. Haslam, J. A. Wright, D. A. O'Reilly, D. J. Sherlock, T. Coleman and N. L. Simmons, *Drug Metab. Dispos.*, 2011, **39**, 2321.

43. S. Tavelin, J. Taipalensuu, F. Hallböök, K. Vellonen, V. More and P. Artursson, *Pharm. Res.*, 2003, **20**, 397.

44. M. Jamei, D. Turner, J. Yang, S. Neuhoff, S.- Polak, A. Rostami-Hodjegan and G. Tucker, *AAPS J.*, 2009, **11**, 225.

45. M. D. Harwood, S. Neuhoff, G. L. Carlson, G. Warhurst and A. Rostami-Hodjegan, *Biopharm*, 2013, **34**, 2.

46. S. Neuhoff, K. R. Yeo, Z. Barter, M. Jamei and D. B. Turner, *J. Pharm. Sci.*, **102**, 3145.

47. M. S. Roberts, B. M. Magnusson, F. J. Burczynski and M. Weiss, *Clin. Pharmacokinet.*, 2002, **41**, 751.

48. M. A. Hediker, M. F. Romero, J. B. Peng, A. Rolfs, H. Takanaga and E. A. Bruford, *Pflugers Arch.*, 2004, **447**(5), 465–468.

49. M. A. Hediger, B. Clemencon, R. E. Burrier and E. A. Bruford, *Mol. Aspects Med.*, 2013, **34**, 95–107.

50. A. H. Schinkel and J. W. Jonker, *Adv. Drug Delivery Rev.*, 2012, **64**, 138.

51. K. M. Morrissey, C. C. Wen, S. J. Johns, L. Zhang, S. M. Huang and K. M. Giacomini, *Clin. Pharmacol. Ther.*, 2012, **92**, 545.

52. M. K. Nøhr, R. V. Juul, Z. I. Thale, R. Holm, M. Kreilgaard and C. U. Nielsen, *Eur. J. Pharm. Sci.*, 2015, **69**, 10.

53. D. T. Twaites and C. M. Anderson, *Br. J. Pharmacol.*, 2011, **164**, 1802.

54. O. C. Jeon, S. R. Hwang, T. A. Al-Hilal, J. W. Park, H. T. Moon, S. Lee, J. H. Park and Y. Byun, *Pharm. Res.*, 2013, **30**, 959.

55. F. T. Okkels, O. R. Jensen, B. Steffansen, 2006, Evolva AG, *WO 06056604A1*, 2006, *Patent application*.
56. Z. Cai, J. Huang, H. Luo, X. Lei, Z. Yang, Y. Mai and Z. Liu, *Drug Target*, 2013, **21**, 574.
57. I. Tamai, Y. Sai, A. Ono, Y. Kido, H. Yabuuchi, H. Takanaga, E. Satoh, T. Ogihara, O. Amano, S. Izeki and A. Tsuji, *J. Pharm. Pharmacol.*, 1999, **51**, 1113.
58. K. Takahashi, K. Yoshisue, M. Chiba, T. Nakanishi and I. Tamai, *J. Pharm. Sci.*, 2015, **10**, 1002.
59. M. Hiratochi, K. Tatani, K. Shimizu, Y. Kuramochi, N. Kikuchi, N. Kamada, F. Itoh and M. Isaji, *Eur. J. Pharmacol.*, 2012, **5**, 690.
60. S. Rebello, S. Zhao, S. Hariry, M. Dahlke, N. Alexander, A. Vapurcuyan, I. Hanna and V. Jarugula, *Eur. J. Clin. Pharmacol.*, 2012, **68**, 697.
61. M. Grube, K. Kock, S. Oswald, K. Draber, K. Meissner, L. Eckel, M. Bohm, S. B. Felix, S. Vogelgesang, G. Jedlitschky, W. Siegmund, R. Warzok and H. K. Kroemer, *Clin. Pharmacol. Ther.*, 2006, **80**, 607.
62. J. Imanaga, T. Kotegawa, H. Imai, K. Tsutsumi, T. Yoshizato, T. Ohyama, Y. Shirasaka, I. Tamai, T. Tateishi and K. Ohash, *Pharmacogenet. Genomics*, 2011, **21**, 84.
63. Y. Shitara, K. Maeda, K. Ikejiri, K. Yoshida, T. Horie and Y. Sugiyama, *Biopharm. Drug Dispos.*, 2013, **34**, 45.
64. N. Nakamichi, H. Shima, S. Asano, T. Ishimoto, T. Sugiura, K. Matsubara, H. Kusuhara, Y. Sugiyama, Y. Sai, K. Miyamoto, K. Tsuji and Y. Kato, *J. Pharm. Sci.*, 2013, **102**, 3407.
65. T. C. Burnette and P. de Miranda, *Drug Metab. Dispos.*, 1994, **22**, 60.
66. M. Posada and D. E. Smith, *Pharm. Res.*, 2013, **30**, 2931.
67. B. Yang, Y. Hu and D. E. Smith, *Drug Metab. Dispos.*, 2013, **41**, 1867.
68. FDA, *Draft Guidance for industry drug interaction studies, study design data analysis and implication for dosing and labeling*, 2012.
69. B. Bretschneider, M. Brandsch and R. Neubert, *Pharm. Res.*, 1999, **16**, 55.
70. L. Wang, C. Wang, Q. Liu, Q. Meng, X. Huo, P. Sun, X. Yang, H. Sun, Y. Zhen, J. Peng, X. Ma and K. Liu, *Eur. J. Pharm. Sci.*, 2014, **63**, 77–86.
71. S. Misaka, F. Müler and M. F. Fromm, *Curr. Opin. Pharmacol.*, 2013, **13**, 847.
72. K. Maeda and Y. Sugiyama, *Mol. Aspects Med.*, 2013, **34**, 711.
73. J. K. Giacomini, S.-M. Huang, D. J. Tweedie, L. Z. Benet, K. L. R. Brouwer, X. Chu, A. Dahlin, R. Evers, V. Fischer, K. M. Hillgren, K. A. Hoffmaster, T. Ishikawa, D. Keppler, R. B. Kim, C. A. Lee, M. Niemi, J. W. Polli, Y. Sugiyama, P. W. Swaan, J. A. Ware, S. H. Wright, S. W. Yee, M. J. Zamek-Gliszczynski and L. Zhang, *Nat. Rev.*, 2010, **9**, 215–236.
74. H. Kushuhara, H. Furuie, A. InanoI, A. Sunagawa, S. Yamada, C. Wu, S. Fukizawa, N. Morimoto, I. Ieiri, M. Morishita, K. Sumita, H. Mayahara, T. Fujita, K. Maeda and Y. Sugiyama, *Br. J. Pharmacol.*, 2012, **166**, 1793.

75. C. A. Lee, M. A. O'Connor, T. K. Ritchie, A. Galetin, J. A. Cook, I. Ragueneau-Majlessi, H. Ellens, B. Feng, M. E. Taub, M. F. Paine, W. Polli, J. A. Ware and M. J. Zamek-Gliszczynski, *Drug Metab. Dispos.*, 2015, **43**, 490.
76. D. J. Seward, A. S. Koh, J. L. Boyer and N. Ballatori, *J. Biol. Chem.*, 2003, **278**, 27473.
77. N. Ballatori, W. V. Christian, J. Y. Lee, P. A. Dawson, C. J. Soroka, J. L. Boyer, M. S. Madejczyk and N. Li, *Hepatology*, 2005, **42**, 1270.
78. C. U. Nielsen, B. Brodin and B. Steffansen, *Molecular Biopharmaceutics. Aspects of Drug Characterization, Drug Delivery and Dosage Form Evaluation*, ed. B. Steffansen, B. Brodin and C. U. Nielsen, Pharmaceutical Press, London, pp. 193–212.
79. P. A. Dawson and S. J. Karpen, *J. Lipid Res.*, 2015, **56**, 1085.
80. X. Zheng, S. Ekins, J. P. Raufman and J. E. Polli, *Mol. Pharm.*, 2009, **6**, 1591.
81. C. Henry, X. Vitrac, A. Decendit, R. Ennamany, S. Krisa and J. M. Merillon, *J. Agric. Food Chem.*, 2005, **53**, 798.
82. Z. Cai, J. Huang, H. Luo, X. Lei, Z. Yang, Y. Mai and Z. Liu, *J. Drug Target*, 2013, **21**, 574.
83. A. Diez-Sampedro, M. P. Lostao, E. M. Wright and B. A. Hirayama, *J. Membr. Biol.*, 2000, **176**, 111.
84. L. K. Gram, G. M. Rist and B. Steffansen, *Mol. Pharm.*, 2009, **6**, 145.
85. L. K. Gram, G. M. Rist, H. Lennernäs and B. Steffansen, *Eur. J. Pharm. Sci.*, 2009, **37**, 378–386.
86. A. S. Grandvuinet and B. Steffansen, *J. Pharm. Sci.*, 2011, **100**, 3817.
87. A. S. Grandvuinet, L. Gustavsson and B. Steffansen, *Mol. Pharm.*, 2013, **10**, 3285.
88. B. Steffansen and A. S. Grandvuinet, *Transporters in Drug Development*, ed. Y. Sugiyama and B. Steffansen, AAPS Press, Springer New York, 2013, p. 23.
89. T. Yokooji, N. Mori and T. Murakami, *J. Pharm. Pharmacol.*, 2011, **63**, 206.
90. M. Tomi, A. Tajima, M. Tachikawa and K.-I. Hosoya, *Biochim. Biophys. Acta*, 2008, **1778**, 2138.
91. S. B. Frølund, *Intestinal Drug delivery via the proton coupled amino acid transporter, PAT: exploring the interaction between dipeptides, dipeptidomimetics and PAT1*, PhD Thesis, 2012, University of Copenhagen, 1–146.
92. M. P. Lostao, B. A. Hirayama, D. D. Loo and E. M. Wright, *J. Membr. Biol.*, 1994, **142**, 161.
93. G. Kottra and H. Daniel, *J. Pharmacol. Exp. Ther.*, 2007, **322**, 829.
94. G. Jedlitschky, I. Leier, U. Buchholz, J. Hummel-Eisenbeiss, B. Burchell and D. Keppler, *Biochem. J.*, 1997, **327**, 305.
95. I. J. Hidalgo, T. J. Raub and R. T. Borchardt, *Gastroenterology*, 1989, **96**, 736.
96. D. W. Loe, K. C. Almquist, S. P. Cole and R. G. Deeley, *J. Biol. Chem.*, 1996, **271**, 9683.

97. E. Bakos, R. Evers, E. Sinkó, A. Váradi, P. Borst and B. Sarkadi, *Mol. Pharmacol.*, 2000, **57**, 760–768.

98. D. T. Twaites and C. M. H. Anderson, *Br. J. Pharmacol.*, 2011, **164**, 1802.

99. G. L. Amidon, H. Lennernas, V. P. Shah and J. R. Crison, *Pharm. Res.*, 1995, **12**, 413.

100. D. H. Omkvist, *Intestinal drug delivery via hPEPT1: in vitro and in silico evaluation of the relationship between substrate structure, binding and translocation*, PhD Thesis, 2011, University of Copenhagen, pp. 1–64.

101. C. U. Nielsen, B. Brodin, F. S. Jørgensen, S. Frokjaer and B. Steffansen, *Expert Opin. Ther. Pat.*, 2002, **12**, 1329.

102. C. U. Nielsen, J. Våbenø, R. Andersen, B. Brodin and B. Steffansen, *Expert Opin. Ther. Pat.*, 2005, **15**, 153.

103. B. Steffansen, C. U. Nielsen, B. Brodin, A. H. Eriksson, R. Andersen and S. Frokjaer, *Eur. J. Pharm. Sci.*, 2004, **21**, 3.

Drug Transporters in the Kidney

GIT WENG CHUNG,* SARAH FAYE BILLINGTON,
SARAH ELIZABETH JENKINSON AND
COLIN DOUGLAS BROWN

Institute for Cellular and Molecular Biosciences, Newcastle University
Medical School, Framlington Place, Newcastle upon Tyne NE2 4HH, UK
*Email: git.chung@newcastle.ac.uk

4.1 Introduction

The kidney plays a primary role in the clearance of a wide range of predominately hydrophilic endogenous compounds and xenobiotics from the
body, particularly molecules that have negligible hepatic clearance. The
importance of the kidney in drug clearance is illustrated by a recent study
which identified that, out of a sample of 330 clinically-relevant drug molecules, around 30% undergo renal clearance.[1] Similarly, the kidney also
plays a key role in the clearance of polar metabolites generated by hepatic
metabolism of hydrophobic drug molecules, and more recently renal elimination has been shown to play an important role in the rapid clearance of
peptide therapeutics from the bloodstream.[2]

Renal clearance primarily consists of the filtration of free drug from the
plasma across the glomerular barrier into the lumen of the nephron followed by elimination of the drug molecule into the urine. For a subset of
drug molecules, the rate of renal clearance is modified either by tubular
absorption and/or by tubular secretion during drug transit through the

RSC Drug Discovery Series No. 54
Drug Transporters: Volume 1: Role and Importance in ADME and Drug Development
Edited by Glynis Nicholls and Kuresh Youdim

Published by the Royal Society of Chemistry, www.rsc.org

nephron. Tubular absorption and tubular secretion of drug molecules is mediated by a series of transport proteins located in specific nephron segments. The role of transporters in determining the absorption, distribution, metabolism and excretion (ADME) of drug molecules has been studied in great depth over the last few years. Indeed, a clear focus on understanding drug handling and drug–drug interactions (DDIs) at the level of membrane transporters has become routine for the development of drugs by most pharmaceutical companies.[3] The importance of renal clearance and the role of transport proteins in determining renal clearance, and as potential sites for DDIs, has led regulatory agencies to issue guidance to investigate the role of key membrane transporters in the renal handling of new molecular entities (NMEs) when renal clearance mediates a significant proportion (more than 25%) of the total NME clearance.[4–7]

The aim of this chapter is to provide an overview of the physiology of the kidney, a review of clinically-relevant transporters that play a key role in determining renal clearance, and also to provide a brief insight into the currently available preclinical *in vitro* models and their advantages and disadvantages in the study of renal drug transport. Since rodents, more specifically rats, are routinely the choice for preclinical studies, transporter expression in these animals is also explored in this chapter.

4.2   The Anatomy of the Kidney

The kidney is a bean-shaped organ located towards the back of the abdomen of many mammals that is highly perfused, receiving around a fifth of cardiac output per minute. At a gross level, the kidney can be subdivided into three distinct regions: the cortex making up the outer third of the kidney; the renal medulla making up the bulk of the inner two-thirds; and the renal pelvis (Figure 4.1A). The renal artery enters the kidney at the hilus located on the medial surface of the kidney along with the ureter and renal vein; within the kidney, blood is distributed to the cortex *via* a series of subdividing blood vessels that finally give rise to a glomerular capillary network. Around 93% of the blood supply is restricted to the cortex with the remaining 7% feeding the medulla *via* vasa recta blood vessels arising from juxtaglomerular nephrons.[8]

The functional unit of the kidney is the nephron. A human kidney contains about 1.25 million nephrons compared with approximately 300 000 nephrons in the rat kidney, a common *in vivo* model.[9,10] A unit of nephron comprises the Bowman's capsule that encases the glomerulus, a proximal tubule, loop of Henle, and the distal and collecting tubule segments (Figure 4.1B). Three kinds of nephrons are found in the kidney, defined by the spatial location of their glomerulus within the cortex: superficial nephrons with very short loops of Henle; mid-cortical nephrons with short loops of Henle; and juxta-medullary nephrons that possess long loops of Henle projecting deep into the medulla. Juxtaglomerular nephrons, which only constitute about 7% of nephrons in humans and rats, have a specialised role in maintaining fluid

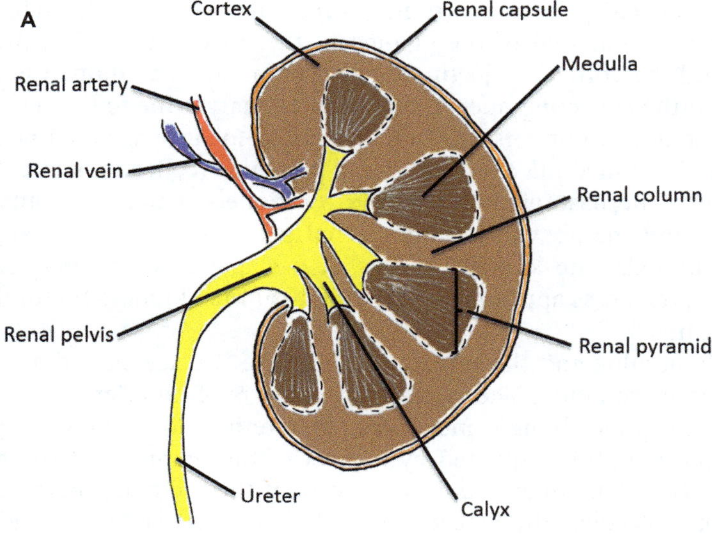

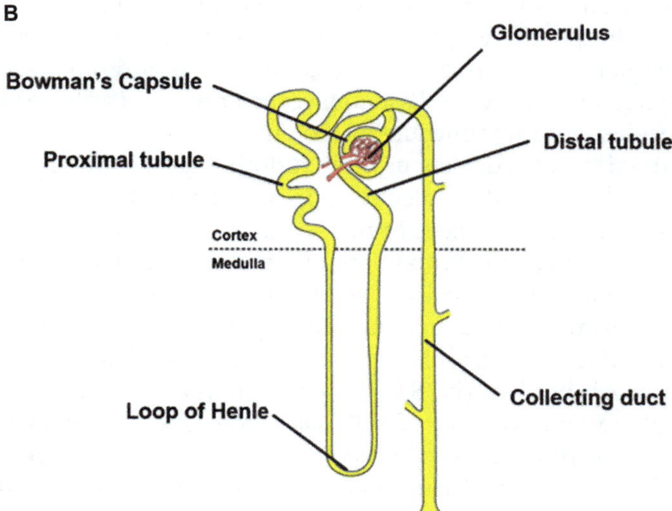

Figure 4.1 Anatomy of the human kidney and nephron. (A) The cortex and medulla are distinct regions visible on the surface of a bisected human kidney. The medulla is divided into several renal pyramids. The apexes of the pyramids have small openings that lead into the calyces of the renal pelvis, which is responsible for draining the urine produced into the bladder *via* the ureter. (B) A unit of nephron comprises the Bowman's capsule, which encases the glomerulus, the proximal tubule, the loop of Henle and the distal tubule. They are found in the cortex of the kidney. In addition to the descending and ascending loops of Henle, collecting ducts are found in the medulla of the kidney.

(B) Reproduced from https://commons.wikimedia.org/wiki/File:Gray1130.svg.

balance. The kidney has a complex three-dimensional (3D) architecture. The cortex contains all of the glomeruli, the proximal tubules, the distal tubule and the cortical collecting ducts. The medulla contains the loop of Henle and the collecting ducts that deliver primary urine to the renal pelvis and the ureter. In humans, the kidney is multi-lobar, which can be seen at the level of the medulla as several renal pyramids (collections of collecting ducts), with the base of each pyramid positioned at the corticomedullary boundary and the apex towards the renal pelvis. In rats, and many other small mammals, the kidney consists of a single lobe of renal pyramid. Otherwise, the gross appearance of the small animal kidneys resembles that of the human.[8,10]

The glomerulus and the proximal tubule are the key sites that regulate renal drug elimination. The glomerulus consists of an afferent arteriole, a glomerular capillary bundle and an efferent arteriole. Blood flow through the glomerulus is tightly regulated by control of the diameter of the afferent arteriole. The glomerular capillary consists of three distinct barriers to filtration of molecules: the fenestrated endothelial cell layer; the underlying collagen matrix; and the filtration slits generated by the interdigitating foot processes of the podocytes (the epithelial cells lining the Bowman's capsule).[11] Together these form a barrier that limits the filtration based on the size and charge of a molecule. Essentially, the barrier is freely permeable to small molecules and impermeable to proteins.

The proximal tubule consists of three morphologically distinct segments (S1, S2 and S3).[12,13] The S1 segment comprises the initial convoluted portion of the proximal tubule. The cells in this segment have a tall brush border on the apical membrane, and the basolateral membrane forms extensive lateral invaginations with adjacent cells. The S2 segment consists of the rest of the convoluted proximal tubule and the initial portion of the straight tubule. The structure of the cells in this segment is similar to that of S1, except for the shorter brush border. The S3 segment comprises the remainder of the proximal tubule and consists of cells with even shorter brush borders—although this varies between species—and fewer basolateral invaginations.[8,10,11] The mitochondria number is also lower in these cells than cells of the other two segments, which suggests less active transport in this region.[13] All three segments are dealt with as a single collective proximal tubule in this chapter.

4.3 Renal Clearance of Xenobiotic Compounds

The renal clearance of xenobiotic compounds is one of several mechanisms essential to protecting the body from toxicity. Foreign chemical substances that undergo renal excretion include drug molecules, toxic contaminants from food, toxins produced by micro-organisms, agrichemicals and heavy metals. Overall, the process of renal clearance is a net result of three key mechanisms: glomerular filtration, tubular secretion and tubular reabsorption.

As blood enters each nephron through afferent arterioles, it passes through glomerular capillaries then leaves *via* efferent arterioles. In the glomerular capillaries, water, urea, glucose, amino acids and other small molecules pass across the basement membrane of the Bowman's capsule into the nephron in a process called ultrafiltration, with red blood cells, proteins and other molecules too large to pass through being left behind in the capillaries. This is a highly efficient process; on average the human kidneys collectively receive around 20% of cardiac output (1000 ml of blood per minute) and form around 120 ml of protein-free ultrafiltrate per minute.[14,15] The ultrafiltrate within the nephron then passes through the proximal tubule, the loop of Henle, the distal convoluted tubule and a series of collecting ducts to form urine.

In terms of renal drug clearance, many drug molecules and their metabolites are excreted in urine solely as a consequence of glomerular filtration. If a xenobiotic is freely filtered from the renal glomerular capillaries into the Bowman's capsule and is neither reabsorbed nor secreted, then its clearance rate (C_x) will equal the glomerular filtration rate (GFR), which is given in eqn (4.1):

$$\mathrm{GFR} = \frac{U_x \times V}{P_x} \tag{4.1}$$

where U_x is the urinary concentration of the molecule, V is the rate of urine production and P_x is the plasma concentration of the molecule.

Several techniques can be used to estimate GFR; clinically, the most common method is based upon the rate of clearance of creatinine, an endogenous metabolite of muscle metabolism.[16,17] At steady state, creatinine production is matched by renal creatinine elimination, which means that an estimate of GFR can be made solely from a measure of plasma levels. Creatinine clearance gives a relatively accurate estimate of GFR and is widely used to monitor renal function. However, the use of creatinine clearance to monitor both GFR and renal function is not without its problems. These arise predominately because, in addition to being freely filtered, a proportion (10–30%) of the total creatinine appearing in the urine arises from the tubular secretion of creatinine in the proximal tubule of the nephron. Therefore, anything that interferes with the tubular secretion of creatinine will reduce the renal clearance of creatinine and result in an increase in plasma creatinine levels, which may be misinterpreted as a decrease in renal function. Clinically, increases in plasma creatinine concentrations have been reported for a number of transporter-mediated drug–creatinine interactions, most notably for drugs that inhibit organic cation transporter 2 (OCT2) and multidrug and toxin extrusion transporters (MATEs) in the renal proximal tubules, which are thought to be key transporters in the renal secretion of creatinine (discussed in more detail in Section 4.4). For instance, cimetidine has been shown to increase creatinine concentration in the plasma by as much as 15% in patients with normal GFR.[18] For these reasons, within drug development, the use of creatinine as a measure of GFR and renal function should be avoided in favour of markers that do not undergo renal secretion. Cystatin C, a low molecular weight protein freely filtered in the kidney,[19] has been suggested as a good alternative marker to creatinine that provides a more accurate measurement of GFR.[20]

4.4   Drug Transporter Expression in the Proximal Tubule

The kidney clears many drug molecules solely by filtration. However, for those where the renal C_x is modified either by tubular secretion and/or tubular absorption, the rate of clearance is defined by eqn (4.2):

$$C_x = \text{Amount filtered} - \text{Amount absorbed} + \text{Amount secreted} \qquad (4.2)$$

This concept is illustrated in Figure 4.2, which shows that the renal clearance of a drug (C_x) is dependent upon its fate within the nephron; for a drug that is freely filtered and is neither absorbed or secreted, then C_x is equal to GFR;

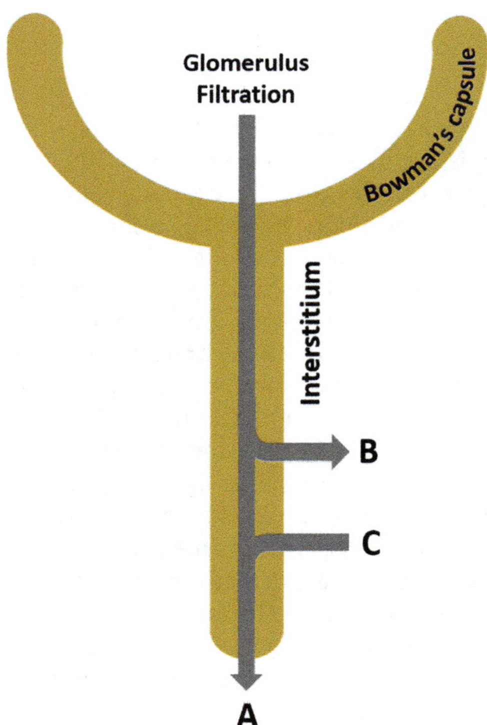

Figure 4.2   Summary of renal clearance. For freely filtered small molecules, renal clearance, C_x, is dependent upon whether the molecule is (A) filtered and not absorbed or secreted by the tubule $(C_x = GFR)$, (B) filtered with additional secretion by the tubule $(C_x > GFR \times P_x)$ or (C) filtered with reabsorption by the tubule $(C_x < GFR \times P_x)$, where GFR is the glomerulus filtration rate and P_x is the plasma concentration.

for a drug that is freely filtered and is neither reabsorbed or secreted, then Cx is equal to GFR. For a drug that undergoes tubular absorption C_x will be less than the filtered load $GFR \times P_x$; and for a drug molecule that undergoes tubular secretion, C_x will be greater than $GFR \times P_x$.

Tubular secretion occurs primarily in the proximal tubule of the nephron and is due to the presence of a number of drug transporter proteins that are asymmetrically distributed to either the basolateral or apical membrane of the epithelial cells. A summary of all of the currently known membrane transporters at the human renal pharmacological barrier is given in Figure 4.3. Whilst all transporters expressed in the proximal tubule cells are vital in fluid, electrolyte and substrate homeostasis, several families of these transporters are also now known to be involved in the transport of drug molecules. In 2010, the International Transporter Consortium (ITC) published a review entitled, "Membrane transporters in drug development" in which transporters considered to be clinically relevant were highlighted.[21] Subsequently, several regulatory agencies published draft regulatory guidance on drug interactions, that included the consideration of the impact of transporter proteins on drug pharmacokinetics.[22–24] The guidances state that if the renal clearance of a NME separately accounts for more than 25% of total drug elimination, then modulation of the renal transporters involved in active secretion may be clinically relevant. The renal drug transporters that should currently be studied in this instance are shown in Table 4.1 and discussed in detail in this chapter.

The uptake of molecules from the circulation across the basolateral membranes of proximal tubular cells is often mediated by the solute carrier (SLC) transporter superfamily. SLC transporters can be grouped into three categories by their transport mechanisms: (i) facilitated diffusion *via* a concentration gradient; (ii) Na^+ utilisation or H^+ co-transport; and (iii) substrate exchange.[25] All SLC transporters act independently of ATP hydrolysis. The main renal xenobiotic uptake transporters include members of the organic anion transporter (OAT), organic anion transporting polypeptide (OATP) and organic cation transporter (OCT) families.

The efflux of molecules across the apical membrane of proximal tubule cells into ultrafiltrate is typically mediated by both the ATP binding cassette (ABC) and SLC transporter superfamilies. The physiological function of ABC transporters is to protect cells against toxic compounds and metabolites by utilising energy from ATP hydrolysis to efflux substrates by active transport, as detailed in the sections below. The main renal efflux transporters known to date include the ABC transporters multidrug resistance protein 1 (MDR1), breast cancer resistance protein (BCRP) and the multidrug resistance associated proteins (MRPs), together with the SLC transporters MATE1 and MATE2-K.

4.4.1 OATs

The OAT isoforms OAT1 (*SLC22A6, Slc22a6*), OAT2 (*SLC22A7, Slc22a7*), OAT3 (*SLC22A8, Slc22a8*) and urate transporter 1 (URAT1; *SLC22A12, Slc22a12*) are

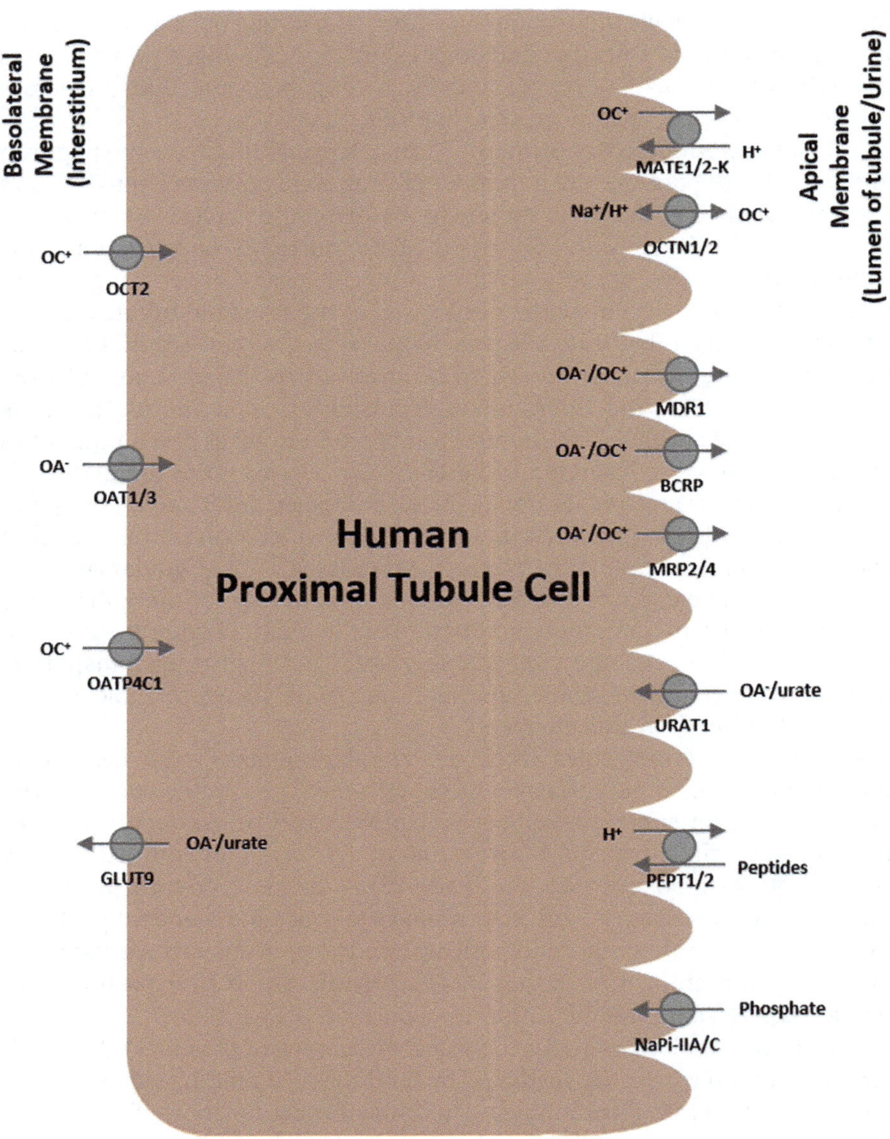

Figure 4.3 Drug transporters of human proximal tubule cells. The figure shows several drug transporters expressed on the membrane of human proximal tubule cells that require consideration in the development of drugs. They include OCT2, OAT1/3, OATP4C1 and GLUT9 on the basolateral membrane, and MATE1/2-K, OCTN1/2, MDR1, BCRP, MRP2/4, URAT1, PEPT1/2 and NaPi-IIA/C on the apical membrane.

found in both human and rat proximal tubule cells. OAT4 (*SLC22A11*) has only been found in humans and primates thus far.[26] In both species, OAT1 and OAT3 are localised to the basolateral membrane and mediate the transport of organic anions and anionic drugs (OA⁻) from blood into the

Table 4.1 Renal transporters that the FDA, EMA and PMDA consider to be clinically relevant in drug development.[a]

Transporter	ITC/FDA	EMA	PMDA
OCT1	−	Consider	Consider
OCT2	+	+	+
OAT1	+	+	+
OAT3	+	+	+
MATEs	Consider	Consider	+
MDR1	+	+	+
BCRP	+	+	+
MRPs	Consider	−	Consider

[a]These transporters play key roles in determining the bioavailability, therapeutic efficacy, and pharmacokinetics of a variety of drug molecules. Information gathered from Giacomini *et al.*, 2010 (ITC);[21] US Food and Drug Administration (FDA), 2012;[24] European Medicines Agency (EMA), 2012;[23] and Japanese Ministry of Health Labour and Welfare (PMDA), 2014.[22]

tubule. OAT2 is present in both the basolateral and apical membranes of human proximal tubule cells, but only the apical membrane in rodents.[27,28] OAT4 and URAT1 are located at the apical membrane.

Rodent OAT1 was the first OAT to be cloned from rat kidney in 1997.[29,30] The gene is located on chromosome 11q12.3 and is paired with the gene for OAT3,[31] which was first cloned from rat brain in 1999.[32] OATs are 542–556 amino acids long with 12 predicted transmembrane spanning domains.[33] They function as antiporters; the uptake of extracellular OA^- across the basolateral membrane is driven by the efflux of intracellular α-ketoglutarate.[34] Thus, OAT1 and OAT3 are functionally coupled to Na^+-driven mono- and di-carboxylate transporters that establish and maintain the intracellular/extracellular gradients of lactate, nicotinate and α-ketoglutarate.

OAT1 and OAT3 substrates, listed in Table 4.2, are generally monovalent or divalent anions that are smaller than 500 Da,[21] although OAT3 can also transport some positively charged drugs such as cimetidine.[35] Studies of OAT1 and OAT3 in expression systems such as *Xenopus laevis* oocytes and transfected cell lines have shown that they have wide overlapping substrate specificities.[36] OAT1 is primarily known for its high affinity transport of *para*-aminohippurate (PAH) from renal proximal tubule cells;[37] OAT3 can also transport PAH but with slightly lower affinity.[38] Studies using $Oat1^{-/-}$ and $Oat3^{-/-}$ knockout mice have led to the identification of many endogenous OAT substrates and allow initial assessments of the influence of each transporter on the renal handling of selected compounds.[39,40] Genetic variants in the OATs have so far not been associated with clinically-significant DDIs or changes in the amino acids transcribed.[41]

DDIs have been observed for drugs transported by OAT1 and OAT3, particularly for those drugs with a narrow therapeutic index. For example, several case studies have reported an increase in serum urate levels during antihypertensive therapy with the low dose diuretics torsemide and hydrochlorothiazide;[42–44] OATs have the ability to antiport diuretics and urate, and an increase in the former in the bloodstream will cause an increase in urate

Table 4.2 SLC transporters and their substrates and inhibitors.[a,b]

	Substrates	Inhibitors
OCT1 (*SLC22A1*)	TEA	Quinine
	MPP^+	Desipramine
	Metformin	Cimetidine
	NMN	Quinidine
		Imipramine
		Amiloride
		Chlorpheniramine
		Diphenhydramine
		Monoamines
OCT2 (*SLC22A2*)	TEA	Quinine
	MPP^+	TPA
	NMN	Imipramine
	Metformin	Amiloride
		Chlorpheniramine
		Diphenhydramine
		Monoamines
		Cimetidine
		Quinidine
		Phenformin
OCT3 (*SLC22A3*)	TEA	Cimetidine
	MPP^+	DMA
	Guanidine	NMN
OCTN1 (*SLC22A4*)	TEA	Cimetidine
	L-Carnitine	Palmitoyl-L-carnitine
	Verapamil	Procainamide
	Pyrilamine	Quinine
		Quinidine
OCTN2 (*SLC22A5*)	TEA	DMA
	L-Carnitine	Cimetidine
	Cephaloridine	Desipramine
		Procainamide
		Clonidine
		Palmitoyl-L-carnitine
		Octanoyl-L-carnitine
		Choline
OAT1 (*SLC22A6*)	PAH	Probenecid
	MTX	Indomethacin,
	Adefovir	Ibuprofen,
	Tenofovir	Carprofen
	Cidofovir	
	cAMP, cGMP	
	PGE_2	
	Urate	
	α-Ketoglutarate	
	β-Lactam antibiotics	
OAT2 (*SLC22A7*)	PAH	Probenecid
	Methotrexate	NSAIDS
	α-ketoglutarate	
	PGE_2	
	Salicylate	
	Acetylsalicylate	

Table 4.2 (*Continued*)

	Substrates	Inhibitors
OAT3 (*SLC22A8*)	E3S	Probenecid
	PAH	Betamipron
	PGE$_2$	Cimetidine
	Pravastatin	Bumetanide
	Taurocholate	DIDS
	OTA	DHEAS
	MTX	NSAIDS
	cAMP	Dicarboxylates
	β-Lactam antibiotics	
OAT4 (*SLC22A11*)	OTA	Probenecid
	E3S	Piroxicam
	PAH	Indomethacin
	Urate	DHEAS
	β-Lactam antibiotics	BSP
		Dicarboxylates
OAT5 (*SLC22A10*)	OTA	Probenecid
	E3S	Dicarboxylates
		BSP
		DHEAS
		Ibuprofen
		Furosemide
URAT1 (*SLC22A12*)	Urate	Probenecid
	Lactate	Benzbromarone
	Orotate	Indomethacin
		Nicotinate
NPT4 (*SLC17A3*)	Urate	Probenecid
	PAH	Indomethacin
	E3S	Piroxicam
	OTA	Bumetanide
		Furosemide
OATP1A2 (*SLCO1A2*)	Methotrexate	
	E3S	
	DHEAS	
OATP4C1 (*SLCO4C1*)	Digoxin	
	Sitagliptin	
	Ouabain	
PEPT1 (*SLC15A1*)	Small peptides *e.g.* Gly-Sar	Cephalexin,
		Cyclacillin
		Ceftibuten
		Cefadroxil
		Valacyclovir
		Alafosfalin
PEPT2 (*SLC15A2*)	Gly-Sar	Cephalexin
		Cyclacillin
		Ceftibuten
		Cefadroxil
		Valacyclovir
		Alafosfalin
		Ampicillin
		Amoxicillin

Table 4.2 (*Continued*)

	Substrates	Inhibitors
MATE1 (*SLC47A1*)	TEA	Cimetidine
	MPP$^+$	Quinidine
	Metformin	Quinine
	Guanidine	Thiamine
	NMN	Amiloride
	Cephalexin	Nicotine
	Procainamide	Corticosterone
	PAH	Verapamil
	E3S	Famotidine
	Acyclovir	
	Ganciclovir	
MATE2 (*SLC47A2*)	TEA	Cimetidine
	MPP$^+$	Quinidine
	Metformin	Verapamil
	Procainamide	Nicotine
	Guanidine	Testosterone
	E3S	Famotidine
	Acyclovir	
	Ganciclovir	

[a]The table shows the substrates and inhibitors of SLC membrane transporters found in the proximal tubule cells.
[b]BSP: bromosulfophthalein; DHEAS: dehydroepiandrosterone sulfate; DIDS: 4,4′-diisothiocyano-2,2′-stilbenedisulfonic acid; DMA: 5-(*N*,*N*-dimethyl)amiloride; E3S: estrone-3-sulphate; MPP$^+$: 1-methyl-4-phenylpyridinium; MTX: methotrexate; NMN: *N*1-methylnicotinamide; NSAIDS: nonsteroidal anti-inflammatory drugs; OTA: ochratoxin-A; PAH: para-aminohippuric acid; PGE$_2$: prostaglandin E$_2$; TEA: tetraethyl ammonium; TPA: tetrapentylammonium.

absorption from the urine, which results in hyperuricaemia. In some cases, the nephrotoxicity of therapeutics can be mitigated by inhibiting the transport of potentially nephrotoxic drugs into proximal tubular cells. For example, cidofovir is an OAT substrate that can cause kidney damage due to its accumulation in the proximal tubular cells, but co-administration with probenecid, an OAT inhibitor, results in a reduced accumulation of cidofovir in the proximal tubule cells and thus reduced nephrotoxicity.[45]

OAT2 was first identified in rat livers in 1994[46,47] and was later found to be highly expressed at the mRNA level in rat kidneys.[48,49] Evidence for similar renal expression of OAT2 in the human kidney was not published until recently.[50] OAT2 interacts with a wide range of nucleobases, nucleosides and nucleotides (see Table 4.2). Three splice variants of OAT2 that result in amino acid changes have been identified, but the clinical consequences of these are yet to be investigated.[51]

As mentioned earlier, it is important to understand which transporters are involved in the tubular secretion of creatinine, as creatinine clearance is used to measure GFR. Recently, OAT2 was implicated in the renal handling of creatinine. Previous evidence suggested that basolateral OCT2 served as the primary means of creatinine uptake into proximal tubular cells. However, this theory was challenged when knockout of renal OCT1/OCT2 expression in mice was found to have no effect on renal clearance of creatinine, but that in mice lacking OCT1/OCT2, the rate of creatinine secretion was sensitive to inhibition in the presence of the OAT substrate PAH, suggesting that OAT transporters may be involved in creatinine transport.[52] A role for OAT2 as a creatinine transporter was confirmed in a screen of the ability of various OAT isoforms to transport creatinine.[48] At physiologically relevant creatinine concentrations, the affinity of creatinine for OAT2 has been shown to be greater than the affinity for OCT2, OCT3, MATE1 and MATE-2K.[28,53] In addition, OAT2 mRNA expression in the kidney is reported to be three-fold higher than expression of OCT2.[50] Collectively, these data suggest that OAT2 plays a major role in the tubular secretion of creatinine. Evidently, OATs influence the renal secretion and clearance of many drugs and can be important in drug development and in the consideration of potential DDIs.

4.4.2 OATPs

The OATP (*SLCO*) family consists of 11 members that are expressed in multiple tissues in both rodents and humans. They transport a wide range of substrates, including xenobiotics and endogenously synthesised compounds such as bile acids.[54] The first member to be identified was rodent OATP1A1 (*Slco1a1*) in rat liver.[55] Indeed, many OATPs are found predominantly in the liver while others are ubiquitously expressed.[54] In general, OATPs mediate the sodium-independent transport of a diverse range of amphiphilic organic compounds with a molecular weight greater than 350 Da.[56,57] The general predicted structure consists of 12 transmembrane spanning domains.[58] The majority of substrates are OA^-, although some OATPs have been reported to also transport neutral and cationic compounds (see Table 4.2).[59] They function as antiporters, the uptake of organic compounds being coupled with the efflux of bicarbonate, glutathione or glutathione-*S* conjugates in an electroneutral exchange.[60,61] The overall direction of transport is dependent upon the local substrate gradients.

In human renal proximal tubule cells, only two OATP superfamily isoforms have been characterised, basolateral OATP4C1 (*SLCO4C1*) and apical OATP1A2 (*SLCO1A2*). In contrast, rodents are thought to express multiple OATP isoforms, including basolateral OATP4C1 (*Slco4c1*), apical OATP1A1 (*Slco1a1*), OATP1A2 (*Slco1a2*), OATP1A3_v1 (*Slco1a3v1*, previously known as Oat-K1) and OATP1A3_v2 (*Slco1a3v2*, previously known as Oat-K2).[62]

OATP4C1 was first cloned from rat kidney in 2004 and based upon northern blot analysis it was thought to be kidney-specific.[63] More recently, however, microarrays have suggested that it is also expressed within the liver, although this has not yet been verified with polymerase chain reaction (PCR) or protein analysis.[64] The human gene is located on chromosome 5q21.[62,65] From rat OATP4C1 localisation studies it is assumed that human OATP4C1 is also located on the basolateral membrane of proximal tubule cells. OATP4C1 is reported to possess multiple substrate recognition sites; so far, two distinct recognition sites for estrone-3-sulfate (E3S) and digoxin have been characterised.[65]

OATP1A2 was first cloned within rat liver, and in humans is located on chromosome 12p12.[62,66] In both human and rat proximal tubular cells, it is expressed at the apical membrane,[67] where it is thought to be responsible for either the reabsorption from or secretion of xenobiotics into the urine. In the kidney, OATP1 expression is stimulated by testosterone and inhibited by oestrogen. As a consequence, kidney OATP1 is less abundantly expressed in female compared with male rats.[68,69] OATP1A3_v1 and OATP1A3_v2 are exclusively expressed in rat kidney.[70,71] OATP1A3_v1, transfected into Lilly Laboratory cell-porcine kidney-1 (LLC-PK1) or Madin–Darby canine kidney (MDCK) cells, was identified as a bi-directional transporter of methotrexate (MTX) across the apical membrane that could be inhibited by anti-inflammatory drugs such as indomethacin, ketoprofen and ibuprofen, and by folic acid (Table 4.2).[72,73] In addition, an inwardly directed gradient of folic acid derivatives was also found to stimulate MTX efflux, suggesting that OATP1A3_v1 could function as an MTX/folic acid derivative exchanger.[73] OATP1A3_v2 also transports MTX across the apical membrane and therefore may play the same role in the kidney as OATP1A3_v1.[74] *In vitro* studies indicate that rodent OATP1A3_v1 and OATP1A3_v2 are multi-specific transporters, mediating the transport of a wide variety of compounds that are also substrates for OATPs such as ochratoxin A (OTA), thyroid hormones and conjugated steroids.[75]

4.4.3 Organic Anion Transporters URAT1, GLUT9 and NPT4

Several organic anion transporters capable of transporting urate in the kidney have recently been identified and characterised. URAT1 (*SLC22A12*) was identified in the apical membrane of proximal tubule epithelial cells in humans.[76] URAT1 in rat kidney was previously identified to be a novel renal specific transporter, but is now acknowledged to be identical to rat URAT1 found in other organs.[77–79] Transfection of oocytes with URAT1 demonstrated that it mediated the transport of urate but was inhibited by uricosuric drugs such as probenecid and benzbromarone.[76,78,79] Studies have also shown that URAT1-mediated transport of urate is electroneutral and can be *trans*-stimulated by Cl⁻ gradients and gradients of lactate, which is transported by the sodium-monocarboxylate transporter.[76,79] Given the high Cl⁻ and lactate concentrations generated within the proximal tubule cell, these conditions would drive the reabsorption of urate. Recently, transport of orotate, a heterocyclic precursor of pyrimidine synthesis, was also shown in human embryonic kidney (HEK293) cells transfected with human URAT1. Transport of orotate is mediated by the same co-transport system as urate since it is dependent on Cl⁻ exchange and stimulated by lactate.[80]

Another important urate transporter has been found in the human kidney. Glucose transporter 9 (GLUT9, *SLC2A9*) is found in the basolateral membrane of proximal tubule cells and is a known glucose/fructose transporter,[81] although the affinity of GLUT9 for urate transport is far higher than for either glucose or fructose, and also higher than URAT1 for urate.[81,82] GLUT9 transport of urate is inhibited by benzbromarone, suggesting that it may possess similar binding sites to URAT1.[82] Uptake of urate by GLUT9 is voltage-driven and hence this transporter is also referred to as URATv1.[83] Its location in the basolateral membrane suggests a possible role as an efflux transporter in proximal tubules that forms part of a urate transport system with URAT1, where urate taken up from the lumen by URAT1 is then secreted into the blood by GLUT9.[83]

The sodium phosphate transporter 4 (NPT4; *SLC17A3*) has been identified recently as an apical membrane transporter in the liver and kidney. It is a multi-specific OAT and its substrates include those of OAT1 and OAT3, including PAH, E3S, OTA and urate. Similar to URATv1, it exists in two distinct isoforms of different lengths, although only the long isoform, NPT4_L, is expressed in the plasma membrane when injected into oocytes.[84] Efflux of urate by NPT4, like URATv1, is voltage-driven and is inhibited by diuretic drugs such as bumetanide and furosemide (see Table 4.2).[84] This results in the prevention of the removal of urate, which could lead to hyperuricaemia. In addition, since NPT4 affinity for OTA is far lower than for OAT1 and OAT3, this may result in accumulation of OTA in tubular cells, leading to nephrotoxicity.[85]

Apical NPT4 may function in parallel to the URAT1/GLUT9 transport system, being part of a urate secretory system. In this model, OAT1 and OAT3 transport urate from the blood into the proximal tubule, which is then effluxed into the urine *via* BCRP, OAT4 and NPT4.[84] Therefore, a balance between urate secretion and reabsorption *via* the kidney may be maintained by two separate transport systems. Genome-wide association studies have also identified single nucleotide polymorphisms (SNPs) in OAT4 and BCRP that are associated with disturbances of urate homeostasis.[86–90]

4.4.4 OCTs

The OCT isoforms OCT2 (*SLC22A2*) and OCT3 (*SLC22A3*) are found in both human and rat proximal tubule cells on the basolateral membrane. OCT1 (*SLC22A1*) has also been identified at the basolateral membrane of rodent proximal tubular cells by northern blot analysis and RNA *in situ* hybridisation,[91,92] although *in situ* hybridisation studies have failed to detect OCT1 in human kidney. OCT1 has been reported to be expressed primarily in the liver in humans, indicating a species difference in renal transporter expression.[93,94]

Rodent OCT1 was the first OCT isoform to be cloned in 1994 from rat kidney.[91] A human orthologue was later cloned in 1997. A second member, OCT2, was identified in rat kidney by homology screening and later cloned in humans.[95,93] OCT3 was independently cloned and identified as the extraneuronal monoamine transporter in 1998. OCTs encode proteins that are 542–556 amino acids long with 12 predicted transmembrane spanning domains.[33] OCT1, OCT2 and OCT3 mediate the passive facilitated diffusion of a broad range of structurally diverse organic cations (OC$^+$s) down their electrochemical gradients.[92,96] Therefore, transport can occur in either direction and is driven not only by the difference in substrate concentration but also by membrane potential. Due to the negative resting membrane potential of the cell maintained by Na^+/K^+ ATPase, transport of OC^+s into the cell is favoured at normal membrane potential.[97] The functional characteristics of these transporters have been studied in expression systems such as *Xenopus laevis* oocytes and transfected cell lines. They have extensively overlapping substrate specificities with transported OC^+s ranging in size from 60 to 350 Da, with at least one positively charged moiety at physiological pH (examples are shown in Table 4.2).[98]

In addition to the OCTs, another subfamily of human OCT transporters (organic cation transporters novel (OCTN)) that are known to interact with several drugs, have been identified in proximal tubule cells.[99] OCTN1 (*SLC22A4*) was the first to be identified in human foetal liver. Its expression ceases in adult liver but it is highly expressed in the kidney.[100] It mediates pH-dependent transport of tetraethyl ammonium (TEA) in transfected HEK293 cells but at a lower affinity than the OCT family of transporters.[100] Expression in *Xenopus laevis* oocytes also showed transport of other OC^+s such as verapamil and pyrilamine, indicating that OCTN1 is a multi-specific OCT.[101] A second transporter, OCTN2 (*SLC22A5*), has also been identified in the human kidney. Similar to OCTN1, it is not expressed in the liver and mediates transport of a broad spectrum of OC^+s.[102,103]

Rat OCTN1 and OCTN2 have also been found to be widely distributed in rat tissues;[103–105] OCTN1 mRNA was detected at the highest level in rat kidney, with moderate levels in the liver, intestine, heart and brain. The mRNA levels of rat renal OCTN1 were also found to increase steadily with the age of the rat.[104] The functions of rat OCTN1 and OCTN2 are similar to those of humans,[101] with rodent OCTN2 having a higher affinity for TEA compared with human OCTN2.[103] Another transporter, OCTN3, has also been found in the rat kidney but not in the human,[106,107] and has the ability to mediate transport of L-carnitine in a Na^+-independent manner.[107]

OCTN1 and OCTN2 are both situated in the apical membrane of proximal tubules,[108,109] in contrast to OCTs, which are basolateral transporters. This provides an indication that OCTNs may be important for the reabsorption of carnitine and OC^+s from urine rather than their elimination. However, due to the bidirectional nature of the OCTN transporters (direction of substrate travel is dependent on the pH of urine),[101] it is hypothesised that they also help in the secretion of OC^+s into the urine depending on the environment in the tubule.[104] Regardless of the direction of transport by OCTNs, they have been shown to mediate the transport of many OC^+s and constitute potential pharmacologic targets in the kidney.

4.4.5 MATEs

Within mammals there are two MATE proteins, MATE1 (*SLC47A1*) and MATE2 (*SLC47A2*). Currently, the MATE2 splice variant, MATE2-K, is the only member of the subfamily in which functional transport activity has been demonstrated.[110] MATE2-K appears to be exclusively expressed in the apical membrane of human renal proximal tubular cells, whilst MATE2 is expressed in intracellular vesicles.[111]

The MATE transporters appear to function in conjunction with basolateral OCT2 to mediate the excretion of OC^+s from the blood into ultrafiltrate.[112,113] Initial characterisation of the transporter when it was first identified within *Vibrio parahaemolyticus* described MATE1 as a secondary active antiporter.[114] It couples the movement of OC^+ out of the cell, against its concentration gradient, using the energetically favourable inward movement of H^+ and Na^+ along their electrochemical gradient. Similar to OCTs, substrates of these transporters include structurally diverse low molecular weight OC^+s. In addition, select OA^-s, such as E3S, acyclovir and ganciclovir, are also reported to be substrates for MATE1.[115] The functional characteristics of MATE transporters have been studied in expression systems such as *Xenopus laevis* oocytes and transfected cell lines, and studies indicate that they have extensively overlapping substrate specificities.[115,116] Examples of endogenous and exogenous substrates are shown in Table 4.2.

In MATE1 and MATE2-K, 11 and 2 non-synonymous SNPs, respectively, have been identified,[117–119] with the mutations G64D and V480M in MATE1 and G211V in MATE2-K causing a complete loss of membrane expression. Thus, these variants may affect the pharmacokinetics of substrates. The allelic frequency of dysfunctional mutations is less than 5% and homozygous carriers have not been found. In addition, the ubiquitously expressed transcription factor specificity protein 1 (SP1) and activator protein 1 (AP-1) are reported to affect MATE1 transcription in both human and rat proximal tubule cells.[120–122] Polymorphisms in both the SP1 and AP-1 transcription factor binding regions have been identified (rs72466470, rs2252281), which result in decreased binding and reduced transcriptional activity. The MATE genes are also located in the commonly deleted region in Smith–Magenis syndrome, a genomic disorder of chromosome 17p11.2.[123,124] The disease is a developmental disorder, with major features including mild to moderate intellectual disability, distinctive facial features, sleep disturbances and behavioural problems. The relevance of MATE1 and MATE2-K in the progression of this disorder is under investigation.

Experiments in *Mate1*$^{-/-}$ knockout mice have identified the role of MATE1 in the pharmacokinetics of many drug molecules. For example, plasma and renal concentrations of metformin were increased and urinary excretion was decreased in *Mate1*$^{-/-}$ mice compared with control mice, indicating that MATE1 plays a role in the tubular secretion of metformin.[125] In heterozygous MATE1 knockout mice, the pharmacokinetics of metformin were not significantly different from wild type, suggesting that MATE1 is not the rate limiting step in the tubular secretion of metformin.[126]

4.4.6 MDR1

MDR1 (also known as P-gp, *ABCB1*) was first cloned from human carcinoma cells,[127] and is highly expressed on the apical membrane of human and rat proximal tubule cells. In humans, MDR1 is encoded by one gene, whereas rats possess two isoforms *Mdr1a* and *Mdr1b*.[128]

The function of apically expressed MDR1 in the kidney is the movement of substrates out of the cell for renal clearance, using ATP hydrolysis to provide energy for active transport against steep concentration gradients. The transporter consists of 1276–1280 amino acids in a tandem duplicated structure, with 12 hydrophobic transmembrane domains, and a high resolution structure of mouse MDR1 has been described.[129] It contains at least three distinct substrate binding sites within the ligand binding domain and two ATP binding motifs.[130] Endogenous and exogenous substrates of human and rat MDR1 have been identified through functional studies in membrane vesicles and transfected cell lines (see Table 4.3). Substrates are preferentially neutral or hydrophobic OC^+ molecules ranging in size from less than 200 to 1900 Da. Cholesterol, an endogenous substrate of MDR1,[131] is known to stimulate basal ATPase activity.[132,133] Furthermore, it has been shown that cholesterol affects the affinity of low molecular weight molecules (350–500 Da) for MDR1,[134] possibly by directly binding or allosterically affecting the substrate binding site to help the recognition of smaller substrates. Many inhibitors of MDR1 contain aromatic ring structures, a tertiary or secondary amino group and have high lipophilicity;[135] they can be categorised into either high affinity substrates that bind non-competitively or efficient inhibitors of ATP hydrolysis.

Various studies have indirectly implicated the involvement of MDR1 in DDIs.[136,137] Clinically, an important substrate of MDR1 in the kidney is digoxin, a cardiac glycoside prescribed to patients suffering from chronic heart failure.[138] Many drugs are co-administered with digoxin; patients suffering from the disease are usually also given other heart medications such as verapamil and quinidine.[139,140] It was noticed that the co-administration of these drugs lowered the renal clearance of digoxin, which was later attributed to the inhibition of MDR1-mediated efflux of digoxin by the other drugs.[140] Digoxin has a narrow therapeutic window and slight changes in its plasma concentration can cause digitalis toxicity.[141] It is therefore necessary to evaluate the risks of drugs taken with digoxin in inducing DDIs *via* the inhibition of MDR1 activity.

4.4.7 BCRP

BCRP (*ABCG2*) was first identified in the breast cancer cell line MCF-7 where it appeared to play a role in multidrug resistance.[142] BCRP is now known to be an efflux transporter that may prevent drugs from penetrating tissues, and can be important in drug disposition and distribution. Screening of human tissues revealed that the transporter is expressed in all major organs, such as the brain, liver and intestines.[143] Studies also confirmed moderate BCRP expression at the mRNA and protein level in normal human kidneys,[142] although levels of BCRP mRNA were found to be up-regulated several fold in renal carcinomas.[144] Expression of functional BCRP was

Table 4.3 ABC transporters and their substrates and inhibitors.[a,b]

	Substrates	Inhibitors
MDR1 (*ABCB1*)	Rhodamine 123	Verapamil
	Hoechst 33342	Vinblastine
	Calcein	Cyclosporin A
	Digoxin	Quinidine
	Daunorubicin	Valspodar
	Doxorubicin	
	Actinomycin	
	Adriamycin	
	Colchicine	
	Mitoxantrone	
	Ivermectin	
BCRP (*ABCG2*)	Hoechst 33342	Nelfinavir
	Doxorubicin	Elacridar
	Daunorubicin	Fumitremorgin C
	Mitoxantrone	
	Topotecan	
MRP1 (*ABCC1*)	Glutathione	Vincristine
	GS-DNP	Vinblastine
	LTC$_4$	BSO
	Doxorubicin	Arsenate
	Daunorubicin	Sulfinpyrazone
	Bilirubin	MK571
MRP2 (*ABCC2*)	GS-DNP	Probenecid
	PGA1	Bromsulphthalein
	LTC4	GF120918
	Ethacrynic acid	Glucuronide conjugates
	Etoposide	Fura-AM
	Doxorubicin	MK571
	Epirubicin	Dipyridamole
	Vinblastine	BSO
	Vincristine	
	PAH	
	Bilirubin	
	Saquinavir	
	Ritonavir	
	Indinavir	
MRP3 (*ABCC3*)	E217βG	MK571
	LTC4	
	SG-DNP	
	DHEAS	
	Bilirubin	
MRP4 (*ABCC4*)	cAMP	MK571
	cGMP	Dipyridamole
	E217βG	Glucuronide Conjugates
	PAH	Probenecid
	Adefovir	Methotrexate
	Tenofovir	PGA1
	Mercaptopurines	PGE1
		Progesterone
		Estramustine

Table 4.3   (*Continued*)

	Substrates	Inhibitors
MRP5 (*ABCC5*)	cAMP	Probenecid
	cGMP	Trequinsin
	GS-DNP	Sildenafil
	Thioguanine	
	PMEA	
	Mercaptopurines	
MRP6 (*ABCC6*)	BQ123	Benzbromarone
	LTC4	Indomethacin
	SG-DNP	
	GS-NEM	
	Etoposide	
	Teniposide	
	Daunorubicin	
	Doxorubicin	

[a]The table shows the substrates and inhibitors of ABC membrane transporters found in the proximal tubule cells.
[b]BSO: DL-buthionine (*S,R*)-sulfoximine; DHEAS: dehydroepiandrosterone sulfate; E217βG: estradiol-17-β-D-glucuronide; GS-DNP: *S*-dinitrophenylglutathione; LTC$_4$: leukotriene C$_4$; PGA$_1$: Prostaglandin A$_1$; PGE$_1$: prostaglandin E$_1$; PMEA: 9-(2-phosphonylmethoxyethyl)adenine; SG-DNP: dinitrophenyl *S*-glutathione.

detected in cultures of primary tubule cells, as Hoechst 33342 efflux from these cells could be specifically blocked by the BCRP inhibitor fumi-tremorgin C.[145] Rat kidneys have constitutively higher mRNA levels of the transporter compared with human.[146]

Both human and rat BCRP are found on the apical membrane of the proximal tubule cells,[145,147] where it acts as an efflux pump by mediating the unidirectional transport of substrates out of the tubule.[148] BCRP substrates are diverse, ranging from drugs such as nitrofurantoin, dipyridamole and cimetidine, to endogenous compounds such as estrones and bile acids.[149–151] BCRP-mediated transport may be inhibited by highly potent and specific inhibitors such as fumitremorgin C or its analogue Ko143.[152] There is an overlap in substrate specificity between BCRP and MDR1. It is thought that MDR1 generally transports hydrophobic compounds whereas BCRP additionally transports hydrophilic conjugated organic anions, particularly sulfates, with high affinity (see Table 4.3).[153]

4.4.8   MRPs

The MRP family of transporters consists of nine structurally related members (MRP1–9), which have a wide tissue and species distribution.[154] All members function as lipophilic anion efflux transporters and are expressed on either the apical or basolateral membrane of epithelial cells.[155] MRP1 (*ABCC1*) was first identified in the drug resistant lung cancer cell line H69AR, suggesting that this transporter plays a role in drug resistance in cancer cells.[156] MRP2 (*ABCC2*), MRP3 (*ABCC3*), MRP4 (*ABCC4*) and MRP5 (*ABCC5*) have subsequently been identified in various cancer cell lines and tissue samples, and all MRPs have been shown to confer resistance to many drugs, including anticancer and antiviral agents.[157]

MRP1 and MRP2 are the most abundant MRPs in the human kidney.[157] MRP3–5 have also been detected in the kidney, but their expression is more abundant in other tissues, for instance MRP3 expression is greater in the liver.[158–160] The mRNA of other MRPs (MRP6–9) has been detected in other tissues, including the kidney, but the functional expression of these transporters has yet to be investigated.[161,162] A similar expression pattern has been observed in rat kidney, with high expression of rat MRP1, MRP2 and MRP4, but low expression of MRP3.[163,164] MRP2 and MRP4 are located in the apical membrane of proximal tubule cells,[165–168] whereas MRP5 and MRP6 are found in the basolateral membrane.[169,170] MRP1, however, is not greatly expressed in the proximal tubule, but is found at higher levels in glomeruli and the basolateral membrane of distal tubule cells.[171]

As with other drug transporters, transfected cells have been used to investigate the function and substrate specificity of many of the MRPs. Mammalian cells transfected with MRP1, MRP2 or MRP6 transport glutathione-*S* conjugates.[159,166,172] MRP2 or MRP4 transfected into SF9 cells mediate ATP-dependent transport of PAH, with MRP4 showing higher affinity than MRP2.[173,174] This is particularly important as PAH is also a well-known substrate for OATs in proximal tubules. MRP4 and MRP5 are able to transport cyclic nucleotides such as cAMP and cGMP, and are the only members of the MRP family that do so.[168,175–177]

MRPs also have an important role to play in multidrug resistance, and are a major elimination pathway for anticancer and antiviral drugs *via* the kidney. For example, anticancer drugs such as daunorubicin and vinblastine are transported by MRP1 and MRP2.[166,172] In contrast, MRP6 shows low level resistance to anticancer drugs compared with MRP1 and MRP2, suggesting that it does not play a major role in drug resistance in the kidney and may function predominantly as an OAT.[159]

4.4.9 Peptide Transporters (PEPT1 and PEPT2)

Peptide transporter 1 (PEPT1) and PEPT2 are a class of transporters responsible for the uptake of short peptides of not more than three amino acid subunits. PEPT1, first cloned from rabbit small intestine,[178] is found in both the intestine and the apical membrane of rat and human proximal tubule cells.[179,180] In contrast, PEPT2 is predominantly expressed in the kidney.[181] Both isoforms are H$^+$ coupled transporters with substrates that include drug molecules such as the β-lactam antibiotics, which have a structural resemblance to small peptides (see Table 4.2).[182]

Uptake of the synthetic dipeptide substrate tracer, glycyl-sarcosine (Gly-Sar), by PEPT1 and PEPT2 was found to be inhibited by peptide-derived antibiotics including cephalosporin,[183–185] the antiviral drug valaciclovir[186] and the antibacterial phosphonodipeptide.[187] All of these compounds inhibited PEPT2-mediated transport of Gly-Sar with greater potency than PEPT1, suggesting that PEPT2 may be a higher affinity peptide transporter. This may also suggest that it is the predominant peptide transporter in the kidney and, thus, may be important for the reabsorption of small peptides and antibiotics from the kidney lumen,[188] and could play a role in the systemic distribution of these compounds.

4.4.10 Phosphate Transporters

Phosphate transporters expressed in proximal tubule cells are also of interest in drug development. Phosphate is a key component in a number of biological processes, and a healthy serum level is maintained by regulating the expression of phosphate transporters involved in the renal reabsorption of glomerulus-filtered phosphate. The three main phosphate transporters found in the apical membrane of proximal tubule cells are members of the SLC superfamily, namely type II sodium dependent phosphate transporters NaPi-IIA (*SLC34A1*) and NaPi-IIC (*SLC34A3*), and phosphate transporter-2 (PiT2; *SLC20A2*). NaPi-IIA/C are regulated by parathyroid hormone (PTH) and fibroblast growth factor 23 (FGF23) in the kidney, with the latter requiring the co-factor, klotho-α, to elicit its effect.[189,190] The presence of PTH or FGF23 has been shown to cause the removal of NaPi-IIA/C *via* endocytosis, and in doing so limits the amount of phosphate that is reabsorbed. The effects of PTH and FGF23 on PiT2, however, are not well-characterised.[191]

The biological processes that phosphate is involved in range from bone mineralisation to protein phosphorylation; these processes will be adversely affected should phosphate transporters be inhibited by other substrates. Hypophosphataemia is observed in up to 32% of patients taking tenofovir and hyperphosphaturia and osteomalacia also manifest in these patients.[192] Recent data from our laboratory suggest that tenofovir is a potent inhibitor of NaPi-IIA/C, which could explain the observed clinical symptoms that occur with tenofovir therapy—tenofovir prevents the reabsorption of filtered phosphate, which leads to low serum levels (hypophosphataemia) and high urine levels (hyperphosphaturia) of phosphate, and as a result phosphate demineralises from bone structures to compensate for the fall in serum levels (osteomalacia).[193]

4.4.11 Receptor-mediated Endocytosis (Megalin and Cubilin)

In addition to transporters, multi-ligand endocytic receptors are also involved in the transport of molecules in the kidney. These endocytic receptors are essential components in the maintenance of nutrient homeostasis in the proximal tubule. Two of the well-characterised endocytic receptors are megalin and cubilin. Megalin is a member of the low density lipoprotein receptor family with a single transmembrane domain, whereas cubilin is a peripheral glycoprotein attached to the phospholipid bilayer of the cell membrane.[194,195] They are both co-expressed in the apical surface of the proximal tubule cells, although megalin is more widely distributed within the membrane.[196–198] It has been proposed that ligand binding to megalin initiates the process of endocytosis.[199] However, some ligands such as transferrin (an iron binding complex) bind primarily to cubilin, and the binding to the receptor mediates the uptake of iron from urine. Despite evidence for preferential binding of substrates to either megalin or cubilin, evidence from studies using megalin-deficient proximal tubule cells suggests that functional expression of both megalin and cubilin is required for a functional endocytic process.[199]

A clinically important class of megalin ligand is the aminoglycosides, of which the broad spectrum antibiotic gentamycin is a prototypic example. Gentamycin is freely filtered and is reabsorbed from the urine into the proximal tubule cell *via* megalin-mediated endocytosis.[200] Clinically, this receptor-mediated accumulation of gentamycin within the proximal tubule cells is thought to be the key process that initiates gentamycin-induced nephrotoxicity; the accumulation of gentamycin causes oxidative stress in the cell and subsequent apoptosis.[201] More recently, megalin/cubilin receptor-mediated endocytosis has been reported to be the initial step in both oligonucleotide and polymycin-induced nephrotoxicity.[201] Importantly, megalin may also mediate the prevention of apoptosis. Studies have shown that the megalin receptor can initiate the uptake of survivin, an inhibitor of apoptosis found abundantly on the apical surface of the proximal tubule cells.[202] It is thus hypothesised that survivin acts as a protective molecule of kidney injury. These discoveries have highlighted the importance of endocytic receptors, and that approaches for renal protection *via* megalin and/or cubilin-mediated endocytosis need to be further investigated.

4.5   *In vitro* Renal Models

Much of the information about the functions and substrate specificity of renal transporters has been investigated using *in vitro* renal models,[4] including kidney slices, *Xenopus laevis* oocytes and various established cell lines.[6] Whilst Chapter 7 provides more detailed information about the various *in vitro* models that are available, this section will focus specifically on kidney systems that are currently used for the study of renal drug transport.

Models of the renal proximal tubule commonly used to measure transport activity in preclinical drug development include expression systems, transfected cells, immortalized animal and human renal cell lines, membrane vesicles, cortical renal slices and primary cells. They can be used for mechanistic studies to determine the rate-limiting step in transepithelial transport, and to identify transporter-based DDIs. To date, most information on renal drug transporters has come from screening of drug molecules in cells transfected with OCT1/OCT2 or OAT1/OAT3,[203] and it is only recently that more holistic physiological models of the proximal tubule are becoming available, which are explored in the sections below. The advantages and disadvantages of each of these techniques are briefly mentioned in this section.

4.5.1 *Xenopus Laevis* Oocyte Expression System

One of the earliest techniques to determine the function of a transport protein was microinjection of the transport protein cRNA into an expression system such as *Xenopus laevis* oocytes.[204] This provides a platform for studying only the transporter of interest, without the influence of other transporters, and the structure and identification of substrates and inhibitors of OCTs and OATs have been elucidated using this technique.[30,91,95] A drawback of this model is its lack of physiological relevance due to the absence of other transporters that may influence the distribution of the molecule under investigation, together with the loss of complexity and interactions between transporters found in more complex models. A marked variation in expression levels and kinetic parameters (such as maximum rate of transport and transporter affinity values) between batches of oocytes has also been reported.[205] Kinetic values derived from oocytes can differ from those from mammalian expression systems due to differences in the plasma lipid membrane.[206] Furthermore, this system limits any use of high throughput drug screens, as it is very labour intensive—each oocyte needs to be individually injected with cRNA.

4.5.2 Transfected and Immortalized Renal Cell Lines

Membrane transporters of interest can also be heterologously over-expressed *via* transfection into immortalized animal and human cell lines. Cultured renal cell lines commonly used to study drug transporter interactions include MDCK, LLC-PK1 and HEK293 cells.[207,208] Cells can also be stably transfected to express multiple transporters. For example, a quintuple multi-transporter renal model of creatinine clearance containing OAT2, OCT2, OCT3, MATE1 and MATE2-K has been produced commercially.[209] Transporters that are expressed stably or transiently in these cell lines can be used to characterise drug transporter-mediated interactions at the level of the kidney.

Immortalized renal epithelial cell lines can also inherently express transporters of interest without the need for transfection. For instance, LLC-PK1 cells express functional MDR1 and have been used to investigate the efflux of prototypic substrates.[210] MDCK cells have been used to investigate a range of drugs (examples include etoposide and irinotecan in MRP2 transfected MDCK cells),[211,212] and HK-2 cells, which are immortalized human proximal tubule cells, have been used to investigate MDR1-mediated efflux of many compounds.[213] In addition, HK-2 cells also showed functional expression of other members of the ABC family, such as MRP2, and have been used as a model of drug-induced nephrotoxicity.[136]

Cell lines do have several limitations—the most important of these is that the vast majority of cell lines have undergone substantial dedifferentiation and have lost most of the basic functions of the original cell type. For example, MDCK cells do not express transporters such as NaPi2, Na^+/glucose cotransporter 1 and 2, or amino acid transporters usually found in the proximal tubule. The same is true for LLC-PK1 cells,[214] which lack expression of the enzyme fructose-1,6-bisphosphatase, rendering them incapable of gluconeogenesis, which is a key metabolic pathway in proximal nephron cells.[215] In addition, LLC-PK1 cells are not responsive to PTH and lack a probenecid-sensitive OAT.[215,216] Recently, it was shown that HK-2 cells also lack many key transporters and are thus a poor model for drug transporter studies.[217] However, a number of immortalized human cell lines such as renal proximal tubule epithelial cells (RPTEC) and conditionally immortalized proximal tubule epithelial cells (ciPTEC) are becoming available.[218,219] The data on these are still limited but initial data suggest that RPTEC cells lack many key transporters typical of the proximal tubule.[220] ciPTECs are better characterised but do not endogenously express OATs.[219] Additionally, slight variations in culture conditions (*i.e.* culture medium, passage frequency, passage protocol) between laboratories have been shown to adversely affect MDR1 expression in MDCK cells, leading to a large variation in reported function.[221]

4.5.3 Cortical Renal Slices

An alternative to immortalized cells is the use of fresh primary tissue. Isolated cortical renal slices represent the closest anatomical *in vitro* model of the kidney. Since around 80% of each renal slice is composed of proximal tubule cells, they have potential to provide good information on renal drug handling, and a study in 1984 used renal slices to demonstrate that cisplatin competitively inhibited uptake of the OCT2 substrate TEA.[222,223] Cross-sections are available from both human and animal kidneys, allowing a direct comparison between species. However, the lumen of the tubules within the slices collapse, which means they provide information only on basolateral uptake, not apical membrane transport or the net direction of movement.[224] Other drawbacks include their heterogeneous population of cells, making discrimination between cell types difficult, their short finite lifespan and the requirement for a skilled experienced worker for kidney dissection.

4.5.4 Primary Proximal Tubular Cells

The need for a robust *in vitro* cell model of drug transport that expresses a full complement of transporters has already been highlighted. This can potentially be achieved by using primary cells derived from intact tissue that, at the time of isolation, express the full complement of drug transporters. Proximal tubular cells can be isolated using a wide range of techniques, including enzymatic tissue digestion, differential sieving, gradient density centrifugation and fluorescence-activated cell sorting. The isolated cells in suspension are physiologically relevant, intact cell models containing all of the clinically important transporters.[222] However, they are only viable for 2–3 hours, after this functional integrity declines rapidly.[225] Furthermore, during the course of an experiment, the polarity of the cells is lost, which results in a marked down-regulation of transporter expression.

In recent years, proximal tubule cells isolated from intact tissue by collagenase digestion and isopycnic centrifugation have been successfully cultured from human and rat kidneys. Cultures are reported to be structurally polarised, with numerous microvilli and tight junctions at the apical side, and preserved the characteristic features of proximal tubule cells, *e.g.* alkaline phosphatase and γ-glutamyl transferase enzyme activity.[226,227] Human primary proximal tubule cells also express the OA^- and OC^+ transporters absent from many immortalized proximal tubule cell lines, in addition to the ABC transporters MDR1 and MRP2.[228,229] The cells have shown prototypic transport of endogenous OA^- and OC^+ such as PAH and creatinine, and xenobiotics such as rosuvastatin.[228–230] Similarly, cultured rat proximal tubule cells have shown sodium-dependent uptake of α-methyl-D-glucopyranosine, ergothionine and carnitine, suggesting expression of Na^+/glucose and OC^+/carnitine transport proteins.[231] Furthermore, several studies have confirmed the suitability of rat and human primary proximal tubular cell cultures as *in vitro* models to study nephrotoxicity. For example, Lash *et al.* found similarities in the biochemical properties of cultures of rat primary proximal tubule cells and native cells when chemical injury was induced, which demonstrated the use of the former as a good model of nephrotoxicity.[228,232–236]

4.6   Species Differences in Renal Handling

A major challenge in drug development is the extrapolation of drug safety information from animals to humans. Approximately 54% of compounds fail at Phase I.[237] For each drug lost at this stage it is estimated that \$740 million is lost in developmental costs and 6.5 years in lead time to the clinic.[237] Overall, this results in a delay in getting molecules to the clinic and an increased cost of successful drug molecules. Rats are routinely the choice for preclinical studies, and the relevance of studies using this species for human biology has to be interpreted with great caution. Although data from animals may be reasonably extrapolated to humans, there are still limitations, not least because of the differences in physiology of renal handling of molecules between the two species.[238,239]

Several studies have made direct comparisons between renal clearance in man and in preclinical species such as rodents, dogs and monkeys. In 2005, Tahara and co-workers found no species differences in the handling of a toolkit of 11 compounds by human, rat and monkey OAT1; similarly, a screen of nine compounds against human and monkey OAT3 showed good correlations in kinetic values but, in rat, three out of the nine compounds did not correlate well with the primate data, suggesting a real species difference in their handling.[240] Similarly, screening of a panel of 36 molecules of diverse chemistries showed a tighter correlation for the handling of molecules between the dog and human kidney with a much poorer correlation between the rat and human kidney.[241]

As previously outlined, only OATP4C1 and OATP1A2 are expressed in the human kidney. In contrast, multiple OATP isoforms are expressed in the rat, including OATP4C1, OATP1A1, OATP1A2, OATP1A3_v1 and OATP1A3_v2.[62] To date, there is little clear information as to the importance of species differences in the expression of OATPs for renal drug handling, although a recent study did show a clear difference in the handling of digoxin between rat and human proximal tubule monolayers that may be ascribed to differences in OATP expression. In human proximal tubule cell monolayers, there was a net secretion of digoxin, mediated by OATP4C1 uptake at the basolateral membrane and MDR1 efflux across the apical membrane. In contrast, in rat proximal tubule monolayers, digoxin exhibited net absorption from the apical to the basolateral membrane, which could be inhibited by the addition of the OATP substrate T3 to the apical membrane, suggesting that expression of apically located OATPs may explain the difference in renal handling of digoxin between human and rat.[242,243]

Differences in renal handling of drug molecules between preclinical species and humans can result in unpredicted DDIs. For example, a clinical DDI between probenecid and the H2 antagonist famotidine in humans was not predicted from rat data; this lack of predictive ability has been attributed to species differences in the handling of these molecules by rat and human OAT3, and by the lack of expression of OCT1 in the human kidney.[244] There is also evidence that transporters present in both species may be differentially regulated by hormones and other molecules, which may further complicate the interpretation of comparisons between humans and other preclinical species.[245]

4.7 Development of Predictive *In vitro* Models of Drug Transport

As discussed, a key deficiency in the drug development process is the availability of applicable preclinical *in vitro* models that can be used to predict toxicological and efficacious outcomes in the clinical setting. Many widely used preclinical *in vitro* models of transport in the kidney are based on transfected human or animal cells. These models express a limited number of human renal transporters, and so do not accurately reflect the situation *in vivo*. In contrast, cultured primary renal proximal tubule cell monolayers appear to maintain a full complement of key renal transporters, resulting in a more physiologically-relevant and thus predictive model of drug handling in the clinical setting.

One of the barriers in the use of primary cell based *in vitro* models is the availability of fresh tissues for cell isolation. To address this, there is ongoing research into the differentiation of induced pluripotent stem cells (iPSCs), which are more accessible than fresh tissues,[246,247] into proximal tubule-like cells with replicable physiological properties to their primary counterparts.[247] Indeed, stem cell markers, such as the transcription factors Nanog and Sox2, were markedly down-regulated at the mRNA level in iPSCs upon treatment with morphogenetic proteins (which is one of the ways to differentiate iPSCs from proximal tubule cells),[248] whereas proximal tubule cell markers (aquaporin 1 and γ-glutamyl transferase, amongst others) were expressed at levels similar to native cells.[247] Their utility as nephrotoxicity models was also investigated; well-characterised nephrotoxins were tested on iPSC-derived proximal tubule cells and demonstrated over 85% accuracy in nephrotoxicity prediction compared with native cells.[249] While an increase in mRNA expression of several transporters has been identified—OAT1, OAT3, OCT2, MDR1 and PEPT1—the full array of key drug transporters associated with the proximal tubule has not been functionally characterised in iPSC-derived proximal tubule cells.[247] This needs to be addressed to ensure that these cells have the scope to be *in vitro* models of renal drug handling as well as nephrotoxicity.

Improvements to currently available *in vitro* models have also been initiated. There is increasing interest in the 3D culture of cells *in vitro* to replicate the physiological condition. For instance, several research groups have tried to incorporate flow of culture medium to primary proximal tubule cells in a bid to mimic the fluid shear stress that blood and tubular fluid exert on these cells. One group observed improved morphology and expression of membrane transporters in these cells when compared with cells grown under static conditions,[250,251] while another demonstrated enhanced transporter activities in proximal tubule cells when grown under fluidic conditions.[252] The concept of flow has led to the engineering of devices that feature continuously perfused chambers inhabited by cell monolayers,[253] and can be performed at such a small scale that the industry has referred to these devices as organ-on-chips.[252,253] The initial design of the kidney-on-a-chip has two perfused channels separated by a proximal tubule cell layer, with the apical channel mimicking the urinary lumen and basolateral channel mimicking the interstitial space.[252,254] A research group tried to culture human primary proximal tubule cells on these devices and used them for nephrotoxicity tests with great success.[252] Development of these devices with other organs have also led to multiple organs-on-a-chip with which one could, in theory, follow the movement of a substrate from the gut to the liver to the kidney,[254,255] replicating the ingestion, metabolism and excretion of a drug.

While the abovementioned flow of media or fluids adds an additional dimension to the cell culture technique, it is not considered a fully 3D culture technique.[256] True 3D cultures rely on matrices or scaffolds upon which cells grow, and as such provide cell-to-cell interactions that promote the development of the correct phenotypes and functions.[257,258] Cells in these cultures will self-assemble over a period of time into microstructures known as organoids, which have the potential to be good *in vitro* models of the tissue the cells are derived from.[257] For example, when mouse proximal tubule cells were isolated and cultured on hyaluronic acid hydrogel (a commonly used extracellular matrix),[257,259,260] tubular organisation of the cells was observed and the organoids were viable for up to 6 weeks in culture.[257] Recapitulation of *in vivo* cellular responses was also evident in the organoids—kidney injury marker-1 proteins and other biomarkers of renal injury were up-regulated in response to cisplatin-induced toxicity, as would be the case in native tissues.[257] A drawback in the use of organoid cultures, however, is difficulty in the functional analysis of entrapped cells. For instance, quantification of cellular transport of substrates may be hampered by the inaccessible luminal contents, which would limit their use in drug transporter studies. A possible solution to this predicament could be the incorporation of the organoids into microfluidic devices. These innovations provide increasingly promising *in vitro* models for studying renal drug handling, which would help facilitate the drug development process.

4.8 Conclusion

The impact that membrane transporters in the proximal tubule cells, and by association the kidney, have on drug transport has been discussed in this chapter. Understanding renal drug transport and the roles the transporters play will provide an insight into drug secretion and reabsorption, and more importantly the mechanisms of renal DDIs and toxicity. The knowledge of drug handling in renal tubular processes can also aid in the development of new medicines, which will undoubtedly involve at least one of the transporters discussed in this chapter.

References

1. B. Feng, J. L. LaPerle, G. Chang and M. V. Varma, *Expert Opin. Drug Metab. Toxicol.*, 2010, **6**, 939–952.
2. L. Pollaro and C. Heinis, *Med. Chem. Commun.*, 2010, **1**, 319–324.
3. P. D. Dobson and D. B. Kell, *Nat. Rev. Drug Discovery.*, 2008, **7**, 205–220.
4. T. I. T. Consortium, *Nat. Rev. Drug Discovery.*, 2010, **9**, 215–236.
5. P. M. Beringer and R. L. Slaughter, *Ann. Pharmacother.*, 2005, **39**, 1097–1108.
6. P. L. Bonate, K. Reith and S. Weir, *Clin. Pharmacokinet.*, 1998, **34**, 375–404.
7. U. F. a. D. Adminstration, ed. U. F. D. Adminstration, Sliver Spring, MD, 2012.
8. K. M. Madsen and C. C. Tisher, in *The Kidney*, ed. B. M. Brenner, Elsevier, Philadelphia, 7th edn, 1976, vol. 1, pp. 3–72.
9. J. R. Nyengaard and T. F. Bendtsen, *Anat. Rec.*, 1992, **232**, 194–201.
10. D. P. Haley and R. E. Bulger, *Am. J. Anat.*, 1983, **167**, 1–13.
11. R. E. Bulger, *Am. J. Anat.*, 1965, **116**, 237–255.
12. M. J. Helbert, S. E. Dauwe, I. Van der Biest, E. J. Nouwen and M. E. De Broe, *Kidney Int.*, 1997, **52**, 414–428.

13. A. B. Maunsbach, *J. Ultrastruct. Res.*, 1966, **16**, 239–258.
14. W. D. McArdle, F. I. Katch and V. L. Katch, *Exercise Physiology: Nutrition, Energy, and Human Performance*, Lippincott Williams & Wilkins, 2010.
15. A. S. Levey, J. Coresh, T. Greene, L. A. Stevens, Y. L. Zhang, S. Hendriksen, J. W. Kusek and F. Van Lente, *Ann. Intern. Med.*, 2006, **145**, 247–254.
16. A. S. Levey, R. D. Perrone and N. E. Madias, *Annu. Rev. Med.*, 1988, **39**, 465–490.
17. L. A. Stevens and A. S. Levey, *Med. Clin. North Am.*, 2005, **89**, 457–473.
18. E. Andreev, M. Koopman and L. Arisz, *J. Intern. Med.*, 1999, **246**, 247–252.
19. M. Horio, E. Imai, Y. Yasuda, T. Watanabe and S. Matsuo, *Clin. Exp. Nephrol.*, 2011, **15**, 868.
20. L. A. Inker, C. H. Schmid, H. Tighiouart, J. H. Eckfeldt, H. I. Feldman, T. Greene, J. W. Kusek, J. Manzi, F. Van Lente, Y. L. Zhang, J. Coresh and A. S. Levey, *N. Engl. J. Med.*, 2012, **367**, 20–29.
21. K. M. Giacomini, S. M. Huang, D. J. Tweedie, L. Z. Benet, K. L. Brouwer, X. Chu, A. Dahlin, R. Evers, V. Fischer, K. M. Hillgren, K. A. Hoffmaster, T. Ishikawa, D. Keppler, R. B. Kim, C. A. Lee, M. Niemi, J. W. Polli, Y. Sugiyama, P. W. Swaan, J. A. Ware, S. H. Wright, S. W. Yee, M. J. Zamek-Gliszczynski and L. Zhang, *Nat. Rev. Drug Discovery*, 2010, **9**, 215–236.
22. Japanese Ministry of Health Labour and Welfare, 2014, Drug interaction guideline for drug development and labeling recommendations.
23. European Medicines Agency, 2012, Guideline on the Investigation of Drug Interactions.
24. US. Food and Drug Adminstration, 2012, Drug interaction studies - Study design, data analysis, implications for dosing and labeling recommendations.
25. H. Kusuhara and Y. Sugiyama, *Drug Metab. Pharmacokinet.*, 2009, **24**, 37–52.
26. A. L. VanWert, M. R. Gionfriddo and D. H. Sweet, *Biopharm. Drug Dispos.*, 2010, **31**, 1–71.
27. G. Burckhardt and B. C. Burckhardt, *Handb. Exp. Pharmacol.*, 2011, 29–104.
28. H. Shen, T. Liu, B. L. Morse, Y. Zhao, Y. Zhang, X. Qiu, C. Chen, A. C. Lewin, X. T. Wang, G. Liu, L. J. Christopher, P. Marathe and Y. Lai, *Drug Metab. Dispos.: Biol. Fate Chem.*, 2015, **43**, 984–993.
29. D. H. Sweet, N. A. Wolff and J. B. Pritchard, *J. Biol. Chem.*, 1997, **272**, 30088–30095.
30. T. Sekine, N. Watanabe, M. Hosoyamada, Y. Kanai and H. Endou, *J. Biol. Chem.*, 1997, **272**, 18526–18529.
31. S. A. Eraly, B. A. Hamilton and S. K. Nigam, *Biochem. Biophys. Res. Commun.*, 2003, **300**, 333–342.

32. H. Kusuhara, T. Sekine, N. Utsunomiya-Tate, M. Tsuda, R. Kojima, S. H. Cha, Y. Sugiyama, Y. Kanai and H. Endou, *J. Biol. Chem.*, 1999, **274**, 13675–13680.

33. H. Koepsell and H. Endou, *Pflugers Arch. – Eur. J. Physiol.*, 2004, **447**, 666–676.

34. E. B. Berkhin and M. H. Humphreys, *Kidney Int.*, 2001, **59**, 17–30.

35. S. H. Wright, *Toxicol. Appl. Pharmacol.*, 2005, **204**, 309–319.

36. A. N. Rizwan and G. Burckhardt, *Pharm. Res.*, 2007, **24**, 450–470.

37. M. Hosoyamada, T. Sekine, Y. Kanai and H. Endou, *Am. J. Physiol.*, 1999, **276**, F122–F128.

38. S. H. Cha, T. Sekine, J. I. Fukushima, Y. Kanai, Y. Kobayashi, T. Goya and H. Endou, *Mol. Pharmacol.*, 2001, **59**, 1277–1286.

39. S. A. Eraly, V. Vallon, D. A. Vaughn, J. A. Gangoiti, K. Richter, M. Nagle, J. C. Monte, T. Rieg, D. M. Truong, J. M. Long, B. A. Barshop, G. Kaler and S. K. Nigam, *J. Biol. Chem.*, 2006, **281**, 5072–5083.

40. V. Vallon, T. Rieg, S. Y. Ahn, W. Wu, S. A. Eraly and S. K. Nigam, *Am. J. Physiol. Renal Physiol.*, 2008, **294**, F867–F873.

41. H. J. Shin, C. H. Lee, S. S. Lee, I. S. Song and J. G. Shin, *Clin. Chim. Acta; Int. J. Clin. Chem.*, 2010, **411**, 99–105.

42. B. Dussol, J. Moussi-Frances, S. Morange, C. Somma-Delpero, O. Mundler and Y. Berland, *J. Clin. Hypertens.*, 2012, **14**, 32–37.

43. W. Löffler, R. Landthaler, J. X. de Vries, I. Walter-Sack, A. Ittensohn, A. Voss and N. Zöllner, *Clin Investig*, 1994, **72**, 1071–1075.

44. C. S. Wilcox, *J. Am. Soc. Nephrol.*, 2002, **13**, 798–805.

45. K. C. Cundy, Z. H. Li and W. A. Lee, *Drug Metab. Dispos.: Biol. Fate Chem.*, 1996, **24**, 315–321.

46. G. D. Simonson, A. C. Vincent, K. J. Roberg, Y. Huang and V. Iwanij, *J. Cell Sci.*, 1994, **107**(Pt 4), 1065–1072.

47. T. Sekine, S. H. Cha, M. Tsuda, N. Apiwattanakul, N. Nakajima, Y. Kanai and H. Endou, *FEBS Lett.*, 1998, **429**, 179–182.

48. S. C. Buist, N. J. Cherrington, S. Choudhuri, D. P. Hartley and C. D. Klaassen, *J. Pharmacol. Exp. Ther.*, 2002, **301**, 145–151.

49. Y. Kobayashi, M. Ohbayashi, N. Kohyama and T. Yamamoto, *Eur. J. Pharmacol.*, 2005, **524**, 44–48.

50. Y. Cheng, A. Vapurcuyan, M. Shahidullah, L. M. Aleksunes and R. M. Pelis, *Drug Metab. Dispos.: Biol. Fate Chem.*, 2012, **40**, 617–624.

51. A. Emami Riedmaier, A. T. Nies, E. Schaeffeler and M. Schwab, *Pharmacol. Rev.*, 2012, **64**, 421–449.

52. C. Eisner, R. Faulhaber-Walter, Y. Wang, A. Leelahavanichkul, P. S. Yuen, D. Mizel, R. A. Star, J. P. Briggs, M. Levine and J. Schnermann, *Kidney Int.*, 2010, **77**, 519–526.

53. E. I. Lepist, X. Zhang, J. Hao, J. Huang, A. Kosaka, G. Birkus, B. P. Murray, R. Bannister, T. Cihlar, Y. Huang and A. S. Ray, *Kidney Int.*, 2014, **86**, 350–357.

54. J. Konig, in *Drug Transporters, Handbook of Experimental Pharmacology 201*, ed. M. F. F. a. R. B. Kim, 2011, pp. 1–28.

55. E. Jacquemin, B. Hagenbuch, B. Stieger, A. W. Wolkoff and P. J. Meier, *Proc. Natl. Acad. Sci. U. S. A.*, 1994, **91**, 133–137.

56. B. Hagenbuch and P. J. Meier, *Biochim. Biophys. Acta*, 2003, **1609**, 1–18.

57. M. Roth, A. Obaidat and B. Hagenbuch, *Br. J. Pharmacol.*, 2012, **165**, 1260–1287.

58. B. Hagenbuch and C. Gui, *Xenobiotica; Fate Foreign Compounds Biol. Syst.*, 2008, **38**, 778–801.

59. X. Bossuyt, M. Muller and P. J. Meier, *J. Hepatol.*, 1996, **25**, 733–738.

60. R. Castro, I. Badagnani, T. R. Taylor, C. M. Brett, C. C. Huang, D. Stryke, M. Kawamoto, S. J. Johns, T. E. Ferrin, E. J. Carlson, E. G. Burchard and K. M. Giacomini, *J. Pharmacol. Exp. Ther.*, 2006, **2**, 521–529.

61. S. Masuda, A. Takeuchi, H. Saito, T. Abe and K. Inui, *J. Pharmacol. Exp. Ther*, 2001, **299**, 261–267.

62. P. Meier and B. Hagenbuch, *Pflugers Arch.: Eur. J. Physiol.*, 2004, **447**, 653–665.

63. T. Suzuki, T. Mikkaichi, T. Onogawa, M. Tanemoto, H. Mizutamari, M. Okada, T. Chaki, S. Masuda, T. Tokui, N. Eto, M. Abe, F. Satoh, M. Unno, T. Hishinuma, K. Inui, S. Ito, J. Goto and T. Abe, *Proc. Natl. Acad. Sci. U. S. A.*, 2004, **101**, 3569–3574.

64. K. Bleasby, J. C. Castle, C. J. Roberts, C. Cheng, W. J. Bailey, J. F. Sina, A. V. Kulkarni, M. J. Hafey, R. Evers, J. M. Johnson, R. G. Ulrich and J. G. Slatter, *Xenobiotica; Fate Foreign Compounds Biol. Syst.*, 2006, **36**, 963–988.

65. T. Mikkaichi, T. Suzuki, T. Onogawa, M. Tanemoto, H. Mizutamari, M. Okada, T. Chaki, S. Masuda, T. Tokui, N. Eto, M. Abe, F. Satoh, M. Unno, T. Hishinuma, K. Inui, S. Ito, J. Goto and T. Abe, *Proc. Natl. Acad. Sci. U. S. A.*, 2004, **101**, 3569–3574.

66. B. Hagenbuch, G. A. Kullak-Ublick, B. Stieger, C. D. Schteingart, A. F. Hofmann, A. W. Wolkoff and P. J. Meier, *Gastroenterology*, 1995, **109**, 1274–1282.

67. W. Lee, H. Glaeser, L. H. Smith, R. L. Roberts, G. W. Moeckel, G. Gervasini, B. F. Leake and R. B. Kim, *J. Biol. Chem.*, 2005, **280**, 9610–9617.

68. Y. Gotoh, Y. Kato, B. Stieger, P. J. Meier and Y. Sugiyama, *Am. J. Physiol. Endocrinol. Metab.*, 2002, **282**, E1245–E1254.

69. R. Lu, N. Kanai, Y. Bao, A. W. Wolkoff and V. L. Schuster, *Am. J. Physiol.*, 1996, **270**, F332–F337.

70. H. Saito, S. Masuda and K. Inui, *J. Biol. Chem.*, 1996, **271**, 20719–20725.

71. S. Masuda, A. Takeuchi, H. Saito, Y. Hashimoto and K. Inui, *FEBS Lett.*, 1999, **459**, 128–132.

72. S. Masuda, H. Saito and K. Inui, *J. Pharmacol. Exp. Therapeut.*, 1997, **283**, 1039–1042.

73. A. Takeuchi, S. Masuda, H. Saito, Y. Hashimoto and K. Inui, *J. Pharmacol. Exp. Ther.*, 2000, **293**, 1034–1039.

74. S. Masuda, K. Ibaramoto, A. Takeuchi, H. Saito, Y. Hashimoto and K. Inui, *Mol. Pharmacol.*, 1999, **55**, 743–752.

75. A. Takeuchi, S. Masuda, H. Saito, T. Abe and K. Inui, *J. Pharmacol. Exp. Ther.*, 2001, **299**, 261–267.

76. A. Enomoto, H. Kimura, A. Chairoungdua, Y. Shigeta, P. Jutabha, S. Ho Cha, M. Hosoyamada, M. Takeda, T. Sekine, T. Igarashi, H. Matsuo, Y. Kikuchi, T. Oda, K. Ichida, T. Hosoya, K. Shimokata, T. Niwa, Y. Kanai and H. Endou, *Nature*, 2002, **417**, 447–452.

77. E. Hu, Z. Chen, T. A. Fredrickson, N. Spurr, S. Gentle, M. Sims, Y. Zhu, W. Halsey, S. Van Horn, J. Mao, G. M. Sathe and D. P. Brooks, *Exp. Nephrol.*, 2001, **9**, 156–164.

78. M. Hosoyamada, K. Ichida, A. Enomoto, T. Hosoya and H. Endou, *J. Am. Soc. Nephrol.: JASN*, 2004, **15**, 261–268.

79. M. Sato, T. Wakayama, H. Mamada, Y. Shirasaka, T. Nakanishi and I. Tamai, *Biochim. Biophys. Acta*, 2011, **1808**, 1441–1447.

80. D. Miura, N. Anzai, P. Jutabha, S. Chanluang, X. He, T. Fukutomi and H. Endou, *J. Physiol. Sci.*, 2011, **61**, 253–257.

81. V. Vitart, I. Rudan, C. Hayward, N. K. Gray, J. Floyd, C. N. A. Palmer, S. A. Knott, I. Kolcic, O. Polasek, J. Graessler, J. F. Wilson, A. Marinaki, P. L. Riches, X. Shu, B. Janicijevic, N. Smolej-Narancic, B. Gorgoni, J. Morgan, S. Campbell, Z. Biloglav, L. Barac-Lauc, M. Pericic, I. M. Klaric, L. Zgaga, T. Skaric-Juric, S. H. Wild, W. A. Richardson, P. Hohenstein, C. H. Kimber, A. Tenesa, L. A. Donnelly, L. D. Fairbanks, M. Aringer, P. M. McKeigue, S. H. Ralston, A. D. Morris, P. Rudan, N. D. Hastie, H. Campbell and A. F. Wright, *Nat. Genet.*, 2008, **40**, 437–442.

82. M. J. Caulfield, P. B. Munroe, D. O'Neill, K. Witkowska, F. J. Charchar, M. Doblado, S. Evans, S. Eyheramendy, A. Onipinla, P. Howard, S. Shaw-Hawkins, R. J. Dobson, C. Wallace, S. J. Newhouse, M. Brown, J. M. Connell, A. Dominiczak, M. Farrall, G. M. Lathrop, N. J. Samani, M. Kumari, M. Marmot, E. Brunner, J. Chambers, P. Elliott, J. Kooner, M. Laan, E. Org, G. Veldre, M. Viigimaa, F. P. Cappuccio, C. Ji, R. Iacone, P. Strazzullo, K. H. Moley and C. Cheeseman, *Plos Med*, 2008, **5**, 1509–1523.

83. N. Anzai, K. Ichida, P. Jutabha, T. Kimura, E. Babu, C. J. Jin, S. Srivastava, K. Kitamura, I. Hisatome, H. Endou and H. Sakurai, *J. Biol. Chem.*, 2008, **283**, 26834–26838.

84. P. Jutabha, N. Anzai, K. Kitamura, A. Taniguchi, S. Kaneko, K. Yan, H. Yamada, H. Shimada, T. Kimura, T. Katada, T. Fukutomi, K. Tomita, W. Urano, H. Yamanaka, G. Seki, T. Fujita, Y. Moriyama, A. Yamada, S. Uchida, M. F. Wempe, H. Endou and H. Sakurai, *J. Biol. Chem.*, 2010, **285**, 35123–35132.

85. P. Jutabha, N. Anzai, K. Hayashi, M. Domae, K. Uchida, H. Endou and H. Sakurai, *J. Pharmacol. Sci.*, 2011, **116**, 392–396.

86. I. A. Bobulescu and O. W. Moe, *Adv. Chronic Kidney Dis.*, 2012, **19**, 358–371.

87. A. Dehghan, A. Köttgen, Q. Yang, S.-J. Hwang, W. H. L. Kao, F. Rivadeneira, E. Boerwinkle, D. Levy, A. Hofman, B. C. Astor, E. J. Benjamin, C. M. van Duijn, J. C. Witteman, J. Coresh and C. S. Fox, *Lancet*, 2008, **372**, 1953–1961.

88. M. A. Hediger, R. J. Johnson, H. Miyazaki and H. Endou, *Physiology*, 2005, **20**, 125–133.

89. M. Lipkowitz, *Curr. Rheumatol. Rep.*, 2012, **14**, 179–188.

90. C. Wallace, S. J. Newhouse, P. Braund, F. Zhang, M. Tobin, M. Falchi, K. Ahmadi, R. J. Dobson, A. C. B. Marçano, C. Hajat, P. Burton, P. Deloukas, M. Brown, J. M. Connell, A. Dominiczak, G. M. Lathrop, J. Webster, M. Farrall, T. Spector, N. J. Samani, M. J. Caulfield and P. B. Munroe, *Am. J. Hum. Genet.*, 2008, **82**, 139–149.

91. V. Gorboulev, D. Gründemann, S. Gambaryan, M. Veyhl and H. Koepsell, *Nature*, 1994, **372**, 549–552.

92. U. Karbach, J. Kricke, F. Meyer-Wentrup, V. Gorboulev, C. Volk, D. Loffing-Cueni, B. Kaissling, S. Bachmann and H. Koepsell, *Am. J. Physiol. Renal Physiol.*, 2000, **279**, F679–F687.

93. J. Ulzheimer, V. Gorboulev, A. Akhoundova, I. Ulzheimer-Teuber, U. Karbach, S. Quester, C. Baumann, F. Lang, A. E. Busch and H. Koepsell, *DNA Cell Biol.*, 1997, **16**, 971–981.

94. L. Zhang, M. J. Dresser, J. K. Chun, P. C. Babbitt and K. M. Giacomini, *J. Biol. Chem.*, 1997, **272**, 16548–16554.

95. M. Okuda, H. Saito, Y. Urakami, M. Takano and K. Inui, *Biochem. Biophys. Res. Commun.*, 1996, **224**, 500–507.

96. Y. Urakami, M. Okuda, S. Masuda, H. Saito and K. I. Inui, *J. Pharmacol. Exp. Ther.*, 1998, **287**, 800–805.

97. S. H. Wright and W. H. Dantzler, *Physiol. Rev.*, 2004, **84**, 987–1049.

98. J. W. Jonker and A. H. Schinkel, *J. Pharmacol. Exp. Ther.*, 2004, **308**, 2–9.

99. L. Pochini, M. Scalise, M. Galluccio and C. Indiveri, *J. Biomol. Screening*, 2013, **18**(8), 851–867.

100. I. Tamai, H. Yabuuchi, J. Nezu, Y. Sai, A. Oku, M. Shimane and A. Tsuji, *Febs Lett.*, 1997, **419**, 107–111.

101. H. Yabuuchi, I. Tamai, J. Nezu, K. Sakamoto, A. Oku, M. Shimane, Y. Sai and A. Tsuji, *J. Pharmacol. Exp. Ther.*, 1999, **289**, 768–773.

102. X. Wu, P. D. Prasad, F. H. Leibach and V. Ganapathy, *Biochem. Biophys. Res. Commun.*, 1998, **246**, 589–595.

103. X. Wu, W. Huang, P. D. Prasad, P. Seth, D. P. Rajan, F. H. Leibach, J. Chen, S. J. Conway and V. Ganapathy, *J. Pharmacol. Exp. Ther.*, 1999, **290**, 1482–1492.

104. A. L. Slitt, N. J. Cherrington, D. P. Hartley, T. M. Leazer and C. D. Klaassen, *Drug Metab. Dispos.*, 2002, **30**, 212–219.

105. X. Wu, R. L. George, W. Huang, H. Wang, S. J. Conway, F. H. Leibach and V. Ganapathy, *Biochim. Biophys. Acta*, 2000, **1466**, 315–327.

106. I. Tamai, R. Ohashi, J. I. Nezu, Y. Sai, D. Kobayashi, A. Oku, M. Shimane and A. Tsuji, *J. Biol. Chem.*, 2000, **275**, 40064–40072.

107. M. M. Cano, M. L. Calonge and A. A. Ilundain, *J. Cell. Physiol.*, 2010, **223**, 451–459.
108. I. Tamai, K. China, Y. Sai, D. Kobayashi, J.-i. Nezu, E. Kawahara and A. Tsuji, *Biochim. Biophys. Acta*, 2001, **1512**, 273–284.
109. I. Tamai, T. Nakanishi, D. Kobayashi, K. China, Y. Kosugi, J.-i. Nezu, Y. Sai and A. Tsuji, *Mol. Pharm.*, 2003, **1**, 57–66.
110. S. Masuda, T. Terada, A. Yonezawa, Y. Tanihara, K. Kishimoto, T. Katsura, O. Ogawa and K. Inui, *J. Am. Soc. Nephrol.*, 2006, **17**, 2127–2135.
111. T. Komatsu, M. Hiasa, T. Miyaji, T. Kanamoto, T. Matsumoto, M. Otsuka, Y. Moriyama and H. Omote, *Int. J. Biochem. Cell Biol.*, 2011, **43**, 913–918.
112. M. Otsuka, T. Matsumoto, R. Morimoto, S. Arioka, H. Omote and Y. Moriyama, *Proc. Natl. Acad. Sci. U. S. A.*, 2005, **102**, 17923–17928.
113. M. J. Dresser, M. K. Leabman and K. M. Giacomini, *J. Pharm. Sci.*, 2001, **90**, 397–421.
114. Y. Morita, A. Kataoka, S. Shiota, T. Mizushima and T. Tsuchiya, *J. Bacteriol.*, 2000, **182**, 6694–6697.
115. Y. Tanihara, S. Masuda, T. Sato, T. Katsura, O. Ogawa and K. Inui, *Biochem. Pharmacol.*, 2007, **74**, 359–371.
116. S. Masuda, T. Terada, A. Yonezawa, Y. Tanihara, K. Kishimoto, T. Katsura, O. Ogawa and K. Inui, *J. Am. Soc. Nephrol.: JASN*, 2006, **17**, 2127–2135.
117. M. Kajiwara, T. Terada, K. Ogasawara, J. Iwano, T. Katsura, A. Fukatsu, T. Doi and K. Inui, *J. Hum. Genet.*, 2009, **54**, 40–46.
118. Y. Chen, K. Teranishi, S. Li, S. W. Yee, S. Hesselson, D. Stryke, S. J. Johns, T. E. Ferrin, P. Kwok and K. M. Giacomini, *Pharmacogenomics J.*, 2009, **9**, 127–136.
119. A. Yonezawa and K. Inui, *Br. J. Pharmacol.*, 2011, **164**, 1817–1825.
120. M. Kajiwara, T. Terada, J. Asaka, K. Ogasawara, T. Katsura, O. Ogawa, A. Fukatsu, T. Doi and K. Inui, *Am. J. Physiol. Renal Physiol.*, 2007, **293**, F1564–F1570.
121. M. Baumann, J. Pontiller and W. Ernst, *Mol. Biotechnol.*, 2010, **45**, 241–247.
122. J. Ha Choi, S. Wah Yee, M. J. Kim, L. Nguyen, J. Ho Lee, J. O. Kang, S. Hesselson, R. A. Castro, D. Stryke, S. J. Johns, P. Y. Kwok, T. E. Ferrin, M. Goo Lee, B. L. Black, N. Ahituv and K. M. Giacomini, *Pharmacogenet. Genomics*, 2009, **19**, 770–780.
123. W. Bi, J. Yan, P. Stankiewicz, S. S. Park, K. Walz, C. F. Boerkoel, L. Potocki, L. G. Shaffer, K. Devriendt, M. J. Nowaczyk, K. Inoue and J. R. Lupski, *Genome Res.*, 2002, **12**, 713–728.
124. R. E. Slager, T. L. Newton, C. N. Vlangos, B. Finucane and S. H. Elsea, *Nat. Genet.*, 2003, **33**, 466–468.
125. M. Tsuda, T. Terada, T. Mizuno, T. Katsura, J. Shimakura and K. Inui, *Mol. Pharmacol.*, 2009, **75**, 1280–1286.
126. K. Toyama, A. Yonezawa, M. Tsuda, S. Masuda, I. Yano, T. Terada, R. Osawa, T. Katsura, M. Hosokawa, S. Fujimoto, N. Inagaki and K. Inui, *Pharmacogenet. Genomics*, 2010, **20**, 135–138.

127. K. Ueda, D. P. Clark, C. J. Chen, I. B. Roninson, M. M. Gottesman and I. Pastan, *J. Biol. Chem.*, 1987, **262**, 505–508.
128. G. Li and J. A. Bush, *Carcinogenesis*, 2002, **23**, 1603–1607.
129. S. G. Aller, J. Yu, A. Ward, Y. Weng, S. Chittaboina, R. Zhuo, P. M. Harrell, Y. T. Trinh, Q. Zhang, I. L. Urbatsch and G. Chang, *Science*, 2009, **323**, 1718–1722.
130. A. B. Shapiro, K. Fox, P. Lam and V. Ling, *Eur. J. Biochem./FEBS*, 1999, **259**, 841–850.
131. A. Garrigues, A. E. Escargueil and S. Orlowski, *Proc. Natl. Acad. Sci. U. S. A.*, 2002, **99**, 10347–10352.
132. A. Rothnie, D. Theron, L. Soceneantu, C. Martin, M. Traikia, G. Berridge, C. F. Higgins, P. F. Devaux and R. Callaghan, *Eur. Biophys. J.: EBJ*, 2001, **30**, 430–442.
133. L. Gayet, G. Dayan, S. Barakat, S. Labialle, M. Michaud, S. Cogne, A. Mazane, A. W. Coleman, D. Rigal and L. G. Baggetto, *Biochemistry*, 2005, **44**, 4499–4509.
134. Y. Kimura, S. Y. Morita, M. Matsuo and K. Ueda, *Cancer Sci.*, 2007, **98**, 1303–1310.
135. R. B. Wang, C. L. Kuo, L. L. Lien and E. J. Lien, *J. Clin. Pharm. Ther.*, 2003, **28**, 203–228.
136. G. Tramonti, N. Romiti, M. Norpoth and E. Chieli, *Renal Failure*, 2001, **23**, 331–337.
137. N. Romiti, G. Tramonti and E. Chieli, *Toxicol. Appl. Pharmacol.*, 2002, **183**, 83–91.
138. K. S. Fenner, M. D. Troutman, S. Kempshall, J. A. Cook, J. A. Ware, D. A. Smith and C. A. Lee, *Clin. Pharmacol. Ther.*, 2008, **85**, 173–181.
139. S. A. Hunt, *J. Am. Coll. Cardiol.*, 2005, **46**, e1–e82.
140. C. Woodland, S. Ito and G. Koren, *Ther. Drug Monit.*, 1998, **20**, 134–138.
141. G. Koren, *Clin. Pharmacokinet.*, 1987, **13**, 334–343.
142. L. A. Doyle, W. Yang, L. V. Abruzzo, T. Krogmann, Y. Gao, A. K. Rishi and D. D. Ross, *Proc. Natl. Acad. Sci. U. S. A.*, 1998, **95**, 15665–15670.
143. R. Allikmets, L. M. Schriml, A. Hutchinson, V. Romano-Spica and M. Dean, *Cancer Res.*, 1998, **58**, 5337–5339.
144. J. E. Diestra, G. L. Scheffer, I. Catala, M. Maliepaard, J. H. Schellens, R. J. Scheper, J. R. Germa-Lluch and M. A. Izquierdo, *J. Pathol.*, 2002, **198**, 213–219.
145. M. Huls, C. D. A. Brown, A. S. Windass, R. Sayer, J. J. M. W. van den Heuvel, S. Heemskerk, F. G. M. Russel and R. Masereeuw, *Kidney Int.*, 2008, **73**, 220–225.
146. Y. Tanaka, A. L. Slitt, T. M. Leazer, J. M. Maher and C. D. Klaassen, *Biochem. Biophys. Res. Commun.*, 2005, **326**, 181–187.
147. M. Huls, J. J. van den Heuvel, H. B. Dijkman, F. G. Russel and R. Masereeuw, *Kidney Int.*, 2006, **69**, 2186–2193.
148. M. Maliepaard, G. L. Scheffer, I. F. Faneyte, M. A. van Gastelen, A. C. L. M. Pijnenborg, A. H. Schinkel, M. J. van de Vijver, R. J. Scheper and J. H. M. Schellens, *Cancer Res.*, 2001, **61**, 3458–3464.

149. E. Kis, T. Nagy, M. Jani, É. Molnár, J. Jánossy, O. Ujhellyi, K. Német, K. Herédi-Szabó and P. Krajcsi, *Ann. Rheum. Dis.*, 2009, **68**, 1201–1207.

150. M. Jani, P. Szabó, E. Kis, É. Molnár, H. Glavinas and P. Krajcsi, *Biol. Pharm.Bull.*, 2009, **32**, 497–499.

151. E. Beéry, Z. Rajnai, T. Abonyi, I. Makai, S. Bánsághi, F. Erdő, I. Sziráki, K. Herédi-Szabó, E. Kis, M. Jani, J. Márki-Zay, K. Gábor Tóth and P. Krajcsi, *Drug Metab. Pharmacokinet.*, 2012, **27**, 349–353.

152. I. S. Haslam, J. A. Wright, D. A. O'Reilly, D. J. Sherlock, T. Coleman and N. L. Simmons, *Drug Metab. Dispos.*, 2011, **39**, 2321–2328.

153. C. MacLean, U. Moenning, A. Reichel and G. Fricker, *Drug Metab. Dispos.*, 2008, **36**, 1249–1254.

154. D. Keppler, *Drug Transporters*, ed. M. F. Fromm and R. B. Kim, Springer Berlin Heidelberg, 2011, vol. 201, pp. 299–323.

155. M. G. Belinsky and G. D. Kruh, *Br. J. Cancer*, 1999, **80**, 1342–1349.

156. S. P. Cole, G. Bhardwaj, J. H. Gerlach, J. E. Mackie, C. E. Grant, K. C. Almquist, A. J. Stewart, E. U. Kurz, A. M. Duncan and R. G. Deeley, *Science*, 1992, **258**, 1650–1654.

157. M. Kool, M. de Haas, G. L. Scheffer, R. J. Scheper, M. J. van Eijk, J. A. Juijn, F. Baas and P. Borst, *Cancer Res.*, 1997, **57**, 3537–3547.

158. M. Kool, M. van der Linden, M. de Haas, F. Baas and P. Borst, *Cancer Res.*, 1999, **59**, 175–182.

159. M. G. Belinsky, Z.-S. Chen, I. Shchaveleva, H. Zeng and G. D. Kruh, *Cancer Res.*, 2002, **62**, 6172–6177.

160. J. König, A. T. Nies, Y. Cui, I. Leier and D. Keppler, *Biochim. Biophys. Acta*, 1999, **1461**, 377–394.

161. E. Hopper, M. G. Belinsky, H. Zeng, A. Tosolini, J. R. Testa and G. D. Kruh, *Cancer Lett.*, 2001, **162**, 181–191.

162. H. Yabuuchi, H. Shimizu, S. Takayanagi and T. Ishikawa, *Biochem. Biophys. Res. Commun.*, 2001, **288**, 933–939.

163. N. J. Cherrington, D. P. Hartley, N. Li, D. R. Johnson and C. D. Klaassen, *J. Pharmacol. Exp. Ther.*, 2002, **300**, 97–104.

164. C. Chen and C. D. Klaassen, *Biochem. Biophys. Res. Commun.*, 2004, **317**, 46–53.

165. T. P. Schaub, J. Kartenbeck, J. König, O. Vogel, R. Witzgall, W. Kriz and D. Keppler, *J. Am. Soc. Nephrol.*, 1997, **8**, 1213–1221.

166. R. Evers, M. Kool, L. van Deemter, H. Janssen, J. Calafat, L. C. Oomen, C. C. Paulusma, R. P. Oude Elferink, F. Baas, A. H. Schinkel and P. Borst, *J. Clin. Invest.*, 1998, **101**, 1310–1319.

167. T. P. Schaub, J. Kartenbeck, J. Konig, H. Spring, J. Dorsam, G. Staehler, S. Storkel, W. F. Thon and D. Keppler, *J. Am. Soc. Nephrol.: JASN*, 1999, **10**, 1159–1169.

168. R. A. M. H. van Aubel, P. H. E. Smeets, J. G. P. Peters, R. J. M. Bindels and F. G. M. Russel, *J. Am. Soc. Nephrol.*, 2002, **13**, 595–603.

169. J. Wijnholds, C. A. A. M. Mol, L. van Deemter, M. de Haas, G. L. Scheffer, F. Baas, J. H. Beijnen, R. J. Scheper, S. Hatse,

E. De Clercq, J. Balzarini and P. Borst, *Proc. Natl. Acad. Sci. U. S. A.*, 2000, **97**, 7476–7481.

170. G. L. Scheffer, X. Hu, A. C. Pijnenborg, J. Wijnholds, A. A. Bergen and R. J. Scheper, *Lab. Invest. J. Tech. Methods Pathol.*, 2002, **82**, 515–518.

171. K. C. Peng, F. Cluzeaud, M. Bens, J. P. Duong Van Huyen, M. A. Wioland, R. Lacave and A. Vandewalle, *J. Histochem. Cytochem.: Off. J. Histochem. Soc.*, 1999, **47**, 757–768.

172. R. Evers, G. J. Zaman, L. van Deemter, H. Jansen, J. Calafat, L. C. Oomen, R. P. Oude Elferink, P. Borst and A. H. Schinkel, *J. Clin. Invest.*, 1996, **97**, 1211–1218.

173. R. A. Van Aubel, J. G. Peters, R. Masereeuw, C. H. Van Os and F. G. Russel, *Am. J. Physiol. Renal Physiol.*, 2000, **279**, F713–F717.

174. P. H. E. Smeets, R. A. M. H. Van Aubel, A. C. Wouterse, J. J. M. W. Van Den Heuvel and F. G. M. Russel, *J. Am. Soc. Nephrol.*, 2004, **15**, 2828–2835.

175. G. Jedlitschky, B. Burchell and D. Keppler, *J. Biol. Chem.*, 2000, **275**, 30069–30074.

176. Z.-S. Chen, K. Lee and G. D. Kruh, *J. Biol. Chem.*, 2001, **276**, 33747–33754.

177. P. R. Wielinga, I. van der Heijden, G. Reid, J. H. Beijnen, J. Wijnholds and P. Borst, *J. Biol. Chem.*, 2003, **278**, 17664–17671.

178. Y. Fei, Y. Kanai, S. Nussberger, V. Ganapathy, F. H. Leibach, M. F. Romero, S. K. Singh, W. F. Boron and M. A. Hediger, *Nature*, 1994, **368**, 563–566.

179. H. Saito, M. Okuda, T. Terada, S. Sasaki and K. Inui, *J. Pharmacol. Exp. Ther.*, 1995, **275**, 1631–1637.

180. R. Liang, Y. Fei, P. D. Prasad, S. Ramamoorthy, H. Han, T. L. Yang-Feng, M. H. Hediger, V. Ganapathy and F. H. Leibach, *J. Biol. Chem.*, 1995, **270**, 6456–6463.

181. W. Liu, R. Liang, S. Ramamoorthy, Y. Fei, M. E. Ganapathy, M. A. Hediger, V. Ganapathy and F. H. Leibach, *Biochim. Biophys. Acta*, 1995, **1235**, 461–466.

182. M. Brandsch, I. Knütter and E. Bosse-Doenecke, *J. Pharma. Pharmacol.*, 2008, **60**, 543–585.

183. M. E. Ganapathy, M. Brandsch, P. D. Prasad, V. Ganapathy and F. H. Leibach, *J. Biol. Chem.*, 1995, **270**, 25672–25677.

184. M. E. Ganapathy, P. D. Prasad, B. Mackenzie, V. Ganapathy and F. H. Leibach, *Biochim. Biophys. Acta*, 1997, **1324**, 296–308.

185. T. Terada, H. Saito, M. Mukai and K. Inui, *Am. J. Physiol. Renal Physiol.*, 1997, **273**, F706–F711.

186. M. E. Ganapathy, W. Huang, H. Wang, V. Ganapathy and F. H. Leibach, *Biochem. Biophys. Res. Comm.*, 1998, **246**, 470–475.

187. J. Neumann, M. Bruch, S. Gebauer and M. Brandsch, *Eur. J. Biochem.*, 2004, **271**, 2012–2017.

188. T. Nakanishi, A. Fukushi, M. Sato, M. Yoshifuji, T. Gose, Y. Shirasaka, K. Ohe, M. Kobayashi, K. Kawai and I. Tamai, *Mol. Pharma.*, 2011, **8**, 2142–2150.

189. J. Gattineni, C. Bates, K. Twombley, V. Dwarakanath, M. L. Robinson, R. Goetz, M. Mohammadi and M. Baum, *FGF23 decreases renal NaPi-2a and NaPi-2c expression and induces hypophosphatemia in vivo predominantly via FGF receptor*, 2009, vol. 1.

190. R. Goetz, Y. Nakada, M. C. Hu, H. Kurosu, L. Wang, T. Nakatani, M. Shi, A. V. Eliseenkova, M. S. Razzaque, O. W. Moe, M. Kuro-o and M. Mohammadi, *Proc. Natl. Acad. Sci. U. S. A.*, 2010, **107**, 407–412.

191. J. Biber, N. Hernando and I. Forster, *Annu. Rev. Physiol.*, 2013, **75**, 535–550.

192. K. Buchacz, J. T. Brooks, T. Tong, A. C. Moorman, R. K. Baker, S. D. Holmberg, A. Greenberg and the HIVOSI, *HIV Medicine*, 2006, **7**, 451–456.

193. S. F. Billington. G. Chung and C. D. Brown, 2014, *Poster in AAPS Annual Meeting*, American Association of Pharmaceutical Scientists, San Diego, California, USA.

194. P. Bork and G. Beckmann, *J. Mol. Biol.*, 1993, **231**, 539–545.

195. R. Raychowdhury, J. L. Niles, R. T. McCluskey and J. A. Smith, , *Science*, 1989, **244**, 1163–1165.

196. D. Kerjaschki and M. G. Farquhar, *Proc. Natl. Acad. Sci. U. S. A.*, 1982, **79**, 5557–5561.

197. B. Seetharam, J. S. Levine, M. Ramasamy and D. H. Alpers, *J. Biol. Chem.*, 1988, **263**, 4443–4449.

198. F. Chatelet, E. Brianti, P. Ronco, J. Roland and P. Verroust, *Am. J. Pathol.*, 1986, **122**, 500–511.

199. R. Kozyraki, J. Fyfe, P. J. Verroust, C. Jacobsen, A. Dautry-Varsat, J. Gburek, T. E. Willnow, E. I. Christensen and S. K. Moestrup, *Proc. Natl. Acad. Sci. U. S. A.*, 2001, **98**, 12491–12496.

200. J. Nagai, H. Tanaka, N. Nakanishi, T. Murakami and M. Takano, *Am. J. Physiol. Renal Physiol.*, 2001, **281**, F337–F344.

201. Y. Quiros, L. Vicente-Vicente, A. I. Morales, J. M. López-Novoa and F. J. López-Hernández, *Toxicol. Sci.*, 2011, **119**, 245–256.

202. T. Jobst-Schwan, K. X. Knaup, R. Nielsen, T. Hackenbeck, M. Buettner-Herold, P. Lechler, S. Kroening, M. Goppelt-Struebe, U. Schloetzer-Schrehardt, B. G. Fürnrohr, R. E. Voll, K. Amann, K.-U. Eckardt, E. I. Christensen and M. S. Wiesener, *Am. J. Physiol. Renal Physiol.*, 2013, **305**, F734–F744.

203. B. Feng, S. Hurst, Y. Lu, M. V. Varma, C. J. Rotter, A. El-Kattan, P. Lockwood and B. Corrigan, *Mol. Pharm.*, 2013, **10**, 4207–4215.

204. J. B. Gurdon, C. D. Lane, H. R. Woodland and G. Marbaix, *Nature*, 1971, **233**, 177–182.

205. A. A. Tokmakov, T. Hashimoto, Y. Hasegawa, S. Iguchi, T. Iwasaki and Y. Fukami, *FEBS J.*, 2013, **281**(1), 104–114.

206. A. L. Goldin, *Methods Enzymol.*, 1992, **207**, 266–279.

207. A. Takeuchi, S. Masuda, H. Saito, T. Abe and K. Inui, *J. Pharmacol. Exp. Ther.*, 2001, **299**, 261–267.

208. K. Kuteykin-Teplyakov, C. Luna-Tortos, K. Ambroziak and W. Loscher, *Br. J. Pharmacol.*, 2010, **160**, 1453–1463.

209. Y. Zhang, M. S. Warren, X. Zhang, S. Diamond, B. Williams, N. Punwani, J. Huang, Y. Huang and S. Yeleswaram, *Drug Metab. Dispos.*, 2015, **43**, 485–489.

210. T. Ohtomo, H. Saito, N. Inotsume, M. Yasuhara and K. I. Inui, *J. Pharmacol. Exp. Ther.*, 1996, **276**, 1143–1148.

211. A. Guo, W. Marinaro, P. Hu and P. J. Sinko, *Drug Metab. Dispos.*, 2002, **30**, 457–463.

212. F. R. Luo, P. V. Paranjpe, A. Guo, E. Rubin and P. Sinko, *Drug Metab. Dispos.*, 2002, **30**, 763–770.

213. M. J. Ryan, G. Johnson, J. Kirk, S. M. Fuerstenberg, R. A. Zager and B. Torok-Storb, *Kidney Int.*, 1994, **45**, 48–57.

214. K. Kuteykin-Teplyakov, C. Luna-Tortós, K. Ambroziak and W. Löscher, *Br. J. Pharmacol.*, 2010, **160**, 1453–1463.

215. G. Gstraunthaler, W. Pfaller and P. Kotanko, *Am. J. Physiol.*, 1985, **248**, C181–C183.

216. W. Pfaller and G. Gstraunthaler, *Environ. Health Perspect.*, 1998, **106**(Suppl. 2), 559–569.

217. S. Jenkinson, G. Chung, E. van Loon, N. Bakar, A. Dalzell and C. Brown, *Pflugers Arch. – Eur. J. Physiol.*, 2012, **464**, 601–611.

218. M. Wieser, G. Stadler, P. Jennings, B. Streubel, W. Pfaller, P. Ambros, C. Riedl, H. Katinger, J. Grillari and R. Grillari-Voglauer, *hTERT Alone Immortalizes Epithelial Cells of Renal Proximal Tubules Without Changing their Functional Characteristics*, 2008.

219. J. Jansen, C. M. S. Schophuizen, M. J. Wilmer, S. H. M. Lahham, H. A. M. Mutsaers, J. F. M. Wetzels, R. A. Bank, L. P. van den Heuvel, J. G. Hoenderop and R. Masereeuw, *Exp. Cell Res.*, 2014, **323**, 87–99.

220. L. Aschauer, L. N. Gruber, W. Pfaller, A. Limonciel, T. J. Athersuch, R. Cavill, A. Khan, G. Gstraunthaler, J. Grillari, R. Grillari, P. Hewitt, M. O. Leonard, A. Wilmes and P. Jennings, *Mol. Cell. Biol.*, 2013, **33**, 2535–2550.

221. J. Bentz, M. P. O'Connor, D. Bednarczyk, J. Coleman, C. Lee, J. Palm, Y. A. Pak, E. S. Perloff, E. Reyner, P. Balimane, M. Brännström, X. Chu, C. Funk, A. Guo, I. Hanna, K. Herédi-Szabó, K. Hillgren, L. Li, E. Hollnack-Pusch, M. Jamei, X. Lin, A. K. Mason, S. Neuhoff, A. Patel, L. Podila, E. Plise, G. Rajaraman, L. Salphati, E. Sands, M. E. Taub, J.-S. Taur, D. Weitz, H. M. Wortelboer, C. Q. Xia, G. Xiao, J. Yabut, T. Yamagata, L. Zhang and H. Ellens, *Drug Metab. Dispos.*, 2013, **41**, 1347–1366.

222. J. W. Lohr, G. R. Willsky and M. A. Acara, *Pharmacol. Rev.*, 1998, **50**, 107–141.

223. J. A. Nelson, G. Santos and B. H. Herbert, *Cancer Treat. Rep.*, 1984, **68**, 849–853.

224. C. K. Atterwill and C. E. Steele, *In vitro Methods in Toxicology*, Cambridge University Press, Cambridge, 1987.

225. D. P. Jones, G. B. Sundby, K. Ormstad and S. Orrenius, *Biochem. Pharmacol.*, 1979, **28**, 929–935.

226. J. S. Kim, H. Z. Hwang, S. H. Yeo, S. W. Ko, S. M. Song, Y. K. Kim, D. J. Kim, H. Y. Oh, H. Y. Choi, M. K. Kim, S. Toru, T. Kohsaka, Y. Kim and D. K. Jin, *Renal Failure*, 2001, **23**, 21–29.

227. L. C. Racusen, C. Monteil, A. Sgrignoli, M. Lucskay, S. Marouillat, J. G. Rhim and J. P. Morin, *J. Lab. Clin. Med.*, 1997, **129**, 318–329.

228. L. H. Lash, D. A. Putt and H. Cai, *Toxicology*, 2006, **228**, 200–218.

229. C. D. Brown, R. Sayer, A. S. Windass, I. S. Haslam, M. E. De Broe, P. C. D'Haese and A. Verhulst, *Toxicol. Appl. Pharmacol.*, 2008, **233**, 428–438.

230. A. Verhulst, R. Sayer, M. E. De Broe, P. C. D'Haese and C. D. Brown, *Mol. Pharmacol.*, 2008, **74**, 1084–1091.

231. T. Nakanishi, A. Fukushi, M. Sato, M. Yoshifuji, T. Gose, Y. Shirasaka, K. Ohe, M. Kobayashi, K. Kawai and I. Tamai, *Mol. Pharm.*, 2011, **8**, 2142–2150.

232. L. H. Lash, D. A. Putt, S. E. Hueni, R. J. Krause and A. A. Elfarra, *J. Pharmacol. Exp. Ther.*, 2003, **305**, 1163–1172.

233. L. H. Lash, D. A. Putt, S. E. Hueni and B. P. Horwitz, *Toxicol. Appl. Pharmacol.*, 2005, **206**, 157–168.

234. L. H. Lash, S. E. Hueni and D. A. Putt, *Toxicol. Appl. Pharmacol.*, 2001, **177**, 1–16.

235. B. S. Cummings, J. M. Lasker and L. H. Lash, *J. Pharmacol. Exp. Ther.*, 2000, **293**, 677–685.

236. B. S. Cummings and L. H. Lash, *Toxicol. Sci.: Off. J. Soc. Toxicol.*, 2000, **53**, 458–466.

237. S. M. Paul, D. S. Mytelka, C. T. Dunwiddie, C. C. Persinger, B. H. Munos, S. R. Lindborg and A. L. Schacht, *Nat. Rev. Drug Discovery*, 2010, **9**, 203–214.

238. J. H. Lin, *Drug Metab. Dispos.*, 1995, **23**, 1008–1021.

239. A. S. Bass, M. E. Cartwright, C. Mahon, R. Morrison, R. Snyder, P. McNamara, P. Bradley, Y.-Y. Zhou and J. Hunter, *J. Pharmacol. Toxicol. Methods*, 2009, **60**, 69–78.

240. H. Tahara, M. Shono, H. Kusuhara, H. Kinoshita, E. Fuse, A. Takadate, M. Otagiri and Y. Sugiyama, *Pharm. Res.*, 2005, **22**, 647–660.

241. S. W. Paine, K. Ménochet, R. Denton, D. F. McGinnity and R. J. Riley, *Drug Metab. Dispos.*, 2011, **39**, 1008–1013.

242. S. F. Billington, G. Chung and C. D. Brown, 2014, *Poster in AAPS Annual Meeting*, American Association of Pharmaceutical Scientists, San Diego, California, USA.

243. R. J. Roman and M. L. Kauker, *Circ. Res.*, 1976, **38**, 185–191.

244. H. Tahara, H. Kusuhara, H. Endou, H. Koepsell, T. Imaoka, E. Fuse and Y. Sugiyama, , *J. Pharmacol. Exp. Ther.*, 2005, **315**, 337–345.

245. G. Seki, M. Nakamura, M. Suzuki, N. Satoh and S. Horita, *World J. Nephrol.*, 2015, **4**, 307–312.

246. L. Suter-Dick, P. M. Alves, B. J. Blaauboer, K.-D. Bremm, C. Brito, S. Coecke, B. Flick, P. Fowler, J. Hescheler, M. Ingelman-Sundberg, P. Jennings, J. M. Kelm, I. Manou, P. Mistry, A. Moretto, A. Roth,

D. Stedman, B. van de Water and M. Beilmann, *Stem Cells Dev.*, 2015, **24**(11), 1284–1296.

247. K. Kandasamy, J. K. C. Chuah, R. Su, P. Huang, K. G. Eng, S. Xiong, Y. Li, C. S. Chia, L.-H. Loo and D. Zink, *Sci. Rep.*, 2015, **5**, 12337.

248. M. Kang and Y.-M. Han, *PLoS ONE*, 2014, **9**, e94888.

249. Y. Li, Z. Y. Oo, S. Y. Chang, P. Huang, K. G. Eng, J. L. Zeng, A. J. Kaestli, B. Gopalan, K. Kandasamy, F. Tasnim and D. Zink, *Toxicol. Res.*, 2013, **2**, 352–365.

250. Y. Duan, N. Gotoh, Q. Yan, Z. Du, A. M. Weinstein, T. Wang and S. Weinbaum, *Proc. Natl. Acad. Sci. U. S. A.*, 2008, **105**, 11418–11423.

251. Y. Duan, A. M. Weinstein, S. Weinbaum and T. Wang, *Proc. Natl. Acad. Sci. U. S. A.*, 2010, **107**, 21860–21865.

252. K.-J. Jang, A. P. Mehr, G. A. Hamilton, L. A. McPartlin, S. Chung, K.-Y. Suh and D. E. Ingber, *Integr. Biol.*, 2013, **5**(9), 1119–1129.

253. A. Bhushan, N. Senutovitch, S. S. Bale, W. J. McCarty, M. Hegde, R. Jindal, I. Golberg, O. Berk Usta, M. L. Yarmush, L. Vernetti, A. Gough, A. Bakan, T. Y. Shun, R. Biasio and D. Lansing Taylor, *Stem Cell Res. Ther.*, 2013, **4**, S16.

254. S. Kim and S. Takayama, *Kidney Res. Clin. Pract.*, 2015, **34**, 165–169.

255. I. Maschmeyer, A. K. Lorenz, K. Schimek, T. Hasenberg, A. P. Ramme, J. Hubner, M. Lindner, C. Drewell, S. Bauer, A. Thomas, N. S. Sambo, F. Sonntag, R. Lauster and U. Marx, *Lab Chip*, 2015, **15**, 2688–2699.

256. E. R. Shamir and A. J. Ewald, *Nat Rev Mol Cell Biol*, 2014, **15**, 647–664.

257. A. I. Astashkina, B. K. Mann, G. D. Prestwich and D. W. Grainger, *Biomaterials*, 2012, **33**, 4700–4711.

258. F. Pampaloni, E. G. Reynaud and E. H. K. Stelzer, *Nat. Rev. Mol. Cell Biol.*, 2007, **8**, 839–845.

259. S. Gerecht, J. A. Burdick, L. S. Ferreira, S. A. Townsend, R. Langer and G. Vunjak-Novakovic, *Proc. Natl. Acad. Sci.*, 2007, **104**, 11298–11303.

260. G. D. Prestwich, *Acc. Chem. Res.*, 2008, **41**, 139–148.

Drug Transporters at the Blood–Brain Barrier

DAVID DICKENS,* STEFFEN RADISCH AND
MUNIR PIRMOHAMED

Department of Molecular and Clinical Pharmacology, University of
Liverpool, Liverpool, UK
*Email: david.dickens@liv.ac.uk

5.1 The Blood–Brain Barrier

5.1.1 Overview

The central nervous system (CNS) is of fundamental importance for the control and regulation of physiological processes. It is an extremely sensitive microenvironment and strict homeostatic regulation is essential in order to maintain neuronal signalling. A key aspect in regulation is the physical separation of the CNS from the rest of the body by means of CNS barriers, which restrict the access and exit of molecules to and from the brain. This provides control over the concentration and composition of ions, neurotransmitters, macromolecules, neurotoxins and nutrients, as well as the entry/exit of xenobiotics.[1]

Three cellular barriers enclose the CNS:

- The blood–brain barrier (BBB), which separates the blood and the brain interstitial fluid;
- The choroid plexus (CP), which is between the blood and ventricular cerebrospinal fluid (CSF);

RSC Drug Discovery Series No. 54
Drug Transporters: Volume 1: Role and Importance in ADME and Drug Development
Edited by Glynis Nicholls and Kuresh Youdim

Published by the Royal Society of Chemistry, www.rsc.org

- The arachnoid epithelium (meningeal barrier), which separates the blood and the subarachnoid CSF.

These three cellular barriers regulate the exchange of compounds and nutrients at the interfaces between the blood and the brain or its fluid spaces.[2]

The BBB can be considered the most important of the three and has been studied extensively for its role in pharmacology and development, mainly because it has the biggest surface area for the passage of compounds from the blood into the brain, and it directly separates the blood from the extracellular fluid (ECF) of CNS neuronal tissue, and hence is in close proximity to neurons. Due to this, the primary focus of this chapter will be on the BBB with the CP briefly discussed.

The BBB is a cellular barrier that comprises specialised brain endothelial cells that form the walls of microcapillaries and can regulate the passage of endogenous substances and xenobiotics into and out of the brain, which maintains brain homeostasis and neuronal signalling (Figure 5.1). The barrier is not static and can be considered a biologically active interface with regulatory actions that involve transport, secretory and enzymatic roles.[3] This can thus present both a challenge and an opportunity for drug development. The present chapter outlines the role of drug transporters at the BBB and CP in drug pharmacology, with a particular focus on the distribution and penetration of drugs into the brain.

5.1.2 BBB in Numbers

The human brain is composed of approximately 86 billion neurons with similar numbers of glial cells that utilise between 15 and 20% of the entire energy of the body.[4] To provide this energy, the human BBB has a large surface area of between 15 and 25 m^2 for the passage of nutrients. In addition, the cell body of a neuron is between 10 and 20 μm from the nearest microcapillary, providing a network that infuses the entire brain.[1] This dense network results in a microvessel length of 600 km, with the diameters of microcapillaries being as small as 7–10 μm. The BBB vasculature is 3% of the total brain volume with the brain endothelial cells comprising 0.1% of the total brain by cell number.[5] The scale of the BBB results in 15–20% of the blood flow from the heart going to the brain. The size and importance of the BBB in maintaining homeostasis of the brain requires regulation with associated cells, so a cellular framework is formed that has been coined the 'neurovascular unit'.[6]

5.1.3 Neurovascular Unit

The BBB is not only composed of brain endothelial cells; other cell types also make up the neurovascular unit. The brain endothelial cells form the physical barrier of the microvessel with contributions from pericytes, astrocytes and neurons[2] (Figure 5.1).

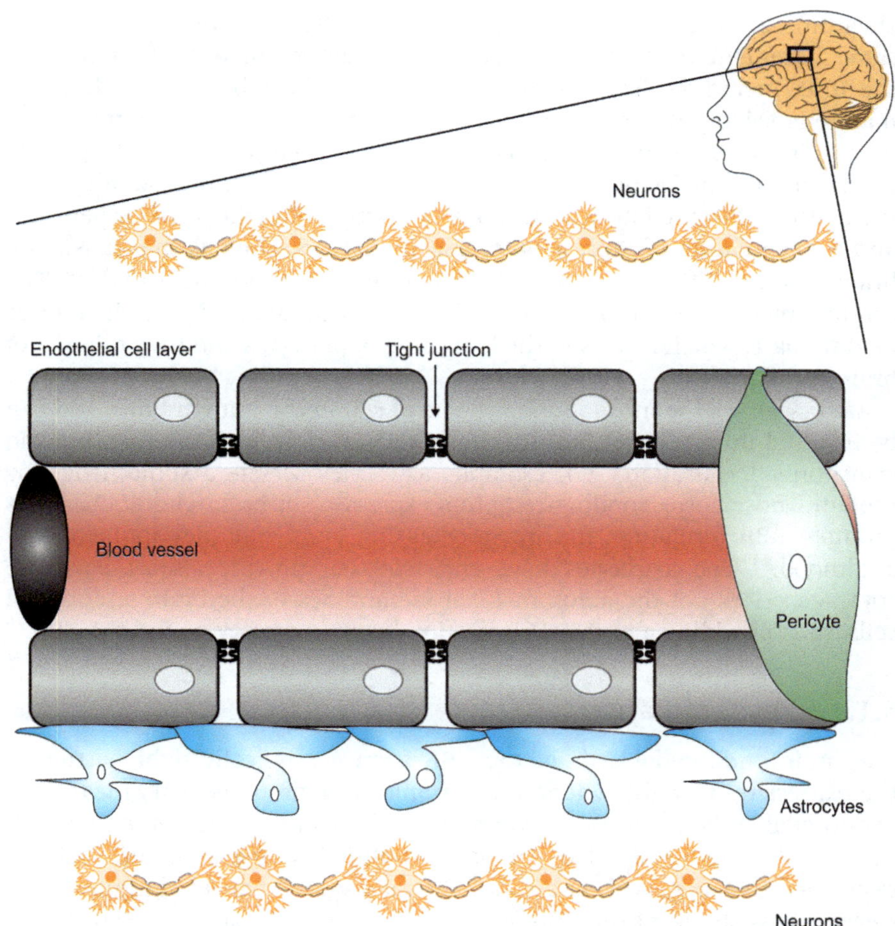

Figure 5.1 The BBB as a neurovascular unit. The BBB forms a cellular framework termed the neurovascular unit.[6] The brain endothelial cells comprise the cell type that forms the physical barrier of the microvessels within the BBB. Pericytes at the BBB are involved in regulating cerebral blood flow while astrocytes regulate BBB functions, including transporter localisation and tight junction proteins.

The brain endothelial cells are a highly specialised endothelial cell type due to their interactions with surrounding CNS cells and tissues during development and maturation. During development, early specialisation of cells into endothelial cells is due at least in part to the excretion of retinoic acid from glial cells.[7] Additionally, neuronal progenitor cells, *via* the WNT/β-catenin signalling pathway, have been proposed to drive the endothelial progenitor cells towards the specialised brain endothelial cell phenotype.[8] Surrounding cell types such as pericytes and astrocytes assist in maintaining this specialised phenotype into adulthood.

Pericytes are a type of perivascular cell that wraps around endothelial cells, with the highest density found in vessels of the neural tissues. As well as sheathing brain endothelial cells, pericytes of the CNS are attached to the same basement membrane.[9] They have a number of important roles at the BBB, including developmental, regulation of cerebral blood flow and maintenance of BBB permeability.[10–13] Brain pericytes contribute to the developing BBB by controlling tight junction formation and by decreasing the expression of factors that are involved in increasing permeability.[10] In the mature BBB, the physical coverage of brain endothelial cells by pericytes and the regulation of expression of specific factors, such as the transmembrane protein MFSD2, control BBB permeability and integrity.[10,13] The cerebral blood flow can be altered following neuronal activity *via* the release of glutamate, which results in the loosening of pericytes and thus dilation of brain capillaries.[12]

Astrocytes are the most abundant type of cell in the human brain and can be involved in regulating BBB functions such as tight junction proteins and transporter localisation.[2] For example, astrocytes secrete a sonic hedgehog protein whose corresponding receptors expressed on brain endothelial cells promote BBB formation and integrity during embryonic development and adulthood.[14] The astrocytes have end feet that provide almost complete coverage (>99%) of the abluminal cell membranes of the brain endothelial cells, thus providing a physical interaction between astrocytes and the BBB.[15]

5.1.4   Physical Barrier

The brain endothelial cells of the BBB form a physically tight barrier by the expression of tight junction and transporter proteins. This means the endothelial cells at the BBB have a specialised phenotype compared with peripheral endothelial cells. This is exemplified by measures of electrical resistance that are used as an indirect measure of barrier tightness. The electrical resistance of the BBB *in vivo* is thought to be at least 1500 Ω cm^2, with values up to 6000 Ω cm^2 recorded, and a recent *in vitro* model of human brain endothelial cells derived from induced pluripotent stem cells (iPSCs) has achieved values as high as 6000 Ω cm^2.[16,17] This is much higher than the resistance observed in peripheral endothelial cells; for example, in small bowel and muscle endothelial capillaries, the resistance values are 2 Ω cm^2 and 20 Ω cm^2, respectively.[18,19] This tightness of the BBB results in a low paracellular permeability.

The tight junctions between the brain endothelial cells are considered essential for both conferring the tightness of the BBB, thus reducing paracellular permeability, and in polarising the brain endothelial cells into a luminal and abluminal membrane.[2] The tight junctions are composed of a complex of proteins (including zonula occludens (ZO)-1, occludin, claudin 3, claudin 5 and claudin 12[1]) that span the intercellular cleft and thereby constitute a physical barrier. They are part of the junctional complex generally involved in cell adhesion. The junctional complex also includes adherens junctions, which are comprised of important structural proteins such as the cadherin–catenin complex. These attach endothelial or epithelial cells and are essential for the formation of tight junctions.[1,20] Adherens junctions are located underneath tight junctions at the most apical part of the junctional complex between the apical and basolateral cell membrane.[21] Tight junctions also contain junctional adhesion molecules (JAMs), which are proposed to be involved in leukocyte cell adhesion.[1,22] In addition, scaffold proteins such as ZO-1, ZO-2 and ZO-3 cluster the spanning proteins and connect them to the actin/myosin cytoskeleton.[23] Tight junctions can also restrict membrane trafficking within one cell, thereby separating the apical from the basolateral membrane domain.[21] Figure 5.2 illustrates the basic composition and arrangement of tight junctions and adherens junctions.

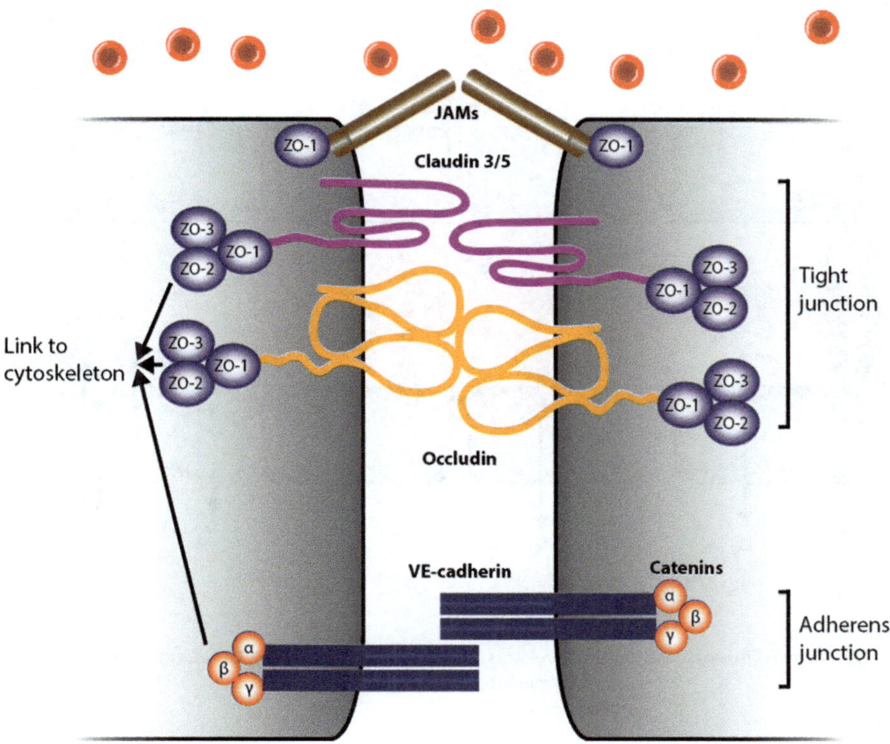

Figure 5.2 Model of BBB tight junctions and adherens junctions. A simplified model illustrating tight junction and adherens junction composition and arrangement. JAMs, claudin 3/5 and occludin are important tight junction proteins with the claudins and occludins spanning and sealing the intercellular cleft between brain endothelial cells. ZOs comprise scaffolding protein clusters and connect tight junction proteins to the actin/myosin cytoskeleton. Vascular endothelial (VE)-cadherin proteins are adhesive junction proteins important for structural integrity and tight junction formation. They are linked to the cytoskeleton by means of the catenin scaffolding proteins.

5.1.5 Transport at the BBB

Compounds can enter the brain by a number of different mechanisms: the paracellular aqueous pathway, transcellular lipophilic pathway, transporter-mediated, receptor-mediated transcytosis and adsorptive transcytosis (Figure 5.3). The focus of this chapter is on the role of transporters in

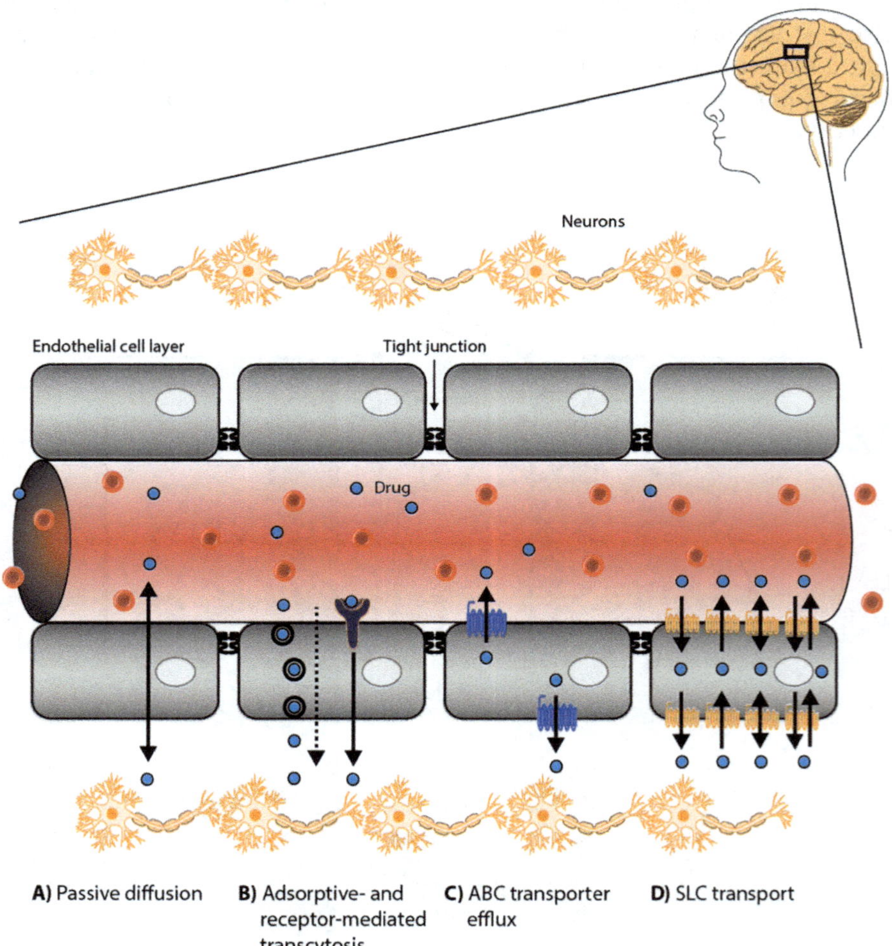

Figure 5.3   Model of BBB transport. Simplified model illustrating important drug transport mechanisms across brain endothelial cell layers. (A) Passive diffusion: in the case of transcellular diffusion, the accepted rule is that the higher the lipophilicity of a compound, the greater the passive diffusion through the BBB. (B) Adsorptive and receptor mediated transcytosis: proteins such as insulin and albumin can be transported through the BBB by specific receptor-mediated endocytosis and transcytosis pathways. (C) ABC transporter efflux: carrier-mediated efflux is a major obstacle for many pharmacological agents as it is a mechanism involved in expelling drugs from the brain. (D) SLC transport: secondary-active and facilitative transporters of the SLC transporter family can be involved in the carrier-mediated uptake and removal of compounds into and out of the brain.

affecting the transcellular pathway by efflux or influx mechanisms, and how this affects drug delivery and/or prevents exposure to the brain. The importance of the transcellular lipophilic pathway in the passage of drugs through cells has been hotly debated in a number of recent reviews and will not be discussed further in this chapter. Interested readers are referred to various in-depth reviews concerning this debate on the relative significance of transporter-mediated and passive diffusion of compounds into cells.[24,25] However, it should be noted that lipophilicity is not always a good guide to CNS penetration, as even highly lipophilic drugs can be prevented from entry by efflux transporter activity.[26] Increased lipophilicity can also lead to disadvantages such as increased non-specific binding.[27,28]

The brain endothelial cells are polarised and express a range of efflux and influx transporters at the apical and basolateral membranes that control the movement of substances through the cells. They are known to have more mitochondria than peripheral endothelial cells (such as skeletal endothelial cells), probably due, at least in part, to the active nature of the barrier and an increased energy requirement for active transport processes.[29,30] Therefore, through the action of transporters, and by providing a physical barrier, the BBB aids in keeping out toxins from the brain while allowing in essential nutrients to maintain brain homeostasis. From a pharmacological perspective, it has been reported that it prevents >98% of small compound drugs and nearly all large-molecule therapeutics (biologicals) from having a pharmacological effect on the brain.[28]

5.2   Modelling of the BBB

Modelling of the BBB by *in vitro* and *in vivo* methods is an important endeavour for predicting transport of drugs at the human BBB, as each cellular barrier has its own novel characteristics and transportisome. There have been recent advances in the study of drug transport in BBB models, and this section will outline the use and relevance of these different *in vivo*, *in vitro* and *in silico* methods. All of the models outlined below have both advantages and limitations, with systems biology/physiologically-based modelling approaches that interpolate transport data from multiple laboratory-based models being particularly useful. For example, the prediction of *in vivo* penetration of CNS drugs can be improved by integrating permeability, P-glycoprotein (P-gp) efflux and drug free fractions in the blood and brain into a mathematical model.[27,31] Alternatively, new molecular entities can be modelled at an early stage using quantitative structure–activity relationships (QSAR) for both biological activity and BBB permeability, taking into account interactions with both the target receptor and BBB transporters[32,33] (covered in more detail in Chapter 7).

5.2.1   Cellular Models of the BBB

Cellular *in vitro* models of the BBB have improved in recent years and can provide a useful tool for investigating drug transport. A number of brain immortalized endothelial cell lines have been developed and include human (hCMEC/D3), rat (RBE4 cells) and mouse (endo3) cell lines. The hCMEC/D3 cell line is a well-characterised human brain endothelial cell line[34,35] that has been used for *in vitro* drug uptake assays by multiple groups to study efflux and influx transporters.[36–40] However, like all cell lines generated to date, these brain endothelial cells have lost some of their unique protein expression pattern outside of their native environment and thus display a more generic endothelial cell phenotype.[41] This leads to a loss of barrier tightness and consequently transcellular permeability studies of the compound of interest are not possible, which remains a major drawback of this cell line. However, cell lines that do provide the possibility to study transcellular permeability include cells such as Caco-2 and MDCK transfected with P-gp, although they are of epithelial origin and so their relevance to the BBB is highly debatable.[42]

Primary brain endothelial cells provide a useful compromise between the cell line approach and *in vivo* experiments.[43] However, they are low to medium throughput due to the fact that only 0.1% of brain cells are endothelial cells, and thus yields are low and extended passaging results in a loss of phenotype. Porcine, bovine and rodent models are favoured sources of primary brain endothelial cells and, in co-culture with astrocytes, high transendothelial electrical resistance (TEER) is achievable—up to 1300 Ω cm^2. This high electrical resistance means a tight barrier is formed and thus the models can be useful as a tool for drug permeability studies investigating transcellular permeability through the BBB.[44]

Recent developments include deriving brain endothelial cells from stem cell sources.[8,17,45,46] The most developed model utilises either iPSCs or embryonic stem cells (ES) to derive cells with a brain endothelial phenotype with a high TEER (around 5000 Ω cm^2), and as such the cells are suitable for transcellular permeability studies.[8,17] This is an exciting area for further research as it could in the longer term lead to an *in vitro* BBB model derived from patient cohorts of responders and non-responders to a particular drug treatment, to investigate whether transport is a variable for drug response.

5.2.2 *In vivo* Models

A variety of *in vivo* models exist to study the transport of drugs at the BBB. These range from non-mammalian organisms such as zebrafish through to rodents and humans. The non-mammalian studies are at an early stage, but proof-of-principle drug transport studies have been performed. In fruit flies (*Drosophila melanogaster*), a P-gp homologue (*Mdr65*) has conserved function at its CNS barrier that confers protection against cytotoxic pharmaceuticals,[47] and evidence from zebrafish suggests that drug transport could also be investigated in this model.[48,49]

Rodents are the most common *in vivo* animal model used to investigate drug transport at the BBB, in whole animal studies or with techniques such as microdialysis or *in situ* perfusion. *In situ* perfusion entails controlling the circulation to the brain by directly infusing saline and drug into the major vessels leading to the brain. The amount of drug in the brain is determined at set time points and, from these measurements, the kinetics and permeability constants of brain uptake are determined.[50] Microdialysis determines the free drug concentration in the brain by the insertion of a probe with a semi-permeable membrane into the rodent brain and is one of the most sensitive methods available for studying BBB transport of drugs *in vivo*.[43] The amount of free drug is the concentration of drug that is unbound while the total is the combined protein-bound and unbound drug concentration. However, the insertion of the probe can lead to damage of the surrounding tissue, including the BBB.[51] In whole animal experiments using rodents, the total brain concentration of a drug can be determined following administration. The use of knockout mice to determine the effect of a specific transporter in the uptake of a compound into the brain has become an important technique for the study of BBB transport. Due to the recent advances in generating and culturing rat ES lines with homologous recombination,[52] knockout rats such as $Mdr1a^{-/-}$ and $Abcg2^{-/-}$ are now being used for drug transport studies, for example in determining the brain to plasma ratios of novel small compound drugs.[53] A hypothesised disadvantage of knockout animals is that compensatory processes could induce other drug transporters to be upregulated following knockout of a particular transporter. However, in a study by Agarwal *et al.* in $Mdr1a/b^{-/-}$ or $Abcg2^{-/-}$ rodent knockout animals, no difference was observed in the expression of the measured proteins (29 protein molecules, including 12 ABCs, 10 SLCs, 5 receptors and 2 housekeeping proteins) in the brain capillary endothelial cells of the single and double knockout animals.[54] Nonetheless, it should also be noted that species differences in transporter expression need to be taken into account when extrapolating data to humans.[55]

Positron emission tomography (PET) is a non-invasive technique that enables the regional brain measurement of radioactivity for a radiolabelled compound to be obtained in both animals and humans[56] (refer to *Drug Transporters: Volume 2 Recent Advances and Emerging Technologies*, Chapter 6 for further details). For example, a study in mice showed that elacridar increased the brain uptake of radiolabelled gefitinib (a tyrosine kinase inhibitor) by around 12-fold compared with control mice receiving gefitinib only.[57] Another avenue is the use of a fluorescent probe substrate and bioluminescence imaging, for example D-luciferin, the endogenous substrate of firefly luciferase, has recently been shown to be a breast cancer resistance protein (BCRP) substrate.[58] This approach requires a transgenic mouse but is the first study to generate and validate a specific BCRP probe for imaging studies.

Sampling from the CSF in humans can be used as a surrogate for the brain ECF concentration[59] and this approach has been used to obtain the unbound drug concentrations of some P-gp substrates in rodents.[60] However, due to the differences in transporter expression and activity at the BBB compared with the CP, the CSF does not always correlate with the brain ECF drug concentrations.[61,62] One reason for this may be that drug transport into the CSF takes place mainly at the CP, with only a small part from the flow of the brain fluid emptying into the CSF[1] (see Section 5.5).

5.3 Efflux Transporters Expressed at the BBB

The family of ATP binding cassette (ABC) transporter genes currently comprises 48 genes (excluding pseudogenes). The official list is available on the HGNC website (http://www.genenames.org/cgi-bin/genefamilies/set/417). The ABC superfamily is further divided into seven subfamilies with designated letters A–G and each gene assigned a unique number following the family root and the subfamily letter, *e.g. ABCB1*. All ABC transporters, corresponding to their name, are characterised by the presence of two ATP binding cassettes, also referred to as the nucleotide binding domains (NBDs). NBDs bind to ATP and the energy derived from ATP hydrolysis is utilised as the driving force for active substrate translocation against an electrochemical gradient. NBDs typically consist of two highly conserved nucleotide binding motifs, referred to as Walker A and Walker B, linked by another highly conserved motif, the ABC signature or C motif.[63] In addition to the NBDs, ABC transporters further comprise two transmembrane domains (TMDs) with varying numbers of transmembrane helices (TMHs). While a functional transporter consists of the core structure of two TMDs and two NBDs, the corresponding gene may only encode a half-transporter, with one TMD and one NBD, and subsequent homo- or hetero-dimerisation at the protein level.[63] Functionally, ABC transporters are involved in the unidirectional, active extrusion of xenobiotics and endogenous substances, such as metabolic products and lipids, and are thus important in cell detoxification systems.[64] A number of recent reviews have suggested that too much focus has been placed on BBB permeability and not enough on understanding BBB transport, however this view is debated.[59,65] Along with this clearance function, many ABC transporters are recognised as mediators of a multidrug resistance phenotype, particularly for cytotoxic anticancer drugs such as vinblastine and doxorubicin.[66] The ABC efflux transporters known to be expressed at the BBB are outlined in Figure 5.4 and discussed further below.

5.3.1 P-gp

P-gp (*ABCB1*, MDR1) is the prototypical efflux transporter. It is an ATP-dependent multidrug efflux transporter that is highly expressed at the BBB and can affect the permeation of its substrate drugs into the brain.[67] P-gp is localised to the luminal side of the brain endothelial cell and as such pumps drugs out of the cells and into the blood. This reduces penetration into the CNS and has led to the recognition of its role as a "gatekeeper". This is because P-gp can act as a protective mechanism against potentially toxic xenobiotics.[26] This gatekeeper role is enhanced by its broad substrate base,

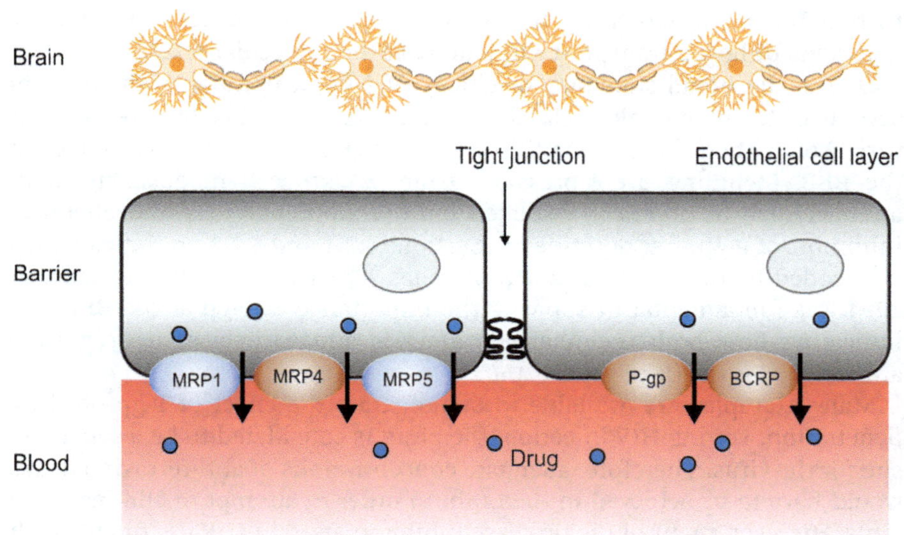

Figure 5.4 Subcellular localisation of efflux transporters at the BBB. Human ABC transporters at the BBB that have been introduced in this chapter with confirmed subcellular localisation are P-gp (*ABCB1*), BCRP (*ABCG2*) and MRP4 (*ABCC4*) (in red). In blue is the proposed localisation of efflux transporters that could be putatively expressed at the BBB, MRP1 (*ABCC1*) and MRP5 (*ABCC5*).

and readers are referred to Schinkel and Fromm for extensive reviews on this topic.[26,68]

The topology and structure of P-gp have been well described, which has led to a functional understanding of its poly-specific drug binding. Utilising X-ray crystallography techniques, the structure of P-gp from a variety of non-human organisms has been solved.[69–72] As ABCB transporters have the same basic core structure of two TMDs and two NBDs, the mechanistic insights gained from these crystal structures are of high value for the whole *ABCB* transporter family. Mouse MDR1A was crystallised in the inward-facing conformation and the two TMDs displayed a pseudo two-fold symmetry to each other, forming a large portal with many hydrophobic amino acid residues localised in the membrane at the assumed substrate binding pocket.[69] Consistent with the predicted P-gp topology, each domain displays a bundle of six TMHs.[69] Mechanistically, the inward-facing structure does not allow substrates to access the binding pocket from the outer membrane layer or the extracellular space. Instead, it has been suggested that substrates enter the binding pocket from within the inner membrane layer, which further stimulates ATP binding to the NBDs and is followed by a conformational change of P-gp to the outward-facing state.[69] ATP hydrolysis has been suggested to be a likely mechanism to disrupt the dimerisation of both NBDs, and to allow the transporter to flip back into the inward-facing conformation.[69,73] This provides an example of how translation of structure to function using structural biology techniques can address mechanistic questions of how a transporter functions at the molecular level.

Due to the broad substrate specificity of P-gp, a number of drugs from several different chemical classes are substrates of this transporter.[68] A variety of methods have been utilised to investigate the importance of P-gp at the BBB, including overexpression using *in vitro* cell models, knockout animals such as $Mdr1a^{-/-}$, isolated brain endothelial cells and chemical inhibitors of P-gp,[68] as described in Sections 5.2.1 and 5.2.2. In rodents, P-gp is encoded by the *Mdr1a* and *Mdr1b* genes. The work by Schinkel *et al.* in 1994 was fundamental in showing the importance of P-gp at the BBB,[26,67] indicating how both the pharmacokinetics and toxicity of a drug were affected by knockout of mouse *Mdr1a*.

Many examples are available to demonstrate the effect of P-gp on drug penetration. During HIV infection, the brain is considered to be a sanctuary site for the virus. Therefore, adequate concentrations of anti-retroviral drugs would have to be achieved in the brain in order to attempt to eliminate the virus. However, many of the protease inhibitors used in HIV treatment, such as saquinavir, indinavir and nelfinavir, are P-gp substrates, which restricts their access to the brain. This contributes to viral persistence and reduced effectiveness. The importance of P-gp for these compounds was shown using the $Mdr1a^{-/-}$ knockout mouse in which, following intravenous injection, the brain concentration of drug was elevated from 7- to 36-fold relative to control mice.[74] To overcome the action of P-gp, a number of approaches have been tried and include the use of inhibitors of P-gp and pro-drug approaches.[75–77]

Imatinib is a tyrosine kinase inhibitor used in the treatment of multiple cancers, including chronic myelogenous leukaemia (CML). Using a number of approaches, including the use of multidrug resistant cell lines, transcellular permeability in P-gp overexpressing cells and $Mdr1a/1b^{-/-}$ knockout mice, imatinib has been shown to be a P-gp substrate.[78–81] In one study, an 11-fold greater uptake of imatinib into the brain was observed in the $Mdr1a/1b^{-/-}$ knockout mouse compared with control mice, providing an explanation for one of the limitations of using imatinib in CML patients, who can show a complete haematological response to the drug but still retain a sanctuary site in the CNS.[78–81]

In contrast, the restriction of BBB drug entry by P-gp can sometimes be of benefit, since it may result in a reduced CNS side effect profile and for some drugs can result in a repositioning of the drug to another clinically therapeutic area. An example of this is domperidone, a dopamine antagonist with limited CNS effects (due to its P-gp mediated transport at the BBB[67]) that can be used in patients with Parkinson's disease in order to minimise adverse effects on the extrapyramidal system, *i.e.* involuntary movements. Similarly, the P-gp substrate loperamide, an opioid, is excluded from the CNS, and as such its peripheral opioid-like effects can be used clinically as an antidiarrhoeal.[26]

Colchicine is a potentially neurotoxic alkaloid used as an anti-gout medication that is excluded from the brain on oral dosing due, at least in part, to its efflux at the BBB by P-gp. Studies measuring colchicine uptake into the brain of rodents in the absence or presence of a P-gp inhibitor (PSC-833), using microdialysis and *in situ* perfusion methodologies, indicate enhanced brain uptake in the presence of PSC-833.[82–84] Notably, when directly injected into the brain of rodents to bypass the effect of P-gp, colchicine induces neurotoxicity, leading to sporadic dementia in the animals.[84]

5.3.2 BCRP

BCRP (*ABCG2*) is a drug efflux transporter that is part of the *ABCG* subfamily. BCRP was discovered and cloned from a multidrug resistance breast cancer cell line and found to extrude a number of chemotherapeutic drugs.[85] Together with P-gp, BCRP is recognised as a mediator for multidrug resistance in cancer, with a complementary and overlapping substrate profile to P-gp.[85] Unlike P-gp, the BCRP gene encodes only one TMD and one NBD with oligomerisation of the protein being critical to produce a functional transporter.[86] As well as being expressed in breast cancer, BCRP has been found to have a functional role in a number of tissues, including the BBB.[87] It displays apical membrane localisation, with immunofluorescence studies and proteomic analysis confirming its brain endothelial cell expression at the luminal membrane.[88–90]

BCRP substrates include both endogenous and xenobiotic compounds. Endogenous BCRP substrates include urate and 17β-estradiol-17-β-D-glucuronide (E217βG), with xenobiotic compounds including the chemotherapeutic drug mitoxantrone, the H2 blocker cimetidine and the HMG-CoA reductase inhibitor pitavastatin.[87] One of the first studies showing the importance of BCRP at the BBB investigated the transport of mitoxantrone in $Mdr1a^{-/-}$ knockout mice in the presence and absence of a dual BCRP and P-gp inhibitor (elacridar).[91] Real-time quantitative reverse transcription polymerase chain reaction (PCR) was used to show a high expression of BCRP in brain endothelial cells compared with the cortex in wild type mice. However, a number of negative studies at the BBB for known BCRP substrates exist,[92] a finding that may be explained at least in part by the overlapping substrate specificity of BCRP and P-gp. BCRP is thought to act in conjunction with P-gp in preventing the entry of some xenobiotics into the brain[93] and, for specific co-substrates, both P-gp and BCRP may need to be inhibited or knocked out for a functional inhibitory effect to be observed. The interplay between efflux transporters at the BBB is discussed in more detail in Section 5.3.5.

For BCRP, a non-synonymous genetic polymorphism (Q141K) at the NBD has in multiple studies been linked to high serum urate levels and thus to gout[94–96] due to reduced urate transport in cells.[94] The variant BCRP protein has been linked to reduced stability at the NBD and thus reduced protein expression.[95] The relevance of this for BCRP-mediated transport at the BBB is unknown but it should be noted that in patients with diffuse large B-cell lymphoma, the Q141K polymorphism was associated with chemotherapy-induced diarrhoea following R-CHOP treatment.[97] R-CHOP comprises rituximab plus cyclophosphamide/doxorubicin/vincristine/prednisone, with cyclophosphamide and doxorubicin, a BCRP substrate drug, highly associated with diarrhoea.[97]

5.3.3 MRP4

MRP4 (*ABCC4*) is part of the multidrug resistance associated protein (MRP; *ABCC*) subfamily and is highly expressed at the BBB.[98] Twelve genes are assigned to the *ABCC* subfamily of membrane transporters designated as *ABCC1–ABCC12* with the encoded proteins being further divided into three classes, namely the MRPs, the sulfonylurea receptors (SURs) and the cystic fibrosis transmembrane conductance regulator (CFTR). The MRP transporters are ATP-dependent transporters with two TMDs and NBDs, and the transport of substrates may be either glutathione (GSH)-independent or GSH-dependent (GSH may either be co-transported with the substrate or it may act as a stimulant[99]). MRP4 was first identified as a protein that conferred resistance to nucleoside-based antiviral drugs in a T-lymphoid cell line.[100] It is known to display overlapping but also quite distinct transport characteristics within the *ABCC* subfamily. A number of substrates of both endogenous and xenobiotic origin have now been identified for MRP4.

Prostaglandins are a unique endogenous substrate of MRP4.[101] They are effluxed by MRP4 and an *in vitro* study determined that nonsteroidal anti-inflammatory drugs (NSAIDs) were inhibitors of this transport process.[101] This could mean that NSAIDs have a dual anti-inflammatory activity in terms of their ability to inhibit both the synthesis and release of prostaglandins from cells. Cyclic nucleotides, signalling molecules that control cell migration, are also endogenous substrates of MRP4. Thus, MRP4 may play a role in the regulation of intracellular cyclic nucleotide levels and subsequently affect the migration of fibroblasts, which could have significance in wound repair.[102] Xenobiotic substrates for MRP4 are numerous and include the antibiotics ceftizoxime and cefazolin, and the immunosuppressant methotrexate.[98]

The expression of MRP4 at the BBB has been shown to be localised to the luminal membrane by immunofluorescence.[103–105] Quantitative mass spectrometry studies have also detected MRP4 protein in human brain endothelial cells.[89] In agreement with these results, *ABCC4* mRNA was also detectable in two other studies utilising isolated human brain microvessels.[106,107] The $Mrp4^{-/-}$ knockout mouse was first utilised in 2004 to show the enhanced accumulation of the anticancer agent topotecan in brain tissue and CSF.[104] It was proposed that the presence of MRP4 in the BBB can confer resistance to topotecan and protect the brain from this chemotherapeutic drug.[104] In a mouse brain endothelial cell line (bEnd.30), MRP4 transport was assessed by utilising siRNA targeting, and inhibition of MRP4 was shown to increase the uptake of its substrate azidothymidine (an antiretroviral drug) into the brain, suggesting the functional importance of MRP4 in this *in vitro* model of the BBB.[108] Interplay between MRP4 and other efflux transporters at the BBB can occur and is discussed in Section 5.3.5.

5.3.4 Putatively Expressed BBB Efflux Transporters

In addition to MRP4, other MRP transporters have been linked to the BBB. One of these is MRP1 (*ABCC1*). MRP1 contains three TMDs and two NBDs. This five domain structure and the cytoplasmic N-terminus of MRP1 is atypical of ABC proteins, which generally have two TMDs and two NBDs.[99] Proteomic analysis of isolated human brain microvessels failed to detect the protein above the limit of quantification, but an immunolocalisation study reported weak BBB expression in the abluminal (basolateral) membrane.[89,105] However, another study detected MRP1 at the luminal membrane of brain endothelial cells.[103] In addition, other studies also confirmed mRNA expression in isolated human brain microvessels.[106,107] However, MRP1 does not appear to have functional activity at the BBB when tested *in vivo*.[109] These negative experiments included work in $Mdr1a^{-/-}$ knockout mice treated with chemical inhibitors of MRP1 (probenecid or MK571) with etoposide as a substrate, or alternatively etoposide, vincristine and doxorubicin as substrates in $Mrp1^{-/-}$ knockout mice.[109]

The MRP5 (*ABCC5*) transporter has also been linked to the BBB. MRP5, like MRP4, is a short MRP containing two TMDs and NBDs. MRP5 is expressed in many human tissues with mRNA expression in the brain and much lower mRNA levels detectable in the liver.[110] No protein expression was found above the detection limit by quantitative mass spectrometry in isolated human brain microvessels,[89] but two independent immunolocalisation studies revealed MRP5 protein expression on the luminal (apical) side of brain endothelial cells.[103,105] In addition, MRP5 mRNA has been detected in isolated human brain microvessels.[106,107] The substrate profile appears to have a distinct overlap with MRP4, particularly with regards to nucleotides.[110] However, its functional role in the BBB for transport of xenobiotics has yet to be fully determined.

5.3.5 Interplay Between Efflux Transporters

The efflux transporters expressed at the BBB can work in tandem to exclude a drug from the brain[93] because of their overlapping substrate specificity. In practice, this means that if one transporter is knocked out or inhibited, then the other efflux transporter(s) may compensate by excluding the compound from the brain. Numerous examples of drugs that are substrates for dual efflux transporters at the BBB now exist, including a number of tyrosine kinase inhibiters used in cancer treatment.

At the mouse BBB, imatinib entry into the brain has been found to be restricted not only by P-gp (see Section 5.3.1) but also by BCRP.[78,111] Using knockout mice, the brain concentration of imatinib in $Abcg2^{-/-}$ mice was similar to that in wild type mice, whereas the concentration in $Mdr1a/1b^{-/-}$ and $Mdr1a/1b^{-/-}$ $Bcrp^{-/-}$ mice was increased more than 3- to 4-fold and 19- to 50-fold, respectively, compared with wild type mice. This clearly shows the synergistic activity of these transporters at the BBB for imatinib transport.

Another example is the drug sunitinib, which has been approved by the US Food and Drug Administration (FDA) for the treatment of renal cell carcinoma and imatinib resistant gastrointestinal stromal tumour. The brain accumulation of sunitinib is restricted by P-gp and BCRP, and can be enhanced by oral elacridar co-administration in knockout mice studies.[112]

The oral availability and brain penetration of the B-RAF mutant inhibitor (V600E) vemurafenib are affected by efflux transporters at the BBB. They can be enhanced by inhibition of both P-gp and BCRP by elacridar treatment, resulting in increased brain accumulation in mice.[113] While studies in knockout mice found that the brain to plasma ratios of vemurafenib were only increased 1.7-fold in $Mdr1a/1b^{-/-}$ mice, and not increased in $Abcg2^{-/-}$ mice, ratios were increased by 21-fold in triple knockout $Mdr1a/1b^{-/-}$ $Abcg2^{-/-}$ mice, clearly showing the importance of both of these transporters in vemurafenib brain penetration. Similarly, URB937, a fatty acid amide hydrolase inhibitor (FAAH), is a dual substrate for P-gp and BCRP, and these efflux transporters restrict its access to the brain.[114] Thus, URB937 is restricted to the periphery, revealing an unexpected role for fatty acid amide hydrolase in pain initiation control outside of the CNS.

This phenomenon of interplay between efflux transporters at the BBB is not just limited to BCRP and P-gp. An example of this is methotrexate, with an effect on brain penetration only being observed when both BCRP and MRP4 are knocked out in rodents.[115] It is also possible for a drug to be a substrate of all three of the main efflux transporters expressed at the BBB, *i.e.* P-gp, MRP4 and BCRP. An example of this is the antitumour camptothecin analogues. Investigations using knockout mice found that these drugs are restricted by MRP4 together with P-gp and BCRP, thus forming a robust cooperative drug efflux system that restricts their entry into the brain.[116] This cooperative drug efflux system is also exemplified by lapatinib, a dual tyrosine kinase inhibitor. Whilst the double knockout $Mdr1a/b^{-/-}$ mice only showed a 3- to 4-fold increase in brain concentration and $Abcg2^{-/-}$ mice had no increase when compared with wild type mice, the $Mdr1a/b^{-/-}$ $Abcg2^{-/-}$ triple knockout animal was found to have a 40-fold increase in brain concentration compared with the wild type mice.[117] This type of work has led to the suggestion that inhibitors of efflux transporters could be used to boost the brain penetration of drugs through inhibition of efflux transporters at the BBB; however, even if suitable inhibitors could be found, a drawback of this approach is that this could lead to toxic side effects in other organs expressing the transporter(s) of interest (see Section 5.6).

5.4 Influx Transporters Expressed at the BBB

Secondary-active and facilitative transporters belong to the superfamily of solute carrier (SLC) transporters. SLCs represent the largest known superfamily of membrane transporters and the official list currently consists of over 300 *SLC* transporter encoding genes within 52 families (*SLC1–SLC52*) and is available from the Human Genome Organisation Gene Nomenclature Committee at the European Bioinformatics Institute (HGNC) website (http://slc.bioparadigms.org/). A corresponding review article for each family has recently been published in *Molecular Aspects of Medicine*[118] and some selected families that have relevance to the BBB are highlighted in more detail in this chapter.

SLC genes are allocated to a subfamily if the coded proteins share at least 20% of the amino acid sequence, and can be divided into further subfamilies.[118] The official nomenclature for each gene starts with *SLC* to indicate the corresponding gene superfamily and is followed by the family number and subfamily letter. Finally, each gene is allocated a unique number. The SLC superfamily comprises a diverse set of transporters involved in the absorption and excretion of a broad spectrum of physiologically important endogenous molecules/ions, but also xenobiotics such as drugs. For example, members of the *SLC2A* family are essential for an adequate sugar supply to the brain and other organs, while the *SLC39A* family is essential for metal ion homeostasis, particularly zinc.[119,120] However, uncertainty exists around which influx transporters are important at the BBB. This is therefore a developing area of research, with a recent and major review on *SLC* transporters calling for systematic research on this important but relatively uncharacterised family of proteins.[121] The current knowledge of the likely expression of certain transporters is summarised in Figure 5.5.

5.4.1 LAT1

LAT1 (*SLC7A5*) was cloned at the end of the 1990s and found to be highly expressed in brain endothelial cells.[106,122,123] It is an amino acid antiporter that pumps a substrate in and then effluxes an internal substrate out of the cell with a 1 : 1 stoichiometry.[124] The amino acids that LAT1 transports are neutral amino acids such as phenylalanine and leucine.[125] LAT1 forms a heterodimer with CD98 (*SLC3A2*, 4F2 heavy chain), which is proposed to aid in the localisation of LAT1 to the plasma membrane.

As well as being a nutrient transporter, a number of drugs have been shown to be substrates for LAT1. These include L-DOPA and gabapentin.[37,126] The knockout mouse for LAT1 is embryonic lethal, which has restricted research on this transporter. However, the recent use of a conditional knockout in T-cells has, at least for T-cell differentiation, shown an essential requirement for LAT1.[127] At the BBB, LAT1 is thought to be localised to both the luminal and abluminal membranes of brain endothelial cells, as demonstrated by immunofluorescence studies.[128,129] The use

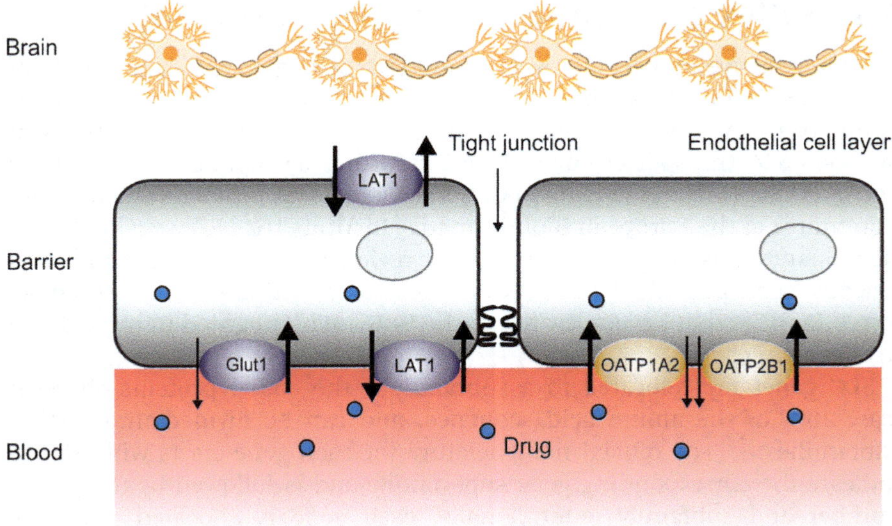

Figure 5.5   Subcellular expression of influx transporters at the BBB. SLC transporters at the BBB that have been introduced in this chapter with confirmed expression are LAT1 (*SLC7A5*) and GLUT1 (*SLC2A1*) (in purple). The putative expression of OATP1A2 (*SLCO1A2*) and OATP2B1 (*SLCO2B1*) is shown in yellow with their proposed subcellular expression; other SLC transporters have also been proposed (see text).

of either competitive inhibitors or siRNA targeted to LAT1 has shown its functional importance in brain endothelial cells of mouse and human origin.[37,130] Due to LAT1 enrichment and functional expression at the BBB, LAT1 is considered a promising target for drug delivery to the brain (see Section 5.7).

5.4.2   Organic Anion Transporting Polypeptide Transporters

The organic anion transporting polypeptide transporters (OATPs) are a class of influx transporters that are part of the SLC superfamily and, more specifically, encoded by the solute carrier organic anion (*SLCO*) gene subfamily. The human OATP family consists of 11 members. Controversy surrounds which, if any of the OATPs are functionally important at the BBB.[131] However, this class of transporter has been shown to have an important role in the influx of organic anions at other tissue types such as the kidney and liver.[131] OATP1A2 (OATP-A, *SLCO1A2*) was the first OATP to be linked to expression in human brain endothelial cells.[132] Since this finding, OATP2B1 (*SLCO2B1*) has also been suggested to be expressed in human brain endothelial cells in paraffin embedded sections.[133] However, in quantitative mass spectrometry studies no OATP protein was detected above the limits of detection in brain endothelial cells.[89] Rodent OATP1A4 (*Slco1a4*) has been suggested to function at the BBB, with its transport function being upregulated in hypoxia, leading to more drug being delivered to the brain.[134,135] The human counterpart to rodent OATP1A4 is unclear, so the relevance of this finding for the human BBB is unknown; however it does share 72% sequence homology at the amino acid level to human OATP1A2.[134] Further research is required to determine which, if any, OATP is functionally important at the human BBB.

5.4.3 Monocarboxylate Transporters

Monocarboxylic acid transporters (MCTs; *SLC16A*) comprise a family of at least eight proton-coupled transporters.[136] MCT1 can transport endogenous substrates such as lactate and drug substrates such as salicylate, atorvastatin, nateglinide, γ-hydroxybutyrate and nicotinic acid.[136] A number of *in vivo* experiments in rodents and *in vitro* studies with brain endothelial cells suggest that the BBB has a monocarboxylate transport system.[136] MCT1 protein expression has been detected in the brain endothelial cells of the BBB with the assumption, due to a lack of further evidence, that this is the main MCT expressed at the BBB.[137] The significance of MCT-mediated transport of drug substrates at the BBB is at present unknown, but a recent study found that MCT1 was downregulated in brain microvessels in patients with temporal lobe epilepsy.[138]

5.4.4 Organic Cation Transporters

Organic cation transporters (OCTs) comprise a class of transporters that are part of the *SLC22* subfamily. OCT1 (*SLC22A1*), OCT2 (*SLC22A2*) and OCT3 (*SLC22A3*) constitute the first subgroup of functionally characterised transporters from the *SLC22A* family. Corresponding to their name, OCTs predominantly recognise organic cations or weakly alkaline molecules that are positively charged at physiological pH.[139] Around 40% of all orally administered drugs exhibit these physicochemical properties[140] and, accordingly, a large number of compounds have been found to interact with OCT1–3, particularly as inhibitors, with broadly overlapping but distinctive specificities.[139]

The translocation of substrates follows a facilitated, bi-directional mechanism down an electrochemical gradient.[139] Drug library screening for OCT1 and OCT2 inhibition has revealed some common physicochemical features of potent inhibitors, including a positive net charge and high lipophilicity.[141] At the mRNA level, OCT1 is most abundantly expressed in the liver but is also detectable in the intestine, kidney, brain and other organs. The protein is localised to the luminal (apical) membrane of brain endothelial cells,[142–145] which could result in OCT1 mediating the uptake of substrates from the blood into the brain. OCT2 mRNA is most significantly detectable in the kidney but also present in the brain, intestine and other tissues.[139] The protein is localised in the luminal (apical) membrane of brain endothelial cells.[144,146] OCT3 transcripts are detectable in various tissues including the liver, kidney, brain and intestine.[139] Protein expression of OCT3 has recently been shown in isolated human brain microvessels, although its subcellular localisation remains to be determined.[106]

The role of specific organic cation transporter(s) in BBB transport is unclear. However, studies have linked different OCTs to the transport of a number of drugs at the BBB. These include the neurotoxin MPTP, lamotrigine, amisulpride and sulpiride.[36,144,147,148] Either multiple different transporters are important for the uptake of these drugs or an uncharacterised BBB specific organic cation-like transporter has functional importance at the BBB. For example, both choline and pyrilamine have been linked to a carriage process mediated by an unidentified cation transporter at the BBB, termed either the pyrilamine or choline transporter.[149,150]

5.4.5 Organic Anion Transporters

The organic anion transporters (OATs), OAT1–3, OAT7, OAT4, URAT1 and OAT10 proteins (*SLC22A6–9* and *SLC22A11–13*) belong to the third and largest subgroup of membrane transporters from the *SLC22A* family. They mediate the transport of organic anions in either direction and are particularly important in the first step of renal excretion.[151] Most, if not all, are secondary-active organic anion exchangers utilising counter-ions such as α-ketoglutarate, lactate and nicotinate.[151] To date, it remains unclear whether any of the OATs/URAT1 are involved in human BBB transport.[152]

A quantitative absolute proteomics study with isolated human brain microvessels attempted to analyse the expression of 114 transporters at the human BBB.[89] None of the investigated organic anion uptake transporters were above the limit of quantification. However, the same negative results were obtained in this study for the entire *SLC22A* family and all 11 human *SLCO* (OATP) transporters.[89] Caution in the interpretation of these results is therefore warranted, as there may be issues with the sensitivity of the method used, although other transporters of the *ABC* family, such as P-gp, BCRP and MRP4, were detected. In contrast, recent work that analysed the mRNA expression of most *SLC* transporters in isolated human brain microvessels found that the majority of the *SLC22A* genes were expressed.[106]

At least in rodents, a candidate OAT at the BBB is OAT3. OAT3 has been found to be localised to the abluminal membrane in brain endothelial cells of mouse and rat origin.[153,154] In one study, an active metabolite of oseltamivir (Ro 64-0802), used in the treatment of influenza virus, was microinjected into the cerebrum of knockout *Oat3*$^{-/-}$ mice. The amount of Ro 64-0802 in the brain was significantly greater in the *Oat3*$^{-/-}$ knockout mice than wild type mice 2 hours after the microinjection, suggesting that OAT3 is a candidate transporter for this compound at the BBB.[155] An additional study utilised a model organic anion drug (dehydroepiandrosterone sulfate) as a probe substrate of OAT3 and investigated its efflux from the mouse brain. The elimination of the probe compound from the brain after microinjection into the cerebral cortex was found to be delayed in *Oat3*$^{-/-}$

knockout mice compared with the controls, suggesting a functional role of OAT3 at the mouse BBB.[156] How this *in vivo* work in mice correlates to human BBB transport is, however, as yet unknown.

5.4.6 Nutrient Transporters

As the BBB is involved in maintaining brain homeostasis, a number of influx transporters that transport endogenous nutrients are expressed. These include LAT1 (see Section 5.4.1), GLUT1 and MFSD2A. GLUT1 (*SLC2A1*) is part of the sugar porter subfamily of the major facilitator superfamily (*MFS*), one of the largest secondary transporter superfamilies. GLUT1 is highly expressed at the BBB with subcellular localisation at both the luminal and abluminal membranes of brain endothelial cells, and has the essential role of transporting glucose into the brain.[89,157–159] Mutations of GLUT1 that affect activity can result in reduced transport of glucose and are associated with neurological syndromes as a result of lack of energy supply to the brain.[160] The crystal structure of the human glucose transporter GLUT1 has been solved recently and this gives insight into its mode of action.[161] GLUT1 has no known drug substrates but interactions with drugs have been observed, for example barbiturates and sodium valproate *in vitro* can reduce GLUT1-mediated transport of glucose.[162,163] However, how or if this could affect the role of the BBB in maintaining brain homeostasis is still unknown. MFSD2A is a novel omega three fatty acid transporter that was recently identified at the mouse BBB,[13,164] and is thought to be a key regulator of BBB function. MFSD2A has no known drug substrates but it has recently been shown to be a putative transporter of tunicamycin, which is used as an *in vitro* chemical tool to induce the unfolded protein response.[165] The recent identification of MFSD2A at the BBB is a good example of how additional influx transporters are being found and how this is an ongoing area of study.

5.5 Transporters Expressed at the CP

CP epithelial cells are the main site of CSF secretion and, along with the arachnoid membrane, form the blood–CSF barrier. In brief, the CP is a vascularised tissue that is located in each ventricle of the brain. The CP epithelial cells form a monolayer of cells that are joined together by tight junctions to form the blood–CSF barrier, which prevents the paracellular passage of molecules from the circulatory system into the CSF.[166] Like BBB endothelial cells, the CP epithelial cells express a variety of transporters. These transporters are involved in maintaining CSF homeostasis such as movement of nutrients or waste products out/into the CSF, and the transport of pharmaceutical drugs in/out of the CSF.[167] The concentration of a drug in the CSF is considered to be the product of transport through the CP epithelium cells and as such does not readily correlate with brain concentrations of a compound.[65] A limited number of gene deletion studies using knockout mice have been used to investigate the significance of transporters at the rodent CP, as outlined in an extensive review.[167] This review details immunohistochemical studies of transporter localisation in knockout mice that have determined a major difference between drug efflux transporter proteins in CP epithelial cells compared with those in the brain endothelial cells of the BBB. P-gp and BCRP in the CP epithelial cells were found to be located on the CSF-facing membrane, while at the BBB these transporters were localised at the blood-facing membrane.[167] This results in the P-gp and BCRP transporters at the BBB pumping substrates out of the brain, while in the CP they pump substrates into the CSF. Such a differential effect of the transporters at the two CNS barriers emphasises the difference in the basic biology and explains, at least in part, why the CSF is not always a good surrogate for the brain concentration of a drug.

5.6   Challenge

Up to 98% of all drugs may fail to cross the BBB at pharmacologically relevant concentrations, which has resulted in a major challenge for the development of CNS drugs to combat neurological disorders and brain tumours.[28] The traditional method has been to use medicinal chemistry approaches to alter the properties of a potential candidate compound in order to enhance its lipid solubility, but this approach has failed to produce many effective treatments for CNS diseases. This is due, at least in part, to efflux transporters such as P-gp transporting lipophilic substrate compounds out of the brain (see Section 5.3).

Alternatively, if a drug is transported by an influx transporter, this could cause issues such as carriage across the BBB and undesired CNS side effects. One example of a CNS adverse effect is with tecadenoson (CVT-510), an A1 adenosine receptor agonist that has undergone small scale clinical trials due to its potent anti-arrhythmic effects in tachycardias.[168] A1 adenosine receptors are expressed both on cardiac and brain tissues, and the drug is known to cause CNS sedative effects in humans.[168,169] A recent study has shown that tecadenoson is a substrate of nucleoside transporters *in vitro* and *in vivo* in mice, with inhibition of the equilibrative nucleoside transporter 1 (ENT1) reducing the BBB transport of the drug.[169]

Cheminformatics approaches have recently suggested that pharmaceutical drugs that have a structural similarity to cellular metabolites and nutrients can have a similar transport profile to their homologous cellular metabolite.[170,171] A recent example of how this concept of metabolite or nutrient likeness may be important, but is sometimes overlooked, is with fedratinib. Fedratinib is a Janus kinase 2 inhibitor that was in a Phase III trial for the treatment of myelofibrosis, but was withdrawn due to the onset of Wernicke's-like encephalopathy in a subset of patients.[172] Fedratinib has structural similarities to thiamine and was subsequently found to interact directly with the thiamine transporter [*SLC19A3*, thiamine transporter 2 (THTR2)], providing a clear explanation for the onset of Wernicke's encephalopathy and an example of a drug–nutrient interaction at the transporter level with a serious adverse reaction.[173] This study utilised Caco-2 cells as an oral absorption model but also found that the free drug concentration of fedratinib in the rodent brain was higher than expected compared with its physicochemical properties. A study from the 1980s showed that thiamine at the BBB has a saturable active transport component.[174] If this transporter-mediated carriage was also the case for fedratinib at the BBB, this could then be an explanation of why this compound was able to penetrate the brain at levels higher than predicted.

It is also known that certain diseases may affect the expression of transporters at the BBB and the tightness of the barrier. For example in Alzheimer's disease, a PET study in humans has shown decreased activity of P-gp at the BBB,[175] and animal models of peripheral pain have indicated increased P-gp expression at the rodent BBB, which reduces morphine penetration into the brain.[176] HIV infection of a limited number of astrocytes at the BBB is known to result in a disrupted BBB and is proposed to be a cause of the HIV-associated cognitive impairment that can occur in infected individuals.[177] The relevance of this affect for antiretroviral drug uptake into this sanctuary site is unknown. In addition, transporter activity at the BBB can be affected by diseases or impairment in other organs in the body, for example liver damage may lead to increased P-gp and MRP2 expression and function at the BBB in rodents.[178] Ongoing research will no doubt shed further light on the changes at the BBB during disease processes and how they might affect drug delivery *via* changes in drug transporters.

5.7 Opportunity

As well as challenges, opportunities for targeting drugs to the CNS also exist due to the expression of drug transporters at the BBB. This includes strategies to overcome the lack of brain penetration by targeting the mechanism of uptake; targeting efflux transporter(s) at the BBB so that a drug is excluded from the brain, thus potentially avoiding CNS side effects; and using transport inhibitors to boost the uptake of a drug into the brain.

An interaction of a drug with an efflux transporter is not necessarily a negative outcome, since this can result in a drug being excluded from the brain and avoiding potentially serious CNS side effects. An example of this is with sedating and non-sedating antihistamines. Cetirizine and hydroxyzine are related structurally, but cetirizine has a non-sedating profile, due to the fact that it is a P-gp substrate and is excluded from the CNS, whereas hydroxyzine is not a P-gp substrate and can be used as a sedative drug.[179]

The use of efflux transporter inhibitors has been proposed as an adjuvant treatment to target drugs to the brain that would otherwise be excluded by P-gp and/or BCRP, with success being achieved in preclinical animal studies. Although this could be of particular relevance to, for example, tyrosine kinase inhibitors, for which both P-gp and BCRP at the BBB have been linked to the lack of brain uptake that restricts their applications in the treatment of brain tumours, this approach has failed to translate into the clinic. Early generation transport inhibitors have failed in clinical trials as adjuvants for cancer therapy, due to either a lack of clinical benefit or high incidence of toxicity,[180] and no P-gp inhibitor has yet received approval from the European regulatory agencies or the US FDA as an adjuvant treatment.[180,181] The situation with third generation inhibitors such as tariquidar or elacridar showed promise in early phase clinical trials,[182,183] but in the case of tariquidar, a Phase III trial in non-small-cell lung cancer was terminated due to safety concerns. An area of basic research that is in its infancy is the modulation of drug transporter expression to achieve more drug penetration to the target site, which could have great potential in the long term.[184]

Targeting influx transporters expressed at the BBB to achieve uptake of the drug into the brain is an opportunity that is attracting current interest. Historically this has happened more by accident than design, a classic example being L-DOPA. Dopamine does not cross the BBB but a prodrug of dopamine, L-DOPA, has high permeability across the BBB and is a substrate of LAT1, a BBB influx transporter.[123,126,185] Subsequently, a number of groups developed compounds that are transported by LAT1, enabling delivery to the brain; these are at an early stage but appear promising.[186,187] Utilising *in silico* modelling, Geier *et al.* were able to predict and experimentally validate four novel ligands of LAT1.[188] With the advance of structural biology in terms of solving the structure of a number of SLC transporters, this opens up the realistic possibility of solving the structure of BBB influx transporters and thus enabling QSAR to be performed to specifically target ligands to the BBB transporter and thus the brain.[161,189] Drawbacks to this methodology are not yet clear but could include drug–nutrient interactions and saturation by the drug of the transporter-mediated process.

5.8 Summary

In this chapter we have outlined the physiology of the BBB and how drug transporters are an important consideration in how drugs are able to penetrate into the brain. This is an exciting area of research that will continue to develop over time, particularly in terms of the importance and identification of influx transporters at the BBB.

References

1. N. J. Abbott, A. A. Patabendige, D. E. Dolman, S. R. Yusof and D. J. Begley, *Neurobiol. Dis.*, 2010, **37**, 13–25.
2. N. J. Abbott, L. Ronnback and E. Hansson, *Nat. Rev. Neurosci.*, 2006, **7**, 41–53.
3. E. Neuwelt, N. J. Abbott, L. Abrey, W. A. Banks, B. Blakley, T. Davis, B. Engelhardt, P. Grammas, M. Nedergaard, J. Nutt, W. Pardridge,

G. A. Rosenberg, Q. Smith and L. R. Drewes, *Lancet Neurol.*, 2008, **7**, 84–96.

4. F. A. Azevedo, L. R. Carvalho, L. T. Grinberg, J. M. Farfel, R. E. Ferretti, R. E. Leite, W. Jacob Filho, R. Lent and S. Herculano-Houzel, *J. Comp. Neurol.*, 2009, **513**, 532–541.

5. A. D. Wong, M. Ye, A. F. Levy, J. D. Rothstein, D. E. Bergles and P. C. Searson, *Front. Neuroeng.*, 2013, **6**, 7.

6. C. Iadecola, *Nat. Rev. Neurosci.*, 2004, **5**, 347–360.

7. M. R. Mizee, D. Wooldrik, K. A. Lakeman, B. van het Hof, J. A. Drexhage, D. Geerts, M. Bugiani, E. Aronica, R. E. Mebius, A. Prat, H. E. de Vries and A. Reijerkerk, *J. Neurosci.*, 2013, **33**, 1660–1671.

8. E. S. Lippmann, S. M. Azarin, J. E. Kay, R. A. Nessler, H. K. Wilson, A. Al-Ahmad, S. P. Palecek and E. V. Shusta, *Nat. Biotechnol.*, 2012, **30**(8), 783–791.

9. E. A. Winkler, R. D. Bell and B. V. Zlokovic, *Nat. Neurosci.*, 2011, **14**, 1398–1405.

10. R. Daneman, L. Zhou, A. A. Kebede and B. A. Barres, *Nature*, 2010, **468**, 562–566.

11. A. Armulik, G. Genove, M. Mae, M. H. Nisancioglu, E. Wallgard, C. Niaudet, L. He, J. Norlin, P. Lindblom, K. Strittmatter, B. R. Johansson and C. Betsholtz, *Nature*, 2010, **468**, 557–561.

12. C. N. Hall, C. Reynell, B. Gesslein, N. B. Hamilton, A. Mishra, B. A. Sutherland, F. M. O'Farrell, A. M. Buchan, M. Lauritzen and D. Attwell, *Nature*, 2014, **508**, 55–60.

13. A. Ben-Zvi, B. Lacoste, E. Kur, B. J. Andreone, Y. Mayshar, H. Yan and C. Gu, *Nature*, 2014, **509**, 507–511.

14. J. I. Alvarez, A. Dodelet-Devillers, H. Kebir, I. Ifergan, P. J. Fabre, S. Terouz, M. Sabbagh, K. Wosik, L. Bourbonniere, M. Bernard, J. van Horssen, H. E. de Vries, F. Charron and A. Prat, *Science*, 2011, **334**, 1727–1731.

15. T. M. Mathiisen, K. P. Lehre, N. C. Danbolt and O. P. Ottersen, *Glia*, 2010, **58**, 1094–1103.

16. A. M. Butt, H. C. Jones and N. J. Abbott, *J. Physiol.*, 1990, **429**, 47–62.

17. E. S. Lippmann, A. Al-Ahmad, S. M. Azarin, S. P. Palecek and E. V. Shusta, *Sci. Rep.*, 2014, **4**, 4160.

18. C. Crone and O. Christensen, *J. Gen. Physiol.*, 1981, **77**, 349–371.

19. S. P. Olesen and C. Crone, *Biophys. J.*, 1983, **42**, 31–41.

20. S. Kurita, T. Yamada, E. Rikitsu, W. Ikeda and Y. Takai, *J. Biol. Chem.*, 2013, **288**, 29356–29368.

21. C. Forster, *Histochem. Cell Biol.*, 2008, **130**, 55–70.

22. C. Weber, L. Fraemohs and E. Dejana, *Nat. Rev. Immunol.*, 2007, **7**, 467–477.

23. C. M. Niessen, *J. Invest. Dermatol.*, 2007, **127**, 2525–2532.

24. P. D. Dobson and D. B. Kell, *Nat. Rev. Drug Discovery*, 2008, **7**, 205–220.

25. K. Sugano, M. Kansy, P. Artursson, A. Avdeef, S. Bendels, L. Di, G. F. Ecker, B. Faller, H. Fischer, G. Gerebtzoff, H. Lennernaes and F. Senner, *Nat. Rev. Drug Discovery*, 2010, **9**, 597–614.

26. A. H. Schinkel, *Adv. Drug Delivery Rev.*, 1999, **36**, 179–194.
27. S. G. Summerfield, K. Read, D. J. Begley, T. Obradovic, I. J. Hidalgo, S. Coggon, A. V. Lewis, R. A. Porter and P. Jeffrey, *J. Pharmacol. Exp. Ther.*, 2007, **322**, 205–213.
28. W. M. Pardridge, *NeuroRx*, 2005, **2**, 3–14.
29. W. H. Oldendorf and W. J. Brown, *Proc. Soc. Exp. Biol. Med.*, 1975, **149**, 736–738.
30. W. H. Oldendorf, M. E. Cornford and W. J. Brown, *Ann. Neurol.*, 1977, **1**, 409–417.
31. S. G. Summerfield, A. J. Stevens, L. Cutler, M. del Carmen Osuna, B. Hammond, S. P. Tang, A. Hersey, D. J. Spalding and P. Jeffrey, *J. Pharmacol. Exp. Ther.*, 2006, **316**, 1282–1290.
32. W. Wang, M. T. Kim, A. Sedykh and H. Zhu, *Pharm. Res.*, 2015, **32**(9), 3055–3065.
33. W. M. Pardridge, *Expert Opin. Ther. Targets*, 2015, **19**(8), 1059–1072.
34. B. B. Weksler, E. A. Subileau, N. Perriere, P. Charneau, K. Holloway, M. Leveque, H. Tricoire-Leignel, A. Nicotra, S. Bourdoulous, P. Turowski, D. K. Male, F. Roux, J. Greenwood, I. A. Romero and P. O. Couraud, *FASEB J.*, 2005, **19**, 1872–1874.
35. B. Weksler, I. A. Romero and P. O. Couraud, *Fluids Barriers CNS*, 2013, **10**, 16.
36. D. Dickens, A. Owen, A. Alfirevic, A. Giannoudis, A. Davies, B. Weksler, I. A. Romero, P. O. Couraud and M. Pirmohamed, *Biochem. Pharmacol.*, 2012, **83**, 805–814.
37. D. Dickens, S. D. Webb, S. Antonyuk, A. Giannoudis, A. Owen, S. Radisch, S. S. Hasnain and M. Pirmohamed, *Biochem. Pharmacol.*, 2013, **85**, 1672–1683.
38. D. Dickens, S. R. Yusof, N. J. Abbott, B. Weksler, I. A. Romero, P. O. Couraud, A. Alfirevic, M. Pirmohamed and A. Owen, *PloS One*, 2013, **8**, e64854.
39. S. Dauchy, F. Miller, P. O. Couraud, R. J. Weaver, B. Weksler, I. A. Romero, J. M. Scherrmann, I. De Waziers and X. Decleves, *Biochem. Pharmacol.*, 2008, **77**(5), 897–909.
40. B. Poller, H. Gutmann, S. Krahenbuhl, B. Weksler, I. Romero, P. O. Couraud, G. Tuffin, J. Drewe and J. Huwyler, *J. Neurochem.*, 2008, **107**, 1358–1368.
41. E. Urich, S. E. Lazic, J. Molnos, I. Wells and P. O. Freskgard, *PloS One*, 2012, **7**, e38149.
42. L. Di, E. H. Kerns, I. F. Bezar, S. L. Petusky and Y. Huang, *J. Pharm. Sci.*, 2009, **98**, 1980–1991.
43. R. Cecchelli, V. Berezowski, S. Lundquist, M. Culot, M. Renftel, M. P. Dehouck and L. Fenart, *Nat. Rev. Drug Discovery*, 2007, **6**, 650–661.
44. A. Patabendige, R. A. Skinner and N. J. Abbott, *Brain Res.*, 2013, **1521**, 1–15.

45. J. Boyer-Di Ponio, F. El-Ayoubi, F. Glacial, K. Ganeshamoorthy, C. Driancourt, M. Godet, N. Perriere, O. Guillevic, P. O. Couraud and G. Uzan, *PloS One*, 2014, **9**, e84179.

46. R. Cecchelli, S. Aday, E. Sevin, C. Almeida, M. Culot, L. Dehouck, C. Coisne, B. Engelhardt, M. P. Dehouck and L. Ferreira, *PloS One*, 2014, **9**, e99733.

47. F. Mayer, N. Mayer, L. Chinn, R. L. Pinsonneault, D. Kroetz and R. J. Bainton, *J. Neurosci.*, 2009, **29**, 3538–3550.

48. R. A. Umans and M. R. Taylor, *Clin. Pharmacol. Ther.*, 2012, **92**, 567–570.

49. A. Fleming, H. Diekmann and P. Goldsmith, *PloS One*, 2013, **8**, e77548.

50. Q. R. Smith and D. D. Allen, *Methods Mol. Med.*, 2003, **89**, 209–218.

51. M. E. Morgan, D. Singhal and B. D. Anderson, *J. Pharmacol. Exp. Ther.*, 1996, **277**, 1167–1176.

52. C. Tong, P. Li, N. L. Wu, Y. Yan and Q. L. Ying, *Nature*, 2010, **467**, 211–213.

53. H. Fuchs, W. Kishimoto, D. Gansser, P. Tanswell and N. Ishiguro, *Drug Metab. Dispos.*, 2014, **42**(10), 1761–1765.

54. S. Agarwal, Y. Uchida, R. K. Mittapalli, R. Sane, T. Terasaki and W. F. Elmquist, *Drug Metab. Dispos.*, 2012, **40**, 1164–1169.

55. X. Chu, K. Bleasby and R. Evers, *Expert Opin. Drug Metab. Toxicol.*, 2013, **9**, 237–252.

56. B. Wulkersdorfer, T. Wanek, M. Bauer, M. Zeitlinger, M. Muller and O. Langer, *Clin. Pharmacol. Ther.*, 2014, **96**, 206–213.

57. K. Kawamura, T. Yamasaki, J. Yui, A. Hatori, F. Konno, K. Kumata, T. Irie, T. Fukumura, K. Suzuki, I. Kanno and M. R. Zhang, *Nucl. Med. Biol.*, 2009, **36**, 239–246.

58. J. Bakhsheshian, B. R. Wei, K. E. Chang, S. Shukla, S. V. Ambudkar, R. M. Simpson, M. M. Gottesman and M. D. Hall, *Proc. Natl. Acad. Sci. U. S. A.*, 2013, **110**, 20801–20806.

59. E. C. de Lange and M. Hammarlund-Udenaes, *Clin. Pharmacol. Ther.*, 2015, **97**, 380–394.

60. T. T. Mariappan, V. Kurawattimath, S. S. Gautam, C. P. Kulkarni, R. Kallem, K. S. Taskar, P. H. Marathe and S. Mandlekar, *Mol. Pharm.*, 2014, **11**, 477–485.

61. M. Friden, S. Winiwarter, G. Jerndal, O. Bengtsson, H. Wan, U. Bredberg, M. Hammarlund-Udenaes and M. Antonsson, *J. Med. Chem.*, 2009, **52**, 6233–6243.

62. E. C. de Lange and M. Danhof, *Clin. Pharmacokinet.*, 2002, **41**, 691–703.

63. M. Dean, A. Rzhetsky and R. Allikmets, *Genome Res.*, 2001, **11**, 1156–1166.

64. J. I. Fletcher, M. Haber, M. J. Henderson and M. D. Norris, *Nat. Rev. Cancer*, 2010, **10**, 147–156.

65. W. M. Pardridge, *J. Cereb. Blood Flow Metab.*, 2012, **32**, 1959–1972.

66. F. J. Sharom, *Pharmacogenomics*, 2008, **9**, 105–127.

67. A. H. Schinkel, J. J. Smit, O. van Tellingen, J. H. Beijnen, E. Wagenaar, L. van Deemter, C. A. Mol, M. A. van der Valk, E. C. Robanus-Maandag, H. P. te Riele, *et al.*, *Cell*, 1994, **77**, 491–502.

68. M. F. Fromm, *Eur. J. Clin. Invest.*, 2003, **33**(Suppl. 2), 6–9.

69. S. G. Aller, J. Yu, A. Ward, Y. Weng, S. Chittaboina, R. Zhuo, P. M. Harrell, Y. T. Trinh, Q. Zhang, I. L. Urbatsch and G. Chang, *Science*, 2009, **323**, 1718–1722.

70. M. S. Jin, M. L. Oldham, Q. Zhang and J. Chen, *Nature*, 2012, **490**, 566–569.

71. J. Li, K. F. Jaimes and S. G. Aller, *Protein Sci.*, 2014, **23**, 34–46.

72. A. B. Ward, P. Szewczyk, V. Grimard, C. W. Lee, L. Martinez, R. Doshi, A. Caya, M. Villaluz, E. Pardon, C. Cregger, D. J. Swartz, P. G. Falson, I. L. Urbatsch, C. Govaerts, J. Steyaert and G. Chang, *Proc. Natl. Acad. Sci. U. S. A.*, 2013, **110**, 13386–13391.

73. G. Tombline, A. Muharemagic, L. B. White and A. E. Senior, *Biochemistry*, 2005, **44**, 12879–12886.

74. R. B. Kim, M. F. Fromm, C. Wandel, B. Leake, A. J. Wood, D. M. Roden and G. R. Wilkinson, *J. Clin. Invest.*, 1998, **101**, 289–294.

75. A. Owen, O. Janneh, R. C. Hartkoorn, B. Chandler, P. G. Bray, P. Martin, S. A. Ward, C. A. Hart, S. H. Khoo and D. J. Back, *J. Pharmacol. Exp. Ther.*, 2005, **314**, 1202–1209.

76. J. Chmielewski and C. Hrycyna, *Ther. Delivery*, 2012, **3**, 689–692.

77. E. F. Choo, B. Leake, C. Wandel, H. Imamura, A. J. Wood, G. R. Wilkinson and R. B. Kim, *Drug Metab. Dispos.*, 2000, **28**, 655–660.

78. S. Bihorel, G. Camenisch, M. Lemaire and J. M. Scherrmann, *J. Neurochem.*, 2007, **102**, 1749–1757.

79. J. Thomas, L. Wang, R. E. Clark and M. Pirmohamed, *Blood*, 2004, **104**, 3739–3745.

80. H. Dai, P. Marbach, M. Lemaire, M. Hayes and W. F. Elmquist, *J. Pharmacol. Exp. Ther.*, 2003, **304**, 1085–1092.

81. N. C. Wolff, J. A. Richardson, M. Egorin and R. L. Ilaria Jr., *Blood*, 2003, **101**, 5010–5013.

82. N. Drion, M. Lemaire, J. M. Lefauconnier and J. M. Scherrmann, *J. Neurochem.*, 1996, **67**, 1688–1693.

83. S. Desrayaud, P. Guntz, J. M. Scherrmann and M. Lemaire, *Life Sci.*, 1997, **61**, 153–163.

84. A. Kumar, N. Seghal, P. S. Naidu, S. S. Padi and R. Goyal, *Pharmacol. Rep.*, 2007, **59**, 274–283.

85. L. A. Doyle, W. Yang, L. V. Abruzzo, T. Krogmann, Y. Gao, A. K. Rishi and D. D. Ross, *Proc. Natl. Acad. Sci. U. S. A.*, 1998, **95**, 15665–15670.

86. J. Xu, Y. Liu, Y. Yang, S. Bates and J. T. Zhang, *J. Biol. Chem.*, 2004, **279**, 19781–19789.

87. R. W. Robey, K. K. To, O. Polgar, M. Dohse, P. Fetsch, M. Dean and S. E. Bates, *Adv. Drug Delivery Rev.*, 2009, **61**, 3–13.

88. Y. Kubo, S. Ohtsuki, Y. Uchida and T. Terasaki, *J. Pharm. Sci.*, 2015, **104**(9), 3060–3068.

89. Y. Uchida, S. Ohtsuki, Y. Katsukura, C. Ikeda, T. Suzuki, J. Kamiie and T. Terasaki, *J. Neurochem.*, 2011, **117**, 333–345.

90. H. C. Cooray, C. G. Blackmore, L. Maskell and M. A. Barrand, *Neuroreport*, 2002, **13**, 2059–2063.

91. S. Cisternino, C. Mercier, F. Bourasset, F. Roux and J. M. Scherrmann, *Cancer Res.*, 2004, **64**, 3296–3301.

92. R. Zhao, T. J. Raub, G. A. Sawada, S. C. Kasper, J. A. Bacon, A. S. Bridges and G. M. Pollack, *Drug Metab. Dispos.*, 2009, **37**, 1251–1258.

93. S. Agarwal, A. M. Hartz, W. F. Elmquist and B. Bauer, *Curr. Pharm. Des.*, 2011, **17**, 2793–2802.

94. O. M. Woodward, A. Kottgen, J. Coresh, E. Boerwinkle, W. B. Guggino and M. Kottgen, *Proc. Natl. Acad. Sci. U. S. A.*, 2009, **106**, 10338–10342.

95. O. M. Woodward, D. N. Tukaye, J. Cui, P. Greenwell, L. M. Constantoulakis, B. S. Parker, A. Rao, M. Kottgen, P. C. Maloney and W. B. Guggino, *Proc. Natl. Acad. Sci. U. S. A.*, 2013, **110**, 5223–5228.

96. L. Zhang, K. L. Spencer, V. S. Voruganti, N. W. Jorgensen, M. Fornage, L. G. Best, K. D. Brown-Gentry, S. A. Cole, D. C. Crawford, E. Deelman, N. Franceschini, A. L. Gaffo, K. R. Glenn, G. Heiss, N. S. Jenny, A. Kottgen, Q. Li, K. Liu, T. C. Matise, K. E. North, J. G. Umans and W. H. Kao, *Am. J. Epidemiol.*, 2013, **177**, 923–932.

97. I. S. Kim, H. G. Kim, D. C. Kim, H. S. Eom, S. Y. Kong, H. J. Shin, S. H. Hwang, E. Y. Lee and G. W. Lee, *Cancer Sci.*, 2008, **99**, 2496–2501.

98. F. G. Russel, J. B. Koenderink and R. Masereeuw, *Trends Pharmacol. Sci.*, 2008, **29**, 200–207.

99. S. P. Cole, *Annu. Rev. Pharmacol. Toxicol.*, 2014, **54**, 95–117.

100. J. D. Schuetz, M. C. Connelly, D. Sun, S. G. Paibir, P. M. Flynn, R. V. Srinivas, A. Kumar and A. Fridland, *Nat. Med.*, 1999, **5**, 1048–1051.

101. G. Reid, P. Wielinga, N. Zelcer, I. van der Heijden, A. Kuil, M. de Haas, J. Wijnholds and P. Borst, *Proc. Natl. Acad. Sci. U. S. A.*, 2003, **100**, 9244–9249.

102. C. Sinha, A. Ren, K. Arora, C. S. Moon, S. Yarlagadda, W. Zhang, S. B. Cheepala, J. D. Schuetz and A. P. Naren, *J. Biol. Chem.*, 2013, **288**, 3786–3794.

103. Y. Zhang, J. D. Schuetz, W. F. Elmquist and D. W. Miller, *J. Pharmacol. Exp. Ther.*, 2004, **311**, 449–455.

104. M. Leggas, M. Adachi, G. L. Scheffer, D. Sun, P. Wielinga, G. Du, K. E. Mercer, Y. Zhuang, J. C. Panetta, B. Johnston, R. J. Scheper, C. F. Stewart and J. D. Schuetz, *Mol. Cell Biol.*, 2004, **24**, 7612–7621.

105. A. T. Nies, G. Jedlitschky, J. Konig, C. Herold-Mende, H. H. Steiner, H. P. Schmitt and D. Keppler, *Neuroscience*, 2004, **129**, 349–360.

106. E. G. Geier, E. C. Chen, A. Webb, A. C. Papp, S. W. Yee, W. Sadee and K. M. Giacomini, *Clin. Pharmacol. Ther.*, 2013, **94**(6), 636–639.

107. M. S. Warren, N. Zerangue, K. Woodford, L. M. Roberts, E. H. Tate, B. Feng, C. Li, T. J. Feuerstein, J. Gibbs, B. Smith, S. M. de Morais, W. J. Dower and K. J. Koller, *Pharmacol. Res.*, 2009, **59**, 404–413.

108. H. Zhang, T. Gerson, M. L. Varney, R. K. Singh and S. V. Vinogradov, *Pharm. Res.*, 2010, **27**, 2528–2543.
109. S. Cisternino, C. Rousselle, A. Lorico, G. Rappa and J. M. Scherrmann, *Pharm. Res.*, 2003, **20**, 904–909.
110. C. D. Klaassen and L. M. Aleksunes, *Pharmacol. Rev.*, 2010, **62**, 1–96.
111. L. Zhou, K. Schmidt, F. R. Nelson, V. Zelesky, M. D. Troutman and B. Feng, *Drug Metab. Dispos.*, 2009, **37**, 946–955.
112. S. C. Tang, J. S. Lagas, N. A. Lankheet, B. Poller, M. J. Hillebrand, H. Rosing, J. H. Beijnen and A. H. Schinkel, *Int. J. Cancer*, 2012, **130**, 223–233.
113. S. Durmus, R. W. Sparidans, E. Wagenaar, J. H. Beijnen and A. H. Schinkel, *Mol. Pharm.*, 2012, **9**, 3236–3245.
114. G. Moreno-Sanz, B. Barrera, A. Armirotti, S. M. Bertozzi, R. Scarpelli, T. Bandiera, J. G. Prieto, A. Duranti, G. Tarzia, G. Merino and D. Piomelli, *Pharmacol. Res.*, 2014, **87C**, 87–93.
115. R. Sane, S. P. Wu, R. Zhang and J. M. Gallo, *Drug Metab. Dispos.*, 2014, **42**, 537–540.
116. F. Lin, S. Marchetti, D. Pluim, D. Iusuf, R. Mazzanti, J. H. Schellens, J. H. Beijnen and O. van Tellingen, *Clin. Cancer Res.*, 2013, **19**, 2084–2095.
117. J. W. Polli, K. L. Olson, J. P. Chism, L. S. John-Williams, R. L. Yeager, S. M. Woodard, V. Otto, S. Castellino and V. E. Demby, *Drug Metab. Dispos.*, 2009, **37**, 439–442.
118. M. A. Hediger, B. Clemencon, R. E. Burrier and E. A. Bruford, *Mol. Aspects Med.*, 2013, **34**, 95–107.
119. B. Thorens and M. Mueckler, *Am. J. Physiol. Endocrinol. Metab.*, 2010, **298**, E141–E145.
120. J. Jeong and D. J. Eide, *Mol. Aspects Med.*, 2013, **34**, 612–619.
121. A. Cesar-Razquin, B. Snijder, T. Frappier-Brinton, R. Isserlin, G. Gyimesi, X. Bai, R. A. Reithmeier, D. Hepworth, M. A. Hediger, A. M. Edwards and G. Superti-Furga, *Cell*, 2015, **162**, 478–487.
122. L. Mastroberardino, B. Spindler, R. Pfeiffer, P. J. Skelly, J. Loffing, C. B. Shoemaker and F. Verrey, *Nature*, 1998, **395**, 288–291.
123. R. J. Boado, J. Y. Li, M. Nagaya, C. Zhang and W. M. Pardridge, *Proc. Natl. Acad. Sci. U. S. A.*, 1999, **96**, 12079–12084.
124. C. Meier, Z. Ristic, S. Klauser and F. Verrey, *EMBO J.*, 2002, **21**, 580–589.
125. Y. Kanai, H. Segawa, K. Miyamoto, H. Uchino, E. Takeda and H. Endou, *J. Biol. Chem.*, 1998, **273**, 23629–23632.
126. H. Uchino, Y. Kanai, D. K. Kim, M. F. Wempe, A. Chairoungdua, E. Morimoto, M. W. Anders and H. Endou, *Mol. Pharmacol.*, 2002, **61**, 729–737.
127. L. V. Sinclair, J. Rolf, E. Emslie, Y. B. Shi, P. M. Taylor and D. A. Cantrell, *Nat. Immunol.*, 2013, **14**(5), 500–508.
128. R. Duelli, B. E. Enerson, D. Z. Gerhart and L. R. Drewes, *J. Cereb. Blood Flow Metab.*, 2000, **20**, 1557–1562.
129. H. Matsuo, S. Tsukada, T. Nakata, A. Chairoungdua, D. K. Kim, S. H. Cha, J. Inatomi, H. Yorifuji, J. Fukuda, H. Endou and Y. Kanai, *Neuroreport*, 2000, **11**, 3507–3511.

130. T. Kageyama, M. Nakamura, A. Matsuo, Y. Yamasaki, Y. Takakura, M. Hashida, Y. Kanai, M. Naito, T. Tsuruo, N. Minato and S. Shimohama, *Brain Res.*, 2000, **879**, 115–121.

131. A. Kalliokoski and M. Niemi, *Br. J. Pharmacol.*, 2009, **158**, 693–705.

132. B. Gao, B. Hagenbuch, G. A. Kullak-Ublick, D. Benke, A. Aguzzi and P. J. Meier, *J. Pharmacol. Exp. Ther.*, 2000, **294**, 73–79.

133. B. Gao, S. R. Vavricka, P. J. Meier and B. Stieger, *Pflugers Arch.*, 2014, **467**(7), 1481–1493.

134. A. Ose, H. Kusuhara, C. Endo, K. Tohyama, M. Miyajima, S. Kitamura and Y. Sugiyama, *Drug Metab. Dispos.*, 2010, **38**, 168–176.

135. B. J. Thompson, L. Sanchez-Covarrubias, L. M. Slosky, Y. Zhang, M. L. Laracuente and P. T. Ronaldson, *J. Cereb. Blood Flow Metab.*, 2014, **34**, 699–707.

136. A. P. Halestrap and M. C. Wilson, *IUBMB Life*, 2012, **64**, 109–119.

137. D. Z. Gerhart, B. E. Enerson, O. Y. Zhdankina, R. L. Leino and L. R. Drewes, *Am. J. Physiol.*, 1997, **273**, E207–E213.

138. F. Lauritzen, N. C. de Lanerolle, T. S. Lee, D. D. Spencer, J. H. Kim, L. H. Bergersen and T. Eid, *Neurobiol. Dis.*, 2011, **41**, 577–584.

139. A. T. Nies, H. Koepsell, K. Damme and M. Schwab, *Handb. Exp. Pharmacol.*, 2011, **201**, 105–167.

140. S. Neuhoff, A. L. Ungell, I. Zamora and P. Artursson, *Pharm. Res.*, 2003, **20**, 1141–1148.

141. H. Koepsell, K. Lips and C. Volk, *Pharm. Res.*, 2007, **24**, 1227–1251.

142. A. T. Nies, E. Herrmann, M. Brom and D. Keppler, *Naunyn Schmiedebergs Arch. Pharmacol.*, 2008, **376**, 449–461.

143. M. V. Tzvetkov, S. V. Vormfelde, D. Balen, I. Meineke, T. Schmidt, D. Sehrt, I. Sabolic, H. Koepsell and J. Brockmoller, *Clin. Pharmacol. Ther.*, 2009, **86**, 299–306.

144. C. J. Lin, Y. Tai, M. T. Huang, Y. F. Tsai, H. J. Hsu, K. Y. Tzen and H. H. Liou, *J. Neurochem.*, 2010, **114**, 717–727.

145. T. K. Han, R. S. Everett, W. R. Proctor, C. M. Ng, C. L. Costales, K. L. Brouwer and D. R. Thakker, *Mol. Pharmacol.*, 2013, **84**, 182–189.

146. H. Motohashi, Y. Sakurai, H. Saito, S. Masuda, Y. Urakami, M. Goto, A. Fukatsu, O. Ogawa and K. Inui, *J. Am. Soc. Nephrol.*, 2002, **13**, 866–874.

147. J. N. Dos Santos Pereira, S. Tadjerpisheh, M. Abu Abed, A. R. Saadatmand, B. Weksler, I. A. Romero, P. O. Couraud, J. Brockmoller and M. V. Tzvetkov, *AAPS J.*, 2014, **16**, 1247–1258.

148. D. Dickens, A. Owen, A. Alfirevic, A. Giannoudis, A. Davies, B. Weksler, I. A. Romero, P. O. Couraud and M. Pirmohamed, *Biochem. Pharmacol.*, 2012, **83**(6), 805–814.

149. T. Okura, A. Hattori, Y. Takano, T. Sato, M. Hammarlund-Udenaes, T. Terasaki and Y. Deguchi, *Drug Metab. Dispos.*, 2008, **36**, 2005–2013.

150. W. J. Geldenhuys and D. D. Allen, *Cent. Nerv. Syst. Agents Med. Chem.*, 2012, **12**, 95–99.

151. A. N. Rizwan and G. Burckhardt, *Pharm. Res.*, 2007, **24**, 450–470.

152. H. Koepsell, *Mol. Aspects Med.*, 2013, **34**, 413–435.
153. S. Mori, H. Takanaga, S. Ohtsuki, T. Deguchi, Y. S. Kang, K. Hosoya and T. Terasaki, *J. Cereb. Blood Flow Metab.*, 2003, **23**, 432–440.
154. S. Ohtsuki, T. Takizawa, H. Takanaga, S. Hori, K. Hosoya and T. Terasaki, *J. Neurochem.*, 2004, **90**, 743–749.
155. A. Ose, M. Ito, H. Kusuhara, K. Yamatsugu, M. Kanai, M. Shibasaki, M. Hosokawa, J. D. Schuetz and Y. Sugiyama, *Drug Metab. Dispos.*, 2009, **37**, 315–321.
156. M. Miyajima, H. Kusuhara, M. Fujishima, Y. Adachi and Y. Sugiyama, *Drug Metab. Dispos.*, 2011, **39**, 814–819.
157. K. Zeller, S. Rahner-Welsch and W. Kuschinsky, *J. Cereb. Blood Flow Metab.*, 1997, **17**, 204–209.
158. I. A. Simpson, S. J. Vannucci, M. R. DeJoseph and R. A. Hawkins, *J. Biol. Chem.*, 2001, **276**, 12725–12729.
159. A. P. Dick, S. I. Harik, A. Klip and D. M. Walker, *Proc. Natl. Acad. Sci. U. S. A.*, 1984, **81**, 7233–7237.
160. J. Klepper, D. Wang, J. Fischbarg, J. C. Vera, I. T. Jarjour, K. R. O'Driscoll and D. C. De Vivo, *Neurochem. Res.*, 1999, **24**, 587–594.
161. D. Deng, C. Xu, P. Sun, J. Wu, C. Yan, M. Hu and N. Yan, *Nature*, 2014, **510**, 121–125.
162. J. Klepper, J. Fischbarg, J. C. Vera, D. Wang and D. C. De Vivo, *Pediatr. Res.*, 1999, **46**, 677–683.
163. H. Y. Wong, T. S. Chu, J. C. Lai, K. P. Fung, T. F. Fok, T. Fujii and Y. Y. Ho, *J. Cell. Biochem.*, 2005, **96**, 775–785.
164. L. N. Nguyen, D. Ma, G. Shui, P. Wong, A. Cazenave-Gassiot, X. Zhang, M. R. Wenk, E. L. Goh and D. L. Silver, *Nature*, 2014, **509**, 503–506.
165. J. H. Reiling, C. B. Clish, J. E. Carette, M. Varadarajan, T. R. Brummelkamp and D. M. Sabatini, *Proc. Natl. Acad. Sci. U. S. A.*, 2011, **108**, 11756–11765.
166. M. P. Lun, E. S. Monuki and M. K. Lehtinen, *Nat. Rev. Neurosci.*, 2015, **16**, 445–457.
167. R. F. Keep and D. E. Smith, *Fluids Barriers CNS*, 2011, **8**, 26.
168. B. B. Lerman, K. A. Ellenbogen, A. Kadish, E. Platia, K. M. Stein, S. M. Markowitz, S. Mittal, D. J. Slotwiner, M. Scheiner, S. Iwai, L. Belardinelli, M. Jerling, R. Shreeniwas and A. A. Wolff, *J. Cardiovasc. Pharmacol. Ther.*, 2001, **6**, 237–245.
169. E. I. Lepist, V. L. Damaraju, J. Zhang, W. P. Gati, S. Y. Yao, K. M. Smith, E. Karpinski, J. D. Young, K. H. Leung and C. E. Cass, *Drug Metab. Dispos.*, 2013, **41**, 916–922.
170. S. O'Hagan and D. B. Kell, *Front. Pharmacol.*, 2015, **6**, 105.
171. S. O'Hagan, N. Swainston, J. Handl and D. B. Kell, *Metabolomics*, 2015, **11**, 323–339.
172. R. Williams, *Expert Opin. Invest. Drugs*, 2015, **24**, 95–110.
173. C. Jamieson, R. Hasserjian, J. Gotlib, J. Cortes, R. Stone, M. Talpaz, J. Thiele, S. Rodig and O. Pozdnyakova, *J. Transl. Med.*, 2015, **13**, 294.

174. J. Greenwood, E. R. Love and O. E. Pratt, *J. Physiol.*, 1982, **327**, 95–103.
175. D. M. van Assema, M. Lubberink, M. Bauer, W. M. van der Flier, R. C. Schuit, A. D. Windhorst, E. F. Comans, N. J. Hoetjes, N. Tolboom, O. Langer, M. Muller, P. Scheltens, A. A. Lammertsma and B. N. van Berckel, *Brain*, 2012, **135**, 181–189.
176. P. T. Ronaldson and T. P. Davis, *Ther. Delivery*, 2011, **2**, 1015–1041.
177. E. A. Eugenin, J. E. Clements, M. C. Zink and J. W. Berman, *J. Neurosci.*, 2011, **31**, 9456–9465.
178. J. Zhang, M. Zhang, B. Sun, Y. Li, P. Xu, C. Liu, L. Liu and X. Liu, *J. Neurochem.*, 2014, **131**, 791–802.
179. J. W. Polli, T. M. Baughman, J. E. Humphreys, K. H. Jordan, A. L. Mote, J. A. Salisbury, T. K. Tippin and C. J. Serabjit-Singh, *J. Pharm. Sci.*, 2003, **92**, 2082–2089.
180. E. Fox and S. E. Bates, *Expert Rev. Anticancer Ther.*, 2007, **7**, 447–459.
181. W. R. Friedenberg, M. Rue, E. A. Blood, W. S. Dalton, C. Shustik, R. A. Larson, P. Sonneveld and P. R. Greipp, *Cancer*, 2006, **106**, 830–838.
182. J. Abraham, M. Edgerly, R. Wilson, C. Chen, A. Rutt, S. Bakke, R. Robey, A. Dwyer, B. Goldspiel, F. Balis, O. Van Tellingen, S. E. Bates and T. Fojo, *Clin. Cancer Res.*, 2009, **15**, 3574–3582.
183. I. E. Kuppens, E. O. Witteveen, R. C. Jewell, S. A. Radema, E. M. Paul, S. G. Mangum, J. H. Beijnen, E. E. Voest and J. H. Schellens, *Clin. Cancer Res.*, 2007, **13**, 3276–3285.
184. D. Miller, *Clin. Pharmacol. Ther.*, 2014, **97**(4), 395–403.
185. W. M. Pardridge and W. H. Oldendorf, *Biochim. Biophys. Acta*, 1975, **401**, 128–136.
186. M. Gynther, K. Laine, J. Ropponen, J. Leppanen, A. Mannila, T. Nevalainen, J. Savolainen, T. Jarvinen and J. Rautio, *J. Med. Chem.*, 2008, **51**, 932–936.
187. L. Peura, K. Malmioja, K. Huttunen, J. Leppanen, M. Hamalainen, M. M. Forsberg, M. Gynther, J. Rautio and K. Laine, *Pharm. Res.*, 2013, **30**, 2523–2537.
188. E. G. Geier, A. Schlessinger, H. Fan, J. E. Gable, J. J. Irwin, A. Sali and K. M. Giacomini, *Proc. Natl. Acad. Sci. U. S. A.*, 2013, **110**, 5480–5485.
189. A. Penmatsa, K. H. Wang and E. Gouaux, *Nature*, 2013, **503**(7474), 85–90.

Drug Transporters in the Lung: Expression and Potential Impact on Pulmonary Drug Disposition

LENA GUSTAVSSON,*[a] CYNTHIA BOSQUILLON,[b] MARK GUMBLETON,[c] TOVE HEGELUND-MYRBÄCK,[d] TAKEO NAKANISHI,[e] DAN PRICE,[c] IKUMI TAMAI[e] AND XIAO-HONG ZHOU[f]

[a] Department of Drug Metabolism, H. Lundbeck A/S, Denmark; [b] School of Pharmacy, University of Nottingham, UK; [c] Cardiff School of Pharmacy and Pharmaceutical Sciences, University of Cardiff, UK; [d] Drug Metabolism and Pharmacokinetics, Respiratory, Inflammation and Autoimmunity iMed, AstraZeneca R&D, Göteborg, Sweden; [e] Department of Membrane Transport and Biopharmaceutics, Faculty of Pharmacy, Kanasawa University, Japan; [f] Department of Bioscience, Respiratory, Inflammation and Autoimmunity iMed, AstraZeneca R&D, Göteborg, Sweden
*Email: LEGU@lundbeck.com

6.1 Introduction

In recent years it has become evident that several drug transporters of the solute carrier (SLC and solute carrier organic anion (SLCO)) and ATP binding cassette (ABC) super families are highly expressed in the lung. Active transport processes may affect the local pulmonary disposition of drugs

RSC Drug Discovery Series No. 54
Drug Transporters: Volume 1: Role and Importance in ADME and Drug Development
Edited by Glynis Nicholls and Kuresh Youdim

Published by the Royal Society of Chemistry, www.rsc.org

administered directly to the lung by the inhaled route as well as the uptake of drugs to the lung from the systemic circulation. The inhaled route of drug delivery provides a unique opportunity to administer drugs directly to the target organ in respiratory diseases. Despite extensive experience and clinically successful therapies in this area, the local pharmacokinetics (PK) in the lung after drug inhalation remain poorly understood. The pharmaceutical industry has focussed on the control of dose size and drug particle deposition in the airways after inhalation. In contrast, the mechanisms of absorption across the pulmonary epithelium as well as the distribution of a drug in different compartments of the lung, once the compound is dissolved, have been less thoroughly investigated. Drug transporters may affect several processes that have an impact on the pharmacological efficacy and toxicity of inhaled drugs, such as the rate and extent of distribution of compounds to the target site, accumulation in specific cell types and retention of the drug in the lung, as well as absorption into the systemic circulation. In addition, drugs may affect the transport of endogenous substances across biological membranes in the lung, thereby contributing to the pharmacological effect profile. Recent reports on the potential impact of organic cation transporters (OCTs) on the disposition of inhaled drugs used in the treatment of asthma and chronic obstructive pulmonary disease (COPD) have highlighted the need to enhance our understanding in this emerging area. In this chapter, the current literature on drug transporter expression in the lung, the functional impact of transporters as well as models used to study pulmonary drug disposition are reviewed.

6.2 The Lung: Anatomy, Morphology and Physiology

6.2.1 Overview

The lung is a complex organ with a variety of different cell types (Figure 6.1), and this is important to take into consideration when assessing the expression and role of pulmonary drug transporters. Based on anatomy, the lung may be divided into the intrapulmonary conducting airways (central compartment), consisting of bronchi and non-respiratory bronchioles, and the distal respiratory tract (peripheral compartment), including the respiratory bronchioles and alveoli (Figure 6.1). There are extensive differences in structure and cell morphology between the central and peripheral regions of the lung. The pulmonary epithelium is much thinner in the alveoli (approximately 0.2 µm thick) compared with the conducting airways (columnar cells approximately 60 µm thick).[1] Furthermore, the surface area is much larger in the alveoli (more than 100 m^2 in the human lung) compared with the bronchial part (a few m^2).[1] The epithelium in the alveoli is highly perfused by an extensive capillary bed, which receives the entire cardiac output to the lung. The large surface area, combined with the high blood perfusion, enable an efficient exchange of oxygen and carbon dioxide between the air and the blood in the distal respiratory tract. Consideration of

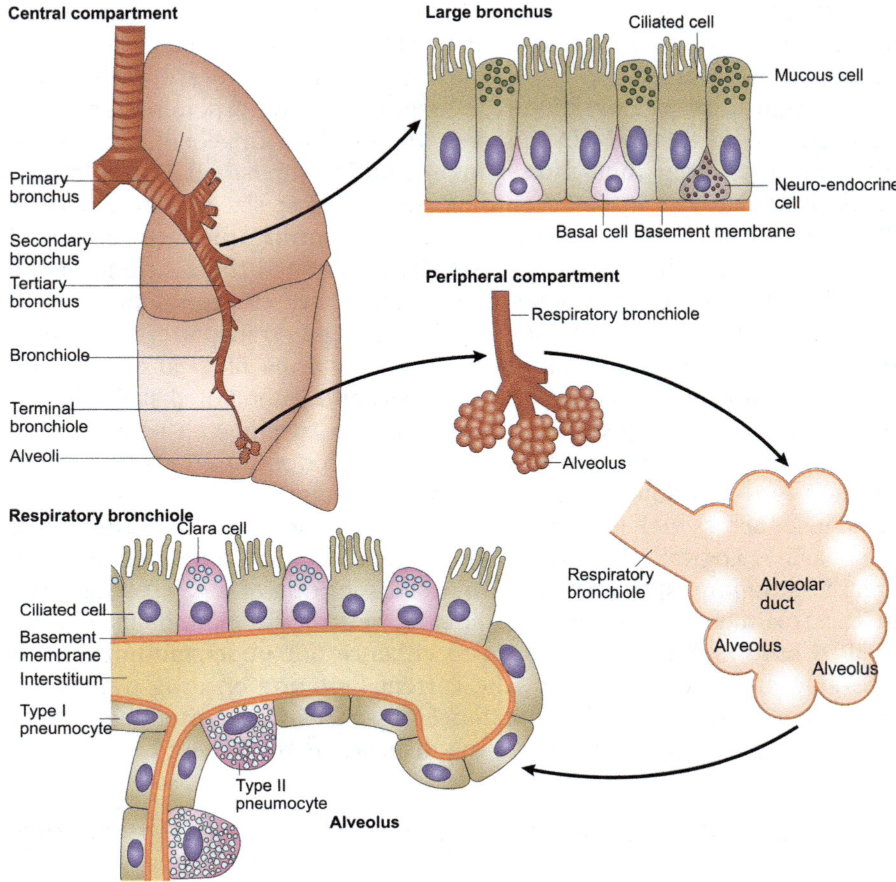

Figure 6.1 Illustration of the lung anatomy and different types of epithelial cells in the intrapulmonary conducting airways (central compartment) and the distal respiratory tract (peripheral compartment).
Reprinted with permission from Macmillan Publishers Ltd.[184]

the differences in morphology between the central and peripheral regions is important when assessing pulmonary drug disposition, since the rate and extent of pulmonary drug absorption and distribution may vary between the different parts of the lung, as well as between different cell types. Likewise, disease may cause structural and biochemical changes that affect drug absorption.

6.2.2   The Healthy Lung

6.2.2.1   *Intrapulmonary Conducting Airways*

The main bronchi start from the bifurcation of the trachea, and then enter each lung together with the pulmonary arteries at the lung hilum. They are divided into two lobar bronchi in the left lung and three lobar bronchi in the

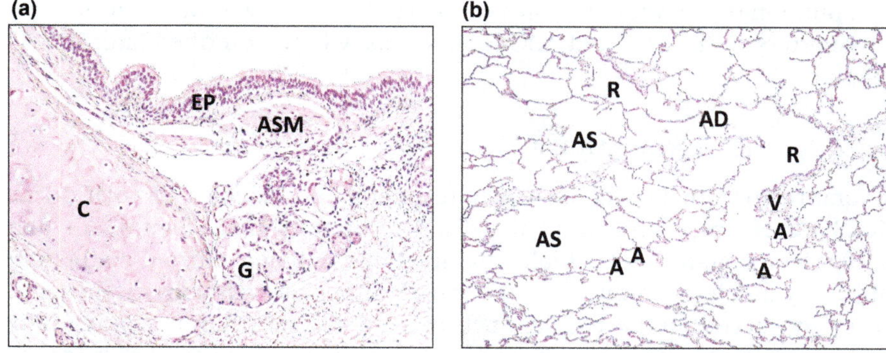

Figure 6.2 Histological images illustrating healthy human lung tissue from the (a) bronchus and (b) distal respiratory tract. Tissues were stained with haematoxylin and eosin (2.5×images). A: alveolus; AD: alveolar duct; AS: alveolar sac; ASM: airway smooth muscle bundle; C: bronchial cartilage; EP: pseudostratified epithelium; G: seromucous glands; R: respiratory bronchiole; V: pulmonary vessel.

right lung. Each of the lobar bronchi divides into variable numbers of segmental bronchi, which further divide into several generations of bronchioles.[2] The basic structure of the conducting airways (Figures 6.1 and 6.2a) consists of:

- *The epithelial mucosa*, which contains columnar ciliated cells interspersed with mucus-secreting goblet cells, basal cells, neuroendocrine cells and Clara cells. All of these cells attach to the basement membrane, but not all of them reach the surface of the airway lumen.
- *The submucosa*, which contains variable amounts of fibrocollagenous tissue, leukocytes and seromucous glands. This is surrounded by a submucosa coat consisting of smooth muscle bands, elastic fibres and partial cartilaginous rings. The seromucous glands and cartilage rings eventually disappear where the bronchi segments divide into bronchioles.

Each cell type in the conducting airways has a specific function. The spontaneous beating of cilia at the surface of the ciliated cells in the epithelial mucosa removes excess mucus in the airways.[2] This has an impact on the pulmonary distribution of drugs, since particulates deposited in the central airway compartment may be eliminated by mucociliary clearance.[3] The basal cells have the potential to differentiate into different cell types in the airways and as such they serve as stem cells.[4] The neuroendocrine cells are thought to play important roles in lung development prenatally and in lung repair and regeneration in adulthood.[4] Clara cells, named after the Austrian histologist Max Clara, are most abundant in terminal bronchioles and secrete a 10 kDa protein (CC10) that inhibits several inflammatory cytokines;[5] they have also been demonstrated to secrete endothelin, which is a powerful broncho- and vaso-constrictor.[6] Several of the cell types in the intrapulmonary conducting airways have been shown to express drug transporters of the SLC and ABC families, as will be described later in this chapter.

6.2.2.2 *Distal Respiratory Tract*

Beyond the terminal bronchioles, the respiratory bronchioles, alveolar ducts, alveolar sacs and alveoli form the distal respiratory tract (Figures 6.1 and 6.2b). The respiratory bronchioles are lined by a cuboidal ciliated epithelium and a very thin layer spiral of smooth muscle. The cilia eventually disappear in the epithelial cells of the alveolar ducts and alveolar sacs. A human adult lung contains about 300–500 million alveoli. Most alveoli open to an alveolar sac or an alveolar duct; however, a few of them open into a respiratory bronchiole.[7] The alveolar wall is rich in blood capillaries, which together with alveolar epithelial cells and their extracellular matrix components make up the centre of the gaseous exchange. The dominant cellular components of the alveoli are described below:

- *Type I pneumocytes* represent about 40% of the alveolar cell population but cover 90% of the surface of the alveolar sacs and alveoli. Their thin cytoplasm forms the cover of the alveolar basement membrane and contributes to the efficient gaseous exchange.
- *Type II pneumocytes* represent about 60% of the alveolar cell population but only populate 5–7% of the alveolar area.[7] They are the main suppliers of the surfactant proteins that make up the air–liquid interface in alveolar walls and are the key components to keep the surface tension of alveoli during breath.[8] The other important functions of type II pneumocytes are as progenitor cells for type I pneumocytes.[9] It has been reported that type II pneumocytes are involved in the remodelling process in the progression of certain diseases, for instance in COPD[10] and idiopathic pulmonary fibrosis (IPF).[11]
- *Alveolar macrophages* are the first line of the lung defence against inhaled bacteria[12] and can migrate freely between air spaces and inter-alveolar septa.[2] They also phagocytise inhaled exogenous particles, including drug particles and endogenous debris.[7] In addition, they secrete various agents including enzymes, for instance collagenase, elastase, lysozyme and different isoforms of matrix metalloproteases (MMPs).[13] The number of alveolar macrophages is augmented in smokers and individuals exposed to dust.[13]

Type I and type II pneumocytes as well as alveolar macrophages express transporter proteins that may affect drug disposition (see Section 6.4).

6.2.3 **The Diseased Lung**

Pathological changes in the diseased lung may affect drug disposition as well as transporter expression. Asthma,[14] chronic bronchitis and COPD[15] are

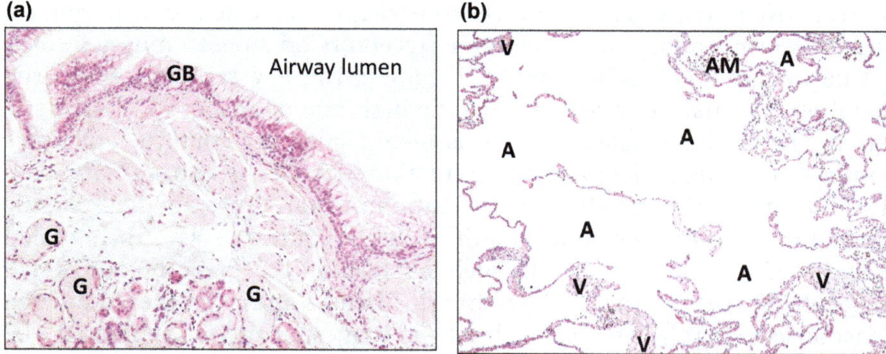

Figure 6.3 Histological images illustrating diseased human lung tissue from the bronchus and distal respiratory tract. Tissues were stained with haematoxylin and eosin. (a) Chronic bronchitis with mucous cell hyperplasia from a COPD bronchus (10×image). (b) Emphysematous lung with enlarged alveolus with increased thickness of alveolar septa as well as increased numbers of alveolar macrophages (2.5× image). A: alveolus; AM: alveolar macrophages; G: seromucous glands; GB: goblet cells; V: pulmonary vessel.

associated with chronic inflammation and extensive morphological and functional changes. These include airway epithelial shedding, goblet cell and submucosal gland hyperplasia, mucus hypersecretion and infiltration of inflammatory cells[12,16] (Figure 6.3a). Another central symptom in asthma and COPD is bronchoconstriction, which is the narrowing of the bronchi and bronchioles of the lung leading to restricted air flow. The restricted air flow may be due to spasmodic contraction of smooth muscles in the airway walls, inflammation or excessive production of mucus, due for example to irritation by allergens or overcooling. Emphysema, commonly observed in COPD, is characterised by an abnormally enlarged air space beyond the terminal bronchioles with destruction of the epithelium[12] (Figure 6.3b). This is associated with abnormal repair and remodelling of lung tissue. Pulmonary infectious diseases and damage caused by airborne pollutants may also affect the morphology, function and drug disposition in the lung.[12]

6.3 Inhalation Therapy and Pulmonary Drug Disposition

Delivering a drug topically to the lung by inhalation is a well-established technique that has specific benefits in terms of target organ selectivity— achieving a high local concentration in the target organ whilst keeping the systemic drug levels low.[17] Consequently, an increased therapeutic index and fast onset are two important advantages of the inhaled route of administration, and inhaled drug strategies have proven clinically successful in the treatment of respiratory diseases such as asthma and COPD.[18] Various inflammatory processes observed in these disorders are commonly treated by corticosteroids, whereas bronchoconstriction is treated with β-agonists and muscarinic antagonists that target receptors on smooth muscle cells.

Due to the huge surface area of the lung, the low capacity of its drug metabolising enzymes and the generally high rate of absorption to the systemic circulation, inhaled delivery is also a potential route to efficiently administer a drug to the systemic circulation without the issues of first-pass metabolism.[1] Although the strategy to administer drugs to the systemic circulation through the lung is still at an exploratory stage, the inhaled delivery of biopharmaceuticals such as insulin for diabetes and small molecules for the treatment of psychiatric diseases and migraines, for which a fast onset is highly beneficial,[17] has been suggested and may potentially be developed in the future.

The fate of the drug from the initial delivery of the drug particle to the lung surface and its appearance in the target compartment or in the systemic circulation consists of several different processes.[3,17,19] The first step is the deposition of the drug and the resulting regional drug distribution, which is dependent on the nature and particle size of the inhaled drug material, the flow rate, and the complex anatomical structure of the lung. It should also be noted that a substantial proportion of the inhaled drug generally ends up in the intestine due to the deposition of a large fraction of the dose in the upper airways and subsequent mucociliary clearance. Computational and *in vitro* models have been developed to predict and optimise the deposition pattern depending on target localisation, *e.g.* central *versus* peripheral lung compartments.[17,19,20] After the particle reaches the surfactant and the lining fluid, it needs to dissolve before the drug molecules can be absorbed across the pulmonary epithelium. For compounds with low solubility, *e.g.* some corticosteroids, the dissolution rate may be the rate limiting step in absorption.[17] Such compounds may be subject to mucociliary clearance and partly removed from their site of deposition.[3]

The physicochemical properties of drugs play a key role in their disposition. In general, the physicochemical space is more diverse for inhaled drugs compared with orally dosed drugs, and in particular there are several inhaled compounds with a higher degree of polarity (Table 6.1).[21–23] So far, there are no defined requirements such as the Lipinski's rule of five for drug absorption in the lung.[22] Since one of the advantages of inhaled treatment of the lung in respiratory diseases is the possibility of keeping the systemic exposure low, it is commonly desirable to develop compounds with poor intestinal absorption. Indeed, many successful inhaled drugs, *e.g.* muscarinic antagonists and β-agonists, are highly hydrophilic and do not readily cross cell membranes (Table 6.1). Because of this, their disposition may be highly dependent on drug transporters for uptake, efflux and vectorial transport across epithelial cell layers. The Biopharmaceutics Drug Disposition Classification System (BDDCS) was developed for orally administered drugs and provides a model to predict the role of transporters in drug disposition and absorption[24] (for more details see Chapter 1). A similar reasoning may be used for inhaled drugs. Drugs that are highly

Table 6.1 Physicochemical properties of selected inhaled drugs and the likely impact of uptake transporters on their disposition.

Drug	Target profile	Ion class	Log D$_{7.4}$	Passive permeability	Verified substrate of	Likely impact of uptake transporters	Ref.
Ipratropium	Antimuscarinic	Quaternary	−0.19	Low	OCT1, OCT2, OCTN2	High	21, 53, 104, 109
Tiotropium	Antimuscarinic	Quaternary	−1.1[b]	Low	OCT1, OCT2	High	21, 109
Albuterol[a]	β-agonist	Base	−1.8	Low	OCT1, OCT2	High	21, 185
Formoterol	β-agonist	Base	0.34	Intermediate	OCT1, OCT3	Intermediate	21, 91
Salmeterol	β-agonist	Base	2.0	High	ND[c,e]	Low	21
Terbutaline	β-agonist	Base	−1.5	Low	OCT1, OCT2	High	21
Budesonide	Corticosteroid	Neutral	2.9	High	ND[d,e]	Low	26
Fluticasone propionate	Corticosteroid	Neutral	>4.2	Intermediate	P-gp (weak)	Intermediate	26

[a]Alternative name: salbutamol.
[b]Calculated.
[c]Tested for OCT1 and OCT2 but not identified as a substrate.
[d]Tested for P-gp but not identified as a substrate.
[e]ND: not detectable.

lipophilic and display a high passive permeability are less likely to be significantly influenced by uptake transporters, *i.e.* SLCs, since their passive diffusion into cells is rapid.[21,25] β-agonists are generally bases with a variability in lipophilicity *e.g.* from albuterol, which has a low log D and only slowly diffuses passively across cell membranes, to salmeterol, which is highly lipophilic and has a high passive permeability[21] (Table 6.1). Most of the muscarinic antagonists, for example ipratropium and tiotropium, which contain a quaternary amine group and show poor passive permeability, and β-agonists are substrates of OCTs, and for those with low lipophilicity, these transporters have a major impact on their cellular uptake.[21] Other groups of inhaled drugs, *e.g.* corticosteroids, are commonly lipophilic. Once dissolved, they rapidly distribute across cell membranes through passive diffusion, and thus, facilitated transport by a transporter protein would most likely not have a major impact. Conversely, ATP-driven transporters may potentially play a role in the disposition of these types of drugs at sub-saturating concentrations. Although some corticosteroids are substrates of P-glycoprotein (P-gp), the transport observed was less pronounced for chemical entities developed for inhaled treatment.[26] It is not known whether inhaled corticosteroids are substrates of other transporters of the ABC family. To add further to the complexity of inhaled drug delivery, combination therapy is commonly used in the inhaled treatment of respiratory diseases. In the treatment of asthma, a glucocorticoid is often administered together with a β-agonist and/or a muscarinic antagonist in order to treat inflammation and bronchoconstriction with the same device. As will be further discussed in Section 6.6.1, some corticosteroids can inhibit OCTs, which may potentially influence the pulmonary disposition of co-administered drugs.

The process of drug transport across the pulmonary epithelial cell barrier is poorly understood and complicated by the regional differences and heterogeneity in cell types. Despite these challenges, there is emerging evidence that drug transporters expressed on pulmonary epithelial cells can play a role in the disposition of inhaled drugs. Thus, an understanding of the impact of pulmonary drug transporters on pharmacological efficacy and potential drug–drug interactions (DDIs) is highly desirable. Physiologically-based PK (PBPK) modelling and simulation is routinely used for the prediction of clinical DDIs of orally dosed drugs, and it would be highly beneficial to develop such a PBPK model for pulmonary drug disposition in the future.[27] Furthermore, an understanding of local drug disposition in the lung is important in order to use rational design in the development of effective and safe inhaled medicines, as well as for providing mechanistic explanations of observations in pharmacological and toxicological models.

6.4 Drug Transporter Families in the Human Lung

As mentioned in the previous section, several transporters are present in lung tissue. In a study of 40 human tissues, Bleasby and co-workers demonstrated that, next to the liver and kidney, the lung was overall one of the tissues with the highest levels of drug transporter gene expression.[28] However, as will be described later in this section, as well as in Section 6.6, some of the drug transporters of importance in the liver, kidney and intestine display a much lower gene expression in the lung. In terms of mRNA expression, two recent studies have investigated drug transporter mRNA expression in samples collected from specific regions of the human lung.[29,30] However, drug transporter expression at the protein level is in general not as well characterised in the lung. Recently, Sakamoto and co-workers published the first studies with liquid chromatography-tandem mass spectrometry (LC-MS/MS)-based quantification of drug transporter proteins in lung tissue from human subjects[31] as well as pulmonary cell lines.[32] An overview of drug transporter expression in human lung tissue is presented in Tables 6.2–6.4.

Based on gene expression data, the SLCs relevant to drug transport that are most highly expressed in the lung are the peptide transporter PEPT2 (*SLC15A2*), the organic cation transporters OCT3 (*SLC22A3*), OCTN1 (*SLC22A4*) and OCTN2 (*SLC22A5*), and the equilibrative nucleoside transporter (ENT1; *SLC29A1*) (Table 6.2). The expression of these transporters in the lung has also been confirmed at the protein level. The organic cation transporters OCT1 (*SLC22A1*) and multidrug and toxin extrusion transporter 1 (MATE1; *SLC47A1*), as well as the amino acid transporters ATB(0+) (*SLC6A14*) and LAT1 (*SLC7A5*) are also present in pulmonary tissues (Table 6.2). On the other hand, organic anion transporter (OAT) 1 and 3 (*SLC22A6* and *8*), which are highly expressed and functionally important for drug transport in the kidney, are not or are barely detectable in the lung.[29,30] In terms of OCT2 (*SLC22A2*), one of the main kidney transporters, there are inconclusive data on its expression in lung tissue. With regard to organic anion polypeptides (OATP; *SLCO*), OATP2A1/prostaglandin transporter (PGT; *SLCO2A1*), OATP2B1 (*SLCO2B1*), OATP3A1 (*SLCO3A1*), OATP4A1 (*SLCO4A1*) and OATP4C1 (*SLCO4C1*) are highly expressed at the gene level (Table 6.3). OATP1B1 (*SLCO1B1*) and SLCO1B3 (*SLCO1B3*), which are important drug transporters in the liver, show no or low gene and protein expression in the lung, respectively. Except for OATP2A1 and OATP2B1, the expression of OATPs in the lung has not been confirmed at the protein level. In this chapter, the SLC and SLCO transporters for which evidence of functional activity relevant to pharmacology and drug disposition exist will be addressed further.

Among the ABC transporters, genes of the multidrug resistance protein 1 (MDR1), also named P-gp (*ABCB1*), and the breast cancer resistance protein (BCRP; *ABCG2*) are expressed in the lung, and their localisation in the lung has also been investigated at the protein level (Table 6.4). Based on mRNA data, several of the multidrug resistance associated proteins (MRPs; *ABCC* family) are expressed in the lung (Table 6.4). The most well studied MRP in the lung is MRP1 (*ABCC1*), and its mRNA and protein expression has been found to change in COPD.[33] Other ABC transporters that are highly expressed in the lung are *ABCA1* and *ABCA3*, which secrete lipids and surfactants out of alveolar type II cells (Table 6.4). Mutations in the *ABCA3* gene have been reported to increase the risk of neonatal respiratory distress

Table 6.2   Expression of selected SLCs in the human lung selected based on their relevance for transport of drugs or pulmonary endogenous substrates.[a]

Transporter	Substrates	mRNA expression[29,30]		Protein expression	
Protein (*gene*)	Examples	Central	Peripheral	LC-MS/MS[31]	Localisation in the lung (IHC)
OCT1 (*SLC22A1*)	Organic cations; ipratropium, tiotropium, β-agonists, acetylcholine[21,90]	+	+	+	Apical; ciliated bronchial epithelial cells[93]
OCT2 (*SLC22A2*)	Organic cations; ipratropium, tiotropium, β-agonists[21,90]	+/−	ND[b]	+	Apical; ciliated bronchial epithelial cells[93] (contradictory data)
OCT3 (*SLC22A3*)	Organic cations; epinephrine, norepinephrine, histamine, β-agonists[90,102]	++/+	++	ND[b]	Basal cells, basolateral membrane of intermediate cells, apical membrane of ciliated cells, airway smooth muscle cells, endothelium[92–94]
OCTN1 (*SLC22A4*)	Ergothioneine[90]	++	+/++	+++	Apical; tracheal epithelium, alveolar macrophages (cytoplasmic)[92]
OCTN2 (*SLC22A5*)	L-carnitine[90]	+/++	+/++	ND[b]	Apical; airway epithelium, alveolar epithelium[92]
OAT1–3 (*SLC22A6–8*)	NA[d]	ND[b]	ND[b]	ND/+ [b]	Low expression or absent
MATE1 (*SLC47A1*)	Similar to OCT1[90,186]	+	+/++	NT[c]	Unknown
MATE2 (*SLC47A2*)	NA[d]	ND[b]	ND[b]	NT[c]	Low expression or absent
PEPT1 (*SLC15A1*)	NA[d]	+/−	ND[b]	ND[b]	Low expression
PEPT2 (*SLC15A2*)	Di- and tri-peptides, peptidomimetics, β-lactam antibiotics[113]	++	++	++	Apical; bronchial epithelium, alveolar type II cells, endothelium[116]
ATB(0+) (*SLC6A14*)	Neutral and cationic amino acids, valacyclovir[143,149]	++	+	NT[c]	Apical; airway ciliated epithelial cells, alveolar epithelial cells[143]
LAT1 (*SLC7A5*)	Amino acids[187]	+/++	+/++	NT[c]	Bronchial epithelial cells, chondrocytes, serous cells, alveolar macrophages[188]
ENT1 (*SLC29A1*)	Nucleosides, adenosine, nucleoside analogues *e.g.* gemcitabine[189]	++	++/+++	NT[c]	Unclear

[a]Relative expression levels based on ref. 29–31: +/−: very low expression; +: low expression; ++: moderate expression; +++: high expression; /: both expression levels have been reported.
[b]ND: not detectable.
[c]NT: not tested.
[d]NA: not applicable.

Table 6.3 Expression of selected SLCOs in the human lung selected based on their relevance for transport of drugs or pulmonary endogenous substrates.[a]

Transporter	Substrates	mRNA expression[29,30]		Protein expression	
Protein (*gene*)	Examples	Central	Peripheral	LC-MS/MS[31]	Localisation in the lung (IHC)
OATP1A2 (*SLC01A2*)	Organic anions, including glucuronide conjugates, bile acids, statins[190]	+	+	+	Unknown
OATP1B1 (*SLCO1B1*)	NA[c]	ND[b]	ND[b]	ND[b]	No expression
OATP1B3 (*SLCO1B3*)	NA[c]	−/+	−/+	+	No or low expression
OATP2A1/PGT (*SLCO2A1*)	Prostaglandins[190]	++	+++	++	Bronchial epithelial cells, alveolar epithelial cells[141]
OATP2B1 (*SLCO2B1*)	Estrone-3-sulfate, prostaglandins, statins[190]	++	++/+++	+	Unknown
OATP3A1 (*SLCO3A1*)	Estrone-3-sulfate, prostaglandins, statins[190]	+/++	+/++	ND[b]	Unknown
OATP4A1 (*SLCO4A1*)	Estradiol 17β-glucuronide, estrone-3-sulfate, bile acids[190]	++	+/++	ND[b]	Unknown
OATP4C1 (*SLCO4C1*)	Estrone-3-sulfate	+/++	+/++	ND[b]	Unknown

[a]Relative expression levels based on ref. 29–31: +/−: very low expression; +: low expression; ++: moderate expression; +++: high expression.
[b]ND: not detectable; NT: not tested.
[c]NA: not applicable.

Table 6.4 Expression of selected ABC transporters in the human lung selected based on their relevance for transport of drugs or pulmonary endogenous substrates.[a]

| Transporter | Substrates | mRNA expression[29,30] | | Protein expression | |
Protein (*gene*)	Examples	Central	Peripheral	LC-MS/MS[31]	Localisation in the lung (IHC)
P-gp/MDR1 (*ABCB1*)	Broad range of drug substrates, chemotherapeutics, corticosteroids[152,191,192]	+/++	+/++	+	Apical; bronchial/bronchiolar epithelium, alveolar type I cells, endothelial cells, alveolar macrophages[154,156–158,193]
BCRP (*ABCG2*)	Chemotherapeutics, tyrosine kinase inhibitors, glucuronide and sulfate conjugates, ciprofloxazine[164,171,194]	+	++	+++	Bronchial epithelium, endothelium, seromucinous glands[158,193]
MRP1 (*ABCC1*)	Organic anions, *e.g.* glutathione, glucuronide and sulfate conjugates, chemotherapeutics, LTC_4[164]	++	++	+++	Basolateral/lateral, bronchial/bronchiolar epithelium, goblet cells[158,169,193]
MRP2 (*ABCC2*)	Similar to MRP1[164]	−/+	−/+	ND[b]	ND[b]
MRP3 (*ABCC3*)	Similar to MRP1, methotrexate[164]	+/++	+/++	+	Unclear
MRP4 (*ABCC4*)	Cyclic nucleotides, nucleoside analogues, LTC_4, prostaglandins[164]	+/++	++	++	Unclear
MRP5 (*ABCC5*)	Cyclic nucleotides, nucleoside analogues, methotrexate[164]	+/++	+/++	+	Unclear
MRP6 (*ABCC6*)	BQ123, anthracyclins[164]	++	++	+	Unclear
CFTR (*ABCC7*)	Chloride channel (defect in cystic fibrosis)[164]	+	+	NT[c]	Apical; bronchial epithelium, seromucinous glands[193,195,196]
ABCA1 (*ABCA1*)	Cholesterol, lipids, surfactant[164]	++	++/+++	NT[c]	Alveolar type II cells[193,197]
ABCA3 (*ABCA3*)	Surfactant[164]	++	++/+++	NT[c]	Alveolar type II cells[193,198]

[a]Relative expression levels based on ref. 29–31: +/−: very low expression; +: low expression; ++: moderate expression; +++: high expression.
[b]ND: not detectable.
[c]NT: not tested.

syndrome.[34] The cystic fibrosis transmembrane conductance regulator (CFTR; *ABCC7*) is a chloride channel that is defective in cystic fibrosis. In this chapter, focus will be on the most highly expressed ABC transporters that are important for drug and drug metabolite transport: P-gp, BCRP and MRP1.

Except for a few ABC and SLC/SLCO transporters, the localisation of drug transporter proteins in the human lung is unclear (Tables 6.2–6.4). The spatial distribution of drug transporters in the different regions of the lung is important to consider given the variety of cell types and regional differences in epithelial cell morphology. In a recent study on well-characterised material from the human lung, OCTN1 and PEPT2 were more highly expressed in the central airways, whereas MATE1, OATP4C1 and OATP2B1 were expressed at higher levels in peripheral lung tissues.[29] Still, the association of protein expression to specific cell types needs to be confirmed for most drug transporters. Given the variety of cell types in the human lung, care should be taken before concluding that transporters with a low expression in a piece of lung tissue are not important, since they actually may be concentrated in a specific cell type, and this cell type could constitute a substantial proportion of the total tissue area (*e.g.* type I pneumocytes).

With regard to drug transporter protein expression in specific cell types, most studies have focused on epithelial cells. In general, immunohistochemistry (IHC) of SLC proteins has indicated an apical localisation of OCT1, OCTN1, OCTN2, PEPT2 and ATB(0+) in epithelial cells in the lung (Table 6.2). Some SLC proteins such as OCT3 are also expressed in other pulmonary cell types, such as smooth muscle cells, endothelial cells and basal cells. With regard to ABC transporters, the P-gp protein was detected on the apical side of epithelial cells whereas MRP1 was observed on the basolateral/lateral membrane (Table 6.4). The localisation of pulmonary transporters in specific cell types, where known, will be described in more detail in the section that discusses their functional importance (Section 6.6).

There are few studies addressing the effect of disease on drug transporter expression. No statistically significant differences were observed in either central or peripheral lung tissue from patients with severe COPD and healthy individuals.[29] It should be noted that the COPD patients in this study were non-smokers for at least 6 months before sampling. However, in a previous study, a correlation between a decrease in MRP1 expression in lung tissue from COPD patients and the severity of disease was reported.[33] In another study, no difference in PEPT2 expression and distribution was found in lung tissue from individuals with cystic fibrosis compared with healthy subjects.[35] It would be expected that certain disease states could cause a change in transporter expression, as has been observed for transporters in other organs. However, it should be noted that there is still a very limited number of studies published and the ones reported include relatively few individuals (around ten or fewer). Thus, currently there are insufficient data to conclude whether respiratory disease states have an effect on the expression of specific transporters in the lung.

6.5 *In vitro* and *In vivo* Models to Study Pulmonary Drug Disposition

6.5.1 Cell Culture Models

In vitro cell models are important tools for screening and for mechanistic studies in drug discovery and development. Although needed for an understanding of pulmonary drug disposition, which is influenced by the anatomy and physiology of the organism, studies in laboratory animals are technically challenging, low throughput and onerous. In addition, there are important ethical aspects that need to be considered, as described by the 3R guide of Replacement, Reduction and Refinement of animal studies.[36] *In vitro* models of the pulmonary absorption barrier consist of cell layers of bronchial or alveolar epithelial origin grown on a permeable artificial membrane mounted into cell culture inserts. Typically, cells are maintained at an air–liquid interface created by removing the culture medium from the apical compartment of the inserts 24 hours to a few days after cell seeding (Figure 6.4). These conditions mimic the physiological environment in the respiratory tract and have been demonstrated to promote cell differentiation into layers morphologically close to the corresponding native epithelium.[37,38] *In vitro* models suitable for vectorial drug transport studies must exhibit absorption barrier properties, which are reflected by the formation of functional tight junctions between cells and the development of a meaningful transepithelial electrical resistance (TEER). Table 6.5 presents an overview of different *in vitro* airway epithelial cell models and their ability to form tight junctions and suitability for culture at an air–liquid interface. In addition, for drug transport screening, it is paramount that they express the range of drug transporters present in the human airway epithelium.

The majority of *in vitro* respiratory absorption systems currently available for drug testing are of bronchial origin.[39,40] Indeed, no convenient, physiologically-relevant and low cost model of the alveolar epithelium has been successfully established to date (2016). For example, although potentially useful for toxicity testing, the human alveolar epithelial cell line A549 has limited applications as a drug permeability screen: its phenotype is close to Type II pneumocytes, a cell population that only represents 5–7% of the

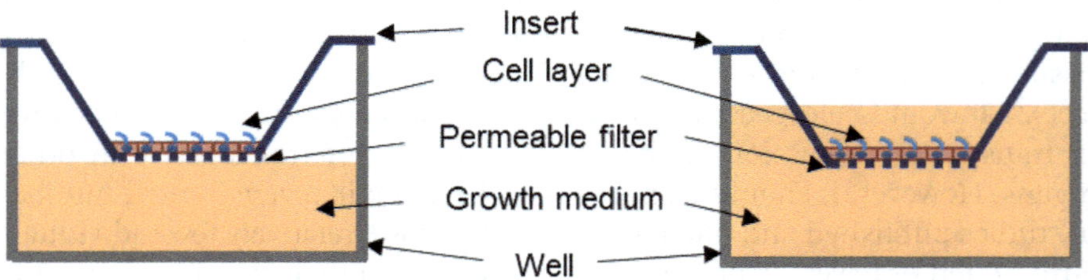

Figure 6.4 Schematic representing airway epithelial cells grown in cell culture inserts in air–liquid (left) or submerged (right) conditions.

Table 6.5 Comparison of airway epithelial cell culture models in terms of their ability to form tight junctions and suitability for culture at an air–liquid interface.

Cell model[a]	Origin	Tight junction formation[b]	Air–liquid interface
Alveolar			
Type I like	Primary cells	Yes	Yes
A549	Cancerous	No	No
TT1	Immortalised	No	No
Bronchiolar			
NCI-H441	Cancerous	Yes	No
Bronchial			
NHBE	Primary cells	Yes	Yes
Calu-3	Cancerous	Yes	Yes
16HBE14o-	Immortalised	Yes	No
VA10	Immortalised	Yes	Yes
NuLi	Immortalised	Yes	Yes
BEAS-2B	Immortalised	No	No

[a]For references, see Section 6.5.1.
[b]As assessed by TEER values >100 Ω cm^2 on Transwell inserts.

alveolar surface area, and it is unable to form electrically tight cell layers when grown on permeable filters.[39,40] A type I pneumocyte cell line TT1 has recently been generated, but again, fails to develop into layers exhibiting TEER values above the background reading (Table 6.5).[41] Modelling the alveolar epithelium *in vitro* therefore relies on primary cultures. Type II pneumocytes are isolated from the lung tissue of humans or various animal species and differentiated into cell layers presenting characteristics of type I pneumocytes together with high TEER values above 2000 Ω cm^2.[42,43] The extensive use of these type I-like cell layers is hindered by the tediousness of the model, the impossibility of passaging type II pneumocytes and the difficulties in obtaining lung tissue samples as a source for these cells. Accordingly, no *in vitro–in vivo* correlation has yet been demonstrated for type I-like pneumocyte layers. Nevertheless, they have been valuable in the elucidation of the transport mechanisms by which therapeutic compounds cross the alveolar epithelium.[44]

The unavailability of an alveolar cell line suited to permeability measurements is perceived as a major limitation in understanding drug disposition in the lung. In this context, the human adenocarcinoma cell line NCI-H441, which presents characteristics of both Clara cells and Type II pneumocytes, has been evaluated very recently as a potential surrogate. The cell line was shown to form layers with well-defined tight junction complexes, high TEER values and low permeability of paracellular markers when cultured under submerged conditions[45] (Figure 6.4 and Table 6.5). On the other hand, it was unable to generate layers sufficiently tight for transport studies when maintained at an air–liquid interface. It is not yet known whether the NCI-H441 cell line will be a valuable tool to predict drug absorption from the peripheral lung.

Several human bronchial epithelial cell culture models have been reported to develop appropriate absorption barrier properties, with each of them presenting advantages and limitations that need to be balanced. The human adenocarcinoma Calu-3 cell line is by far the most extensively used due to its availability from cell banks, maintenance of phenotype over several passages and ease of culture in low cost serum-containing media. Due to the cancerous origin of the cell line, its phenotype is different from bronchial epithelial cells. Nevertheless, Calu-3 cell layers cultured in air–liquid conditions for 10–25 days exhibit characteristic features of the bronchial epithelium, including mucus production, and reproducibly generate TEER values >400 Ω cm^2.[37] Furthermore, permeability data collected in this cell line have been shown to correlate with *in vivo* rat pulmonary absorption parameters.[46]

In contrast, normal human bronchial epithelial (NHBE) cells present a phenotype closer to cells *in vivo*. Similarly to Calu-3 cells, they can be purchased commercially and produce electrically tight, mucus-secreting layers in air–liquid culture.[38] However, they lose their ability to differentiate after three to four passages and are relatively expensive to grow since they require serum-free conditions. Cultures have also been reported to be poorly reproducible, even when cells originate from the same donor.[47,48]

In an attempt to combine the benefits offered both by cancerous cell lines and primary cells, a few human-derived cell lines have been generated through immortalisation of normal bronchial cells. Amongst these, layers of 16HBE14o- cells grown in submerged conditions for 6 days produced drug permeability values that correlate with the absorption rate constant in isolated perfused rat lungs.[49] However, the cell line is unable to differentiate into tight layers morphologically representative of the bronchial epithelium under air–liquid conditions.[50] In contrast, the recently developed VA10 cell line provides a model anatomically close to the native epithelium, exhibiting barrier properties similar to those of Calu-3 and NHBE layers at an air–liquid interface.[51] A major limitation of this model is that the cell layers require 28 days in culture in a medium containing a serum substitute to achieve complete differentiation. Another potentially interesting model is the NuLi cell line, which also generates high TEER values when cultured in air–liquid conditions.[52] Unfortunately, this has not yet been evaluated for drug permeability measurements. Finally, the BEAS-2B cell line has been used to investigate the impact of transporters on the cellular uptake of inhaled drugs,[53] but is unable to form tight junctions[48] and, therefore, is unsuitable for vectorial drug transport studies.

It is noteworthy that besides human cell culture systems, the spontaneously immortalised rat bronchial/bronchiolar RL65 cell line has been shown to be a promising model to bridge the gap between human *in vitro* and rat *in vivo* models.[54] Culture conditions need nevertheless to be further optimised before this can be used as a drug permeability screen.

Pulmonary cell culture models have not yet been thoroughly characterised in terms of transporter expression and functionality. Data are particularly scarce for differentiated air–liquid systems, *i.e.* those valuable for drug transport studies. A semi-quantitative analysis of *ABC*, *SLC* and *SLCO* transporter gene expression in undifferentiated NHBE, Calu-3, 16HBE14o-BEAS-2B and A549 cells grown on tissue culture treated plastic surfaces as well as freshly isolated alveolar type II pneumocytes and type I-like layers has been undertaken.[55] This revealed variations in transporter levels amongst different cells and with time in culture. Of more relevance to *in vitro* permeability measurements, a quantitative micro-array analysis of the same transporter families in differentiated layers of NHBE and Calu-3 cells at low or high passage numbers was recently published.[38] Similar expression levels were obtained in primary cells and the cell line with the exception of P-gp, which was overexpressed in Calu-3 layers. Importantly, transporter expression was shown to be stable over passaging in the cell line.

Protein expression and functionality studies have mainly focussed on P-gp and, to a lesser extent, the OCT and PEPT families.[56] P-gp has been detected on the apical surface of all human respiratory cell culture models suitable for permeability studies, *i.e.* layers of NHBE,[38] Calu-3,[47] 16HBE14o-,[57] VA10,[51] NCI-H441[45] and type I cells,[58] as well as the rat RL65 cell line,[54] with, however, no functional activity observed in VA10 and RL65 cell layers.[51,54] Due to differences in culture conditions and the use of non-specific substrates and inhibitors, contradictory data regarding the transporter activity in Calu-3 cells have been reported.[56] Functionality in 21 day old layers of both Calu-3 and NHBE cells was nevertheless confirmed recently by performing a UIC2 antibody shift assay in the presence of PSC833, a potent P-gp inhibitor.[47]

OCT activity has been detected *via* uptake of the cationic fluorescent dye 4-[4-(dimethylamino)-styryl]-*N*-methylpyridinium (ASP+) in undifferentiated A549, Calu-3, 16HBE14o-[59] and NCI-H441[45] cells, as well as on the apical side of 21 day old air–liquid interfaced Calu-3 layers.[60] Similarly to NHBE layers, the latter were shown to express OCT1, OCT3, OCTN1 and OCTN2, but not OCT2.[60]

Both PEPT1 and PEPT2 were shown to be expressed apically in Calu-3 layers grown in air–liquid conditions[61] while only PEPT2 was present in NHBE layers.[62] As predicted, the transepithelial transport of the dipeptide glycylsarcosine (GlySar) was mediated by PEPT2 in primary cultures,[62] but unexpectedly by PEPT1 in Calu-3 layers.[61]

Very little information is currently available on the functionality of other drug transporters in respiratory *in vitro* models.[56] MRP1 has been localised on the basolateral side of both NHBE[63] and Calu-3 layers.[64] However, functional activity could not be ascertained in the latter due to the chemical probe employed being a substrate for both MRP1 and P-gp.[64] Finally, one study reported the presence of BCRP activity in undifferentiated Calu-3 cells.[65] On the other hand, transcripts for the transporter were not detected in air–liquid interfaced layers.[47]

Although promising, none of the respiratory *in vitro* models described above have been thoroughly assessed for permeability screening or drug–transporter interaction studies. Transporter expression and functionality data reported in the literature overall lack consistency, likely due to variations in culture conditions. There is indeed increasing evidence that passage number and time in culture affect transporter levels.[60,66] Standardisation and establishment of *in vitro–in vivo* correlations are therefore required in order to gain confidence in their usefulness in the preclinical development of drugs targeting the lung tissue.

6.5.2 Isolated Perfused Lung *Ex vivo*

The isolated perfused lung model (IPL) was first developed in the middle of the 20th century as a technique for studying lung function in preclinical animal models. During the 1970s there was a movement to develop the model for investigations of drug absorption and it is now a well-characterised and utilised model, used by many groups to study pulmonary PK.[67] The model has also been used previously to look directly at the effect of drug transporters in the lung, in particular for P-gp, where it provided the first evidence of a role for efflux proteins in lung absorption from an intact lung.[68,69]

In essence, the technique is purely the separation of the lungs from the systemic circulation by creating artificial perfusion through the pulmonary vasculature, usually with artificial media. Lungs are then maintained with artificial circulation and ventilation during the experiment[67] (Figure 6.5). Compounds may be administered to the IPL as dry powder or nebulised solution/suspension by different dosing techniques developed to mimic inhaled dosing in the clinic.[67,70] The absorption profile may then be explored after repeated sampling of perfusate and subsequent compound concentration analysis.

Although the majority of pulmonary drug absorption studies continue to be carried out using cell culture based systems, the specific roles of lung structure and cellular heterogeneity are lost in these mainly monoculture systems. The major advantage of an IPL technique is that it allows the

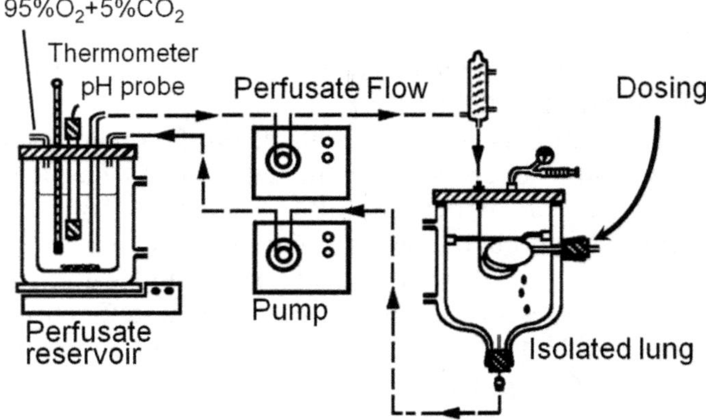

Figure 6.5 Schematic illustration of the IPL.
Reprinted with permission from Elsevier.[67]

investigator to assess the pulmonary absorption of a molecule in a system where structural and cellular integrity, cell–cell interactions and biochemical activity are all present.

By separating the lungs from the systemic circulation, investigators are also able to study the effects of pulmonary PK/pharmacodynamics (PD) without the confounding effects of systemic absorption, distribution, metabolism and excretion (ADME) observed with *in vivo* models, whilst still allowing compounds of interest to be delivered *via* physiologically relevant routes. The isolation of the organ also allows sampling at multiple time points (thus reducing the number of animals used and inter-animal variation) and has the benefit of making mass balance calculations straightforward at the termination of the experiment. This allows the IPL techniques to lend themselves well to studies of acute toxicity, metabolism, local drug disposition, potential for DDIs and indeed drug delivery.

The major limitation for IPL studies comes from the finite viability of the preparation; this is species specific but, at most, experiments can run for 5 hours at 37 °C and far less in small species such as mice, in which 1 hour experiments have been achieved.[68] This severely limits the model in studies of slow processes where an *in vivo* approach is more appropriate, especially for the mouse model; however, drug absorption studies of low molecular weight molecules tend to be rapid and well within the viability window of an IPL. The technique also requires time consuming and complex methodology and apparatus, as well as operator training, and hence is less suitable for a screening based approach. Despite these limitations, the possibility of being able to specifically study pulmonary drug disposition in isolation makes this an important model, in particular for mechanistic studies.

The initial IPL drug absorption studies were performed using rat and guinea pig lungs, and a wide range of inhaled drugs have been put through this early model.[71] Work in the 1980s expanded upon this by specifically developing IPL models to study uptake using first-order kinetic models, identifying compounds that did not behave as would be predicted by an entirely passive absorption profile.[72] For example, insulin absorption was susceptible to non-absorptive losses from mucociliary clearance and metabolism, which was confirmed by altering the deposition profile of insulin within the IPL model.[73] The study also showed that dosing concurrently with metabolic inhibitors increased absorption by preventing loss to proteolytic degradation.

The relevance of the model for the study of pulmonary drug delivery, particularly for more rapidly absorbed low molecular weight drugs, has led to its frequent use, and it is well documented that the absorption profiles obtained correlate well with physicochemical properties, and *in vitro* and *in vivo* absorption.[74] The isolated lung methodology has also been applied in the perfusion of a resected human lung lobe, comparing the onset of action and intrinsic activity of the long and short acting β2-agonists.[75] In another study, a delay in the pulmonary absorption of selected inhaled β2-agonists caused by substrate competition for OCTN transporters was observed.[76]

The IPL system has been used to look specifically at the impact of drug transporters on pulmonary PK in the intact lung. References and examples of these publications can be found in the relevant drug transporter sections later in the chapter. These experiments reveal the ability of the models to identify the role of transporter proteins, both in active uptake and efflux. Taken together, they provide good evidence of a role for drug transporters within the intact lung.

6.5.3 *In vivo* Models

In vivo studies of pulmonary drug disposition enable assessment of the impact of all physiological variables that may affect the fate of the molecule. Although transgenic animals in which one or several transporters have been knocked out or overexpressed are available from commercial sources, there are very few *in vivo* studies published demonstrating an influence of pulmonary drug transporters on drug disposition. This may not be surprising given the complexity in measuring drug disposition locally within the lung. Overall, lung PK may be investigated by measuring the systemic drug concentration after inhaled drug exposure to animals, when the free fraction in plasma can be assumed to be in equilibrium with the free fraction in the lung tissue. However, although plasma may be the only available site of measurement, it is not a good surrogate for measuring pulmonary tissue concentrations of drugs with transporter dependent disposition[77] and this is a major challenge for evaluations of drug transporter interactions in clinical studies. Moreover, after inhaled drug delivery, the plasma concentration may not accurately reflect the concentration in lung tissue for drugs with poor solubility.[17] Evaluation of the disposition of a drug, and potentially its metabolites, in the lung is therefore commonly based on the analysis of excised lung tissue.[19]

Given the heterogeneous architecture of the lung, transporter dependent drug disposition may lead to local differences in drug concentration within the organ. Thus, analysis of the whole lung tissue may not lead to the appropriate estimation of drug concentration in relation to the target site or regions of drug toxicity. Non-invasive imaging techniques such as positron emission tomography (PET) and single-photon emission computed tomography (SPECT) have proven useful in defining the localisation of drugs *in vivo* and revealing the impact of transporters in other organs, *e.g.* the brain.[78–80] One important advantage with PET and SPECT imaging is that they may be used in both the preclinical and clinical setting. Despite their usefulness in areas such as neuroscience, the experience of PET and SPECT imaging in pulmonary drug distribution is limited, with only a few published examples.[19,81,82] Imaging mass spectrometry (IMS) is another promising technique used in drug discovery and development to investigate the distribution of compounds.[83] Importantly, in contrast to techniques using radiolabelled material, IMS enables the individual analysis of both the parent compound and metabolites. Matrix assisted laser desorption ionisation (MALDI) IMS has been used to successfully localise drugs in the rat lung after inhaled treatment.[84] Recently, MALDI IMS was used to monitor the localisation of ipratropium in human bronchial biopsies obtained by fibre-optic bronchoscopy,[85] in which a spatial resolution down to 30 µM was achieved.[85] Thus, IMS is a technique that may aid our understanding of the impact of drug transporters on pulmonary drug disposition.

The deposition of drugs in the lungs is commonly influenced by the choice of administration technique. For example, intratracheal administration of a solution results in deposition of all of the dose in the lungs, but the deposition may be patchy.[86] In contrast, inhalation administration often results in a more even distribution than intratracheal administration,[86] but the deposition depends on several factors such as particle size and breathing pattern. Aerosol generation and administration also require advanced equipment and specific technical skills, and there are several devices to generate aerosols for administration of the test compound in solution, suspension or as a dry powder.[19,87,88] Frequently, a large fraction of the dose ends up in the gastrointestinal tract, which may further confound the interpretation of the PK. These are important experimental factors to consider when studying the distribution of drugs and metabolites in the lung following inhaled drug delivery.

6.6 Drug Transporters and Their Potential Impact on Inhaled Drug Disposition, Efficacy and Toxicity

6.6.1 OCTs of the *SLC22A* Family

Polyspecific OCTs are involved in the PK and tissue disposition of many substrate drugs, including β2-adrenergic and anti-cholinergic bronchodilators (Table 6.1).[21,89,90] There are two major subclasses of carriers responsible for organic cation transport that belong to the SLC22 family: OCT1–3, and OCTN1 and 2. OCT1–3 facilitate polyspecific organic cation transport in a Na^+ independent manner, whereas OCTN1 and 2 are known to recognise and transport ergothioneine and carnitine, respectively, in a Na^+ dependent manner. In addition, the carnitine transporter 2 (CT2; *SLC22A16*, also known as OCT6) has been characterised as a Na^+ independent carnitine/betaine transporter that recognises monocharged cations, and is functionally similar to mouse OCTN3.[91] To date, pulmonary expression of OCTs and OCTNs has been studied by several researchers (Table 6.2).[29–31,92,93] Using IHC, OCT1 and 2 were shown to be localised at the apical membranes of tracheal and bronchial ciliated epithelial cells in humans and rodents, and OCT2 was the major isoform in human airway epithelium.[93] However, the level of OCT2 mRNA expression may be very low[29,30] and data on OCT2 expression in the lung remain contradictory. For OCT3, the protein expression is dependent upon the type of cells, being localised in the plasma membranes of basal cells, the basolateral membrane of intermediate cells and the apical membrane of ciliated cells in the human lung.[93] In addition, OCT3 is also expressed in bronchial smooth muscle cells.[94] Although some reports have indicated that mRNA expression of any OCT isoform was barely detectable in human lung tissues,[92] more recent studies have demonstrated that OCT1 and 3 are consistently expressed in the lung.[29,30] In the case of OCTNs, abundant mRNA expression of both isoforms was found in human airway epithelial cells and immunoreactivity for both of them was detected at the luminal membranes of the airway and alveolar epithelial cells.[92] More recently, quantitative protein analysis indicated moderate expression of OCT1 and 2, and remarkably high expression of OCTN1, but no notable expression of OCT3 and OCTN2 in human tissue from whole lung and primary cultures of human epithelial cells from the trachea, bronchi and alveoli.[31] Thus, expression of OCT and OCTN in human lungs is disputed. At present, little is known about the cellular distribution of CT2 in the lung.

OCTs recognise cationic neurotransmitters, including acetylcholine, dopamine, histamine, epinephrine and serotonin,[90,95–97] as well as steroid hormones,[90,98] and are involved in the local disposition of these biologically active substances. Acetylcholine mediates bronchoconstriction, stimulates the proliferation of primary cultured bronchial epithelial cells[99] and reduces net influx of Na^+ into cells,[100] and thus OCTs may be important in regulating pulmonary functions. Using bronchial epithelial cells[93] and $Oct1/2^{-/-}$ knockout mice, it has been shown that OCT1 and 3 mediate the release of acetylcholine,[101] which can be trans-inhibited by extracellular glucocorticosteroids such as budesonide and beclomethasone.[89] The corticosteroid sensitivity of OCTs' activity is of pharmacological interest because they potentiate β2-adrenergic bronchodilation (compounds that are increasingly used in combination inhalers to treat asthma). Although a number of mechanisms for this effect are involved, interaction of corticosteroids with OCTs has been postulated as one of the non-genomic actions of inhaled corticosteroids.[94,101–103] Among the three OCT isoforms, OCT3 has the highest affinity for corticosteroids[89] and is shown to be expressed in alveolar epithelial cells[93] and bronchial smooth muscle.[94] OCT3 mRNA suppression reduced cationic formoterol uptake and the inhibitory effect of corticosterone.[103] Accordingly, these results suggest that inhaled corticosteroids lowered clearance by inhibiting OCT (OCT3) mediated β2-agonist transport, thereby contributing to the prolonged action of β2-agonists.

Endogenous substrates of OCTN1 and 2 are more limited than those of OCT; therefore, the physiological significance of these transporters is puzzling, given their robust expression in airway epithelial cells in humans[92] and mice.[104] Electron microscopic immunocytochemistry in mice demonstrates that both OCTN1 and 2 are expressed in tracheal ciliated epithelial cells.[104] Because motile cilia of the respiratory tract beat in coordinated waves and function to keep the lower tract sterile by escalating secreted mucus toward the laryngopharynx, OCTN may play an essential role in energy production to sustain cilial beating by supplying epithelial cells with carnitine to push fatty acids into beta-oxidation.[91,105,106] Like OCTs, OCTNs also function as polyspecific OCTs, and their substrates include synthetic compounds and clinically important drugs, such as quinidine and pyrilamine.[90,105] Corticosteroid insensitive uptake of the cationic fluorophore ASP^+ by primary cultured human airway epithelial cells, in which OCTN1 and 2 but small amounts of OCTs were expressed, was inhibited by cationic β_2-agonists such as salbuterol and formoterol, but not by ergothioneine, a specific substrate of OCTN1. Therefore, OCTN2 is considered to contribute to cation transport, and to be involved in the delivery of inhaled cationic bronchodilators to the tracheal and bronchial epithelial cells.[92] Although ASP^+ is a good fluorescent probe to evaluate OCTNs, it should be noted that another group suggested that ASP^+ is translocated across the pulmonary epithelium by apically expressed OCT2 and basolaterally expressed OCT3.[59]

In line with evidence that OCT/OCTN are involved in the translocation of drugs across cell membranes in the lung, the transporter mediated permeation of anti-cholinergic bronchodilators such as ipratropium and tiotropium across airway epithelial cells has also been investigated. Systemic exposure of these cationic drugs is superior using inhaled administration compared with oral dosing. It has also been suggested that a carrier mediated process is involved in their urinary excretion.[107,108] Ipratropium uptake by human epithelium derived BEAS-2B cells, in which abundant mRNA expression of both OCTNs was confirmed,[53,55] was dependent upon temperature and substrate concentration, but independent of extracellular Na^+ concentration.[53] Addition and suppression of *OCTN2* in cell culture models caused a substantial increase and reduction in cellular uptake of ipratropium, respectively. These results suggest that OCTN2 is responsible for pulmonary absorption of ipratropium.[53] Tracheal tissue accumulation of ipratropium, however, was only partially inhibited by co-intratracheal injection with carnitine (specific to OCTN2) or cimetidine (inhibitor of OCTs), but not with ergothioneine.[104] Ipratropium and its structurally related drug, tiotropium, are also substrates of OCT1 and 2;[21,109] therefore, multiple transporters could be involved in the pulmonary absorption of ipratropium. More recently, OCTN2 targeting prodrugs, L-carnitine ester derivatives of prednisolone, were developed in order to increase the availability and slow down the release of prednisolone.[110] In the study, one of the prodrugs tested *in vitro*, PDSC, suppressed lipopolysaccharide (LPS) induced IL-6 production by BEAS-2B cells. Accordingly, OCTN2 could be a pharmacological target for asthma treatment. Based on these observations, OCTs have gained more attention in terms of molecular determinants for efficacy and pulmonary/systemic disposition of inhaled drugs.

In summary, OCTs and OCTNs are physiologically and pharmacologically important membrane transporters expressed in the airway epithelium. Observations regarding their expressions based on conventional assays (*e.g.* reverse transcription polymerase chain reaction and antibody based assays) vary among specimens used, and are partly inconsistent with recent quantitative analyses using LC-MS/MS (Table 6.2); therefore, their expression profile needs to be more carefully examined to understand inter-individual and cell type differences. In addition, little information is available on the molecular mechanisms for regulation of gene expression of these transporters; hence it is an important task to clarify such mechanisms for a better understanding of their physiological and pharmacological significance in the lung.

6.6.2 Peptide Transporters of the *SLC15A* Family

Peptide transporters of the *SLC15A* family are responsible for the cellular uptake of di- and tri-peptides in the body.[111] PEPT2 is highly expressed in the lung whereas PEPT1 is absent or expressed at a low level (Table 6.2). PEPTs are proton dependent oligopeptide transporters that are dependent on the membrane potential and extracellular pH for their activity,[112] with a pH optimum for translocation of substrates across membranes of 4.5–6.5.[112] PEPT2 has a broad substrate specificity[113] and, in general, it is a high affinity transporter with low capacity.[115] Interestingly, PEPT2 has been demonstrated to transport the peptidomimetic β-lactam antibiotics of the cephalosporin and penicillin classes. Other groups of substrates include antiviral drugs, *e.g.* valacyclovir, angiotensin converting enzyme inhibitors, *e.g.* enalapril and fosinopril, as well as antineoplastic drugs, *e.g.* bestatin.[115] PEPT2 is broadly expressed in different tissues.[28]

IHC studies have shown that PEPT2 is expressed on the apical membrane of epithelial cells in bronchi and type II alveolar cells as well as in endothelial cells of small vessels (Table 6.2).[35,116] A similar distribution was found in mouse, rat and human lungs.[35,116] In addition to mRNA and protein measurements, the functional activity of PEPT2 was demonstrated by measuring the uptake of a fluorescent probe into mouse lung *ex vivo* and in cells isolated from human lung. A dipeptide conjugated to a fluorophore, D-Ala-L-Lys-AMCA, was taken up in mouse lung *ex vivo* into the same cell types that were stained positive for PEPT2.[35,116] Similarly, D-Ala-L-Lys-AMCA was distributed into bronchial epithelial cells and type II alveolar cells expressing PEPT2.[35,116] In both cases, cephalosporins and dipeptides, which are PEPT2 substrates, inhibited the uptake of the fluorescent probe. PEPT2 expression has also been demonstrated in rat alveolar macrophages and a potential role in the transport of *S*-nitrosothiol, a product of nitric oxide (NO), has been described.[117]

The inter-individual variation of PEPT2 expression has not been extensively investigated. PEPT2 is polymorphic and two variants, *PEPT2*1* and *PEPT2*2*, that differ in their substrate affinity and pH dependence in the transport of a dipeptide substrate have been identified.[118] Still, it is not known what impact this difference in function has on transport *in vivo*.

This high expression of PEPT2 in the lung opens up the opportunity for targeted inhaled drug delivery of PEPT2 substrates to specific pulmonary cell types or as a first step in the delivery of peptide drugs to the systemic circulation.[112] Antibiotics are commonly used in the treatment of recurrent pulmonary bacterial infections in cystic fibrosis and COPD, and using PEPT2 for targeted drug delivery of anti-infectious drugs to their site of action has been proposed.[112] However, such an approach still has to be validated in a clinically relevant setting.

6.6.3 Other Transporters of the *SLC* and *SLCO* Families

In terms of sensitivity of lung cells to drugs, organic anion transporters have been postulated as a pharmacological target and investigated in the lung. Although information is limited, two major classes of transporter for organic anions are described: OATs and OATPs. Current evidence suggests that expression of OAT1–4 is barely detectable in the lung (Table 6.2),[28,119–121] whereas OATP2A1/PGT, 2B1, 3A1, 4A1 and 4C1 are expressed in human lungs,[28–30,122–125] although their cellular distributions remain unclear[56] (Table 6.3). Furthermore, OATP2B1, 3A1 and 4A1 were also shown to be expressed in human airway epithelial Calu-3, 16HBE14o- and A549 cell lines.[55] The OATP family consists of over 40 members in the human, rat and mouse (reviewed in ref. 126–128), and they have been characterised as a Na^+ independent transport systems for a wide variety of both endogenous and exogenous organic anions, including bile salts, bilirubin and hormone conjugates, as well as a number of drugs in clinical use.[129–131] In terms of their role in the lung, their functions have been studied mainly in terms of the efficacy of substrate drugs using *in vitro* lung cell culture models. For example, recent studies indicated that OATP2B1 may be involved in amiodarone induced pulmonary toxicity in A549 cells.[132] In research on the drug sensitivity of mouse lung cells, Ohbayashi *et al.* reported that alveolar epithelial cells were more sensitive to methotrexate induced cytotoxicity and apoptosis than lung fibroblast cells.[133] Moreover, methotrexate treatment of mice *in vivo* resulted in a fibrotic response accompanied by cell dysfunction of the alveolar epithelium.[133] More recently, the same researchers found that mouse OATP4C1, which can transport methotrexate, was differentially upregulated in mouse alveolar epithelia compared with fibroblasts, and they concluded that expression of OATP4C1 is one of the determinants for methotrexate induced toxicity in the lung.[134]

OATPs not only regulate the cellular disposition of exogenous but also endogenously active substances such as eicosanoids. At least 5 out of the 11 functional human OATP members (OATP1B1, 2A1, 2B1, 3A1 and 4A1) recognise prostaglandins (PGs) as substrates.[125,135] Among these, OATP2A1/PGT was characterised as a SLC for PGE_1, PGE_2, PGF_{2a} and PGD_2, and even thromboxane B_2, with a higher affinity for PGE_2, when it was isolated from adult human kidneys.[135] OATP2A1/PGT mediated cellular uptake of PGE_2 is driven by an exchange of organic anions such as lactate,[136] and is considered to be an initial step for degradation of PGE_2 by cytoplasmic 15-hydroxy PG;[137] therefore, OATP2A1/PGT is now generally accepted to serve as a clearance system for extracellular PGs, particularly PGE_2. At the same time, OATP2A1/PGT is also suggested to be involved in the release of PGE_2 from different types of cells.[136,138,139] Since OATP2A1/PGT expression has been detected in both mouse type II alveolar epithelial cells by IHC[140] and in human bronchial epithelial BEAS-2B cells,[141] OATP2A1/PGT may play a role in regulating pulmonary disposition of PGs. The mRNA expression of a proinflammatory cytokine, IL-6, was associated with an increased release of PGE_2 from BEAS-2B cells treated with LPS.[141] Since OATP2A1/PGT inhibitors reduced the release of PGE_2 under the same conditions, OATP2A1 could be involved in the production of proinflammatory cytokines, such as IL-6. Furthermore, it was shown that bleomycin induced fibrosis is aggravated in $Oatp2a1^{-/-}$ knockout mice.[142] Therefore, these observations suggest that OATP2A1/PGT adjusts physiological functions of eicosanoids, which mediate inflammatory reactions, by regulating their local disposition in the airways. This suggests that agents modifying OATP2A1/PGT function might be promising candidates as anti-inflammatory drugs. In addition to PGs, OATPs are also important carriers for cholate, conjugated hormones and small peptides;[129–131] however, their physiological significance remains unclear in the lung. Future work in this area is warranted to reveal the subcellular distribution and functions of OATPs.

In the lung, it is also important to consider transport of airway mucus, a fluid comprising approximately 90% water plus protein, carbohydrate and lipids, which plays a role in protecting the lung from inhaled pathogens, particles and toxins. Excessive intraluminal mucus is a key component of common airway diseases such as asthma and chronic bronchitis; therefore, the elimination rate of protein from airways should be balanced with secretion of protein from epithelia. There are two classes of protein clearance from the airway; the mucociliary transport of proteins in an upward direction and transverse transport of proteins or their degradation products across the airway epithelium. Although it remains unclear how these degraded products are absorbed into the epithelium, it is known that SLC transporters specialised for transporting small peptides, amino acids and saccharides are expressed in the epithelial cells. In addition to the oligopeptide transporter PEPT2 described in Section 6.6.2, only a few classes of amino acid transporters, including $B^{0,+}$ ($ATB^{0,+}$)[143] and L (L, y + L and y+) systems,[144] have been indicated to be functionally expressed in airway epithelial cells. $ATB^{0,+}$, a Na^+ and Cl^- coupled transport system for neutral and cationic amino acids in humans[145] and mice,[114,146] has been shown by IHC to be expressed predominantly at the apical membranes of ciliated epithelial cells in the airways and in alveolar epithelial cells (Table 6.2).[143] $ATB^{0,+}$ has broader substrate specificity compared with other amino acid transporters, and its substrates include D-isoforms of amino acids (*e.g.* alanine, serine, methionine, leucine and tryptophan),[147] carnitine,[148] cationic amino acid related nitric oxide synthase (NOS) inhibitors,[114] valacyclovir[149] and 1-methyltryptophan.[150] Therefore, it is hypothesised that $ATB^{0,+}$ not only contributes to protein clearance by removal of amino acids from airway lumen, but also to pulmonary disposition of substrate drugs or synthetic agents administered into the body. Based on studies from other organs,[151] it is currently hypothesised that efficient translocation of amino acids is achieved by influx *via* $ATB^{0,+}$ expressed at the apical membranes and efflux *via* y + L systems such as y + LAT1 (*SLC7A6*) and y + LAT2 (*SLC7A7*) in airway or alveolar epithelial cells. This vectorial transport by cooperation is considered an efficient route for pulmonary absorption of arginine derived NOS inhibitors, such as N^{ω}-nitro-L-arginine and N^{G}-monomethyl-L-arginine, because these NOS inhibitors are substrates of $ATB^{0,+}$.[114]

Another group of amino acid carriers are the system L transporters, represented by LAT1 in the lung (Table 6.2). In rat alveolar epithelial cells, *Lat1* was co-expressed with *Pept2*, and was proposed to be involved in the uptake of NO and some derivatives by type II alveolar epithelial cells.[144] NO is a gaseous signal molecule, important not only for protein denitrosylation but also for treatment of pulmonary hypertension; therefore, LAT1 could be an essential determinant of NO metabolism and efficacy of NO related drugs.

In summary, amino acid transporters and organic anion transporting polypeptides are expressed in the human airway and may regulate pulmonary disposition of their endogenous substrates and substrate drugs. Accordingly, these transporters could be considered as molecular targets for pharmacology or drug delivery for inhaled drugs.

6.6.4 MDRs: P-gp

P-gp is encoded by the *MDR1/ABCB1* gene in humans, and the *Mdr1a/Abcb1a* and *Mdr1b/Abcb1b* genes in rodents. It is amongst the most highly studied of the ABC transporters with a large volume of published work in a range of barriers, including the intestine, kidney, liver and blood–brain barrier (BBB). P-gp substrates cover a broad range of therapeutic classes and physicochemical properties and are generally lipophilic or amphipathic,[152] enabling entry into the plasma membrane, a necessity for interaction with the substrate binding site of the transporter (Table 6.4).

In 2006, microarray analysis of transporter mRNA expression across a range of species and organs was performed.[28] In all species, the expression of P-gp witnessed in the lung was in the lower quartile, a far lower level than that seen in the intestine or liver. However, the analysis was performed on a whole organ homogenate and as such the heterogeneous nature of the lung tissue could have served to dilute the mRNA transcript level, particularly if only a few cell types exhibited P-gp expression. It does seem likely, however, that the expression levels in the lung epithelia are not as high as those seen in the intestine. It is interesting to note that in the rodent species the level of expression of *Mdr1b* was higher than *Mdr1a* in the lung, whereas in all other studied organs *Mdr1a* was the more abundant.[153] This observation has also been made in *in vitro* cell models, specifically those of primary isolated alveolar type I cell lines.[154]

A number of groups since the early 1990s has investigated the spatial expression of P-gp within the lung using IHC analyses. All observed P-gp expression was within the lung epithelia, but there are discrepancies with the localisations dependent on the specific techniques and antibodies used (Table 6.4).

Initially, positive staining for P-gp was demonstrated on the luminal surface of bronchial and bronchiolar epithelial cells in frozen tissue samples of human lung.[155,156] Alveolar macrophages were positively stained for P-gp whereas the staining of alveolar epithelia was variable depending on the antibody used.[156] Positive staining for P-gp in the bronchial epithelia has been confirmed by other groups.[154,157,158] Alveolar staining has also been confirmed more recently in intact alveolar epithelia of rats[154] as well as human tissue samples.[58] The presence of the P-gp protein by western blot analysis was also demonstrated in the respective isolated primary alveolar cell lines.[58,154]

Functional activity of P-gp has been demonstrated *in vitro* in cell models. Polarised transport of the well-characterised P-gp substrate rhodamine 123 was observed in primary isolated human alveolar epithelial cells. The transport was both saturable and inhibited by the P-gp inhibitor verapamil, providing good evidence for P-gp mediated efflux.[58] A similar study was performed in rat primary cells in which the accumulation of rhodamine 123 was increased by co-incubation with verapamil.[154] Later work using the human immortalised NHBE and Calu-3 cell lines indicated polarised transport of digoxin in both cell types, although transport was less polarised than in the intestinal Caco-2 cells.[159] In all cell lines, the polarised transport was abolished by co-dosing with GF120918, a selective P-gp/BCRP inhibitor. P-gp mediated efflux from Calu-3 cells has also been demonstrated using the fluoroquinolone antibiotic moxifloxacin,[160] again using a P-gp inhibitor (PSC-833). Moreover, polarised transport of rhodamine 123 in the bronchial epithelial cell line 16HBE14o- was inhibited by co-dosing with verapamil.[57]

Taken together, these studies provide compelling evidence that pulmonary cells grown *in vitro* express functional P-gp, and may give insights into the PK/PD of drugs delivered to the lung or indeed drugs with sites of action in the lung that are delivered systemically. Questions are raised, however, about the expression of transporter proteins in these *in vitro* cell lines in comparison to intact pulmonary epithelia and whether they are truly models of the pulmonary barrier. To this end, studies in intact lungs are necessary to elucidate any potential role for the P-gp transporter.

Although P-gp is one of the most well studied of the pulmonary drug transporters, data in an intact lung system are scarce. In 2003, the effect of P-gp on the accumulation of idarubicin in lung tissue after dosing *via* the circulation in the IPL was examined.[69] It was found that in experiments with the P-gp inhibitors cinchonine and rutin, the accumulation of idarubicin in lung tissue was increased. This implies that the P-gp inhibition occurred at the endothelial barrier, preventing efflux of idarubicin back into the blood, a finding supported by a similar study[161] in a rabbit model with rhodamine 6G performed a year later.

This hypothesis appears to be in agreement with experiments investigating the absorption of drugs from the lung into blood using both *in vivo* and isolated lung models.[74] In both cases, the absorption of the P-gp substrate losartan was above 95% and it was assumed that there was no attenuating effect of P-gp efflux on absorption from the lung lumen. However, the experiments were not designed to investigate efflux and, therefore, the concentrations used may have resulted in saturation of the P-gp transporter. Manford *et al.* provided further evidence for the lack of effect of P-gp efflux from the lumen to blood using digoxin as a substrate in an isolated rat lung.[162] The group used spontaneous *Mdr1a* deficient mice, which maintained their *Mdr1b* expression, and showed that there was no difference in the rate and extent of digoxin absorption between the deficient and control mice. Contrary to this a study, in a similar isolated rat lung model, an increase in absorption of the P-gp substrate rhodamine 123 from the lung to the blood when co-dosed with GF120918 was observed.[163]

Subsequently, a study was published investigating the effects of pulmonary P-gp on a larger selection of substrates in rat and mouse IPLs,[68] and corroborates the results of both of the previous two studies. In both rat and mouse models, an increase in the absorption of rhodamine 123 was observed when co-dosed with GF120918, an observation not seen with digoxin, where the inhibitor did not alter the absorption profile. The same discrepancy was then shown using a *Mdr1a/Mdr1b* knockout mouse model, suggesting it was an effect involving P-gp and not a difference due to an off-target effect of the inhibitor. A similar pattern was seen using the known P-gp substrate loperamide, which showed increased absorption with the inhibitor, whereas absorption of the P-gp substrate saquinavir was not altered by co-dosing with GF120918.

In summary, the studies discussed above provide persuasive evidence for a role for P-gp in lung tissue. Its presence has been observed in a variety of models from *in vitro* cell cultures to human sections at both the protein and mRNA levels, and functionality has now been observed in IPLs in both the rat and mouse. P-gp efflux has been shown to reduce the absorption of a set of substrates from the airways to the blood, yet the transporter appears to have no effect on other well characterised substrates such as digoxin. Despite this, it is apparent that P-gp can affect drug absorption after pulmonary delivery and as such needs to be considered when developing and delivering drugs that are known P-gp substrates, as it could potentially impact their PK/PD within the body.

6.6.5 MRPs: MRP1

The MRPs are members of the *ABCC* subfamily of ABC transporters, which consists of at least 13 separate transporters. Of these, MRP1 is amongst the most well studied and functions primarily as a co-transporter of glutathione, glucuronic acid or sulfate conjugated drugs, usually at the basolateral membrane of cells. It was initially identified due to its role in the chemotherapeutic resistance of anticancer drugs, but it also plays a role in the PK of many other drug classes, with relatively broad substrate specificity.[164]

In the lung, the protein is of interest because it has been reported that MRP1 reduces the oxidative stress generated by smoking and therefore may play a therapeutic role in COPD.[165,166]

In microarray work, MRP1 expression was identified as moderate to high (50–75%) in the human and rodent lung,[28] a finding corroborated by other mRNA based studies (Table 6.4)[29,30,167] and protein staining of intact lungs. IHC of wax embedded lung sections demonstrated expression of MRP1 in a healthy lung, localised around the apical region of the cytoplasm of bronchial epithelial cells.[168] The same study also showed strong expression of the protein using western blot analyses. Two later studies confirmed the presence of MRP1 using IHC techniques, but instead of localisation below the cilia, these studies reported localisation of MRP1 on the basolateral membrane,[158,169] a finding in common with its protein expression in other tissues.

In 2009, a study investigating the expression of a wide range of transporters in *in vitro* cell models of respiratory cell lines[55] found evidence of a high level of MRP1 expression in all cell lines studied. This *in vitro* activity has been confirmed using functional studies by measuring the reduction in efflux of the substrate carboxydichlorofluorescein in the presence of the MRP inhibitor MK-571.[170]

Despite a large number of studies using *in vitro* cell lines to examine the possible roles of MRPs in lung tissue, data from an intact lung system are not available and further work is required in this area. However, the data currently available provide strong evidence for the presence of MRP1 in the lung, particularly at the basolateral membrane of bronchial epithelia.

6.6.6 BCRP

BCRP is a member of the ABC transporters expressed on the apical membranes of epithelial cells where it acts as an efflux transporter for a range of exogenous chemicals from anticancer drugs to ion channel blockers (Table 6.4).[164,171]

Data for BCRP activity within the lung are sparse, with fewer studies performed than for P-gp. In 2007, the relative levels of transporter protein mRNA in many organs across a range of species was studied.[28] For BCRP, the expression levels were in the moderate to higher quartile (50–75%) in human lungs. Further investigations were reported across a range of species, including dog and rodent, although expression levels were lower in rodent lungs.

This relatively high level of mRNA in the lung is also represented by the expression of BCRP protein in both Calu-3 and 16HBE14o- cell models,[55] although the protein is much more highly expressed in the 16HBE14o- cell line, in which it appears overexpressed.

In terms of protein expression levels, results are contradictory and apparently vary according to the stains and techniques used. IHC techniques showed expression of BCRP in alveolar pneumocytes (cell type unspecified) with little or no staining of the bronchial epithelia,[172] whereas earlier work had demonstrated staining of the bronchial epithelial cells and endothelial capillaries but not of alveolar pneumocytes.[158] It is, however, likely that given the staining seen and the high levels of mRNA present in both lung homogenate and *in vitro* lung cell lines, BCRP is present in the lung. At the time of writing (2016), there are no published data on the assessment of functional activity of BCRP in the lung.

6.7 Distribution of Drugs from the Systemic Circulation

Besides their potential impact on the disposition of drugs administered by the inhaled route, lung transporters could also play an important role in the distribution of drugs from the systemic circulation into the pulmonary tissue and epithelial lining fluid (ELF), thus affecting their efficacy or toxicity in the lungs. Key transporters have not yet been identified. However, the few studies published so far point towards a possible involvement of the ABC and/or OATP families.[173–177]

There is increasing evidence in the literature that several anti-infectious agents are actively transported into the ELF following systemic administration. For instance, the macrolide antibiotics clarithromycin, azithromycin and telithromycin, commonly used to treat pulmonary infections, accumulate in the ELF of rats after oral administration.[176,177] Interestingly, the drugs are transported to a greater extent in the basolateral to apical (B–A) than apical to basolateral (A–B) direction in Calu-3 layers, with an apparent efflux ratio in line with their ELF *versus* plasma concentrations.[177] This suggests that an active transport mechanism present in the airway epithelium drives their distribution from the systemic circulation into the ELF. In foals, whose transporter sequence and regulation are similar to those in humans, clarithromycin concentrations in the ELF after oral dosing were enhanced following chronic co-medication with the anti-tuberculosis agent rifampicin.[175] Although rifampicin is a well-known P-gp inducer, expression levels of the transporter in broncho-epithelial cells were unchanged upon repeated administration.[175] Therefore, the involvement of P-gp in clarithromycin accumulation in ELF is unlikely. In contrast, rifampicin caused upregulation of OATP2B1 in the bronchial epithelium, suggesting a potential role of this uptake transporter. However, clarithromycin was later reported not to be a substrate for OATP2B1, OATP1A2, OATP1B1 or OATP1B3,[178] implying that the transporter that facilitates its distribution in ELF after systemic administration is still to be identified.

Similarly to macrolides, higher ELF *versus* plasma concentrations were measured in rats after both aerosol delivery and intravenous injection of the fluoroquinolones ciprofloxacin, moxifloxacin and grepafloxacin.[174] Permeability studies in Calu-3 layers indicated substantially higher B–A transport for the three drugs[173] and PK modelling supported the role of efflux transporters in the ELF accumulation of these drugs.[174] Further studies are nevertheless required to identify the transporter(s) involved.

Although very preliminary, a few studies have indicated that severe pulmonary toxicity of some systemic drugs could result from their accumulation in lung cells following transporter mediated uptake. As mentioned in Section 6.6.3, the antiarrhythmic agent amiodarone, which causes life-threatening lung adverse effects in 5–10% of patients, is taken up in the alveolar epithelial cell line A549 by an active mechanism.[132] Inhibition and siRNA knockdown studies have demonstrated that OATP2B1 is the main transporter responsible for accumulation in this cell line. Similarly, the anticancer drug methotrexate, which is associated with lung fibrosis, is preferentially internalised by mouse type II pneumocytes rather than mouse lung fibroblasts, presumably because the latter express a broader range of MRP transporters.[134] The drug's cellular uptake was indeed enhanced in both cell types in the presence of the MRP inhibitor MK571, although to a larger extent in the fibroblasts.

6.8   Transporter Regulation in the Lung

Whilst some knowledge of the regulation of renal, hepatic and intestinal transporters has been gained recently, information on physiological, pathological, pharmacological or environmental factors affecting transporter expression in the lung is still extremely limited. Interestingly, in a very small study (three males and four females) of transporter expression in lungs, the majority of drug transporter proteins were expressed at higher levels in female donors,[31] suggesting a role of sex hormones in their pulmonary regulation. A broader investigation is therefore needed to confirm this initial observation.

The effect of smoking on lung transporter regulation has only been reported in the literature for P-gp and MRP1. In both cases, expression was similar in smokers, ex-smokers and non-smokers.[157,169] MRP1 was however downregulated in COPD patients and the lowest expression levels were observed in patients with a severe form of the disease.[33]

It is not yet known whether chronic medications, administered either by the inhaled or systemic route, impact pulmonary transporter levels in humans. Nevertheless, a two-fold increase in P-gp pulmonary protein expression was measured in rats following a single oral[179] or intraperitoneal[180] dose of the glucocorticoid dexamethasone, as well as repeated gavages over 4 days.[179] The liver was the only other organ where upregulation of the transporter was observed, while downregulation occurred exclusively in the kidneys.[179] This strongly suggests that different transporter regulation mechanisms operate in different tissues. In foals, chronic administration of rifampicin, which is commonly used as a P-gp and MRP2 inducer, did not modify the expression of the two efflux pumps in the lung tissue but did reduce OATP2A1 levels and increase those of OATP2B1.[175]

Of possible relevance to asthma, dysregulation of OCT1, 2 and 3 was observed in rodent lungs after an acute allergen challenge with ovalbumin.[181] Variations in transporter expression levels were inconsistent between rat and murine tissues, implying inter-species differences in OCT regulation mechanisms. A related study in air–liquid interfaced Calu-3 layers subsequently demonstrated exposure to the inhaled allergen the house dust mite, as well as the proinflammatory stimulant LPS, resulted in fold increases in the expression of the five OCT members.[182] This indicates that allergic inflammation of the airways might modify OCT levels in the epithelium, with a potential impact on the distribution of inhaled bronchodilators in lung tissue. OCTN1 and 2 were also shown to be upregulated in Calu-3 layers by fenofibrate or rosiglitazone, agonists of the peroxisome proliferator activated nuclear receptor α (PPARα) and PPARγ, respectively.[183]

6.9 Summary and Concluding Remarks

It is evident that several transporters of the SLC, SLCO and ABC families known to translocate drugs across cell membranes are expressed in the lung and that the transporter expression profile is different in the lung compared with other organs. The most comprehensive studies of pulmonary drug transporter expression so far are based on mRNA analysis and, with a few exceptions, there are still only limited data on protein abundance. Recent development of mass spectrometry based techniques to quantify transporter proteins constitutes an important complement to IHC techniques. Moreover, there are limited, and often conflicting, data regarding the localisation of transporter proteins to specific cell types in the lung as well as their subcellular distribution. The spatial resolution of drug transporter expression is particularly important to consider in the lung because of its complex anatomical structure and the variety of cell types with fundamentally different morphologies and functions across the different pulmonary regions.

There is, however, some evidence of the functional impact of transporters on pulmonary drug disposition. One of the most compelling examples is the expression of OCTs on the apical side of bronchial epithelial cells, where they have been reported to influence the first step in pulmonary absorption of muscarinic antagonists across the pulmonary epithelial cell barrier. Other examples include peptide and amino acid transporters that are highly expressed on pulmonary epithelial cells. PEPT2 has been suggested for targeted delivery of drugs, *e.g.* antibiotics and antivirals, to cell types with high expression of this transporter. Although inconclusive, evidence also exists that P-gp is functionally important in the lung and influences the disposition of some of its substrates by hindering absorption from the airways. Evidence of the functional importance of drug transporters in the lung are based mainly on *in vitro* cell and IPL models, but no single model has yet been validated for use in a routine pharmaceutical setting. A major challenge in this area is the translation of *in vitro* and *ex vivo* observations to *in vivo* and even more so to the clinical situation.

Compared with orally administered drugs, compounds developed for inhaled drug delivery often include polar molecules that do not readily cross cell membranes. As an example, muscarinic receptor antagonists commonly administered by inhalation in the treatment of asthma and COPD are permanent cations with poor passive permeability and very low oral bioavailability. Consequently, drug transporters may be particularly important for the disposition of the muscarinic antagonists and other classes of inhaled drugs with similar physicochemical properties, as they need transporters to pass cell membranes. In addition, inhaled drugs may also affect the transport of endogenous compounds, *e.g.* PGs, amino acids, peptides and biogenic amines, and contribute to the pharmacological effect. As an example, the non-genomic effects of corticosteroids may be attributed to the inhibition of OCTs in smooth muscle cells.

In conclusion, drug transporters in the lung may be important in the disposition of both inhaled and orally/systemically administered drugs, as well as affecting the translocalisation of endogenous transporter substrates. Consequently, inter-individual variability in pharmacological responses and toxicity may be influenced by pharmacogenetics and/or transcriptional regulation of pulmonary transporter expression, as well as DDIs. Knowledge is limited within this area and, in particular, a better understanding of the impact of functional transporter activity in specific regions/cell types of the lung on the *in vivo* local pulmonary drug disposition is required. Translation of *in vitro* and *ex vivo* data to *in vivo* is not possible with our existing knowledge and consequently no accurate predictions of the impact of lung transporters on drug disposition can be made in the clinical situation. This remains an important emerging research area that in the future may enable enhanced rational design of drugs for the treatment of respiratory diseases and increase our potential to understand mechanisms of pulmonary toxicity.

6.10   Contributions by the Authors

Although this book chapter has been written in collaboration, the major contributions are as follows: X.-H. Zhou: lung anatomy and morphology; C. Bosquillon: *in vitro* cell models, distribution from systemic circulation and regulation of transporters; T. Nakanishi and I. Tamai: OCTs and other SLC/SLCO transporters; D. Price and M. Gumbleton: IPL and ABC transporters; L. Gustavsson and T. Hegelund-Myrbæck: inhalation therapy and pulmonary drug disposition, *in vivo* models, drug transporter families in the human lung and PEPT2; L. Gustavsson was the coordinator of the contributions and the local editor.

References

1. J. S. Patton and P. R. Byron, *Nat. Rev. Drug Discovery*, 2007, **6**, 67.
2. A. Stevens and J. Lowe, Respiratory system, in *Human Histology*, ed. A., Stibbe, Elsevier Mosby, 3rd edn, 2005.

3. B. Olsson, E. Bondesson, L. Borgstrom, E. Edsbacker, S. Eirefelt, K. Ekelund, L. Gustavsson and T. Hegelund-Myrback, *Controlled Pulmonary Drug Delivery*, ed. H., Smyth and A., Hichey, Springer, New York, 2011, p. 21.

4. D. Warburton, L. Perin, R. Defilippo, S. Bellusci, W. Shi and B. Driscoll, *Proc. Am. Thorac. Soc.*, 2008, **5**, 703.

5. H. Wang, Y. Liu and Z. Liu, *Curr. Opin. Allergy Clin. Immunol.*, 2013, **13**, 25.

6. C. J. Johnston, G. W. Mango, J. N. Finkelstein and B. R. Stripp, *Am. J. Respir. Cell Mol. Biol.*, 1997, **17**, 147.

7. B. Young and J. W. Heath, *Wheather's Functional Histology, A Text and Colour Atlas*, 4th edn, 2014.

8. A. V. Andreeva, M. A. Kutuzov and T. A. Voyno-Yasenetskaya, *Am. J. Physiol.: Lung Cell Mol. Physiol.*, 2007, **293**, L259.

9. H. Fehrenbach, *Respir. Res.*, 2001, **2**, 33.

10. E. M. Vlachaki, A. V. Koutsopoulos, N. Tzanakis, E. Neofytou, M. Siganaki, I. Drositis, A. Moniakis, S. Schiza, N. M. Siafakas and E. G. Tzortzaki, *Chest*, 2010, **137**, 37.

11. G. A. Margaritopoulos, I. Giannarakis, N. M. Siafakas and K. M. Antoniou, *Panminerva Med.*, 2013, **55**, 109.

12. B. Corrin, A. G. Nicholson, M. Burke and A. Rice, *Pathology of the Lungs*, Churchill Livingstone, 2011.

13. T. Hussell and T. J. Bell, *Nat. Rev. Immunol.*, 2014, **14**, 81.

14. Global Strategy for Asthma Management and Prevention, 2015, http://www.ginasthma.org/documents/4.

15. Global Strategy for the diagnosis, management, and prevention of Chronic Obstructive Pulmonary Disease, 2015, http://www.goldcopd.org/guidelines-global-strategy-for-diagnosis-management.html.

16. P. J. Barnes, J. M. Drazen and S. I. Rennard, *Asthma and COPD: Basic Mechanisms and Clinical Management*, 2008.

17. P. Backman, H. Adelmann, G. Petersson and C. B. Jones, *Clin. Pharmacol. Ther.*, 2014, **95**, 509.

18. A. J. Hickey, *J. Pharm. Sci.*, 2013, **102**, 1165.

19. B. Forbes, B. Asgharian, L. A. Dailey, D. Ferguson, P. Gerde, M. Gumbleton, L. Gustavsson, C. Hardy, D. Hassall, R. Jones, R. Lock, J. Maas, T. McGovern, G. R. Pitcairn, G. Somers and R. K. Wolff, *Adv. Drug Delivery Rev.*, 2011, **63**, 69.

20. B. Olsson, L. Borgstrom, H. Lundback and M. Svensson, *J. Aerosol Med. Pulm. Drug Delivery*, 2013, **26**, 355.

21. R. Hendrickx, J. G. Johansson, C. Lohmann, R. M. Jenvert, A. Blomgren, L. Borjesson and L. Gustavsson, *J. Med. Chem.*, 2013, **56**, 7232.

22. T. J. Ritchie, C. N. Luscombe and S. J. Macdonald, *J. Chem. Inf. Model.*, 2009, **49**, 1025.

23. A. Tronde, B. Norden, H. Marchner, A. K. Wendel, H. Lennernas and U. H. Bengtsson, *J. Pharm. Sci.*, 2003, **92**, 1216.

24. C. Y. Wu and L. Z. Benet, *Pharm. Res.*, 2005, **22**, 11.

25. K. Sugano, M. Kansy, P. Artursson, A. Avdeef, S. Bendels, L. Di, G. F. Ecker, B. Faller, H. Fischer, G. Gerebtzoff, H. Lennernaes and F. Senner, *Nat. Rev. Drug Discovery*, 2010, **9**, 597.
26. A. Crowe and A. M. Tan, *Toxicol. Appl. Pharmacol.*, 2012, **260**, 294.
27. H. Jones and K. Rowland-Yeo, *CPT: Pharmacometrics Syst. Pharmacol.*, 2013, **2**, e63.
28. K. Bleasby, J. C. Castle, C. J. Roberts, C. Cheng, W. J. Bailey, J. F. Sina, A. V. Kulkarni, M. J. Hafey, R. Evers, J. M. Johnson, R. G. Ulrich and J. G. Slatter, *Xenobiotica*, 2006, **36**, 963.
29. T. Berg, T. Hegelund-Myrback, M. Olsson, J. Seidegard, V. Werkstrom, X.-H. Zhou, J. Grunewald, L. Gustavsson and M. Nord, *Pharmacol. Res. Perspect.*, 2014, **2**, e00054.
30. J. Leclerc, N. E. Courcot-Ngoubo, C. Cauffiez, D. Allorge, N. Pottier, J. J. Lafitte, M. Debaert, S. Jaillard, F. Broly and J. M. Lo-Guidice, *Biochimie*, 2011, **93**, 1012.
31. A. Sakamoto, T. Matsumaru, N. Yamamura, Y. Uchida, M. Tachikawa, S. Ohtsuki and T. Terasaki, *J. Pharm. Sci.*, 2013, **102**, 3395.
32. A. Sakamoto, T. Matsumaru, N. Yamamura, S. Suzuki, Y. Uchida, M. Tachikawa and T. Terasaki, *J. Pharm. Sci.*, 2015, **104**, 3029.
33. M. van der Deen, H. Marks, B. W. Willemse, D. S. Postma, M. Muller, E. F. Smit, G. L. Scheffer, R. J. Scheper, E. G. de Vries and W. Timens, *Virchows Arch.*, 2006, **449**, 682.
34. J. A. Wambach, D. J. Wegner, K. Depass, H. Heins, T. E. Druley, R. D. Mitra, P. An, Q. Zhang, L. M. Nogee, F. S. Cole and A. Hamvas, *Pediatrics*, 2012, **130**, e1575.
35. D. A. Groneberg, P. R. Eynott, F. Doring, Q. T. Dinh, T. Oates, P. J. Barnes, K. F. Chung, H. Daniel and A. Fischer, *Thorax*, 2002, **57**, 55.
36. J. Tannenbaum and B. T. Bennett, *J. Am. Assoc. Lab. Anim. Sci.*, 2015, **54**, 120.
37. C. I. Grainger, L. L. Greenwell, D. J. Lockley, G. P. Martin and B. Forbes, *Pharm. Res.*, 2006, **23**, 1482.
38. H. Lin, H. Li, H. J. Cho, S. Bian, H. J. Roh, M. K. Lee, J. S. Kim, S. J. Chung, C. K. Shim and D. D. Kim, *J. Pharm. Sci.*, 2007, **96**, 341.
39. B. Forbes and C. Ehrhardt, *Eur. J. Pharm. Biopharm.*, 2005, **60**, 193.
40. J. L. Sporty, L. Horalkova and C. Ehrhardt, *Expert Opin. Drug Metab Toxicol.*, 2008, **4**, 333.
41. E. H. van den Bogaard, L. A. Dailey, A. J. Thorley, T. D. Tetley and B. Forbes, *Pharm. Res.*, 2009, **26**, 1172.
42. K. J. Elbert, U. F. Schafer, H. J. Schafers, K. J. Kim, V. H. Lee and C. M. Lehr, *Pharm. Res.*, 1999, **16**, 601.
43. A. Steimer, M. Laue, H. Franke, E. Haltner-Ukomado and C. M. Lehr, *Pharm. Res.*, 2006, **23**, 2078.
44. M. Bur, H. Huwer, C. M. Lehr, N. Hagen, M. Guldbrandt, K. J. Kim and C. Ehrhardt, *Eur. J. Pharm. Sci.*, 2006, **28**, 196.

45. J. J. Salomon, V. E. Muchitsch, J. C. Gausterer, E. Schwagerus, H. Huwer, N. Daum, C. M. Lehr and C. Ehrhardt, *Mol. Pharm.*, 2014, **11**, 995.

46. N. R. Mathia, J. Timoszyk, P. I. Stetsko, J. R. Megill, R. L. Smith and D. A. Wall, *J. Drug Target*, 2002, **10**, 31.

47. V. Hutter, D. Y. Chau, C. Hilgendorf, A. Brown, A. Cooper, V. Zann, D. I. Pritchard and C. Bosquillon, *Eur. J. Pharm. Biopharm.*, 2014, **86**, 74.

48. C. E. Stewart, E. E. Torr, N. H. Mohd Jamili, C. Bosquillon and I. Sayers, *J. Allergy*, 2012, **2012**, 943982.

49. F. Manford, A. Tronde, A. B. Jeppsson, N. Patel, F. Johansson and B. Forbes, *Eur. J. Pharm. Sci.*, 2005, **26**, 414.

50. C. Ehrhardt, C. Kneuer, J. Fiegel, J. Hanes, U. F. Schaefer, K. J. Kim and C. M. Lehr, *Cell Tissue Res.*, 2002, **308**, 391.

51. B. E. Benediktsdottir, A. J. Arason, S. Halldorsson, T. Gudjonsson, M. Masson and O. Baldursson, *Pharm. Res.*, 2013, **30**, 781.

52. J. Zabner, P. Karp, M. Seiler, S. L. Phillips, C. J. Mitchell, M. Saavedra, M. Welsh and A. J. Klingelhutz, *Am. J. Physiol. Lung Cell Mol. Physiol.*, 2003, **284**, L844.

53. T. Nakamura, T. Nakanishi, T. Haruta, Y. Shirasaka, J. P. Keogh and I. Tamai, *Mol. Pharm.*, 2010, **7**, 187.

54. V. Hutter, C. Hilgendorf, A. Cooper, V. Zann, D. I. Pritchard and C. Bosquillon, *Eur. J. Pharm. Sci.*, 2012, **47**, 481.

55. S. Endter, D. Francombe, C. Ehrhardt and M. Gumbleton, *J. Pharm. Pharmacol.*, 2009, **61**, 583.

56. C. Bosquillon, *J. Pharm. Sci.*, 2010, **99**, 2240.

57. C. Ehrhardt, C. Kneuer, M. Laue, U. F. Schaefer, K. J. Kim and C. M. Lehr, *Pharm. Res.*, 2003, **20**, 545.

58. S. Endter, U. Becker, N. Daum, H. Huwer, C. M. Lehr, M. Gumbleton and C. Ehrhardt, *Cell Tissue Res.*, 2007, **328**, 77.

59. J. J. Salomon, S. Endter, G. Tachon, F. Falson, S. T. Buckley and C. Ehrhardt, *Eur. J. Pharm. Biopharm.*, 2012, **81**, 351.

60. M. Mukherjee, D. I. Pritchard and C. Bosquillon, *Int. J. Pharm.*, 2012, **426**, 7.

61. H. B. Sondergaard, B. Brodin and C. U. Nielsen, *Pflugers Arch.*, 2008, **456**, 611.

62. P. M. Bahadduri, V. M. D'Souza, J. K. Pinsonneault, W. Sadee, S. Bao, D. L. Knoell and P. W. Swaan, *Am. J. Respir. Cell Mol. Biol.*, 2005, **32**, 319.

63. A. R. Torky, E. Stehfest, K. Viehweger, C. Taege and H. Foth, *Toxicology*, 2005, **207**, 437.

64. K. O. Hamilton, E. Topp, I. Makagiansar, T. Siahaan, M. Yazdanian and K. L. Audus, *J. Pharmacol. Exp. Ther.*, 2001, **298**, 1199.

65. D. K. Paturi, D. Kwatra, H. K. Ananthula, D. Pal and A. K. Mitra, *Int. J. Pharm.*, 2010, **384**, 32.

66. M. Haghi, P. M. Young, D. Traini, R. Jaiswal, J. Gong and M. Bebawy, *Drug Dev. Ind. Pharm.*, 2010, **36**, 1207.

67. M. Sakagami, *Adv. Drug Delivery Rev.*, 2006, **58**, 1030.

68. G. Al-Jayyoussi, D. F. Price, D. Francombe, G. Taylor, M. W. Smith, C. Morris, C. D. Edwards, P. Eddershaw and M. Gumbleton, *J. Pharm. Sci.*, 2013, **102**, 3382.

69. O. Kuhlmann, H. S. Hofmann, S. P. Muller and M. Weiss, *Anticancer Drugs*, 2003, **14**, 411.

70. P. Ewing, S. J. Eirefelt, P. Andersson, A. Blomgren, A. Ryrfeldt and P. Gerde, *J. Aerosol Med. Pulm. Drug Delivery*, 2008, **21**, 169.

71. A. Ryrfeldt and E. Nilsson, *Biochem. Pharmacol.*, 1978, **27**, 301.

72. P. R. Byron, N. S. Roberts and A. R. Clark, *J. Pharm. Sci.*, 1986, **75**, 168.

73. Y. Pang, M. Sakagami and P. R. Byron, *Eur. J. Pharm. Sci.*, 2005, **25**, 369.

74. A. Tronde, B. Norden, A. B. Jeppsson, P. Brunmark, E. Nilsson, H. Lennernas and U. H. Bengtsson, *J. Drug Target*, 2003, **11**, 61.

75. M. Gnadt, B. Trammer, B. Kardziev, M. K. Bayliss, C. D. Edwards, M. Schmidt and P. Hogger, *Eur. J. Pharm. Biopharm.*, 2012, **81**, 617.

76. M. Gnadt, B. Trammer, M. Freiwald, B. Kardziev, M. K. Bayliss, C. D. Edwards, M. Schmidt, G. Friedel and P. Hogger, *Pulm. Pharmacol. Ther.*, 2012, **25**, 124.

77. K. M. Giacomini, S. M. Huang, D. J. Tweedie, L. Z. Benet, K. L. Brouwer, X. Chu, A. Dahlin, R. Evers, V. Fischer, K. M. Hillgren, K. A. Hoffmaster, T. Ishikawa, D. Keppler, R. B. Kim, C. A. Lee, M. Niemi, J. W. Polli, Y. Sugiyama, P. W. Swaan, J. A. Ware, S. H. Wright, S. W. Yee, M. J. Zamek-Gliszczynski and L. Zhang, *Nat. Rev. Drug Discovery*, 2010, **9**, 215.

78. S. Mairinger, T. Erker, M. Muller and O. Langer, *Curr. Drug Metab.*, 2011, **12**, 774.

79. B. Wulkersdorfer, T. Wanek, M. Bauer, M. Zeitlinger, M. Muller and O. Langer, *Clin. Pharmacol. Ther.*, 2014, **96**, 206.

80. T. Wanek, S. Mairinger and O. Langer, *J. Labelled Compd. Radiopharm.*, 2013, **56**, 68.

81. W. A. van, B. Maas, P. Doze, R. H. Slart, H. W. Frijlink, W. Vaalburg and P. H. Elsinga, *Chest*, 2005, **128**, 3020.

82. N. Matusiak, W. A. van, D. Rozeveld, A. J. van Oosterhout, I. H. Heijink, R. Castelli, H. S. Overkleeft, R. Bischoff, R. A. Dierckx and P. H. Elsinga, *Mol. Imaging Biol.*, 2015, **17**, 680.

83. B. Prideaux and M. Stoeckli, *J. Proteomics*, 2012, **75**, 4999.

84. A. Nilsson, T. E. Fehniger, L. Gustavsson, M. Andersson, K. Kenne, G. Marko-Varga and P. E. Andren, *PLoS One*, 2010, **5**, e11411.

85. G. Marko-Varga, A. Vegvari, M. Rezeli, K. Prikk, P. Ross, M. Dahlback, G. Edula, R. Sepper and T. E. Fehniger, *Clin. Transl. Med.*, 2012, **1**, 8.

86. K. E. Driscoll, D. L. Costa, G. Hatch, R. Henderson, G. Oberdorster, H. Salem and R. B. Schlesinger, *Toxicol. Sci.*, 2000, **55**, 24.

87. E. Selg, P. Ewing, F. Acevedo, C. O. Sjoberg, A. Ryrfeldt and P. Gerde, *J. Aerosol Med. Pulm. Drug Delivery*, 2013, **26**, 181.

88. A. Tronde, G. Baran, S. Eirefelt, H. Lennernas and U. H. Bengtsson, *J. Aerosol Med.*, 2002, **15**, 283.

89. H. Koepsell, B. M. Schmitt and V. Gorboulev, *Rev. Physiol. Biochem. Pharmacol.*, 2003, **150**, 36.

90. H. Koepsell, K. Lips and C. Volk, *Pharm. Res.*, 2007, **24**, 1227.

91. I. Tamai, R. Ohashi, J. Nezu, H. Yabuuchi, A. Oku, M. Shimane, Y. Sai and A. Tsuji, *J. Biol. Chem.*, 1998, **273**, 20378.

92. G. Horvath, N. Schmid, M. A. Fragoso, A. Schmid, G. E. Conner, M. Salathe and A. Wanner, *Am. J. Respir. Cell Mol. Biol.*, 2007, **36**, 53.

93. K. S. Lips, C. Volk, B. M. Schmitt, U. Pfeil, P. Arndt, D. Miska, L. Ermert, W. Kummer and H. Koepsell, *Am. J. Respir. Cell Mol. Biol.*, 2005, **33**, 79.

94. G. Horvath, E. S. Mendes, N. Schmid, A. Schmid, G. E. Conner, M. Salathe and A. Wanner, *J. Allergy Clin. Immunol.*, 2007, **120**, 1103.

95. A. Amphoux, V. Vialou, E. Drescher, M. Bruss, C. C. Mannoury la, C. Rochat, M. J. Millan, B. Giros, H. Bonisch and S. Gautron, *Neuropharmacology*, 2006, **50**, 941.

96. A. E. Busch, S. Quester, J. C. Ulzheimer, V. Gorboulev, A. Akhoundova, S. Waldegger, F. Lang and H. Koepsell, *FEBS Lett.*, 1996, **395**, 153.

97. A. E. Busch, U. Karbach, D. Miska, V. Gorboulev, A. Akhoundova, C. Volk, P. Arndt, J. C. Ulzheimer, M. S. Sonders, C. Baumann, S. Waldegger, F. Lang and H. Koepsell, *Mol. Pharmacol.*, 1998, **54**, 342.

98. D. Grundemann, J. Babin-Ebell, F. Martel, N. Ording, A. Schmidt and E. Schomig, *J. Biol. Chem.*, 1997, **272**, 10408.

99. I. Wessler, C. J. Kirkpatrick and K. Racke, *Pharmacol. Ther.*, 1998, **77**, 59.

100. Y. Saito, T. Ozawa and A. Nishiyama, *J. Membr. Biol.*, 1987, **98**, 135.

101. W. Kummer, S. Wiegand, S. Akinci, I. Wessler, A. H. Schinkel, J. Wess, H. Koepsell, R. V. Haberberger and K. S. Lips, *Respir. Res.*, 2006, **7**, 65.

102. G. Horvath, Z. Sutto, A. Torbati, G. E. Conner, M. Salathe and A. Wanner, *Am. J. Physiol. Lung Cell Mol. Physiol.*, 2003, **285**, L829.

103. G. Horvath, E. S. Mendes, N. Schmid, A. Schmid, G. E. Conner, N. L. Fregien, M. Salathe and A. Wanner, *Pulm. Pharmacol. Ther.*, 2011, **24**, 654.

104. T. Nakanishi, Y. Hasegawa, T. Haruta, T. Wakayama and I. Tamai, *J. Pharm. Sci.*, 2013, **102**, 3373.

105. I. Tamai, *Biopharm. Drug Dispos.*, 2013, **34**, 29.

106. H. Yabuuchi, I. Tamai, J. Nezu, K. Sakamoto, A. Oku, M. Shimane, Y. Sai and A. Tsuji, *J. Pharmacol. Exp. Ther.*, 1999, **289**, 768.

107. K. Ensing, R. A. de Zeeuw, G. D. Nossent, G. H. Koeter and P. J. Cornelissen, *Eur. J. Clin. Pharmacol.*, 1989, **36**, 189.

108. D. Turck, W. Weber, R. Sigmund, K. Budde, H. H. Neumayer, L. Fritsche, K. L. Rominger, U. Feifel and T. Slowinski, *J. Clin. Pharmacol.*, 2004, **44**, 163.

109. T. Nakanishi, T. Haruta, Y. Shirasaka and I. Tamai, *Drug Metab. Dispos.*, 2011, **39**, 117.

110. J. X. Mo, S. J. Shi, Q. Zhang, T. Gong, X. Sun and Z. R. Zhang, *Mol. Pharm.*, 2011, **8**, 1629.

111. I. Rubio-Aliaga and H. Daniel, *Xenobiotica*, 2008, **38**, 1022.

112. D. A. Groneberg, A. Fischer, K. F. Chung and H. Daniel, *Am. J. Respir. Cell Mol. Biol.*, 2004, **30**, 251.

113. M. Brandsch, I. Knutter and E. Bosse-Doenecke, *J. Pharm. Pharmacol.*, 2008, **60**, 543.

114. T. Hatanaka, T. Nakanishi, W. Huang, F. H. Leibach, P. D. Prasad, V. Ganapathy and M. E. Ganapathy, *J. Clin. Invest.*, 2001, **107**, 1035.

115. A. Biegel, I. Knutter, B. Hartrodt, S. Gebauer, S. Theis, P. Luckner, G. Kottra, M. Rastetter, K. Zebisch, I. Thondorf, H. Daniel, K. Neubert and M. Brandsch, *Amino Acids*, 2006, **31**, 137.

116. D. A. Groneberg, M. Nickolaus, J. Springer, F. Doring, H. Daniel and A. Fischer, *Am. J. Pathol.*, 2001, **158**, 707.

117. M. V. Brahmajothi, N. Z. Sun and R. L. Auten, *Am. J. Respir. Cell Mol. Biol.*, 2013, **48**, 230.

118. J. Pinsonneault, C. U. Nielsen and W. Sadee, *J. Pharmacol. Exp. Ther.*, 2004, **311**, 1088.

119. S. H. Cha, T. Sekine, H. Kusuhara, E. Yu, J. Y. Kim, D. K. Kim, Y. Sugiyama, Y. Kanai and H. Endou, *J. Biol. Chem.*, 2000, **275**, 4507.

120. S. H. Cha, T. Sekine, J. I. Fukushima, Y. Kanai, Y. Kobayashi, T. Goya and H. Endou, *Mol. Pharmacol.*, 2001, **59**, 1277.

121. M. Hosoyamada, T. Sekine, Y. Kanai and H. Endou, *Am. J. Physiol.*, 1999, **276**, F122.

122. H. Adachi, T. Suzuki, M. Abe, N. Asano, H. Mizutamari, M. Tanemoto, T. Nishio, T. Onogawa, T. Toyohara, S. Kasai, F. Satoh, M. Suzuki, T. Tokui, M. Unno, T. Shimosegawa, S. Matsuno, S. Ito and T. Abe, *Am. J. Physiol. Renal Physiol.*, 2003, **285**, F1188.

123. K. Fujiwara, H. Adachi, T. Nishio, M. Unno, T. Tokui, M. Okabe, T. Onogawa, T. Suzuki, N. Asano, M. Tanemoto, M. Seki, K. Shiiba, M. Suzuki, Y. Kondo, K. Nunoki, T. Shimosegawa, K. Iinuma, S. Ito, S. Matsuno and T. Abe, *Endocrinology*, 2001, **142**, 2005.

124. G. A. Kullak-Ublick, M. G. Ismair, B. Stieger, L. Landmann, R. Huber, F. Pizzagalli, K. Fattinger, P. J. Meier and B. Hagenbuch, *Gastroenterology*, 2001, **120**, 525.

125. I. Tamai, J. Nezu, H. Uchino, Y. Sai, A. Oku, M. Shimane and A. Tsuji, *Biochem. Biophys. Res. Commun.*, 2000, **273**, 251.

126. J. Konig, *Handb. Exp. Pharmacol.*, 2011, **201**, 1.

127. T. Nakanishi and I. Tamai, *Drug Metab. Pharmacokinet.*, 2012, **27**, 106.

128. T. Nakanishi and I. Tamai, *Biopharm. Drug Dispos.*, 2014, **35**, 463.

129. B. Hagenbuch and P. J. Meier, *Biochim. Biophys. Acta*, 2003, **1609**, 1.

130. B. Hagenbuch and P. J. Meier, *Pflugers Arch.*, 2004, **447**, 653.

131. E. Jacquemin, B. Hagenbuch, B. Stieger, A. W. Wolkoff and P. J. Meier, *Proc. Natl. Acad. Sci. U. S. A.*, 1994, **91**, 133.

132. S. Seki, M. Kobayashi, S. Itagaki, T. Hirano and K. Iseki, *Biochim. Biophys. Acta*, 2009, **1788**, 911.

133. M. Ohbayashi, M. Suzuki, Y. Yashiro, S. Fukuwaka, M. Yasuda, N. Kohyama, Y. Kobayashi and T. Yamamoto, *J. Toxicol. Sci.*, 2010, **35**, 653.

134. M. Ohbayashi, C. Yamamoto, A. Shiozawa, N. Kohyama, Y. Kobayashi and T. Yamamoto, *J. Toxicol. Sci.*, 2013, **38**, 103.

135. R. Lu, N. Kanai, Y. Bao and V. L. Schuster, *J. Clin. Invest.*, 1996, **98**, 1142.

136. B. S. Chan, S. Endo, N. Kanai and V. L. Schuster, *Am. J. Physiol. Renal Physiol.*, 2002, **282**, F1097.

137. T. Nomura, R. Lu, M. L. Pucci and V. L. Schuster, *Mol. Pharmacol.*, 2004, **65**, 973.

138. S. K. Banu, J. A. Arosh, P. Chapdelaine and M. A. Fortier, *Proc. Natl. Acad. Sci. U. S. A.*, 2003, **100**, 11747.

139. B. S. Chan, J. A. Satriano, M. Pucci and V. L. Schuster, *J. Biol. Chem.*, 1998, **273**, 6689.

140. H. Y. Chang, J. Locker, R. Lu and V. L. Schuster, *Circulation*, 2010, **121**, 529.

141. Y. Shirasaka, M. Shichiri, T. Kasai, Y. Ohno, T. Nakanishi, K. Hayashi, A. Nishiura and I. Tamai, *J. Endocrinol.*, 2013, **217**, 265.

142. T. Nakanishi, Y. Hasegawa, R. Mimura, T. Wakayama, Y. Uetoko, H. Komori, S. Akanuma, K. Hosoya and I. Tamai, *PLoS One*, 2015, **10**, e0123895.

143. J. L. Sloan, B. R. Grubb and S. Mager, *Am. J. Physiol. Lung Cell Mol. Physiol.*, 2003, **284**, L39.

144. O. M. Granillo, M. V. Brahmajothi, S. Li, A. R. Whorton, S. N. Mason, T. J. McMahon and R. L. Auten, *Am. J. Physiol. Lung Cell Mol. Physiol.*, 2008, **295**, L38.

145. J. L. Sloan and S. Mager, *J. Biol. Chem.*, 1999, **274**, 23740.

146. T. Nakanishi, R. Kekuda, Y. J. Fei, T. Hatanaka, M. Sugawara, R. G. Martindale, F. H. Leibach, P. D. Prasad and V. Ganapathy, *Am. J. Physiol. Cell Physiol.*, 2001, **281**, C1757.

147. T. Hatanaka, W. Huang, T. Nakanishi, C. C. Bridges, S. B. Smith, P. D. Prasad, M. E. Ganapathy and V. Ganapathy, *Biochem. Biophys. Res. Commun.*, 2002, **291**, 291.

148. T. Nakanishi, T. Hatanaka, W. Huang, P. D. Prasad, F. H. Leibach, M. E. Ganapathy and V. Ganapathy, *J. Physiol.*, 2001, **532**, 297.

149. T. Hatanaka, M. Haramura, Y. J. Fei, S. Miyauchi, C. C. Bridges, P. S. Ganapathy, S. B. Smith, V. Ganapathy and M. E. Ganapathy, *J. Pharmacol. Exp. Ther.*, 2004, **308**, 1138.

150. S. Karunakaran, N. S. Umapathy, M. Thangaraju, T. Hatanaka, S. Itagaki, D. H. Munn, P. D. Prasad and V. Ganapathy, *Biochem. J.*, 2008, **414**, 343.

151. R. Pfeiffer, G. Rossier, B. Spindler, C. Meier, L. Kuhn and F. Verrey, *EMBO J.*, 1999, **18**, 49.

152. T. R. Stouch and O. Gudmundsson, *Adv. Drug Delivery Rev.*, 2002, **54**, 315.

153. J. M. Brady, N. J. Cherrington, D. P. Hartley, S. C. Buist, N. Li and C. D. Klaassen, *Drug Metab. Dispos.*, 2002, **30**, 838.

154. L. Campbell, A. N. Abulrob, L. E. Kandalaft, S. Plummer, A. J. Hollins, A. Gibbs and M. Gumbleton, *J. Pharmacol. Exp. Ther.*, 2003, **304**, 441.

155. C. Cordon-Cardo, J. P. O'Brien, J. Boccia, D. Casals, J. R. Bertino and M. R. Melamed, *J. Histochem. Cytochem.*, 1990, **38**, 1277.

156. P. van der Valk, C. K. van Kalken, H. Ketelaars, H. J. Broxterman, G. Scheffer, C. M. Kuiper, T. Tsuruo, J. Lankelma, C. J. Meijer and H. M. Pinedo, *Ann. Oncol.*, 1990, **1**, 56.

157. E. Lechapt-Zalcman, I. Hurbain, R. Lacave, F. Commo, T. Urban, M. Antoine, B. Milleron and J. F. Bernaudin, *Eur. Respir. J.*, 1997, **10**, 1837.

158. G. L. Scheffer, A. C. Pijnenborg, E. F. Smit, M. Muller, D. S. Postma, W. Timens, P. van der Valk, E. G. de Vries and R. J. Scheper, *J. Clin. Pathol.*, 2002, **55**, 332.

159. M. Madlova, C. Bosquillon, D. Asker, P. Dolezal and B. Forbes, *J. Pharm. Pharmacol.*, 2009, **61**, 293.

160. J. Brillault, W. V. De Castro, T. Harnois, A. Kitzis, J. C. Olivier and W. Couet, *Antimicrob. Agents Chemother.*, 2009, **53**, 1457.

161. D. L. Roerig, S. H. Audi and S. B. Ahlf, *Drug Metab. Dispos.*, 2004, **32**, 953.

162. F. Manford, Y. Riffo-Vasquez, D. Spina, C. P. Page, A. J. Hutt, V. Moore, F. Johansson and B. Forbes, *J. Pharm. Pharmacol.*, 2008, **60**, 1305.

163. D. Francombe *et al.*, *Proc. Respir. Drug Delivery*, 2008, **II**, 461.

164. J. I. Fletcher, M. Haber, M. J. Henderson and M. D. Norris, *Nat. Rev. Cancer*, 2010, **10**, 147.

165. S. E. Budulac, D. S. Postma, P. S. Hiemstra, L. I. Kunz, M. Siedlinski, H. A. Smit, J. M. Vonk, B. Rutgers, W. Timens and H. M. Boezen, *Respir. Res.*, 2010, **11**, 60.

166. S. P. Cole, *Annu. Rev. Pharmacol. Toxicol.*, 2014, **54**, 95.

167. T. Langmann, R. Mauerer, A. Zahn, C. Moehle, M. Probst, W. Stremmel and G. Schmitz, *Clin. Chem.*, 2003, **49**, 230.

168. M. J. Flens, G. J. Zaman, P. van der Valk, M. A. Izquierdo, A. B. Schroeijers, G. L. Scheffer, P. van der Groep, M. de Haas, C. J. Meijer and R. J. Scheper, *Am. J. Pathol.*, 1996, **148**, 1237.

169. J. M. Brechot, I. Hurbain, A. Fajac, N. Daty and J. F. Bernaudin, *J. Histochem. Cytochem.*, 1998, **46**, 513.

170. T. Lehmann, C. Kohler, E. Weidauer, C. Taege and H. Foth, *Toxicology*, 2001, **167**, 59.

171. M. Jani, C. Ambrus, R. Magnan, K. T. Jakab, E. Beery, J. K. Zolnerciks and P. Krajcsi, *Arch. Toxicol.*, 2014, **88**, 1205.

172. P. A. Fetsch, A. Abati, T. Litman, K. Morisaki, Y. Honjo, K. Mittal and S. E. Bates, *Cancer Lett.*, 2006, **235**, 84.

173. J. Brillault, W. V. De Castro and W. Couet, *Antimicrob. Agents Chemother.*, 2010, **54**, 543.

174. A. V. Gontijo, J. Brillault, N. Gregoire, I. Lamarche, P. Gobin, W. Couet and S. Marchand, *Antimicrob. Agents Chemother.*, 2014, **58**, 3942.

175. J. Peters, W. Block, S. Oswald, J. Freyer, M. Grube, H. K. Kroemer, M. Lammer, D. Lutjohann, M. Venner and W. Siegmund, *Drug Metab. Dispos.*, 2011, **39**, 1643.

176. K. Togami, S. Chono and K. Morimoto, *Biopharm. Drug Dispos.*, 2011, **32**, 389.

177. K. Togami, S. Chono and K. Morimoto, *Pharmazie*, 2012, **67**, 389.
178. J. Peters, K. Eggers, S. Oswald, W. Block, D. Lutjohann, M. Lammer, M. Venner and W. Siegmund, *Drug Metab. Dispos.*, 2012, **40**, 522.
179. M. Demeule, J. Jodoin, E. Beaulieu, M. Brossard and R. Beliveau, *FEBS Lett.*, 1999, **442**, 208.
180. R. J. Dinis-Oliveira, J. A. Duarte, F. Remiao, A. Sanchez-Navarro, M. L. Bastos and F. Carvalho, *Toxicology*, 2006, **227**, 73.
181. K. S. Lips, A. Luhrmann, T. Tschernig, T. Stoeger, F. Alessandrini, V. Grau, R. V. Haberberger, H. Koepsell, R. Pabst and W. Kummer, *Life Sci.*, 2007, **80**, 2263.
182. M. Mukherjee, D. I. Pritchard and C. Bosquillon, *Clin. Exp. Allergy*, 2011, **41**, 1832.
183. M. Mukherjee, M. L. Latif, D. I. Pritchard and C. Bosquillon, *J. Pharmacol. Toxicol. Methods*, 2013, **68**, 184.
184. S. Sun, J. H. Schiller and A. F. Gazdar, *Nat. Rev. Cancer*, 2007, 7, 778.
185. C. Ehrhardt, C. Kneuer, C. Bies, C. M. Lehr, K. J. Kim and U. Bakowsky, *Pulm. Pharmacol. Ther.*, 2005, **18**, 165.
186. K. Damme, A. T. Nies, E. Schaeffeler and M. Schwab, *Drug Metab. Rev.*, 2011, **43**, 499.
187. Y. Kanai and H. Endou, *Curr. Drug Metab.*, 2001, **2**, 339.
188. K. Nakanishi, H. Matsuo, Y. Kanai, H. Endou, S. Hiroi, S. Tominaga, M. Mukai, E. Ikeda, Y. Ozeki, S. Aida and T. Kawai, *Virchows Arch.*, 2006, **448**, 142.
189. J. D. Young, S. Y. Yao, J. M. Baldwin, C. E. Cass and S. A. Baldwin, *Mol. Aspects Med.*, 2013, **34**, 529.
190. A. Obaidat, M. Roth and B. Hagenbuch, *Annu. Rev. Pharmacol. Toxicol.*, 2012, **52**, 135.
191. K. Dilger, M. Schwab and M. F. Fromm, *Inflammatory Bowel Dis.*, 2004, **10**, 578.
192. B. I. Florea, I. C. van der Sandt, S. M. Schrier, K. Kooiman, K. Deryckere, A. G. de Boer, H. E. Junginger and G. Borchard, *Br. J. Pharmacol.*, 2001, **134**, 1555.
193. M. van der Deen, E. G. de Vries, W. Timens, R. J. Scheper, H. Timmer-Bosscha and D. S. Postma, *Respir. Res.*, 2005, **6**, 59.
194. T. Ando, H. Kusuhara, G. Merino, A. I. Alvarez, A. H. Schinkel and Y. Sugiyama, *Drug Metab. Dispos.*, 2007, **35**, 1873.
195. N. Kalin, A. Claass, M. Sommer, E. Puchelle and B. Tummler, *J. Clin. Invest.*, 1999, **103**, 1379.
196. E. Puchelle, D. Gaillard, D. Ploton, J. Hinnrasky, C. Fuchey, M. C. Boutterin, J. Jacquot, D. Dreyer, A. Pavirani and W. Dalemans, *Am. J. Respir. Cell Mol. Biol.*, 1992, 7, 485.
197. M. Agassandian, S. N. Mathur, J. Zhou, F. J. Field and R. K. Mallampalli, *Am. J. Respir. Cell Mol. Biol.*, 2004, **31**, 227.
198. G. Yamano, H. Funahashi, O. Kawanami, L. X. Zhao, N. Ban, Y. Uchida, T. Morohoshi, J. Ogawa, S. Shioda and N. Inagaki, *FEBS Lett.*, 2001, **508**, 221.

Section II: Preclinical Models in Current Use within the Pharmaceutical Industry

The Characteristics, Validation and Applications of In silico and In vitro Models of Drug Transporters

PRADEEP SHARMA,*[a] MOHAMMED I. ATARI,[b] ROBERT ELSBY,[c,†] SIMON THOMAS,[b] SIMONE STAHL,[a] CONSTANZE HILGENDORF[d] AND KATHERINE FENNER[a]

[a] DMPK DSM, iMED and Early Development, AstraZeneca R&D, 310 Darwin Building, Cambridge Science Park, Milton Road, Cambridge CB4 0WG, UK; [b] Cyprotex Discovery Limited, 15 Beech Lane, Macclesfield, Cheshire SK10 2DR, UK; [c] DMPK DSM, iMED and Early Development, AstraZeneca R&D, Alderley Park, Macclesfield, Cheshire SK10 4TG, UK; [d] DMPK DSM, iMED and Early Development, AstraZeneca R&D, Pepparedsleden 1, Mölndal 431 83, Sweden
*Email: Pradeep.Sharma@astrazeneca.com

7.1 Introduction

Transporters play important roles in numerous physiological processes. They govern the disposition of endogenous as well as exogenous (xenobiotic) molecules and as such have been shown to have a substantial impact on the

---

†Current address: Cyprotex Discovery Limited, Biohub, Alderley Park, Macclesfield, Cheshire, SK10 4TG, UK.

RSC Drug Discovery Series No. 54
Drug Transporters: Volume 1: Role and Importance in ADME and Drug Development
Edited by Glynis Nicholls and Kuresh Youdim

drug discovery and development process. The study of drug transporters involves a multitude of different types of studies, such as the determination of binding affinities, transport rate, inhibition of transport, regulation of transporter proteins, expression and/or distribution patterns of novel putative carrier proteins, molecular mechanistic studies of substrate carrier interactions, and structure–activity relationships of transporters. Also, quantification in native organs, up-scaling of absolute abundance and prediction of the influence of transporters on absorption, distribution, metabolism and elimination (ADME) parameters is one of the most practically demanding areas of research in transporter science. The most basic aspect amongst all of these research areas is the selection of suitable model(s) for studying active transport. Various *in silico, in situ, in vitro* and *in vivo* model systems are available to accomplish this wide variety of tasks. In recent years, *in silico* (computational) methods have been increasingly used as an alternative or supplement to *in situ, in vitro* and *in vivo* models to answer questions relating to efficacy and safety. In addition, with recent improvements in computational capabilities, *in silico* models can play a major role in the entire drug development process from the preclinical early discovery stage up to late stage clinical development. Robust *in silico* models are generally time- and cost-effective high-throughput screening methods with the advantage of reducing experimental efforts in the screening phase of drug discovery, capable of evaluating compound structures that have not even been synthesised.

In silico models are useful preliminary screens to study transporter properties but full characterisation of these properties involves the use of *in vitro* models as drug molecules progress to the later stages of drug discovery and development. The *in vitro* models that may be used for performing high-throughput transporter assays at the early development or preclinical stages provide several benefits compared with *in vivo* methods, including the need for no or fewer animals, generally less variability in the dataset and the need for less of the often scarce test material. Different *in vitro* systems offer various levels of correlation with the *in vivo* situation and hence results from experiments using *in vitro* models need to be carefully evaluated for prediction of the *in vivo* outcome.

The current philosophy of drug development is moving towards a 'fail early–fail cheaply' paradigm. Therefore, in many cases, *in vitro* ADME approaches are being applied to drug candidates earlier in development and are increasingly being linked with high throughput automation and analysis. Figure 7.1 describes the choice of *in vitro* transporter models for characterising ADME properties of candidate drugs along the value chain in an industrial pharmaceutical setting. A multitude of *in vitro* models have been reported in the literature for drug transporter studies, ranging from simple membrane systems to complicated isolated perfused organs or tissues. In general, *in vitro* models for investigating DMPK properties should meet the following criteria:

- *Accuracy:* the results obtained from a model should have good concordance with the corresponding *in vivo* property (*e.g.* clearance) using

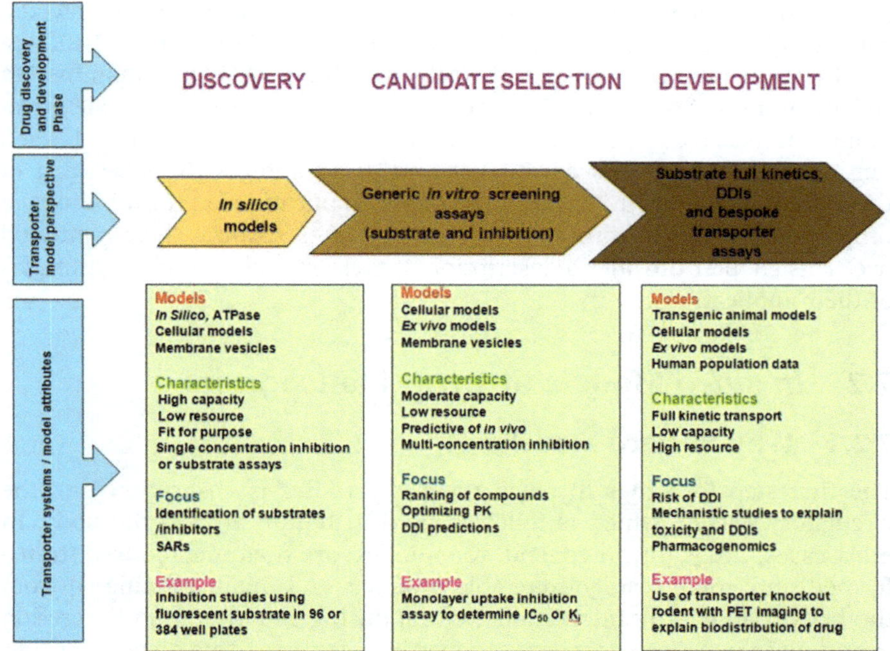

Figure 7.1 *In silico* and *in vitro* models along the value chain in drug discovery and development.

appropriate scaling factors while taking into account other elimination pathways and inter-individual differences in transporter expression. This requires validation of the method with standard compounds with known animal or human performance, taking into account other criteria such as robustness, precision and reproducibility (see below). Due to the promiscuous nature of most transporters with regard to their substrate specificity, the search for highly specific probes and inhibitors is still a burgeoning area of research.

- *Cost:* for early screening, the chosen model should be capable of high-throughput screening of compounds and should not be a labour-intensive procedure to perform. In late stage drug development, thoroughly validated complex models are used to provide detailed transporter kinetics or to provide mechanistic insight into observed atypical pharmacokinetics.
- *Robustness:* the model should be adaptable to the varied chemical structures and physicochemical properties of new chemical entities.
- *Precision:* the results should be reproducible and the execution protocol should be harmonised between different laboratories.

The major types of *in vitro* models for drug transport studies comprise isolated and perfused organs or tissues; primary cell cultures (*e.g.* hepatocytes); sandwich cultures of primary hepatocytes; immortalised cell lines (*e.g.* Caco-2, MDCK, LLC-PK1); genetically engineered cell lines transfected with single transporters (*e.g.* MDCKII-MRP2, LLCPK1-MDR1, HEK293-OATP1B1) or multiple transporters (MDCKII-MRP2/OATP1B1); mammalian cell membrane vesicles; and insect (*e.g. Spodoptera frugiperda*) cell membranes transfected with transporter proteins. As shown in Table 7.1, each one of these models has advantages and disadvantages. Because each of these techniques has its own requirements, applications and limitations, a proper insight into the process being studied often requires their combined use. This chapter provides a description of these models and gives examples of their applications.

7.2   *In silico* Models of Drug Transporters

7.2.1   Why *In silico* Modelling?

The first step for drugs to reach their target sites is absorption into the circulatory system, which is followed by distribution around the body. In most cases, drugs (or, in general, xenobiotics) are then metabolised (transformed) into more readily excretable products and finally eliminated from the body.[1] Drug–drug interactions (DDIs) during absorption and excretion can result in unexpected toxicity.[2] Properties relating to drug disposition and toxicity, *i.e.* absorption, distribution, metabolism, excretion and toxicity (ADMET), parameters can be mediated or influenced by drug transporters, rendering knowledge of transporters indispensable in understanding drug pharmacokinetics and pharmacodynamics. Indeed, detailed three-dimensional (3D) information about the structure, function and mechanism of transport proteins would have a great potential value in contributing to rational drug design, particularly in the drug discovery phase.

The literature contains extensive information on the functional and biochemical attributes of drug transporters and considerable work has been done to establish their protein structures (including the ATP binding cassette (ABC) and solute carrier (SLC) superfamilies). However, the use of a combined structural, functional and mechanistic approach to understand transporter interactions has been limited to date, mainly because of the poor availability of high resolution transporter structures, and it remains challenging to express and crystallise such proteins.[3,4] That said, the number of membrane proteins of known 3D structure, determined by X-ray crystallisation or nuclear magnetic resonance (NMR) spectroscopy, has grown exponentially in the past 30 years. For interested readers, an up-to-date list of membrane protein structures is maintained by the Stephen White Laboratory at the University of California at Irvine (see http://blanco.biomol.uci.edu/mpstruc/).

The absence of a specific experimental 3D structure can nonetheless be addressed using *in silico* modelling tools[5] to integrate and understand the biochemical and biophysical properties of drug transporters. Such tools can be integrated with whole body physiologically-based pharmacokinetic (PBPK)

Table 7.1 List of *in vitro* models with characteristics and applications.[a]

Model/description/ examples	Advantages	Disadvantages	*In vitro* parameters generated	Application	Ref.
ATPase assay membranes					
Measures the ability of a drug to stimulate ATP hydrolysis. *e.g.* P-gp expressing *Spodoptera frugiperda* (Sf9) membranes	Rapid, high-throughput assay, recovery is not an issue, no extractions/ separations, simple colorimetric or radiometric assay	Cannot be used for SLC uptake transporters, non-functional assay, cannot differentiate substrates/inhibitors, some substrates could interfere with the ATPase activity at a high concentration range	ATPase activity (nmol min^{-1} mg^{-1} of protein) determined as the difference between the amount of inorganic phosphate released from ATP in the absence and presence of vanadate	Screening tool to identify ligands of ABC transporters (P-gp, BCRP)	61, 62
Membrane vesicles					
Membrane vesicles are sub-cellular fractions prepared from organs or cells. Transport of substrate into the lumen of closed membranes expressing transporters is measured *e.g.* Purified membrane vesicles from HEK293 cells expressing MRP2	Rapid, high-throughput functional assay (measures actual transport of substrate), good for low permeability compounds, can be used for highly cytotoxic compounds, high transporter expression, possible to 'titrate' transporter expression, purified system for accurate kinetic studies due to direct access to the active site, can be used	Unsuitable for highly permeable compounds, poor recovery of lipophilic compounds due to non-specific binding, batch-to-batch variability, contamination with other cellular components, heterogeneity with respect to size and orientation, labour-intensive techniques	Transporter activity (pmol min^{-1} mg^{-1} of protein), K_m, IC_{50}	Identification of substrates and inhibitors of ABC and SLC transporters, study of transporter kinetics, study the effect of SNPs and transporter mechanisms, CMVs have been extensively used to investigate cholestatic (interference in bile secretion) potential of xenobiotics	53, 60, 213, 214

Table 7.1 (*Continued*)

Model/description/examples	Advantages	Disadvantages	*In vitro* parameters generated	Application	Ref.
	to yield information regarding driving forces (ion gradients, temperature-dependence and ATP-dependence), devoid of metabolising enzymes, versatile range of applications for multitude of transporters in various organs and species				
Primary cell cultures Cells obtained from native tissues and grown as suspension or plated cultures are used for substrate and/or inhibition assays *e.g.* Hepatocytes, proximal renal tubule cells	High similarity to the physiology of organism, possible with all major organs of interest (liver, kidney, brain, *etc.*), possible with different species and humans, cryopreservation of primary cells for longer storage and availability, can be used to make more complicated *in vitro* models	Low or medium throughput, expression of multiple transporters complicates elucidation of involvement of individual transporters, loss of expression on isolation, cryopreservation and culture, cost- and technique-intensive requiring maintenance lab and	Transporter activity $(\text{pmol}^{-1}\ \text{min}^{-1}$ million cells), intracellular concentrations, passive and active transport contribution, K_m, IC$_{50}$, K_p	Prediction of organ clearances, building PBPK or IVIVE, identification of substrates and inhibitors of transporters, study of transporter kinetics	114, 116, 120, 225

			equipment, cells may contain metabolising enzymes and complicate interpretations due to metabolism		
Primary cultures in special configuration					
Primary cells oriented in specific configuration matrices and/or co-cultures with different cell types to study transporters *e.g.* SCH, bioreactors, endothelial cells in co-cultures with glioma and neuronal cells	Mimic physiology of complicated organ systems such as the brain, liver and kidney, regenerate cytoarchitecture and cellular polarity, effect of co-cultures in heterogeneous system can be studied	Low throughput, cost and technique intensive, long culture times, involvement of multiple transporters could be confounding, cost and technique intensive requiring maintenance lab and equipment	Transporter activity ($\mathrm{pmol}^{-1}\,\mathrm{min}^{-1}$ million cells), intracellular concentrations, specific parameters such as biliary clearance, intracellular concentrations, K_m, IC$_{50}$, K_p	Prediction of BBB, kinetics, prediction of organ clearances, prediction of cholestatic potential, building PBPK or IVIVE, identification of substrates and inhibitors of transporters, study of transporter kinetics	148, 151, 224
Immortalised cell lines					
Immortalised cell lines are perpetually dividing cell lines from primary or early passage cells or from transgenic animals that carry an immortalising gene *e.g.* HEK293	High-throughput, readily available and unlimited supply of cells, cryopreservation possible, well defined and stable, can be standardised and exchanged between labs, plateable in inserts and amenable to high-throughput assays, can be genetically modified to express specific transporter of interest	Cells do not represent exact phenotype of native cells in which transporter is present *in vivo*, transporter activity or expression may be unstable, recovery of compounds could be an issue, high variability across labs, long culture times	Transepithelial transport (permeability, nm s^{-1}), K_m, IC$_{50}$	BCS classification of drugs, efflux ratios, Identification of substrates and inhibitors of transporters, study of transporter kinetics	98, 101, 110

Table 7.1 (*Continued*)

Model/description/ examples	Advantages	Disadvantages	*In vitro* parameters generated	Application	Ref.
Transgenic cell lines or genetically modified cells					
Genetic transfection of transporter into host cell using cDNA Heterologous expression involves expression into host cell of different species, *e.g.* CHO-OAT1 cells Homologous expression is expression in host cell of the same species, *e.g.* HEK293-OATP1B1 cells	High transporter expression compared with primary and immortalised cells, multiple transfections possible to get polarised transport and to study multiple transporters	Non-physiological expression of transporters, endogenous expression in test system requires use of 'control' or vector transfected cell lines as an additional experimental run adding complexity and costs, poorly permeable compounds cannot be studied for efflux transporters, highly permeable compounds cannot be studied for uptake transporters, kinetic parameters (K_m and V_{max}) may be different from native cell lines because of difference in expression levels and partitioning into cell membrane bilayers	Transporter activity ($pmol^{-1}\ min^{-1}$ million cells), transepithelial transport (permeability, $nm\ s^{-1}$), intracellular concentrations, specific parameters such as biliary clearance, intracellular concentrations, K_m, IC_{50}, K_p	Efflux ratios, identification of substrates and inhibitors of transporters, study of transporter kinetics, prediction of organ clearances, building PBPK or IVIVE	71, 72, 87, 227

cRNA injected oocytes					
e.g. PEPT1 expressed on oocytes	Efficient expression of transporters	Experimental expertise needed, seasonal variation possible and assay performed at non-physiological temperatures	Transporter activity (pmol^{-1} min^{-1} oocytes), K_m and inhibition parameters (IC$_{50}$)		91, 92
Precision-cut tissue slices					
Ultrathin sectioning of organ systems by specially designed tissue slicers, which produces multi-laminated tissue that is a miniaturised representative bio-specimen of a macroscopic organ *e.g.* Liver, kidney, intestine, brain, lung slices	Versatile range of applications for transporters in various organs and species, cryopreservation of liver slices for storage and subsequent usage reported, preservation of the different cell types and the 3D organisation	Low throughput, poor penetration into deeper cells, low viability, difficult to oxygenate deeper cell layers, cost-extensive and requires expertise, cannot be used for transepithelial transport, presence of metabolising enzymes complicates interpretations due to metabolism	Cell uptake (pmol^{-1} min^{-1} mg of protein)	Liver slices to study the hepatotoxicity and biotransformation of compounds, rat kidney slices have been used for renal uptake studies	163, 164, 166
Isolated perfused organ systems or tissue chambers					
Isolation of organ system or tissue and maintenance *in vitro* by perfusion and oxygenation in special chambers *e.g.* Perfused liver or everted intestinal sac	Retention of anatomical and morphological features, excluding extra-hepatic influences (observed during *in vivo* studies), opportunity to manipulate medium	Low throughput, cost-intensive and requires expertise, only feasible with preclinical species, not in the spirit of the 3Rs of animal usage, complexities of the	Different parameters are generated depending upon the organ system	Hepatic disposition primarily mediated by canalicular transporters, interference in bile excretion or hepatotoxicity caused by xenobiotics due to	190–193

Table 7.1   (*Continued*)

Model/description/ examples	Advantages	Disadvantages	*In vitro* parameters generated	Application	Ref.
	flow, composition and temperature as well as perfusion direction, multiple samples can be taken from both the inflow and outflow medium, allowing pharmacokinetic analysis by compartment modelling	whole organ limit the ability to study individual uptake and excretion mechanisms		competitive inhibition or induction of transporters, mechanistic studies for elucidation of DDIs and hepatic enzyme–transporter interplay, everted intestinal sac and Ussing chamber techniques have been employed to study mechanistic biopharmaceutical rational behind DDIs	

[a]BCS: Biopharmaceutic Classification of Drugs; CMV: canalicular membrane vesicle; IVIVE: *in vitro* to *in vivo* extrapolation; K_p: partition ratio into organ/cells; PBPK: physiologically-based pharmacokinetic; SNP: single nucleotide polymorphism.

models to explore the *in vivo* consequences in the species of interest, as discussed in Section 7.5.2.4 of this chapter and in Chapter 9.

7.2.2 Transporter-based Methods

The aim of transporter-based techniques is to predict the 3D structure of the target protein using either homology modelling (Section 7.2.2.1), fold recognition modelling (Section 7.2.2.2) or *ab initio* modelling (Section 7.2.2.3), where further docking studies can be applied to study interactions between drug transporters and their substrates or inhibitors.

7.2.2.1 *Homology Modelling*

Homology (comparative) modelling is a knowledge-based prediction approach[6] whereby a protein (target) with an unknown structure is aligned with one or more homologous protein sequences with experimentally determined structures (templates) that are available in a Protein Data Bank (PDB).[7] This technique is based on the principle that homologue proteins have similar structures. Among the two approaches for 3D structure prediction described in this section, homology modelling is considered the easiest one. In general, all homology modelling tools are multistep processes that can be summarised by the same seven steps:

(1) Template recognition and initial alignment
(2) Alignment correction
(3) Backbone generation
(4) Loop modelling
(5) Side-chain modelling
(6) Model optimisation
(7) Model validation

In the first step, a PDB, usually that of the Research Collaboratory for Structural Bioinformatics (RCSB; see www.rcsb.org/),[8] is queried using sequence alignment programmes such as Basic Local Alignment Search Tool (BLAST)[9] or any other alignment search method.[10] Successful homology modelling is considered feasible if the target and template sequences have greater than 30% of identical residues in common.[6] The second step, alignment correction, is critical to any successful modelling method since the majority of model errors result from sequence misalignments.[11] In recent years, significant progress has been made in the development of more sophisticated methods, such as the programme CLUSTAL OMEGA,[12] to overcome such errors and arrive at a better alignment. In the third step (*i.e.* backbone generation[10]), the actual model building starts; that is, the coordinates of those template residues that appear in alignment with the model sequence are copied to the target. In addition, only the backbone coordinates (N, C$\alpha^{\ddagger}$, C and O) can be copied if two residues differ and, if they are the same, the side chain can be included. The loop modelling step deals with gaps in the alignment between the model and template sequences. These gaps (loops) can either be in the model sequence (deletions) or in the template sequence (insertions). To overcome deletions, residues are omitted from the template and, therefore, a hole is created in the model that must be closed. In the case of insertions, the continuous backbone from the template is cut and the missing residues are inserted. Both cases suggest conformational changes in the backbone, and the accuracy of loop modelling is vital in determining the effectiveness of homology models for studying transporter–ligand interactions.[13] Two main approaches have been widely used in loop modelling:

(a) Knowledge-based (querying a PDB)[14]
(b) Energy-based, where an energy function is used to judge the quality of a loop[15]

Side-chain modelling (step 5) is the prediction of the target protein side-chain conformation for a given backbone structure. Practically, side-chain placements are obtained from knowledge-based searches using libraries, such as the rotamer libraries,[16] and/or rotamer optimisation methods, for example the clash-detection guided iterative search with rotamer relaxation (CIS-RR) approach.[17] Model optimisation of the structures (step 6) can be achieved by running molecular dynamics simulation of the model[18] to mimic the true folding process. Model validation (final step) is an essential process to check the accuracy of the generated 3D structures, since errors can propagate from former processes. Two types of model validation are available: theoretical validation in which homology modelling software is used to assess the quality of the model; and experimental validation in which the 3D model is tested against real biological results. Validated homology models have been commonly used both to predict binding affinities of drugs to transporters and to examine binding sites.[19,20] Examples of homology modelling software tools are given in Table 7.2.

7.2.2.2 Fold Recognition Methods

Fold recognition (or threading) is a method of protein modelling that entails the assignment of tertiary structures to protein sequences, without access to homologous proteins with known structures.[21] That is, fold recognition enables the identification of proteins with known structures that share similar folds (or fragments of folds) with the new protein (target) sequences. The target sequences are constructed using the identified protein structures as templates. Fold recognition methods (Table 7.2) aim to compute how well

---

‡Cα is the carbon atom to which the amine group and the carboxylic acid group are connected in the amino acid.

Table 7.2 A non-exhaustive list of servers and software tools used in transporter based modelling.

Software	URL	Transporter based method
I-TASSER	http://zhanglab.ccmb.med.umich.edu/ I-TASSER/	*Ab initio* modelling
LOOPP	http://cbsuapps.tc.cornell.edu/loopp. aspx	Fold recognition modelling
MODELLER	http://salilab.org/modeller/	Homology modelling
MUSTER	http://zhanglab.ccmb.med.umich.edu/ MUSTER/	Fold recognition modelling
PHYRE2	http://www.sbg.bio.ic.ac.uk/phyre2/ html/page.cgi?id=index	Homology modelling
ROBETTA	http://www.robetta.org/	*Ab initio* modelling
ROSETTA	https://www.rosettacommons.org/	*Ab initio* modelling
SPARKS^X	http://sparks-lab.org/yueyang/server/ SPARKS-X/	Fold recognition modelling
SWISS-MODEL	http://swissmodel.expasy.org/	Homology modelling

each potential protein structure would fit a particular sequence rather than how well each sequence fits a structure. This is achieved by comparing each target sequence against a database of potential fold templates using energy potentials and/or other similarity scoring functions. The template with the highest similarity score (or lowest energy score) is assumed to best fit the fold of the target protein.

7.2.2.3 Ab initio *Modelling*

The above two methods in transporter-based modelling, namely, homology and fold recognition, build protein structures by alignment with solved (known) template structures.[22] When target proteins have known homologues, high resolution models can be built using the template-based methods.[23] If structure homologues (templates) are absent or exist but cannot be identified from a PDB library, the models can be constructed from first principles. This technique is known as *ab initio* or *de novo* modelling, and has the ability to calculate the 3D structure of a target protein from its primary sequence. Additionally, it can aid in the understanding of how proteins fold in nature using physicochemical principles.[24] Successful predictions for proposed 3D structures of proteins using *ab initio* modelling depend on the following three factors:[24]

(1) A precise energy function (such as physics- and knowledge-based energy functions) upon which the native structure of the target corresponds to the most thermodynamically stable state when compared with all possible decoy structures;
(2) A powerful search method (such as molecular dynamics simulation[25]) that can rapidly identify low energy states *via* a conformational search;

(3) Selection of models from a pool of decoy structures that are structurally close to the native state using model quality assessment programmes (MQAPs).[26]

Presently, the precision of *ab initio* models is low and successes have been restricted to small membrane proteins, due to the computational complexities associated with larger protein molecules. Consequently, modelling transporter proteins using *ab initio* methods (Table 7.2) remains challenging due to the generally large size of transmembrane transporters.

7.2.3 Compound-based Methods

The transporter-based methods (Section 7.2.2) have proven useful in providing structural information at the atomic level. In addition, these methods aid in understanding the interaction between molecules (substrates or inhibitors) and transporters to predict drug transport properties. However, transporter-based techniques are limited by the availability of precise templates and computational power, hence, transporter structures may not always be amenable to transporter-based modelling. As an alternative, compound-based methods, including 3D quantitative structure–activity relationship (QSAR) modelling (Section 7.2.3.1) and pharmacophore modelling (Section 7.2.3.2), do not require previous knowledge of the 3D structures of drug transporters and are frequently used in drug discovery.

7.2.3.1 Compound Descriptor Methods and 3D-QSAR Modelling

In computational chemistry, molecular descriptors are quantities that characterise the properties of molecules. Accordingly, a drug can be described by a set of molecular descriptors. Molecular descriptors are classified into two main groups: experimental measurements (such as logP, molar refractivity and dipole moment) and theoretical molecular descriptors, which are calculated from a representation of the molecular structure.[27] Different groups of theoretical descriptors can be calculated from different representations of the molecular structure. Molecular descriptors derived from the chemical formula are known as zero-dimensional (0D) descriptors (*e.g.* molecular weight and atom type count). One-dimensional (1D) descriptors represent the structural fragments of a molecule (*e.g.* number of hydrogen bond donors/acceptors). Two-dimensional (2D) descriptors are derived from algorithms applied to the topological representation of a molecule (*e.g.* Balaban index and Wiener index). 3D descriptors depend on the conformations of the compound, that is, such descriptors are determined by the structure of the molecule (*e.g.* geometrical descriptors and steric descriptors). Four-dimensional (4D) descriptors are derived from lattice representations of a molecule (*e.g.* molecular interaction fields).[27,28] There are numerous software tools available for calculating molecular descriptors that are widely used in the pharmaceutical industry, examples include Dragon™ (http://www.talete.mi.it/), a package developed by Talete srl, and RDKit™ (http://www.rdkit.org/), open-source cheminformatics software. Table 7.3 shows a number of transporter substrates with their corresponding molecular descriptors calculated using RDKit™; such descriptors, and others associated with related, or similar, properties, have been shown to be important in a number of models relating to the *in silico* identification of P-glycoprotein (P-gp) substrates.[29]

Molecular descriptors are used as input (predictor) variables for QSAR modelling.[30] This is a methodology that seeks statistically meaningful relationships between predictor variables (*i.e.* the molecular descriptors) and an activity (such as a transport rate) across a range of compounds.[31] Such a relationship is expressed in a model that predicts the target activity as a function of the descriptors. The purpose of the model is to make predictions of the target activity value for novel compounds, *i.e.* for compounds that were not involved in model training. This ability to extrapolate to new compounds is the *raison d'être* of the model, and care must be taken to avoid generating a model that predicts well for the training set compounds, but not for other compounds. This adverse effect is known as overfitting.[28] Techniques from the fields of machine learning and statistical pattern recognition are used during model building to reduce the risk of it occurring, and its detrimental impact on model usability.[32] One major benefit of applying compound descriptor methods is the ability to predict transmembrane transport and inhibition when the protein structure is unknown.[29,33]

3D-QSAR models include all QSAR methods that correlate macroscopic target properties (such as protein binding affinity or rate of transport) with calculated atom-based descriptors obtained from the 3D representation of the molecular structures.[34] That is, 3D-QSAR models utilise information regarding the 3D molecular structure of molecules to explain the relationships between 3D molecular interactions, based on aligned training set compounds and their measured biological properties.[35] Such models can

Table 7.3 A select number of transporter substrates with their corresponding molecular descriptors calculated using RDKit™.[a]

Compound name	SlogP	SMR	TPSA	AMW	NumHBD	NumHBA	NumAromaticRings
Digoxin	2.2181	192.6108	203.06	780.949	6	14	0
Paclitaxel	3.7357	217.6901	221.29	853.918	4	14	3
Pravastatin	2.4404	111.4432	124.29	424.534	4	6	0
Doxorubicin	0.0013	131.7544	206.07	543.525	6	12	2
Sulfasalazine	1.8002	102.411	137.82	398.4	3	7	2
Valsartan	4.1617	121.3865	112.07	435.528	2	5	3
Bosentan	4.2039	144.6573	145.65	551.625	2	10	4
Pitavastatin	4.5181	117.0294	90.65	421.468	3	4	3
Ciprofloxacin	1.5833	88.479	74.57	331.347	2	5	2
Ranitidine	1.459	84.4008	83.58	314.411	2	7	1

[a]AMW: molecular weight; NumAromaticRings: number of aromatic rings; NumHBA: number of hydrogen bond acceptors; NumHBD: number of hydrogen bond donors; SMR: molar refractivity; SlogP: logP, TPSA: topological polar surface area.

then be used to predict the activity of other compounds. The 3D-QSAR models (Table 7.3) are, in general, applied to aid in designing compounds with high or low transporter binding affinity, depending on the specific requirement. Among the 3D-QSAR techniques widely used in the pharmaceutical industry are comparative molecular analysis (CoMFA) and comparative molecular similarity indices analysis (CoMSIA).[36,37]

7.2.3.2 *Pharmacophore Modelling*

A pharmacophore is a molecular framework that represents the important structural features that are required for a certain biological activity from a series of ligands with a similar mechanism of action.[38] Pharmacophore features include steric and electrostatic characteristics or hydrogen bonding capabilities, examples include hydrogen bond donors/acceptors, hydrophobic regions, aromatic ring centroids, acidic centres (negatively charged at physiological pH) and basic centres (positively charged at physiological pH).[39,40] The pharmacophore is then defined by these features with their corresponding positions within the molecule of interest. Pharmacophore models are based on the assumption that structurally-diverse molecules establish a similar interaction mechanism when binding to a receptor.[41] Therefore, when comparing a set of known transporter substrates/inhibitors with a set of known non-substrates/non-inhibitors, the mutual set of pharmacophore features that are essential for binding to the transporter active site can be identified. Consequently, one can determine whether molecules are substrates/inhibitors or non-substrates/non-inhibitors based on having the same pharmacophore features. Pharmacophore models are not able to predict the strength of binding to a particular transporter active site and cannot distinguish between molecules of the same class. However, this approach (Table 7.4) is useful for medicinal chemists, especially in identifying the key ligand–transporter interactions[42,43] without knowing the structure of the transporter.

Table 7.4   A non-exhaustive list of servers and software tools used in substrate-based modelling.

Software	URL	Substrate-based method
CATALYST	http://accelrys.com/	Pharmacophore modelling
Cyprotex auto-QSAR	http://www.cyprotex.com/insilico	QSAR modelling
GOLPE	http://www.miasrl.com/	QSAR modelling
SYBYL®-X Suite	http://www.certara.com/products/molmod/sybyl-x/	QSAR modelling
PharmaGist	http://bioinfo3d.cs.tau.ac.il/PharmaGist/	Pharmacophore modelling
Pharmer	http://smoothdock.ccbb.pitt.edu/pharmer/	Pharmacophore modelling

This section has summarised the different *in silico* techniques that are commonly used in the pharmaceutical industry to study drug transporters. Two main categories are discussed, namely, transporter- and compound-based methods. Computational (*in silico*) techniques continue to improve and thus provide information for prioritising *in vitro* assays. This way, competition for resources becomes relatively easier in the drug discovery process.

7.3 *In vitro* Models of Transporters

7.3.1 Membrane-based Models: Transport Assays Utilising Vesicles and the ATPase Assay

Transport assays that make use of membrane vesicles are suitable for studying members of the SLC as well as the ABC transporter family, and can be employed to study transporter substrates as well as inhibitors. Two classical assay formats can be used to evaluate transporter activity using membrane preparations: the ATPase assay, which allows the study of the interaction of compounds with ABC transporters; and the vesicular transport assay.

The vesicular transport assay generally utilises inside-out (inverted) membrane vesicles, which can be prepared from various sources through a series of homogenisation and centrifugation steps. They can be used to quantify the accumulation of a probe substrate into the vesicle (Figure 7.2A). During the preparation, inside-out (inverted) and right-side-out vesicles are generated together with open membrane lamellae; these fractions can be separated using centrifugation techniques in dextran density gradients.[44] However, separation of inside-out and right-side-out vesicles is usually not necessary because the vesicles with the "wrong" orientation do not contribute to the reaction analysed. For ABC transporters, for example, the substrate and ATP binding sites, usually located within the cell, are accessible to substrates, inhibitors and co-factors (*e.g.* glutathione for some multidrug resistance proteins (MRPs)) in inside-out vesicles. These inside-out vesicles overcome the issue of inaccessibility of the transporter to the otherwise poorly permeable co-factor (ATP) and test compounds (Figure 7.2A). Right-side-out vesicles can be employed to study transporter processes of non-ABC uptake transporters where the driving force is known (*e.g.* Na$^+$, H$^+$);[45] examples include uptake of taurocholate *via* sodium-taurocholate co-transporting polypeptide (NTCP),[46] *para*-aminohippurate *via* organic anion transporter (OAT), or dipeptide transport in intestinal and renal brush border membranes.[47]

The amount of substrate taken up into the vesicles is quantified after separation of the vesicles from the incubation buffer though a filtration step.[48–50] Substrates that carry a radiolabelled tracer are often used for such tests, but other endpoints utilise fluorescence or liquid chromatography-mass spectrometry (LC-MS) based detection of non-radiolabelled substrates.

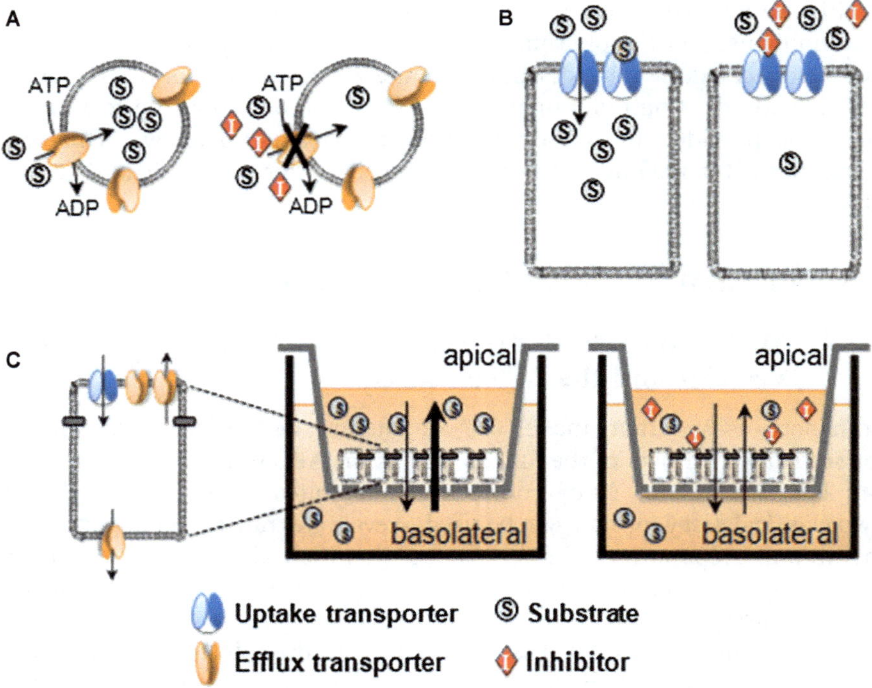

Figure 7.2   Principles of exemplar transporter assays. (A) Activity of transporters can be determined using membrane vesicles, the example illustrated shows ATP-dependent transport *via* an ABC transporter. The amount of substrate (S) taken up into the vesicle in an ATP-dependent manner can be quantified. In the presence of an inhibitor (I) the amount of substrate taken up into the vesicle is reduced. (B) Activity of SLC uptake transporters can be measured, *e.g.* in transfected cells overexpressing the protein of interest, by determining the cellular concentration of a probe substrate in transporter-expressing cells compared with control (mock) transfected cells. Inhibition of uptake activity results in reduced cellular substrate concentrations. (C) Transwell culture formats allow the study of probe substrate fluxes where substrates will show directional transport with different flux rates in the apical to basolateral (A-to-B) direction compared with the basolateral to apical (B-to-A) direction. If inhibitors, *e.g.* of efflux transporters, are present in the apical compartment the B-to-A to A-to-B ratio will be reduced.

It is important to be aware that compounds that are highly permeable may return false negative results in vesicle substrate assays. This is due to their high nonspecific binding and/or passive diffusion through the membrane, with leakage of the test compounds out of the vesicles effectively counteracting any accumulation. In addition to evaluation of substrate potential, the vesicle model is also suited to investigate inhibition of transporter activity. In this format, endogenous molecules [*e.g.* leukotriene C4 (LTC4) or estradiol-17β-glucuronide (E17βG) for multidrug resistance-associated protein 2 (MRP2), and taurocholate for bile salt export pump (BSEP)] or clinically-relevant co-medications (*e.g.* digoxin for P-gp and methotrexate for breast cancer resistance protein (BCRP)) are typically used as the probe substrate. Control conditions for either assay format utilise incubations where ATP is replaced with AMP (which does not support active transport), or where vesicles prepared from mock-transfected cells are used. One of the advantages of this approach is that it is well suited for the investigation of interactions with an individual transporter protein, for example if vesicles are prepared from an overexpression system.

Overall, membrane vesicle assays enable the direct investigation of concentration–transport relationships, determination of apparent IC_{50} values, estimation of key kinetic parameters (*e.g.* K_m, K_i and V_{max})[§] and investigation of the mode of inhibition. In addition, the approach can be adapted to a multi-well plate format, allowing convenient and rapid generation of data with substantial numbers of compounds, an important consideration in early discovery. Other practical advantages of membrane vesicles are that they can be generated in large quantities and be stored frozen (liquid nitrogen or at -80 °C) without loss of activity. A detailed description of the vesicle methodology is provided in recent reviews and protocols are available.[48–50]

Transporter studies utilising membrane vesicles can be undertaken using apical or basolateral plasma membrane vesicles prepared from polarised cells in tissues such as the liver, kidney or intestine.[51–53] These methods have been employed since the 1970s, investigating a range of solute uptake and efflux transport processes, *e.g.* amino acid, ion, sugar, cation and anion transport.[44] Whilst preparation of such membranes is technically challenging in order to obtain pure fractions, they have the advantage that they express multiple transporters at physiologically relevant levels in a native membrane environment. They enable the study of more complex transporter interactions such as the *trans*-inhibition of rodent BSEP by E17βG, which also requires the presence of MRP2.[54] However, preparations obtained from native tissues need to be carefully characterised *via* assessment of the presence/absence of marker proteins, to ensure that no contamination with the opposite membrane domain has occurred. Examples of marker enzyme activities include alkaline phosphatase, γ-glutamyltransferase or 5′-mononucleotidase for apical membranes, and Na^+/K^+ ATPase or glucagon-stimulating adenylate cyclase for basolateral membranes.

Due to the recent advances in molecular biology and recombinant techniques, vesicles are now more often prepared from heterologous expression systems.[55] Typical examples are insect cells such as High Five or cells derived from *Spodoptera frugiperda* (*e.g.* *Sf*9 or *Sf*21) after transduction with

---

[§]The Michaelis–Menten constant (K_m) is the substrate concentration that results in half of the maximal transport rate (V_{max}) in transporter-mediated transport. The inhibition constant (IC_{50}) is the inhibitor concentration that inhibits transport rate to half of initial transport rate. K_i is the absolute inhibition constant and equal to the concentration of competing substrate in a competition assay that would occupy 50% of transporters if no substrate were present. The K_i value can be extrapolated from the IC_{50} using the Cheng–Prusoff equation.[49,217,230]

baculoviruses that carry the gene of interest.[56] Insect cells are easy to manipulate, the required technologies are well-established and large quantities of membrane can be isolated. An alternative approach is to use vesicles prepared from transfected mammalian cells (*e.g.* HEK),[57] although these usually give a lower yield of vesicles compared with insect cells. Mammalian cells provide a more relevant membrane lipid environment than insect cell membranes and their higher cholesterol content has been found to result in higher activities for some transporters. In most cases this does not affect the K_m value, although a recent example showed a change in kinetic behaviour for E17βG transport by MRP2.[58,59] Other studies comparing insect and mammalian cell-derived vesicles, or insect vesicles prepared with and without cholesterol supplementation, did not observe a significant difference when evaluating a range of test compounds for transport and transporter interactions with MRP2, BCRP or BSEP.[58,60]

Since ABC transporter activity is coupled to ATP hydrolysis as the driving force, interactions may also be assessed indirectly by quantification of ATPase activity. The ATPase assay is suitable for high-throughput applications and does not require the use of radioactivity. During the hydrolysis of ATP to ADP, inorganic phosphate is released, which can be measured by a simple colorimetric reaction.[50] *Ortho*-vanadate is a specific inhibitor of ABC-type ATPases. Therefore, determination of the vanadate-sensitive proportion of the measured ATPase activity helps to distinguish between enzymatically-mediated and spontaneous and/or background ATP hydrolysis. It should be noted, however, that vanadate sensitivity may also be a result of the activity of endogenous ATPases present in the membrane preparation and, therefore, it may not be specific to the transport process investigated. In order to differentiate ABC transporter-mediated ATPase activity from the activity of other ATPases that may be present in the preparation, inhibitors of other ion-transporting ATPases can be employed (*e.g.* ouabain or ethylene glycol tetraacetic acid (EGTA) for Na^+/K^+ ATPase, sodium azide for F-type ATPase, and oligomycin for H^+ ATPases) to determine their effect on the total ATPase activity measured in the incubation.

Although the stimulation of ATPase activity indicates that a molecule interacts with an ABC transporter within the membrane, the assay cannot be used to identify transporter substrates directly.[61–63] In addition, different stoichiometrics between molecules of ATP hydrolysed and substrates transported have been reported, and stimulation of ATPase activity can also occur independently of substrate translocation, as demonstrated for BCRP and multidrug resistance protein 1 (MDR1).[64,65] Thus, if an interaction of a molecule is identified in the ATPase assay during the early discovery phase, it is recommended to confirm transport *via* other experimental approaches. As with the vesicle-based assay, the ATPase assay can also be used to identify inhibitors. In this format, ATPase activity is stimulated *via* incubation with a known activator such as a high turnover substrate and the effect of a putative inhibitor on the maximal induced ATPase activity is measured. Whilst for some transporter substrate/inhibitor pairs there is good agreement between

ATPase and vesicle data, for other pairs there is less agreement[58,66,67] and it may therefore be advisable to study an interaction of interest using different methodologies to confirm the data.

7.3.2 Cell-based Models: Genetically Modified Cells

The usefulness of immortalised cell lines and primary cultures (Sections 7.3.3 and 7.3.4) can be limited by their low expression of the required transporter protein(s). However, the prospects of using established cell lines for transporter studies became much brighter with the advent of genetically engineered cells (also termed 'cDNA expression systems', 'recombinant cell systems', 'genetically modified organisms' (GMOs) or 'transgenic cell lines'). There are three main steps in the construction of genetically engineered cell lines: (i) the generation of a full length complementary DNA (cDNA; a double stranded DNA synthesised from a messenger RNA (mRNA) template in a reaction catalysed by the enzyme reverse transcriptase) for the protein to be expressed; (ii) the selection of an expression vector that is compatible with the host cell; and (iii) the transfer of the expression vector containing the cDNA into the host cell. The cDNAs for many of the known drug transporters have already been cloned and are becoming increasingly available from commercial sources. If the cDNAs of specific isoforms are unavailable, full length cDNAs can be synthesised by applying standard reverse transcriptase polymerase chain reaction (RT-PCR) procedures.[68] When the transfected cDNA is integrated into the genome of the expressing cell, so that the expression of the transporter(s) remains effective over several passages, then it is called a 'stable' transfection. This provides a constant supply of cells over a long period of time. However, the expression of transporters is known to decrease after several passages (*i.e.* sub-culturing of cells from one lot to another). Recently, there has been an increase in the use of 'transiently' transfected cell lines where the inserted nucleic acid exists in the cell only for a limited period of time and is not integrated into the genome. Cell lines transiently transfected with transporters are often prepared *in situ* and used to avoid constant maintenance and sub-culturing of cell lines in laboratories.

Several types of biological systems (bacteria, yeast, insect cells, mammalian cells and oocytes) are available for transfection. When the transfected transporter gene and host cell expression system are from the same species it is called a homologous expression system, and if they are different it is called heterologous expression (Table 7.1). One of the most important considerations in the selection of an expression system is the 'background' transporter expression and, for heterologous systems, the post-transcriptional changes in the expressed transporter protein that may occur from the action of the host cell machinery. In addition, the expression system should possess high transfection efficiency, a capacity to express the transporter of interest (for example, sorting of transporters to membranes), stability (transporter structure retained on membrane insertion) and maintenance of transport activity of the expressed transporter(s) over several passages (no loss of activity when sub-cultured repeatedly).[69,70]

One of the most useful systems for determining the biological effects of xenobiotics during drug development are genetically transfected cell lines. They should be characterised for both genotypical (transporter mRNA levels, protein expression and localisation by western blotting) and phenotypical features (functional substrate and transporter assays using prototypical substrates/inhibitors).[71,72] A selection of common, genetically transfected cell models in current use are described below. Since regulatory agencies recommend[73–75] that transport studies are conducted in an *in vitro* system in which the human *in vivo* transporter function is preserved, cell lines genetically transfected with human transporters of interest are a very useful *in vitro* model.

7.3.2.1   Cell Lines of Human Origin: HEK293 and HeLa Cells

HEK293 cells were generated by transformation of human embryonic kidney cell cultures with sheared adenovirus 5′ DNA, and were first described in 1977.[76] The major advantage of this cell line is the minimal expression of endogenous transporters and metabolising enzymes, making them a good host for expressing exogenous transporters.[71] Consequently, these cell lines have been extensively used to express uptake transporters such as organic anion transporting polypeptides (OATPs; OATP1B1/1B3), OATs (OAT1/3), organic cation transporters (OCTs; OCT1/2/3) and multidrug and toxin extrusion proteins 1/2K (MATE1/2K).

HeLa cells, used within various disciplines of medical research, were the first human immortalised cells and were isolated from cervical tumour cells during surgery on a female patient named Henrietta Lacks (hence the name HeLa) in 1951.[77] Although use of this cell line is not very common in transporter science, there are some reports of its use for expressing uptake transporters such as OCT1 and OATP1B3.[78] Uptake studies in these single uptake transporter expressing cell lines can be used to determine intracellularly accumulated drug, as shown in Figure 7.2B.

7.3.2.2   Non-human Cell Lines: MDCKII, LLC-PK1 and CHO Cells

The MDCKII (Madin Darby Kidney cells type 2) cell line is a polarised cell line with tight junctions and separation of apical and basolateral membranes that enable the transport of molecules across the cell. MDCKII has been shown to sort transporters to apical and basolateral membranes depending upon their native existence. For example, efflux transporters such as MDR1 are sorted to the apical membrane and uptake transporters such as OATP1B1 are sorted to the basolateral membrane. This offers the additional advantage to MDCKII cell lines of the capability to express multiple transporters.

Thus, double-transfected MDCKII cell lines, MDCKII-OATP1B1/MRP2[79] and MDCKII-OATP1B3/MRP2,[80] as well as quadruple-transfected MDCKII cell lines, MDCKII-OATP1B1/1B3/2B1/MRP2,[81] have been reported and enable the study of transport involving multiple transporters.

MDCKII cells transfected with efflux transporters, for example MDR1, MRP1, MRP2, MRP3 and MRP5, have been used extensively to characterise the transport of compounds such as anticancer drugs, short chain lipids and human immunodeficiency virus (HIV) protease inhibitors.[79,82] As shown in Figure 7.2C, these cell lines are plated on inserts in a multi-well format and transport of the drug from the apical to basolateral side and *vice versa* is studied. The difference in transport rates in both directions is determined, as described in Section 7.4, to study the kinetics of efflux transporters.

MDCKII cell lines suffer from the drawback of highly expressed endogenous transporters. Therefore, transport studies involving MDCKII cell lines expressing human transporters require an additional study with non-transfected wild-type MDCKII cell lines to estimate the basal transport of drugs by endogenous transporters.[83]

MDCKII cell lines transfected with the MDR1 efflux transporter have been investigated as a permeability screen for the blood–brain barrier (BBB).[84] These cell lines show high transepithelial resistance (TEER) of 1800–2200 Ω cm^2, as would be expected for the BBB.

Chinese hamster ovary (CHO) cells, derived from the ovary of the Chinese hamster, and LLC-PK1 (Lilly Laboratories Culture–Pig Kidney Type 1), derived from epithelial cells of porcine kidney proximal tubules, are also widely used for transfection of uptake and efflux transporters. Similar to HEK293 and MDCKII cell lines, CHO cell lines have cell machinery compatible with post-translational modifications (*e.g.* glycosylation), as might be needed by some transporter proteins (*e.g.* P-gp). They have been successfully used to study OAT1, OATP1B1, OCT2 and peptide transporters (PEPT1/2).[85,86]

7.3.2.3   cRNA Injected Oocytes Expressing Transporters

The oocyte (or egg cell) from the South African frog *Xenopus laevis* is frequently used as a functional expression system, mainly because of its ability to efficiently translate foreign genetic information into functional proteins. This, combined with their ability to assemble oligomeric receptor/channel complexes and insert them into the plasma membrane, has greatly stimulated the cloning of transport proteins.[87] Renal transporters, such as OAT1, OAT3, OCT1 and OCT2, peptide transporters, and ABC transporters, such as BCRP, have been extensively characterised in the *Xenopus* oocyte system.[88,89] Availability of commercial kits and the feasibility of immediate use (with no incubations for scale up, as is the case for cell cultures) have further augmented their use for single transfected transporter assays. However, generation of transfected oocytes does involve extensive experimental work due to the need for an individual injection into each oocyte. There have also been some preliminary reports of endogenous transport activity, such as bile salt secretion,[90] although this has not been fully confirmed. On the other hand, there are reports citing the use of oocytes to characterise bile acid transport.[91,92] Finally, intact *Xenopus* oocytes may suffer from limitations due to unstirred layers and intracellular barriers, and the use of membrane vesicles has been suggested as a useful alternative.

7.3.3   Cell-based Models: Immortalised Cell Lines

7.3.3.1   Caco-2 Cells

Caco-2, a human colon adenocarcinoma cell line that undergoes spontaneous enterocytic differentiation in culture, are polarised cells with well-established tight junctions and tight monolayers. This makes them an excellent model for studying transmembrane processes. The Caco-2 cell model is one of the most extensively characterised cell models in the field of drug permeability and absorption studies,[93] and the expression and functionality of transporters is well documented.[71,94] Caco-2 cells express many uptake (PEPT1, OATP2B1, and monocarboxylate transporters (MCTs)) and efflux (P-gp, BCRP, MRP2 and MRP4) transporters, and have been utilised in specific transporter studies of ABC efflux transporters and SLC transporters for peptide, glucose or bile salt transport.[71,95,96]

Additionally, Caco-2 cultures are widely used to observe grapefruit or dietary effects on carrier-mediated absorption and pH-dependent active transport of drugs, including targeting of drugs towards PEPT1 using pro-drug approaches.[97] Transporter assays using Caco-2 cell lines normally comprise plating cells on plastic inserts and measuring transcellular transport across the monolayer (Figure 7.2C). These assays have been thoroughly validated for use in the pharmaceutical industry for regulatory submission.[98] Usually, 14–21 days are required for Caco-2 cell monolayers to be mature for permeability and transporter studies, in order to allow formation of a tight monolayer and expression of transporters on the membranes. After this time, a defined window of up to 1 week is advisable to conduct transporter studies as relative transporter expression levels change during prolonged maintenance of a monolayer culture. While accelerated culture conditions are described, their behaviour is different from classical cultures and transporter functionality needs to be carefully monitored.[96]

Modifications to the Caco-2 cell system to introduce cytochrome (CYP) activity (*e.g.* the Caco-2/TC7 sub-line) allowed investigation of the interplay of enzymes and transporters in Caco-2 monolayers, due to the co-existence of P-gp and CYPs.[99] The wide spectrum of transporter expression in these cells allows the study of the overall effects of multiple transporters in Caco-2, or transporters and CYP enzymes in Caco-2/TC7. The investigation of single transport mechanisms in this cell line can be done with specific probe substrates or highly specific inhibition techniques.

With the introduction of specific gene knock-down technologies (small interfering RNA, single hairpin RNA and zinc-finger technology), Caco-2 sub-lines with single or double transporter functional knockouts have been generated to study the impact of single transporters on membrane transport and overall permeability.[100] Comparison of native and knock-down cells has also allowed the specific identification of transporters contributing to efflux in Caco-2 cultures.[100]

7.3.3.2 Miscellaneous Cell Lines

While the Caco-2 cell model with its tight monolayers made it well suited for studies of transmembrane transport across the intestinal wall, organ-representative models for other epithelia, *e.g.* the kidney or BBB, were more problematic, as there are no stable cancer cell lines with a full complement of transporters that mimic these barriers. Through transduction of rat brain endothelial (RBE) cultures with viral immortalising genes or from primary cultures of brain endothelial cells of SV40 T-antigen expressing transgenic rats and mice, numerous immortalised cell systems have been generated and characterised.[101,102] Consistently, the endothelial morphology and gene expression pattern is retained in these cells, but expression levels or a low TEER prevent their direct use as transendothelial permeability models for BBB or blood–cerebrospinal fluid barrier (BCSFB) transfer. However, specific investigations into transporter function, regulation or pharmacology have been successfully reported.[103,104] Porcine brain endothelial cells (PBECs) exhibit high TEER compared with brain endothelial cell cultures from other species, implying the presence of well-developed tight junctions.[105] Conditionally immortalised proximal tubule epithelial cells (ciPTEC) have been described in the literature, and the first reports on transporter functionality and their application in drug–transporter interaction studies are emerging.[106,107] Similarly, the immortalised human renal proximal tubule cell line RPTEC/TERT1 expresses a range of cationic and anionic transporter proteins and demonstrates functional activity for cation transport and P-gp.[108] The human HepaRG cell line was established from cells isolated from a liver tumour of a female patient suffering from hepatocarcinoma and hepatitis C infection.[109] HepaRG cells can be differentiated into hepatocyte- or biliary-like cells dependent on culture conditions. Polarised expression of basolateral (*e.g.* OATP1B1, OATP2B1, OCT1 and MRP3) and canalicular (*e.g.* P-gp, BSEP and MRP2) transporter proteins has been demonstrated in HepaRG cells using immunocytochemistry. Furthermore, the cell line also shows uptake and canalicular efflux transport activities of prototypical substrates for NTCP, OATP, OCT, BSEP and MRP2.[110]

7.3.4 Cultures of Primary Cells

7.3.4.1 Suspension Cultures of Primary Cells

Freshly isolated and cryopreserved hepatocytes in suspension are a frequently used model to study protein-mediated drug uptake and clearance.

Cryopreserved cells can be obtained either from individual donors or pooled from several donors, the latter being a good surrogate to represent a population and limit differences in individual expression levels or genetic polymorphisms. The viability of cells in suspension declines over time and, therefore, experiments are usually limited to a time frame of a few hours after isolation or resuscitation. Despite enhanced ease of use, a limitation with cryopreserved cells is the decreased activity of uptake transporters compared with freshly isolated cells,[111] which in part can be explained by internalisation of transporter proteins shortly after isolation. This may impact predictions of metabolic clearance for molecules that depend on hepatic uptake.[112]

One classical approach for transporter studies is to determine the appearance of a substrate in the cells.[49] After incubation of cells with a test substance for varying time periods (usually from seconds to a few minutes; model- and substrate-dependent), the cells are separated rapidly from the incubation medium through the use of a cell harvester, centrifugation or a spin through an oil layer ('oil-spin method').[113] Cell pellets can then be lysed and the amount of compound taken up can be quantified *via* liquid scintillation counting or LC-MS. For the oil-spin method, a test compound is incubated with cells for a specific time after which the suspension is centrifuged through a layer of oil. The cells pass through the oil but the incubation solution, containing free test compound, is excluded. An application of the oil spin technique is the assessment of active hepatic uptake of a variety of different drug compounds with a range of active uptake clearance rates, demonstrating the good dynamic range of the method.[114,115] Such studies also provide insight into metabolism–transporter interplay and improve our understanding of the hepatic disposition of drug molecules.[116] In principle, the oil-spin method can be applied to any single cell suspension, for example uptake into transfected cell lines, primary renal proximal tubule cells and cardiomyocytes.[117–119] It is important to note that uptake studies usually require very short incubation times in order to be able to determine initial uptake rates, hence incubation times should be identified and optimised for each experimental system and substrate combination. Extending incubation times beyond the initial linear uptake phase may result in data being confounded due to possible contributions from metabolism or efflux processes. Approaches to determine the passive component of drug uptake in such studies are challenging and can vary from incubations at 4 °C (which also alters membrane fluidity), use of pharmacological inhibitors (which have limited specificity and may be cytotoxic) or incubation without certain co-factors (*e.g.* Na^+, which may impact cell physiology and viability). If radiolabelled substrates are employed, another alternative is the use of high concentrations of unlabelled substrate to inhibit the carrier-mediated proportion of the transport process investigated. Identification of the contribution of individual transporter proteins to the overall uptake can be challenging due to the aforementioned lack of specific inhibitors.

An alternative approach to the oil-spin method for the study of hepatic uptake processes is the loss-from-media assay. Here, the amount of compound left in the supernatant is quantified, following incubation for a range of short time periods (0.5–6 min) and subsequent centrifugation of the hepatocyte suspension. However, it may be difficult to quantify small differences in the amount of compound disappearing from the incubation medium. Nevertheless, for a number of compounds, a sufficient decrease in medium concentrations was measurable in order to determine *in vitro* uptake clearance, which seems to correlate well with *in vivo* clearance, in particular for acidic and zwitterionic drugs.[120] Both the oil-spin method and loss-from-media assay have been discussed in the recent International Transporter Consortium (ITC) whitepaper on *in vitro* methodologies.[49,114,116]

7.3.4.2 Cultures of Plated Primary Cells

A variety of primary cell types isolated from different species and organs can be used in monolayer cultures. With the improvements made in protocols and procedures, cryopreservation of primary cells provides a good and convenient alternative to freshly isolated cells, although changes in transporter activities (*e.g.* reduction in the expression of NTCP, OATPs and OCTs) can be observed during cryopreservation.[121–123] Certain cells might require the presence of foetal bovine serum (FBS) during plating and the cell culture plates/flasks might need to be coated with extracellular matrix proteins such as collagen to improve adherence and plating efficiency.[123] For example, rat hepatocytes have shorter attachment times when plated in the presence of serum.[124] Similarly, attachment rates of foetal small intestinal epithelial cells are better on protein matrices such as fibronectin, laminin or collagen type IV.[125] High cell viability at the time of plating is a key criterion for successful cultures that demonstrate good functional activity. Once plated, cultures are often maintained in serum-free hormonally defined media, as is the case for hepatocytes or renal tubule cells, which need to be optimised for each cell type. This is necessary to avoid de-differentiation of cells, which can occur in the presence of serum.[126] Primary cells are considered to be more physiologically relevant, *e.g.* compared with transfected cells lines, as they express most of the native transporter proteins. As cells from a variety of species can be obtained, the study of species differences in transporter interactions is greatly facilitated, as are investigations of the effect of exogenous stimuli on the expression, localisation and activity of individual transporter proteins, *e.g.* Ca^{2+} or protein kinase signalling, or the effects of cytokines.[127] However, it is important to be aware that levels of functional activity of transporters in cultured cells may still be quite different from *in vivo* activity levels.

7.3.4.3 Cultures of Plated Hepatocytes

Short term plated hepatocytes have been used successfully to study uptake, metabolism and biliary efflux of drugs. For a small test set of four compounds, rat *in vivo* hepatic clearance was predicted with only approximately two-fold error from the *in vitro* measurements in this test system.[128] Whilst cell polarisation of cell membrane domains is lost after hepatocyte isolation, the short plating period of about 90 min in this study was sufficient for the cells to form a loosely confluent layer with a cobblestone appearance, although it is unclear if this time was sufficient to establish tight junctions and full cell polarity. However, *in vivo* biliary clearances were predicted well from *in vitro* efflux measurements, indicating that this culture format retained some functional efflux activity.[128]

One of the challenges of *in vitro* hepatocyte cultures generally is that the cells dedifferentiate over a short time period of a few days. As a result, CYP and transporter activities are altered[129] (a number of important biotransformation enzymes and transporter proteins are down-regulated), therefore limiting their applicability to transporter research. Dedifferentiation is initiated during the isolation procedure as a result of the loss of cytoarchitecture and activation of ischaemia perfusion injury, which trigger molecular changes leading to a proliferative phenotype.[123,130] Specialised culture formats such as sandwich-cultured hepatocytes (SCH) can somewhat extend the culture period (see Section 7.3.5); nevertheless, changes in expression levels of transporters also occur in these more complex models.[129]

7.3.4.4   *Kidney Epithelial Cell Cultures*

Membrane transporters expressed in the proximal tubule of the kidney play a role in the secretion of waste products as well as maintenance of nutrient homeostasis through reabsorption of key molecules such as glucose, amino acids and albumin. Cultures of primary renal tubule cells can be generated from a range of species, including human, monkey, rat and mouse. They can be established from freshly isolated cells or, in the case of human cells, are also available cryopreserved from commercial sources. Cells can be maintained in monolayers when plated on collagen-coated cultureware, but more often are cultured in a transwell configuration on permeable filter supports, to allow directional transepithelial flux studies (Figure 7.2C). For the latter, test compounds are added either to the basolateral or apical compartment and the appearance of the test compound is determined in the corresponding compartment. Good cultures exhibit high TEER values and the characteristic cobblestone appearance, together with expression of the junctional complex protein ZO-1. Functional expression of transporters such as OCT2, OAT1, OAT3, P-gp, BCRP and MRPs has been demonstrated in human tubule cultures, resulting in the secretion of organic cations or anions such as creatinine or *p*-aminohippurate, respectively. Such studies indicate that transporter proteins are expressed in the correct apical or basolateral membrane domain.[131] Likewise, uptake of α-methyl-glucose across the apical membrane demonstrates that reabsorptive processes can also be recapitulated *in vitro*.[131] Furthermore, similar experiments in transwell cultures of rat primary tubule cells indicate that reabsorptive processes involving sodium glucose linked transporter (SGLT), PEPT and organic cation transporter novel (OCTN) can be studied.[132] Co-culture with endothelial cells has been demonstrated to improve proliferation and long term maintenance of human renal tubule cells, as well as increasing mRNA expression of certain transporters such as OCT1, fructose transporter (GLUT5), SGLT2, and of particular note OAT1 and OAT3.[133] In addition to drug transport studies, renal *in vitro* cultures are also suitable for studying transporter-mediated nephrotoxicities.[134]

7.3.4.5   *Trophoblast Cultures*

Transporter proteins fulfil an important function at the foeto–placental barrier by limiting mother-to-foetus transport of potentially harmful xenobiotics and mediating exchange of nutrients and waste products. Trophoblast cells, the functional units of the placenta, can be obtained for *in vitro* cultures of human origin from first trimester or term placentas after natural or caesarean delivery.[135,136] Such cells are used mostly for academic research and are typically undifferentiated and represent the early stages of gestation. They are multinucleated, do not proliferate and spontaneously form syncytia (a single cell or cytoplasmic mass containing several nuclei). Therefore, a limitation of these cultures is the lack of tight monolayer formation, as the cells form aggregates with large intercellular spaces, which make polarised transport studies impossible. Formation of tight junctions can be achieved if trophoblasts are cultured with specialised multiple seeding and differentiation techniques on semi-permeable filters, which leads to the formation of multiple overlapping layers of syncytialised cells.[136] Trophoblasts cultured for a few days *in vitro* show mRNA expression of multiple transporters, including members of the OATP, OCT, OAT, MRP and MCT families.[137] ABC transporter functional activity of P-gp, MDR3, BCRP and MRPs has been demonstrated in primary trophoblast cells using fluorescent substrates such as calcein-AM and Hoechst 33342.[138] A primary culture of trophoblast cells isolated from Wistar rat placenta maintains BCRP and P-gp activity for several passages in culture and is also suitable for cryopreservation.[135,139]

7.3.4.6   *Cultures of Stem Cell-derived Cells*

Stem cell-derived cells are seen as a promising alternative cell source to primary cells as they have the potential to provide a continuous and unlimited supply of cultures devoid of donor-to-donor and preparation variability. They can be employed in *in vitro* models in the same way as primary cells or cell lines, *e.g.* in monolayer, transwell or spheroid cultures. For most of these models, only limited data on transporter expression and activity are available to date. mRNA expression of rodent MRP2, comparable to adult mouse liver, was demonstrated in spheroids of murine embryonic stem cell (ESC)-derived hepatocyte precursor cells.[140] Furthermore, human ESC and induced pluripotent stem cell (iPSC) derived hepatocytes demonstrate mRNA and protein expression of transporters such as OATP1B1, NTCP, BSEP and MRP2, and encouragingly, also demonstrate transport activity for E17βG and taurocholic acid, albeit at lower levels than in plated primary cryopreserved hepatocytes.[141] Similarly, mRNA and protein expression have been demonstrated in human ESC-derived renal proximal tubular-like cells for a range of transporters, including P-gp, OAT1, OAT3, OCTN2, PEPT1, GLUT1 and SGLT2.[142,143] With respect to gut models, iPSC-derived intestinal cells have similar or even higher expression of intestinal marker proteins (*e.g.* villin, E-cadherin and mucin-2) and demonstrate high TEER with low permeability in transwell cultures. However, not all iPSC-derived intestinal cell lines exhibit these characteristics.[144]

Human cord blood-derived haematopoietic stem cells can be differentiated into brain-like endothelial cells (BLECs), which exhibit BBB properties when co-cultured with pericytes. BLECs express a range of amino acid and glucose influx, and P-gp, BCRP and MRP efflux transporters. In addition, cellular retention of vincristine in the presence of the inhibitors verapamil or elacridar indicated functional P-gp transporter activity.[145] Furthermore, this model demonstrated good BBB properties as *in vitro* unbound brain to plasma concentration ratios correlated well with *in vivo* ratios.[145] Human pluripotent stem cell-derived endothelial cells, when co-cultured with astrocytes, also acquire BBB properties and demonstrate polarised transporter expression as well as P-gp-mediated rhodamine and doxorubicin transport activity.[146]

7.3.5 Specialised Culture Formats

7.3.5.1 Sandwich-cultured Hepatocytes

Hepatocytes cultured between two layers of extracellular matrix (ECM; *e.g.* collagen) in a "sandwich" configuration are a model system that allows the study of both uptake and efflux processes (Figure 7.3). In this culture format, hepatocytes regain their cell polarisation over several days, which eventually leads to the formation of bile canalicular networks, which are vital for directional transport studies. Methods for SCH derived from freshly isolated or cryopreserved primary hepatocytes of a range of species, including human, dog, monkey and rat, have been established.[129,147,148] In addition to their expression of functionally-active basolateral and apical transporter proteins, SCH express some phase 1 and 2 drug metabolising enzymes.[129]

In the presence of calcium (or magnesium), the tight junctions keep the canalicular space between hepatocytes closed, whereas removal of calcium from the medium opens up the canalicular spaces. This allows the assessment of biliary excretion of compounds through parallel incubations in the presence and absence of calcium. Two principal parameters that can be determined in this model are the biliary excretion index (BEI) and the *in vitro* biliary clearance.[147] The BEI is the amount of test compound accumulated in cells plus bile ($+Ca^{2+}$) minus the amount in cells only ($-Ca^{2+}$) divided by the amount of compound in cells plus bile. *In vitro* biliary clearance is the

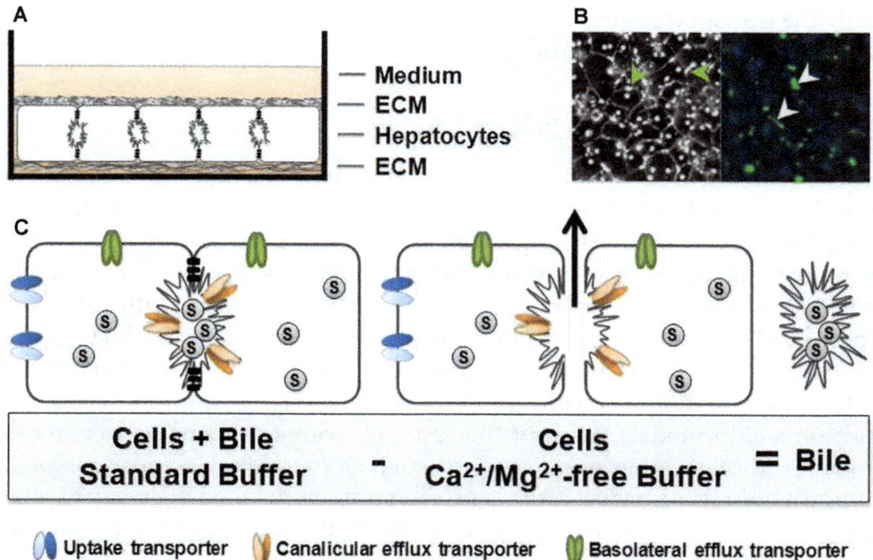

Figure 7.3 (A) In the sandwich culture model, hepatocytes are cultured between a sandwich of ECM; typically, cells are plated onto collagen I-coated plates and then overlaid with collagen or Matrigel. (B) In this configuration, hepatocytes polarise and form canalicular spaces (arrowheads) over several days of culture as can be seen from the phase contrast picture of human hepatocytes on day 7 (left; unpublished data). Hepatocytes demonstrate efflux of the MRP2 substrate CDF into these canalicular spaces (right; nuclei are stained with Hoechst dye; unpublished data). (C) Through incubation of hepatocytes in Ca^{2+}/Mg^{2+} containing and Ca^{2+}/Mg^{2+} free buffers, the amount of substrate (S) excreted into the canaliculi can be determined.

difference in uptake in the presence and absence of Ca^{2+} divided by the compound medium concentration and the incubation time. This model has been used extensively to investigate *in vitro* biliary clearance of a range of endogenous and exogenous substances. In rat SCH, *in vitro* biliary clearance parameters correlated well with *in vivo* biliary clearance in bile duct cannu-lated rats.[148] Similarly, clearance of piperacillin, sestamibi and mebrofenin in human SCH *in vitro* correlated with *in vivo* data.[149,150] Using a modified uptake and efflux protocol in SCH, the contribution of basolateral efflux can also be investigated, as demonstrated, for example, for rosuvastatin.[151]

SCH are also amenable to other mechanistic investigations to identify the contribution of individual transporter proteins. This can be achieved through the use of specific inhibitors, by using hepatocytes isolated from knockout animals or by knockdown using RNA interference techniques.[152] However, protein knockdown in cultured primary hepatocytes using RNA interference is challenging because it is difficult to reach high transfection efficiency with conventional transfection reagents. SCH can also be used to study the transporter-mediated as well as transporter-independent effects of drug molecules in hepatotoxicity.[129,147]

7.3.5.2 Complex Cell Culture Models

Recent years have seen a rise in the development of complex cell systems, with the aim of developing more physiological, *i.e.* organotypic, models that recapitulate cell functions to quasi *in vivo*-like conditions. This is achieved by re-introducing certain features into the culture system that are usually provided by the microenvironment *in vivo* such as contact with other cell types, ECM or exposure to dynamic culture conditions such as flowing medium. A myriad of approaches is available, ranging from micropatterned cultures over scaffolds, organoids, flow-through systems, mini-bioreactors to body-on-a-chip models. Most of the research seems to focus on hepatocyte models, although other organ systems such as the kidney are also targeted. A number of recent review articles provide a more detailed overview of such models.[123,153–155] Here, we present a few selected examples where initial data on transporter expression or activity have been obtained.

It is more and more recognised that cultures consisting of simple monolayers, or of a single cell type, lack important interactions and stimuli provided in the *in vivo* environment. Such limitations can be addressed in 3D culture formats and increasingly in co-culture models, where additional cell types are present, are being explored.

In micro-patterned cultures, cells are patterned onto special ECM domains of microscale dimensions. In one liver model (HepatoPac™), hepatocytes are surrounded by fibroblast cells. In this format, rat hepatocytes maintain their phenotype for at least 4 weeks, form canaliculi that efflux the marker 5-(and-6)-carboxy-2′,7′-dichloro-fluorescein (CDF) and demonstrate higher uptake rates of E17βG and taurocholate compared with SCH.[156] The model also allows co-culture with Kupffer cells, which enables the study of immune-mediated processes.

An example of a 3D co-culture of the intestinal mucosa consists of Caco-2 and stromal cells (fibroblasts and immunocytes) seeded indirectly onto transwells in a collagen matrix. These co-cultures exhibited improved TEER values compared with monocultures, formed a mucus-like layer, and expressed P-gp and BCRP. In addition, there was a good correlation of *in vitro* permeability values *versus* the fraction absorbed in humans for compounds with a range of permeabilities.[157]

In a more complex hepatocyte transwell model, which is perfused and provides haemodynamic stimuli using a cone-and-plate viscometer-based technology, MRP2 expression and canalicular network formation was much improved compared with static cultures.[158]

3D spheroids are microtissues consisting of a single or a mix of several cell types that can be generated through self-aggregation in a hanging-drop culture platform (*e.g.* GravityPLUS™) or in ultra-low adhesion plates. Liver spheroids can consist of hepatocytes or hepatocyte-like cells only (*e.g.* HepG2, HepaRG and primary hepatocytes) or a mixture of hepatocytes and non-parenchymal cells (*e.g.* Kupffer cells) and are typically stable over several weeks. Liver spheroids have been shown to express P-gp and BSEP at the canalicular membrane domain.[159] Furthermore, HepaRG cells cultured as 3D spheroids demonstrate efflux of the MRP2 substrate CDF into bile canalicular structures.[160]

Bioreactors and microfluidic devices (*e.g.* organ-on-a-chip) are examples of the most sophisticated culture models to date for representation of the situation *in vivo*. Culture conditions (*e.g.* nutrient supply, removal of waste products, oxygenation, pH and temperature) are automatically controlled and shear stress, through the flowing medium, provides an important mechanical stimulus. Often, such models are applied to tissue engineering applications but they may find their way into wider use in the future. Currently, such models require a high number of cells and have limited applications for drug transport and toxicity studies because of high binding of lipophilic compounds to the device components. In a kidney model consisting of a microfluidic channel and cells on a porous ECM-coated membrane, human proximal tubule cells experience low-level fluid shear stress that mimics that of the tubule. In this organ-on-a-chip system, tubule cells express MRP4 on their apical membrane; they also demonstrate more *in vivo*-like OCT2 dependent cisplatin toxicity, as well as P-gp efflux activity, compared with static conditions.[161] In bioreactors, ESCs differentiate into renal proximal tubule cells with *in vivo*-like characteristics and show mRNA expression of transporters such as OCTN2, PEPT1 and P-gp, but not OAT1/3.[142] Similarly, fresh or cryopreserved human hepatocytes demonstrate basolateral expression of OATP1B1 protein as well as activity *via* uptake of E17βG in liver bioreactors for up to 7 days.[162]

Whilst these novel models still face many challenges and require further characterisation and validation, as most are still in the proof-of-concept stage, they hold great potential to provide a significant improvement over conventional models.

7.3.6 Precision-cut Tissue Slices

Ultrathin sectioning of fresh organs by specially-designed tissue slicers (*e.g.* Krumdieck tissue slicer) produces tissue sheets that retain complete cells in their 3D environment, but with increased surface area and accessibility to each single cell. Various techniques have been reported for precise organ slicing, and their potential use in drug metabolism and pharmacokinetic studies has been discussed.[163] While solid organs such as the liver, kidney and brain are relatively straightforward to slice, intestinal and lung tissues require stabilisation to make them amenable to slicing.[164]

7.3.6.1 *Liver Slices*

Preparation of 100–300 μm thick slices from fresh liver or liver lobes allows the preparation of large numbers of similar pieces that permit a number of compounds or conditions to be studied to explore uptake mechanisms. Within a slice, the hepatocytes of the inner cell layers have been shown to receive an adequate supply of oxygen and substrates, and slices can be maintained for at least 24 h.[163,164] Liver slices are well accepted as an *in vitro* system to study the hepatotoxicity and biotransformation of compounds as well as the induction of CYP mRNAs. Slices from different animal species can be compared to investigate species differences and advances in cryo-preservation techniques have allowed the successful use of slices after cryostorage. Less advantageous from a transporter perspective is that there is no separation of biliary and portal exposure, as the whole slice is incubated in substrate. However, there are reports of their usage to study induction or repression of transporters,[165] uptake[166] and excretion of bile salts.[167]

7.3.6.2 Kidney Slices

Comprehensive studies on renal uptake of substrates (temperature/oxygen dependence, effect of metabolic inhibitors) using renal slices were reported long before transporter science had emerged as a distinct discipline. Precision-cut renal cortical slices have been used for periods from 2 up to 96 h, although it has been documented that during slicing the lumens of tubular regions may collapse.[168] Kidney slices have been used to evaluate the renal uptake of compounds from the basolateral membrane[169] and to assess gender-based differences in the expression of transporters.[170] The uptake of compounds into kidney slices is much lower than the renal intrinsic uptake clearance *in vivo*, which may be partly due to slow or incomplete diffusion from the surface of the slice.

Freshly isolated renal tissue fragments (glomeruli or tubules) have been used to assess physiology and potential nephrotoxicity, and also the transport of xenobiotics.[171] Epifluorescence microscopy and video image analysis have been used to measure the uptake of the fluorescent anthracycline daunomycin by intact killifish (*Fundulus heteroclitus*) renal proximal tubules.[172] Renal tubules of killifish have been employed extensively to study intracellular mechanisms of transporter physiology,[173] transport of compounds,[174] localisation of renal transporters and transporter-mediated nephrotoxicities.[175,176]

7.3.6.3 Brain Slices

Brain precision-cut slices are an established *in vitro* model to study the transporters involved in normal physiological mechanisms of the brain. Accumulation of mostly fluorescent transporter substrates can be followed microscopically, and the increase and decrease of intracellular accumulation in the presence of different inhibitors can be studied.[177] The most widespread use of brain slices is the rodent brain slice method[178,179] to obtain accurate estimates of the free distribution ratio of compounds into brain tissues and, in context with *in vivo* exposure data, elucidate the influence of drug efflux or uptake transporters on the free brain concentration.

7.3.7 Isolated Perfused Organ Systems and Tissue Chambers

Isolating the organ of interest to make it accessible for detailed investigations of its role in active transport processes is a classical pharmacological approach (*e.g.* organ baths). *In situ* and *ex vivo/in vitro* techniques have been developed for the liver, kidney and intestine as well as lung perfusion models. As with all other methods that are based on isolated organs, normal morphology with different cell types and 3D architecture is closely maintained, while other parameters such as vascular flow, protein content or sequential treatment with different conditions can be controlled in the same organ preparation.

7.3.7.1 Isolated Perfused Liver

The isolated perfused liver allows direct access to blood vessels and bile ducts as in- and out-flow routes to study carrier-mediated transport in the liver over time, alongside formation of metabolites and assessment of their appearance and disappearance into different fluids. Different setups of re-circulating media or single-pass perfusion can be optimised to maximise metabolite formation or study interactions between different treatment schemes. Only a limited number of experiments can be performed with a perfused liver preparation during a 3–4 h window of acceptable viability.[180,181] Despite this, there are successful examples of studies of hepatic disposition-mediated by canalicular transporters, and interference with bile excretion or hepatotoxicity caused by xenobiotics.[182,183] The complexity of the whole intact organ allows the assessment of interdependencies of sequential or competing processes of metabolism and transport. Numerous mechanistic studies to elucidate DDIs and to investigate hepatic enzyme–transporter interplay with isolated perfused rat liver have been reported.[184–187] One needs to bear in mind however that the complexity of the system also limits the ability to specifically investigate individual uptake and excretion mechanisms, as specific and potent inhibitors for single transport proteins are lacking. However, this limitation may be partially overcome by the use of livers from transgenic animals (*e.g.* MRP2 deficient TR$^-$ and Eisai hyperbilirubinaemic rats).

7.3.7.2 Isolated Perfused Kidney

Isolated perfused kidney models to investigate transporters are based on the fundamental methods developed over a century to study renal handling of fluids and homeostasis.[188–190] Maintenance of kidney viability is a persistent challenge to ensure optimal oxygenation, and perfusion is technically demanding and tedious. Over the past decades, the isolated perfused rat kidney has been used to study numerous aspects of renal drug disposition. When water fluxes are monitored thoroughly to account for true transport rates, this *in vitro* model allows the elucidation of the overall contributions of renal transport mechanisms to drug excretion. Therefore, isolated kidney perfusion studies can provide a bridge between *in vitro* findings and *in vivo* disposition. Major applications of this *in vitro* model include characterising renal transporter-mediated excretion mechanisms[191] and screening for clinically-significant drug interactions.[192]

7.3.7.3 Intestinal Segments

Preparation of intestinal sacs or tissue sheets from different regions of the whole length of the intestine is possible for small animals. Everted intestinal sacs are tissue preparations that allow the study of absorption of nutrients and drugs from a bathing solution into the interior buffer of the sac. The major advantage of this method is that multiple regions of the intestine can be studied from the same animal, allowing investigation of regional differences in membrane properties, transporter and enzyme activity, and also pharmacological studies or formulation comparisons. However, the viability of the tissue preparation is very limited, oxygenation of the tissue being a key issue; also, the underlying muscular layer in everted sacs presents an extra barrier to the absorption of drugs.[193] Limited studies have been done using this model for transporter-related investigations.

In Ussing chamber studies this *muscularis* layer is removed and thin intestinal segments are mounted on a specially designed diffusion cell.[194] The Ussing chamber technique allows the study of both rat and mouse intestinal permeation, transport and metabolism. Additionally, human material can be used efficiently through donations from surgical resections. In contrast to other *ex vivo* techniques, Ussing chambers allow continuous monitoring of tissue viability to ensure appropriate quality for permeation and transport data.

Everted intestinal sac and Ussing chamber techniques have been employed to study the mechanistic biopharmaceutical rationale behind DDIs, P-gp inhibition studies, region-dependent modulation of intestinal permeability by drug efflux transporters, interplay of transporters and metabolic enzymes, and effect of pharmaceutical excipients on drug transporter activities in the intestine. Intestinal segments are often the first choice to study the distribution and localisation of drug transporters throughout the different regions of the gastro-intestinal tract.[195,196] A combination of the Ussing chamber method with *in situ* intestinal loop studies (single pass perfusion of intestinal segments) can give detailed insight into the involvement of transporters in the intestinal secretory clearance of drugs.[197] The Loc-I-Gut device is a special sterile polyvinyl tube used for single-pass perfusion of drug solutions through intestinal segments in humans or large animals, mostly to determine segmental permeabilities in different regions.[198] This method had been used to elucidate the interplay of efflux transporters and/or first-pass metabolism in drug absorption. For example, the technique was used to demonstrate increased intestinal permeability of verapamil, with a jejunal concentration reflecting saturation of an active efflux transporter (P-gp) at higher drug concentrations.[199]

7.4 Validation, Variability and Recommendations for Experimental Design of *In vitro* Assays

Preliminary *in silico* screening of inhibitors can be done in early discovery to gauge the level of DDI liability for a series of discovery compounds yet to be synthesised.[200,201] However, the study of DDI potential for a new drug candidate in the late discovery and/or development phase usually begins with *in vitro* studies to determine whether a drug is a substrate or inhibitor of transporters. The assays used in early discovery are screening assays capable of high-throughput screening of transporter inhibitors and substrates. Validation of these early discovery transporter assays has been extensively reported in the literature.[202–205] The results of such *in vitro* studies are subsequently used to inform the nature and extent of clinical studies that may be required to assess potential interactions (United States Food and Drug Administration (US FDA), European Medical Agency (EMA) and Japanese Pharmaceuticals and Medical Devices Agency (PMDA) draft DDI guidance).[73–75] From a pharmaceutical industry perspective, patient safety is paramount and thus any decision to proceed or, crucially even more important, not to proceed with clinical interaction studies based on (in part) the weight of evidence from *in vitro* studies, mean that utilised methodologies must be thoroughly characterised and validated in order to ensure confidence in the *in vitro* data generated. The subject of transporter assay validation has been described in several publications.[49,60,72,98,111,206,207] The following sections describe best practices for validation of transporter assays adopted by some pharmaceutical companies (with subtle inter-company differences) for regulatory submission, based on regulatory guidance and these publications.

The process of assay validation is used to verify that a test system responds appropriately to known substrates and inhibitors of the particular transporter under scrutiny. Additionally, validation is designed to identify and understand experimental variability in order to define how assays should be conducted and robustly quality controlled in future studies to provide consistency of results. For a cell-based *in vitro* transporter assay, inter- and intra-laboratory experimental variability in transporter function (and derived kinetic parameters such as K_m and IC_{50}) may be a consequence of factors such as cell origin, culture medium and feeding regimen, passage number, initial well seeding density, filter size and composition of cell culture inserts, monolayer age, transporter buffer composition and pH, and transport experimental protocol.[72,98,203,208,209,206] The latter two factors are equally important considerations as a source of variability of transporter function in membrane vesicle based assays, alongside vesicle batch and vesicular protein concentration.[60] In addition to biological variability, further variability due to non-specific binding to plasticware (filter inserts, companion plates, incubation plates and rapid filtration plates) or chemical instability during the incubation may be present in both cell- and vesicle-based assays as a consequence of the experimental setup.[60,72,98,207]

Thus, the validation procedure aims to experimentally test the elements above that are most appropriate to the anticipated methodological direction of an assay, in order to determine optimal conditions for transporter function (both substrate and inhibition) that are robust, reproducible and fit for purpose for the intended use of the assay within the drug development process. Furthermore, by reducing variability through validation, the predictive ability of the assay is improved. Recommendations and considerations for validating different *in vitro* transporter assay types are described in this section.

7.4.1 Recommendations for Experimental Design

Recent regulatory guidance and ITC whitepapers have provided general strategies and recommendations regarding preferred experimental approaches for investigating transporter-mediated interactions. All underlie the importance of utilising appropriately validated experimental test systems that have been well characterised with known substrates and inhibitors, and where due consideration has been given to the choice of concentrations, time points, appropriate calculation methods, and selection of probe substrate or inhibitor to allow extrapolation to the clinic.[49,73–75,210] Furthermore, it is good practice not only to test known substrates and inhibitors exhibiting a range of affinities for a transporter (low, medium and high K_m or IC_{50}), but also known non-substrates or non-inhibitors (*e.g.* propranolol or probenecid for P-gp, respectively) to ensure that the test system is able to distinguish between positives and negatives. The existence of multiple binding sites has been reported for some transporters, but the exact clinical significance of transport through these and the impact on drug discovery and development is not yet understood.[211]

7.4.2 General Considerations for Validating Transporter Substrate Assays

7.4.2.1 Choice of Positive Control Substrate

A positive control substrate (at a suitably low concentration to avoid saturation, *e.g.* 1 μM) should give sufficient transport activity in the test system (defined by the uptake rate ratio or efflux ratio) above the endogenous background activity observed in the control condition—mock transfected or parental cells (*e.g.* MDCKII) or control vesicles, for cell- and vesicle-based transporter expression systems, respectively. For hepatocyte uptake assays, the active transport activity can be determined by correcting for the passive component observed in the presence of uptake transporter inhibitors. A minimum uptake rate ratio or efflux ratio of ≥ 2 may be acceptable for some transporters, but a substantially higher signal-to-noise ratio is preferred. A similar magnitude of efflux ratio would be considered a minimum for positive control transport activity in derived cell lines such as Caco-2. Additionally, the transport of the positive control substrate should be inhibited by a prototypical inhibitor of the transporter being studied. For substrate assays, the chosen positive control substrate need not necessarily be a clinically-relevant substrate as its purpose in the assay is solely to demonstrate that the transporter is adequately functionally expressed in the *in vitro* test system, and that the system is thus capable of detecting substrates of the particular transporter. However, from an ease of validation perspective, where possible, it is often advantageous to choose the intended clinically-relevant probe substrate for the corresponding transporter inhibition assay as the positive control substrate for the transporter substrate assay, as this would help to reduce the experimental burden.

7.4.2.2 Chemical/Metabolic Stability and Non-specific Binding

The positive control substrate and the known (from the literature) substrates being tested should be chemically/metabolically stable and exhibit minimal non-specific binding under the conditions of the assay. This is important because in cell-based transporter substrate assays, chemical instability of a drug and/or non-specific binding to the incubation plate could result in an underestimation of the rate of drug transport as the concentration of the drug in the incubation (that provides the driving force for the interaction with the transporter) is reduced. The same rationale applies to vesicle-based substrate assays, which have the added complication of potential non-specific binding to the filtration plate, which could artificially enhance apparent transport activity in the control incubation condition, thereby reducing the signal-to-noise ratio.[60] Collectively, the different characteristics above may result in the identification of false negatives in an assay. In bidirectional transport assays, these may also impact the accuracy of the efflux ratio, potentially leading to false positives if they specifically led to an underestimation of apical to basolateral (A-to-B) apparent permeability (P_{app}). Additionally, it is of paramount importance that substrates are metabolically stable within the *in vitro* test system as metabolism of a drug may result in the generation of an intracellular sink for the transport of the drug, which may result in a false positive if classification was based on a media-loss approach. To minimise such outcomes, the impact of any confirmed non-specific binding in an assay may be appropriately reduced through the use of a low concentration of bovine serum albumin in the incubation plate[212] (*e.g.* 0.1% w/v present on both sides of a polarised cell monolayer, or 1% w/v as a pre-treatment step) or, in the case of vesicle filtration plates, through either pre-incubating the wells of the plate with a high concentration (100 µM) of unlabelled substrate,[60,213] or having a high concentration of unlabelled substrate present in the stop buffer.[214] Of course, the latter two options are only appropriate if radiolabelled substrates are being utilised. If unlabelled substrates are being used for vesicle studies, then an alternative approach to minimise non-specific binding may be to filter vesicles through a gel matrix by centrifugation instead of using filtration plates.[49]

Assessment of chemical stability or non-specific binding can be performed by incubating donor solution vials or blank wells of an incubation plate (and filtration plate if applicable), respectively, with a single low concentration of substrate (*e.g.* 1 µM) under the same incubation time and conditions of the assay, and analytically comparing this sample to that of a time zero sample. Analytical assessment will differ depending upon whether radiolabelled or unlabelled substrates are being utilised. For radiolabelled experiments, liquid scintillation counting, to give counts per minute, is usually used to determine the extent of non-specific binding through calculation of percentage recovery. However, this method cannot be used to assess chemical or metabolic stability as the detection of radioactivity is unable to distinguish between parent molecule and any degraded products. Rather, it is recommended that radioflow high performance liquid chromatography (HPLC) is used to determine radiochemical purity of the substrate to qualitatively assess stability in the assay.[60,72,98] For experiments using unlabelled substrates, assessment of chemical stability and non-specific binding can both be achieved by using the concentration determined by LC-MS to calculate percentage recovery. In bidirectional transport assays, using polarised cell monolayers, assessment of chemical stability and non-specific binding may also be performed directly (rather than separately) within the assay incubation through determination of the mass balance of the substrate (percentage recovery). Typically, a compound would be considered to be chemically stable and exhibit minimal non-specific binding in an assay if the determined recoveries (or radiochemical purities) were ≥80% of the initial (time zero) value.

7.4.2.3 Cell Seeding Density or Protein Linearity

The transport (uptake rate or P_{app}) of the positive control substrate at a single concentration (*e.g.* 1 µM) should ideally be determined in triplicate incubations (wells) over a range of cell seeding densities or vesicle protein amounts (*e.g.* 10, 25, 50, 75 and 100 µg) to optimise transporter function under linear conditions. The positive controls are run in all assays and their transporter activity assessed based on acceptance criteria (see Section 7.4.2.7), which act as a quality control for the cells.

7.4.2.4 Time Linearity

The optimised cell density or vesicular protein amount determined in the experiment above is utilised to assess the transport (uptake rate or P_{app}) of the anticipated lowest and highest concentration of the positive control substrate to be used in kinetic experiments. Using the known (from the literature) substrates, triplicate incubations are performed over a range of incubation time points (*e.g.* 0.5, 1, 2, 3, 5, 10 and 20 min for cell uptake/vesicle assays, or 30, 60, 90 and 120 min for bidirectional transport assays) to optimise the transporter function to be linear with time. This is important as the time course of uptake may deviate quickly from linearity as a result of rapid accumulation of the substrate inside cells or vesicles. Selection of an early time point in the initial linear phase is crucial for accurate determination of kinetic parameters and this time point could be test system dependent for the same substrate.

7.4.2.5 Concentration-Dependence of Transport Activity

The transport (uptake rate or P_{app}) of the positive control substrate should be determined using optimised incubation conditions over a range of at least six concentrations in order to precisely determine an apparent K_m. Ideally, three of the six concentrations should be below and three should be above the K_m value to obtain a good fit of Michaelis–Menten kinetics. For cell uptake and vesicle assays, the corrected (for background control) transporter-mediated uptake rate should be calculated. Incubations should be performed in triplicate on at least three separate experimental occasions in order to facilitate assessment of intra- and inter-day variability. The determined mean apparent K_m for the positive control substrate should be compared with values in the literature to assess test system suitability. In those test systems where it may be experimentally difficult to accurately obtain a true K_m for a positive control substrate, such as polarised cell monolayers for investigating efflux transporters,[209,215] it may be acceptable to solely confirm that the substrate demonstrates concentration-dependent efflux in order to assess system suitability.

Concentration-dependent transport (uptake rate ratio or efflux ratio) of known (from the literature) substrates and a non-substrate should also be assessed using the optimised incubation conditions determined above, typically at three concentrations (*e.g.* 1, 10 and 100 µM). Incubations should be performed in triplicate on three separate experimental occasions in order to assess intra- and inter-day variability. Correct identification of known (from the literature) substrates as transported substrates in the assay, and confirmation of the absence of transport of the non-substrate, can be used to authenticate test system suitability.

For efflux transporter assays, if a known (from the literature) substrate is not correctly identified as being transported by the test system being validated, then it is important to consider the physicochemical properties (*e.g.* intrinsic permeability) of that compound to understand whether the erroneous result is a consequence of limitations inherent to the test system. For example, in polarised cell-based assays, the inability of a compound to cross the basolateral membrane barrier due to its poor intrinsic permeability would prevent it from gaining access to the apical membrane efflux transporter and may result in it being classified as a 'non-substrate'. Conversely, a false negative in a vesicle-based substrate assay is probably due to a compound having high intrinsic permeability such that it exhibits extensive diffusion and thus cannot be trapped within the vesicle lumen. This may be more likely for P-gp substrates and some lipophilic BCRP substrates.[215]

7.4.2.6   *Cell Passage and Number of Days Post-seeding*

For cell-based transporter assays, it is important to determine transport function (uptake rate ratio, efflux ratio and/or K_m) over different cell passage numbers, as well as different days post-seeding, to assess whether the functional expression of the transporter changes with time.[72,98] These properties can be evaluated from the concentration-dependence experiments of the positive control substrate, with the different occasions reflecting different cell passages, in order to set limits on the range of cell passage numbers (*e.g.* passage 25–40 for P-gp in Caco-2) and days post-seeding that can be employed in future assays to ensure acceptable functional transporter activity.

7.4.2.7   *Setting Acceptance Criteria*

If the degree of variability and reproducibility of the inter-day transport data determined from the validation is considered acceptable (typically for *e.g.* precision ≤30%) across the range of substrates studied, then any future assessment of a drug candidate as a substrate of a transporter using the validated assay need not be repeated on triplicate occasions; rather, just performed once (at four concentrations to reflect EMA guidelines), saving both time and resources.[60,72,98]

In order to quality control the transporter function of the *in vitro* test system during routine future assay use, the inter-day transport data determined for the positive control substrate during the validation should be used to derive appropriate acceptance criteria. Usually, such criteria manifest as a combined minimum transporter-expressing system uptake rate and uptake rate ratio, or a combined minimum B-to-A P_{app} and efflux ratio, for a single concentration (*e.g.* 1 µM) of the positive control substrate. The various calculations are described in the equations below.

$$\text{Minimum uptake rate}$$

$$= \text{mean inter-day uptake rate in transporter cells/vesicles}$$

$$- (3 \times \text{standard deviation}) \tag{7.1}$$

$$\text{Minimum uptake rate ratio} =$$

$$\frac{\text{mean inter-day uptake rate in transporter cells/vesicles} - (3 \times \text{standard deviation})}{\text{mean inter-day uptake rate in control cells/vesicles} + (3 \times \text{standard deviation})} \tag{7.2}$$

$$\text{Minimum B-to-A } P_{\text{app}}$$

$$= \text{mean inter-day B-to-A } P_{\text{app}} \text{ in transporter cells/Caco-2}$$

$$- (3 \times \text{standard deviation}) \tag{7.3}$$

Minimum efflux ratio $=$

$$\frac{\text{mean inter-day B-to-A } P_{\text{app}} \text{ in transporter cells/Caco-2} - (3 \times \text{standard deviation})}{\text{mean inter-day B-to-A } P_{\text{app}} \text{ in transporter cells/Caco-2} + (3 \times \text{standard deviation})}$$

$$\tag{7.4}$$

This minimum efflux ratio is appropriate for Caco-2 based assays. However, for MDCKII or LLC-PK1 transporter transfected cell lines, this minimum efflux ratio should be divided by that determined for the parental cell line to give the minimum corrected efflux ratio. In the equations above, the product of three times the standard deviation is approximately equivalent to the 99% confidence interval when the experimental occasion number is less than about 20. As the experimental number increases above 20 and the degrees of freedom approach infinity, then statistically the accuracy of the standard deviation increases as the t-distribution value is reduced. Consequently, the 99% confidence interval then equates to 2.58 times the standard deviation. Post validation, during the routine generation of positive control substrate data, the validity of the derived acceptance criteria for a validated assay should be continuously monitored and the criteria parameters re-calculated once the assay has been performed on more than 20 occasions.

For polarised cell monolayer assays, additional acceptance criteria should be defined for cell monolayer integrity to ensure that the permeability data determined for the compound are acceptable. Criteria should ideally take the form of a minimum TEER value (usually performed as a pre-assay assessment) and a maximum tolerated permeability value (or percentage mass transfer) of a paracellular permeability marker such as mannitol or Lucifer yellow, either co-incubated or post-assay.

7.4.3 General Considerations for Validating Transporter Inhibition Assays

7.4.3.1 *Choice of Probe Substrate*

Ideally, the chosen *in vitro* probe substrate should also be suitable for use as an *in vivo* probe, and should be clinically relevant such that significant DDIs have been observed in the clinic due to inhibition of its transport by the particular transporter under scrutiny (*e.g.* digoxin is the most used probe substrate for P-gp due to its narrow therapeutic index).[216] Additionally, as described earlier for the positive control in substrate assays, the probe substrate should show sufficient transport activity in the test system (defined by the uptake rate/uptake rate ratio or B-to-A P_{app}/efflux ratio) above the endogenous background activity observed in the control condition, *i.e.* mock transfected or parental cells (*e.g.* MDCKII), or control vesicles, for cell or vesicle-based transporter expression systems, respectively. For hepatocyte uptake assays, the active transport activity can be determined by correcting for the passive component observed in the presence of uptake transporter inhibitors (*e.g.* 50 µM rifamycin SV). The ideal probe substrate should also be chemically stable, exhibit minimal non-specific binding (or be adequately managed for, *e.g.* by pre-incubation with the substrate in assays using membrane vesicles) and be "selective" in the test system for the transporter being investigated. The latter attribute is not really a concern for over-expressed single transporter expression systems, but is more critical in cell lines such as Caco-2 that express multiple efflux transporters.

7.4.3.2 Choice of Probe Substrate Concentration

The linearity of probe substrate transport with regard to cell seeding density/ protein linearity and time should be determined and used to derive its concentration dependence (K_m), as described above for validating substrate assays (if such experiments have not already been performed as part of the substrate validation work, or because the probe substrate chosen is different to the positive control substrate). Once the K_m of the probe substrate for the transporter is known, an appropriate concentration for use in inhibition experiments can be chosen as discussed below.

When determining IC_{50} values it is important to consider that for a competitive inhibitor the IC_{50} value is dependent upon probe substrate concentration [S], as described by the Cheng–Prusoff equation.[49,217,230] Consequently, for extrapolation purposes, the IC_{50} should be converted to the absolute inhibition constant (K_i), a parameter that reflects the affinity of the inhibitor for the substrate binding site on the transporter and is independent of substrate concentration. Practically, if a probe substrate is used at a single concentration well below its K_m for the transporter (ideally ten-fold lower; subject to the absence of any non-specific binding or bioanalytical sensitivity issues), then the determined IC_{50} value for an inhibitor in the assay will approximate to the K_i value (assuming competitive inhibition), thereby removing the need to perform any conversion when extrapolating to the clinical situation. Additionally, the extrapolated K_i value should allow more robust comparison of inhibitory potency data across laboratories, which may use different assay methodologies with regard to probe substrate and concentration.[49] It is important to note that this approach of estimating K_i that has been adopted by the pharmaceutical industry assumes that all observed inhibition is competitive in nature, which may not always be the case, and is used as a "short cut" to avoid the need to conduct precise experiments to determine the true K_i value (and inhibition type) by investigating a range of inhibitor concentrations on a range of probe substrate concentrations. The non-competitive mechanisms of transporter inhibitions and their clinical significance are not widely reported in the literature.

7.4.3.3   Choice of Positive Control Inhibitor

Preferably, the chosen positive control inhibitor should inhibit transporter-mediated probe substrate transport in the test system with a low IC_{50} or K_i value (*e.g.* 1–10 μM)[218] and, if possible, allow extrapolation to known clinical DDIs. It should also be chemically stable and exhibit minimal non-specific binding in the test system; this could either be demonstrated experimentally or inferred to be the case based upon a favourable comparison of the determined IC_{50} to values in the literature. The positive control inhibitor need not necessarily be "selective" for the transporter under investigation as long as selectivity (if appropriate for the test system) resides with the probe substrate.

7.4.3.4   Pre-incubation with Inhibitor

Time-dependent inhibition has been reported for some known inhibitors in cell-based uptake transporter inhibition assays, resulting in a left shift in the inhibition curve and a more potent determined IC_{50} value following pre-incubation with the inhibitor compared with pre-incubation with transporter buffer alone.[49,215,219] This phenomenon may reflect a combination of both the typically short incubation times (often 1–2 min) for uptake transporter assays, possibly coupled with the intrinsic permeability properties of the compound, and the potential mechanism of inhibition (*e.g.* if uncompetitive at an intracellular site on the transporter protein), such that a pre-incubation step allows the inhibitor sufficient time to accumulate to a maximum level at its site of action. However, if an inhibitor is competitive in nature, then pre-incubation simply reflects an effect of non-specific binding of the inhibitor within the test system (resulting in lower actual concentrations compared with the nominal concentrations used to fit the IC_{50} curve). This will give the appearance of increased inhibitory potency due to the higher incubation concentrations giving greater inhibition. Consequently, it is recommended to study inhibitory potency in the presence and absence of an inhibitor pre-incubation step (for a selection of inhibitors) during the validation of an uptake transporter inhibition assay in order to direct how future routine assays should be conducted.

Despite also having short incubation times, by design, vesicle-based inhibition assays already incorporate a pre-incubation step with the inhibitor prior to the initiation of the assay with ATP.

Whilst it is common practice, there is no real requirement to perform a pre-incubation step with the inhibitor in polarised cell monolayer efflux assays as time-dependent inhibition is less likely to manifest due to the much longer incubation time (usually a single time point of 90–120 min for MDCKII/LLC-PK1 cells). Indeed, it has been shown that pre-incubating (equilibrating) with transporter buffer alone prior to co-incubating the probe substrate with the inhibitor can result in the determination of an accurate IC_{50} that is predictive of clinical DDI potential[220] and, for known inhibitors, is also comparable to the range of values determined from methodologies that utilise an inhibitor pre-incubation step.[209]

7.4.3.5   *Concentration-dependence of Inhibition of Probe Substrate Transport*

The transport (uptake rate or P_{app}) of the probe substrate should be determined using the optimised incubation conditions established for determination of its K_m, in the absence (vehicle) and presence of a range of at least six concentrations of the positive control inhibitor, or known (from the literature) inhibitors, and a non-inhibitor, in order to determine inhibitory potency (IC_{50}). Subject to practical constraints, inhibitor concentrations should be relevant to anticipated or known maximal clinical exposures at the various target sites of inhibition (*e.g.* intestinal lumen, hepatic inlet, plasma and intracellular). Typical default concentrations for systemic transporters could be 0.3, 1, 3, 10, 30 and 100 µM, whereas it may be appropriate to evaluate higher concentrations (*e.g.* 300 µM) to cover intestinal transporters such as P-gp and BCRP. Incubations should be performed in triplicate on at least three separate experimental occasions in order to facilitate assessment of intra- and inter-day variability. Correct identification of the positive control inhibitor and known (from the literature) inhibitors as "true" inhibitors in the assay with favourable comparison of their determined mean IC_{50} values to values in the literature, coupled with confirmation of the absence of inhibition by a non-inhibitor, is used to confirm the suitability of the test system.

For cell uptake and vesicle assays, two assay formats can be considered: (1) the use of both transporter-expressing cells/vesicles and mock transfected cells/control vesicles in order to calculate the corrected transporter-mediated uptake rate that may then be converted to the percentage of control transport activity; or (2) the use of only transporter-expressing cells/vesicles, where the uptake activity observed at the highest concentration of the positive control inhibitor (that represents 100% inhibition of the transporter) is utilised to define the passive (background) component in order to convert the observed uptake rate for an inhibitor to the percentage of control transport activity.

For polarised cell monolayer efflux inhibition assays, regulatory guidance indicates that two assay formats may be considered, namely bidirectional (A-to-B and B-to-A P_{app}) transport of the probe substrate to give an efflux ratio, or unidirectional B-to-A flux (P_{app}) of the probe substrate.[73,218] However, the international P-gp IC_{50} working group recommended that, when using polarised cell monolayers, efflux transporter inhibition should only be assessed by the use of unidirectional B-to-A flux (P_{app}) of the probe substrate.[209] The reason for this recommendation was the finding that the efflux ratio-based approach resulted in variable IC_{50} values that were several fold lower than those values derived from the unidirectional (B-to-A) or net secretory flux equations.[208,209] Generally, these assays are conducted in transporter-expressing cells only (*e.g.* MDCK-MDR1 or Caco-2) and the B-to-A P_{app} observed at the highest concentration of positive control inhibitor (that represents 100% inhibition of the transporter) is utilised to derive the passive P_{app} for either defining the bottom plateau of the inhibition curve (if plotting non-transformed P_{app} data), or for transforming the data to the percentage of control transport activity prior to mathematical curve fitting.

7.4.3.6 Cell Passage

For cell based transporter assays, it is important to determine inhibition of transport function (IC_{50}) over different cell passage numbers to assess whether the functional response to transporter inhibition changes with time.[72,98] Limits can then be set on the range of cell passage numbers (*e.g.* passage 25–40 for P-gp in Caco-2) that can be employed in future assays to ensure acceptable inhibition of functional transporter activity.

7.4.3.7 Bioanalytical Considerations

If the transport of the probe substrate within the inhibition assay is quantified by LC-MS, then it is important to assess the potential for ion suppression of the analyte within the mass spectrometer by the test inhibitor. Ion suppression may have the potential to give rise to a false positive classification due to the production of artificial "inhibition" that is independent of the biological test system. This can be evaluated by quantifying an appropriate low concentration of analyte (usually equivalent to that determined inside cells/vesicles or receiver compartments in vehicle incubations) in the absence and presence of the highest incubation concentration of test inhibitor (*e.g.* 100 μM). If the determined concentrations are within 20% of each other, then no significant interference has occurred and any observed inhibition is considered true. Use of a stable-label internal standard of the probe substrate in the bioanalytical method would remove the requirement to perform this ion suppression check.

7.4.3.8 Setting Acceptance Criteria

If the degree of variability and reproducibility of the inter-day inhibition (IC_{50}) data determined from the validation are considered acceptable (typically for *e.g.* precision ≤30%) across the range of inhibitors studied, then any future assessment of a drug candidate as an inhibitor of a transporter using the validated assay need not be repeated on triplicate occasions; rather, just performed once, saving both time and resources.[60,72,98]

In order to quality control the inhibitory transporter function of the *in vitro* test system during routine future assay use, both the inter-day transport data determined for the probe substrate (in vehicle incubations) and the inter-day IC_{50} data determined for the positive control inhibitor during the validation can be used to derive appropriate acceptance criteria. Usually, such criteria are manifested as: (1) a minimum uptake rate (or uptake rate ratio), or a minimum B-to-A P_{app} value, observed in vehicle control incubations to establish that a sufficient transport window for defining inhibition exists; and (2) a predefined IC_{50} range that the determined IC_{50} value of the positive control inhibitor should fall within when run alongside drug candidates.[60,72,98] These parameters are determined by eqn (7.1–7.4) described in Section 7.4.1. Additionally, the acceptable IC_{50} range of positive control inhibitors is determined by the following equations:

$$\text{Lower value} = \text{mean inter-day } IC_{50} - (3 \times \text{standard deviation}) \quad (7.5)$$

$$\text{Upper value} = \text{mean inter-day } IC_{50} + (3 \times \text{standard deviation}) \quad (7.6)$$

7.5 *In vitro* Parameters and Calculations for Kinetics and Predictions

7.5.1 Kinetic Parameters Derived from *In vitro* Models

To ensure accurate information is derived from *in vitro* experiments, the method used to estimate the kinetic parameters needs to be considered. A review article[215] discusses many of the important considerations in some detail and is a source of further information. A summary of possible parameters that can be derived from each *in vitro* experiment is given in Table 7.1.

7.5.1.1 *Efflux Transporter Assays*

The accurate estimation of kinetic parameters from *in vitro* transporter experiments can be challenging, due to the complex nature of even the simplest type of experiment. For example, even in the case of a simple inhibition study of an efflux transporter in a vesicle system, where compounds effectively directly interact with the transporter, multiple factors need to be considered. These include the potential for the probe substrate to passively permeate across the vesicle and also the potential for any compound to sequester into the lipid bilayer membrane. Generally, the principles of Michaelis–Menten kinetics are applied to these types of experiments, using controls (either non-transfected vesicles or the omission of essential cofactors such as ATP) to correct for any passive distribution.

Arguably the most common *in vitro* system to study efflux is one where cell monolayers are utilised; either parental cell lines, such as MDCK transfected with a single efflux transporter (*e.g.* MDR1), or immortalised cell lines, such as Caco-2 cells where affinity to one single efflux transporter dominates for a particular substrate. Simple Michaelis–Menten kinetics have been applied to this system,[221] however this approach in many cases is not appropriate

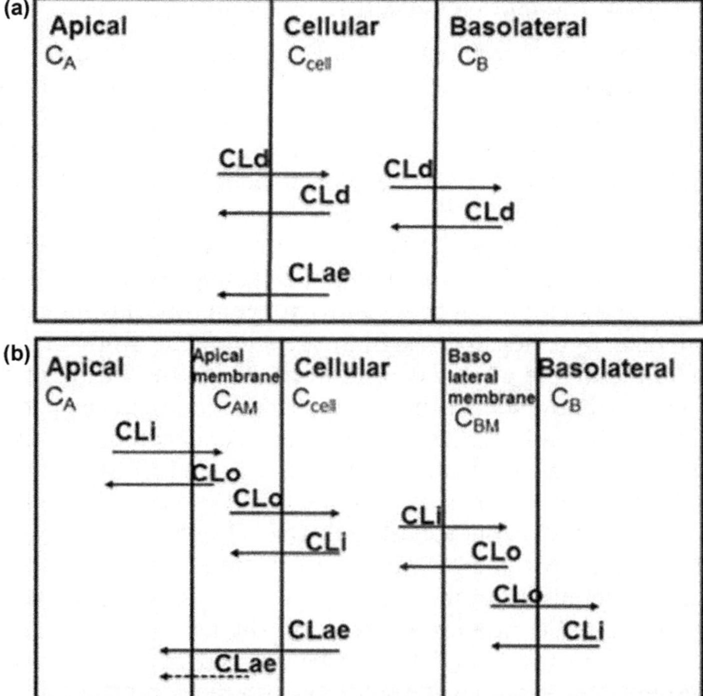

Figure 7.4 Compartmental models to predict transporter kinetics. (a) A three compartment (3C) model was developed with apical, cellular and basolateral compartments. Diffusional clearance (CLd) with apical efflux clearance (CLae) was modelled. (b) A five compartment model was developed with apical, apical membrane, cellular, basolateral membrane and basolateral compartments. Diffusional clearances in (CLi) and out (CLo) of the membranes, with apical efflux clearance (CLae), were modelled. Apical efflux was modelled either out of the cell or out of the apical membrane.
Reproduced with permission from Springer Science and Business Media (copyright 2014).[229]

because the efflux is not directly proportional to the applied drug concentration.[222] This led to the application of compartmental modelling to efflux experiments, with the simplest approach shown in Figure 7.4, where the transfer of a compound between the three compartments of the experiment (basolateral, apical and cell) is described.

To further describe efflux experiments in cell monolayers, more complex compartmental models can be applied, for example two membrane (basolateral and apical) compartments can be added to give a five compartment model (Figure 7.4). This then describes the partition of a compound between media and membrane compartments, and can also describe the potential for active influx to occur, along with the entry of the substrate to the active site of the efflux transporter *via* the membrane.

7.5.1.2   *Influx Transporter Assays*

The simplest form of experiment to study the kinetics of uptake transport is in non-polarised cells (*e.g.* HEK cells) singly transfected with a solute transporter such as OATP1B1; however, the impact of passive permeability needs to be considered using a non-transfected cell. Kinetic parameters to describe the saturable influx can be derived *via* a Michaelis–Menten based approach.[85]

For more complex systems, such as hepatocytes, in which metabolism and transport processes co-exist, dynamic modelling approaches are required to characterise the influx transport. A compartmental model can be applied to describe uptake into hepatocytes in suspension. The work of Menochet *et al.*[223] is one example of this type of work, where the fitting of multiple processes concurrently, such as both active uptake and passive permeability between the buffer and cellular compartments, was completed using data generated over a range of substrate concentrations. This model negates the need for the subtraction of passive uptake using data from separate experiments; however, any effect on the substrates by efflux transporters could not be considered. For compounds where metabolism was known to be significant (*e.g.* telmisartan and repaglinide) the two compartment model was further extended to account for metabolism, as shown in Figure 7.5.

Similar compartmental models can be applied to more sophisticated *in vitro* systems, for example the addition of efflux transport in SCH. As described in Section 7.3.5.1, hepatocytes in this model regain their polarisation, and thus efflux from the hepatocyte can also be considered.[224] Detailed analysis of this compartmental model and other *in vitro* models is outside the scope of this chapter, but is reviewed in recent publications[215] and Chapter 9.

Intracellular binding of the drug in hepatocytes is an important parameter to consider for accurate prediction of *in vivo* clearance. An extension of the compartment model described above was successfully applied by Nordell *et al.*[225] to data generated in human hepatocytes, to include intracellular and membrane binding along with metabolism, to improve the accuracy of prediction of metabolic intrinsic clearance for 11 actively transported compounds.

7.5.2   *In vitro* **Parameters in DDI Predictions**

The study of the potential for a drug candidate to perpetrate a transporter-mediated DDI *versus* a co-administered (victim) drug (comedication) in the clinic begins with the initial *in vitro* evaluation of whether the drug is an inhibitor of transporters. Through the use of various basic static equations, the resulting K_i (or IC_{50} if the probe $[S] \ll K_m$ in the inhibition assay; see the earlier validation section) is put into context with the predicted (by pharmacokinetic modelling) or known inhibitor concentration at the site of inhibition, in order to inform the likelihood of a DDI in the clinic. This can

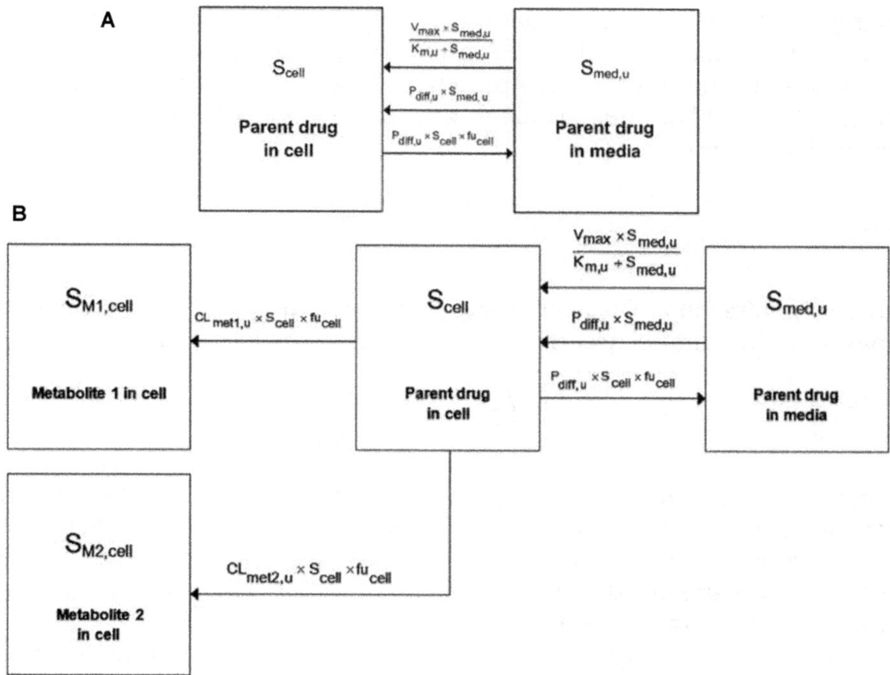

Figure 7.5 (A) Two-compartment model describing the change in cell and media concentration of the parent drug over time because of active uptake ($K_{m,u}$ and V_{max}), passive diffusion ($P_{diff,u}$) and intracellular binding (fu_{cell}) in a plated rat hepatocyte assay. S_{cell} and $S_{med,u}$ represent the total cell and unbound media concentrations, respectively. (B) Extended mechanistic two-compartment model describing the interplay of active uptake ($K_{m,u}$ and V_{max}), passive diffusion ($P_{diff,u}$), intracellular binding (fu_{cell}) and metabolism ($CL_{met1,u}$ and $CL_{met2,u}$) in plated rat hepatocyte assay. $S_{M1,cell}$ and $S_{M2,cell}$ represent the total concentration of the metabolites in the cell. In the case of telmisartan, a single metabolite (metabolite 1) was taken into account for the modelling of its uptake and metabolism. Reproduced with permission from Menochet *et al.* (2012).[223]

guide the clinical programme on the need for any additional clinical studies that may be required to confirm the interaction potential *in vivo*. In addition, if such risk assessments indicate that a drug might have the potential to perpetrate a DDI *in vivo*, then from a patient safety perspective, this information could be used to inform the design of clinical protocols, for example with regard to what comedications may need to be excluded/monitored, or the need for dose adjustment.

7.5.2.1 Basic Static Equations

The basic static equations recommended by regulatory authorities and the ITC for DDI risk assessment are described in detail in Chapter 9, but generally they determine the safety margin ratio between the inhibitor drug concentration at the interacting site (or a surrogate concentration) and its K_i for the transporter, with different cut-offs/thresholds used to flag risk.[73,74,210] A slight modification to this theme is the R value equation calculated as follows:

$$R = 1 + \left(\frac{f_u \times I_{\text{in. max}}}{\text{IC}_{50}} \right) \qquad (7.7)$$

where f_u is fraction of plasma protein binding and $I_{\text{in.max}}$ is the estimated maximum inhibitor concentration at the inlet to the liver and is equal to:

$$C_{\max} + \left(K_a \times \text{dose} \times \frac{F_a F_g}{Q_h} \right) \qquad (7.8)$$

where the $C_{\max}$ is the maximum systemic plasma concentration of inhibitor, *dose* is the inhibitor dose, $F_a F_g$ is the fraction of the dose of inhibitor that is absorbed, k_a is the absorption rate constant of the inhibitor and Q_h is the estimated hepatic blood flow (*e.g.* 1500 ml min^{-1}). If $F_a F_g$ and k_a values are unknown, use 1 and 0.1 min^{-1}.[73,210]

This represents the theoretical fold change in systemic exposure that may occur if a drug were to inhibit a hepatic uptake transporter, with the potential for a DDI *in vivo* being likely if $R \geq 1.25$ (*i.e.* exposure falls outside of bioequivalence). However, it is important to note that the R value risk assessment will inherently overestimate the DDI liability attributed to inhibition of a single transporter, as it assumes that the transporter is solely responsible for the hepatic uptake of the transported victim drug (*i.e.* it contributes 100% towards the disposition pathway). This is not usually the case, as demonstrated for the statins.[226] In addition, the R value is not an *in vitro* to *in vivo* extrapolation (IVIVE) tool that gives an overall predicted fold change in systemic exposure of a victim drug due to a DDI caused by a perpetrator, as it does not take into account other interactions or disposition pathways (enzyme or transporter) that are critically important to the victim drug (which also may or may not be inhibited by the perpetrator). Rather, the R value is merely a qualitative means of flagging a potential DDI risk attributed to inhibition of a single transporter for further investigation/consideration.[227]

7.5.2.2 Mechanistic Static Equations

For IVIVE purposes, more in-depth DDI risk assessments that predict the theoretical change in exposure (fold change in area under the curve (ΔAUC)) due to an inhibitor may be performed using mechanistic static equations such as that described by Zamek-Gliszczynski *et al.* (2009)[228] (adapted from the Rowland–Matin equation for CYP inhibition), which takes into consideration the relative contribution (fraction excreted, f_e) of a particular transporter to a clearance pathway:

$$\text{Fold } \Delta AUC = \frac{1}{\left(\dfrac{f_e}{1 + \dfrac{I}{K_i}}\right) + (1 - f_e)} \qquad (7.9)$$

where I is the unbound maximum hepatic inlet concentration or unbound C_{max} depending upon the location of the transporter under consideration.

The relationship between the magnitude of the $[I]:K_i$ ratio, f_e and predicted fold change in exposure is shown in Figure 7.6. For a transporter pathway that contributes 50% to a clearance process (*i.e.* $f_e = 0.5$), then complete inhibition of that transporter will result in a maximum two-fold increase in the AUC. Large increases in exposure due to inhibition are observed when a transporter is the predominant contributor to a clearance pathway, at $f_e > 0.8$.

An added benefit to this mechanistic static approach to IVIVE is that when parallel transport pathways contribute to a particular disposition route (*e.g.* hepatic uptake mediated *via* OATP1B1 and OATP1B3), the f_e value for each individual transporter can be used as an additive parameter to predict the contribution of inhibiting multiple transporters to any overall increase in exposure.[215] The same outcome can be achieved for sequential disposition pathways (*e.g.* OATP1B1 hepatic uptake followed by hepatic CYP3A4 metabolism) where f_e (or f_m for CYP) values become multiplicative parameters.[226]

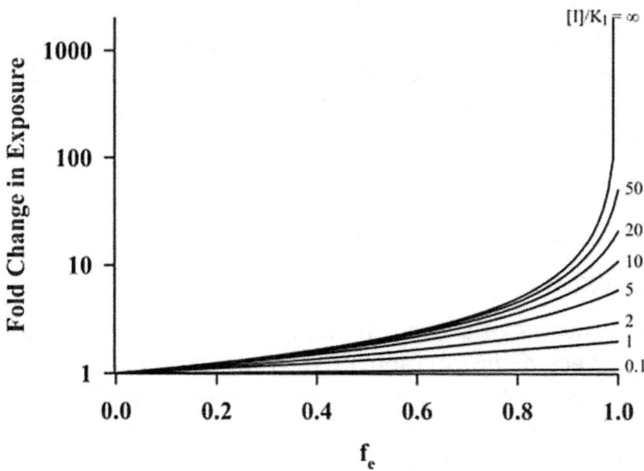

Figure 7.6 Relationship between the $[I]:K_i$ ratio for an inhibitor and the fold change in exposure of a transported substrate as a function of the fraction excreted (f_e) by the transporter.
Reproduced with permission from Zamek-Gliszczynski *et al.*[228]

However, currently, it is difficult to utilise this mechanistic static equation for performing assessments of how intestinal efflux transporter inhibition may impact absorption (fraction absorbed, F_a), and therefore exposure, due to uncertainties in accurately estimating $[I]$. For example, the theoretical maximum intestinal luminal concentration $[I_2]$ (or if lower, the simulated fasted intestinal fluid soluble concentration) may be the appropriate concentration to use for assessing inhibition of intestinal P-gp or BCRP if the inhibitory interaction site on the transporter protein is accessible from the intestinal lumen or the lipid bilayer of the brush border membrane. Conversely, it is also theoretically possible that the intracellular enterocyte concentration $[I_{\mathrm{gut\,max}}]$,[216] derived from dose and enterocyte blood flow, might be the appropriate concentration to use if access to the inhibitory interaction site is only achievable from inside the cell (*e.g.* for the more hydrophilic inhibitors of BCRP compared with P-gp). Consequently, due to these uncertainties, a conservative approach for IVIVE could be to assume that there is complete inhibition of the efflux transporter when the recommended $[I_2]:K_i$ ratio indicates the potential for an interaction (*i.e.* a ratio ≥ 10), and that this will result in complete (100%) absorption ($F_a = 1$). Thus, the predicted fold ΔAUC for the victim drug can be calculated by dividing the resulting predicted overall absorption value (F_aF_g; when $F_a = 1$) by the usual absorption value (F_aF_g) in the absence of perpetrator drug.[226]

This IVIVE approach, employing the mechanistic static equation above and tailored to specific comedications (based upon an understanding of their critical disposition pathways), was recently used to successfully identify the transporter/enzyme inhibition mechanisms underlying a range of observed clinical DDIs with statins.[226] Consequently, this conservative approach (as it assumes constant interaction with the inhibitor throughout the dosing interval) to holistic DDI risk assessment is suggested to be a useful initial tool in the drug development process to predict a maximum extent of change in exposure (AUC) of common comedications due to DDIs by drug candidates, thereby aiding clinical trial design and, ultimately, patient recruitment and safety.

7.5.2.3 Identifying Critical Disposition Pathways and Assigning f_e Values

Pivotal to the success of any mechanistic static or PBPK modelling prediction is the correct determination of $f_{e(m)}$ for each composite critical disposition pathway of the victim drug, be it absorptive (F_a and/or fraction metabolised in enterocytes), hepatic uptake, hepatic metabolism and/or active renal elimination. Towards this goal, data obtained from both clinical pharmacogenetic studies and human mass balance studies, combined with knowledge of the underlying mechanisms responsible for the observed clinical DDIs with the victim drug, can all be assimilated to identify the pathways that are the most critical to the disposition of a drug in humans.

Data obtained from transporter knockout mouse models may also help in this process, but it is important to bear in mind caveats around possible compensatory up-regulation of other transporter/enzyme proteins due to the deletion of a gene.[215]

Once the critical disposition pathway to the overall clearance of a drug is known, then in order to assign an f_e value to a transporter identified as being important to that pathway, it is essential to understand the proportion of clearance that is mediated by active transport *versus* passive permeability. For hepatic uptake transporters, the proportion of active transport can be ascertained using data from *in vitro* hepatocyte uptake studies in the presence and absence of an inhibitor. For renal uptake transporters, the ratio of active tubular secretion to total renal clearance can be determined by subtracting the renal clearance due to passive glomerular filtration (fraction unbound $\times$ glomerular filtration rate) from total renal clearance to give active secretory clearance, which can then be expressed as a percentage. For example, the total renal clearance of pravastatin is on average 400 ml min^{-1} with glomerular filtration clearance equal to 60 ml min^{-1}, giving active secretion as 340 ml min^{-1}, representing 85% of the renal pathway. Consequently, the f_e value for OAT3 (the transporter responsible for the active secretion of pravastatin) can be calculated as being equal to 0.4 (0.47×0.85).

The f_e parameter, in conjunction with $[I]:K_i$ in relation to dose, can be considered as a prerequisite for predicting the extent of any potential increase in exposure that might be observed with an inhibitor of a particular transporter. It is also very useful for the determination of the maximum theoretical fold change in exposure that would be expected if a pathway was to be completely (100%) inhibited in a DDI or was genetically redundant. This theoretical maximum is calculated using the equation, *fold* $\Delta AUC = 1/(1 - f_e)$, for each individual disposition pathway for a drug and provides useful information for deciphering the likely overall mechanism(s) of a clinical DDI in scenarios where the observed change in exposure is far greater than the theoretical maximum expected for each inhibitor type alone.

7.5.2.4   PBPK Modelling

Often, the mechanistic static approach to IVIVE can be supplemented with a dynamic PBPK modelling approach that integrates drug-dependent parameters (*e.g.* concentration–time profile at the interaction site, derived *in vitro* kinetic constants K_m and/or K_i) with physiological and system parameters, in order to further refine overall DDI risk assessment and predictions.[215] As with mechanistic static approaches, the predictive ability of any PBPK modelling approach is critically dependent on the quality and accuracy of the data properties associated with the compounds being simulated; indeed, accuracy of *in vitro* parameters and correct identification of the clinically relevant disposition pathways are crucial. However, as the topic is beyond the scope of this chapter, interested readers are referred to Chapter 9 for further information.

7.6 Summary

In silico and *in vitro* models are important tools that are used to generate data for predictions of drug pharmacokinetics and DDIs. There is a wide variety of *in silico* and *in vitro* models available to do various types of transporter studies, and depending upon the compound properties and purpose of study, it is possible to select the most suitable model. Validation of *in silico* models by experimental data to enhance IVIVE and development of complex *in vitro* 3D models and specialised co-cultures representing organotypical physiology is a major focus of future research.

Conflict of Interest

Pradeep Sharma, Constanze Hilgendorf, Simone Stahl and Katherine Fenner are employees of AstraZeneca, and Mohammed I. Atari, Robert Elsby (former employee of AstraZeneca) and Simon Thomas are employees of Cyprotex Discovery Limited. Cyprotex auto-QSAR is a proprietary software platform of Cyprotex Discovery Limited.

References

1. C. Chang and P. W. Swaan, *Eur. J. Pharm. Sci.*, 2006, **27**, 411–424.
2. G. Szakács, A. Váradi, C. Ozvegy-Laczka and B. Sarkadi, *Drug Discovery Today*, 2008, **13**, 379–393.
3. I. Moraes, G. Evans, J. Sanchez-Weatherby, S. Newstead and P. D. Stewart, *Biochim. Biophys. Acta*, 2014, **1838**, 78–87.
4. A. Ravna, G. Sager, S. Dahl and I. Sylte, in *Transporters as Targets for Drugs*, ed. S. Napier and M. Bingham, Springer-Verlag, Berlin, Heidelberg, 2009, vol. 23, pp. 15–51.
5. M. I. Atari, M. J. Chappell, R. J. Errington, P. J. Smith and N. D. Evans, *Comput. Methods Progr. Biomed.*, 2011, **104**, 93–103.
6. M. Wiltgen and G. P. Tilz, *Wien. Med. Wochenschr.*, 2009, **159**, 112–125.
7. H. Berman, K. Henrick, H. Nakamura and J. L. Markley, *Nucleic Acids Res.*, 2007, **35**, D301–D303.
8. P. W. Rose, A. Prlić, C. Bi, W. F. Bluhm, C. H. Christie, S. Dutta, R. K. Green, D. S. Goodsell, J. D. Westbrook, J. Woo, J. Young, C. Zardecki, H. M. Berman, P. E. Bourne and S. K. Burley, *Nucleic Acids Res.*, 2015, **43**, D345–D356.
9. E. S. Donkor, N. T. K. D. Dayie and T. K. Adiku, *J. Bioinf. Sequence Anal.*, 2014, **6**, 1–6.
10. V. K. Vyas, R. D. Ukawala, M. Ghate and C. Chintha, *Indian J. Pharm. Sci.*, 2012, **74**, 1–17.
11. G. Lunter, A. Rocco, N. Mimouni, A. Heger, A. Caldeira and J. Hein, *Genome Res.*, 2008, **18**, 298–309.

12. F. Sievers, A. Wilm, D. Dineen, T. J. Gibson, K. Karplus, W. Li, R. Lopez, H. McWilliam, M. Remmert, J. Söding, J. D. Thompson and D. G. Higgins, *Mol. Syst. Biol.*, 2011, **7**, 539.

13. J. Lee, D. Lee, H. Park, E. A. Coutsias and C. Seok, *Proteins*, 2010, **78**, 3428–3436.

14. H. P. Peng and A. S. Yang, *Bioinformatics*, 2007, **23**, 2836–2842.

15. A. Shehu and L. E. Kavraki, *Entropy*, 2012, **14**, 252–290.

16. P. Francis-Lyon and P. Koehl, *Proteins*, 2014, **82**, 2000–2017.

17. Y. Cao, L. Song, Z. Miao, Y. Hu, L. Tian and T. Jiang, *Bioinformatics*, 2011, **27**, 785–790.

18. E. R. Lindahl, *Methods Mol. Biol.*, 2008, **443**, 3–23.

19. A. W. Ravna, I. Sylte and G. Sager, *Theor. Biol. Med. Modell.*, 2009, **6**, 20.

20. A. W. Ravna and I. Sylte, *Methods Mol. Biol.*, 2012, **857**, 281–299.

21. L. J. McGuffin, in *Computational Structural Biology: Methods and Applications*, ed. T. Schwede and M. C. Peitsch, World Scientific Publishing Co. Pte. Ltd., Singapore, 2008.

22. S. Wu, J. Skolnick and Y. Zhang, *BMC Biol.*, 2007, **5**, 17.

23. G. Helles, *J. R. Soc.Interface*, 2008, **5**, 387–396.

24. L. Jooyoung, S. Wu and Y. Zhang, in *From Protein Structure to Function with Bioinformatics*, ed. D. J. Rigden, Springer Science + Business Media B.V, Netherlands, 2009.

25. Y. Shan, E. T. Kim, M. P. Eastwood, R. O. Dror, M. A. Seeliger and D. E. Shaw, *J. Am. Chem. Soc.*, 2011, **133**, 9181–9183.

26. D. Fischer, *Curr. Opin. Struct. Biol.*, 2006, **16**, 178–182.

27. R. Todeschini and V. Consonni, *Handbook of Molecular Descriptors*, Wiley-VCH, Weinheim, Germany, 2000.

28. R. J. Dimelow, P. D. Metcalfe and S. Thomas, in *Encyclopedia of Drug Metabolism and Interactions*, ed. A. V. Lyubimov, John Wiley & Sons, Inc., Hoboken, NJ, 2012.

29. S. Thomas, and R. J. Dimelow, in *Drug Metabolism Prediction*, ed. J. Kirchmair, Wiley-VCH Verlag GmbH & Co. KGaA, Weinheim, Germany, 2014.

30. A. Z. Dudek, T. Arodz and J. Gálvez, *Comb. Chem. High Throughput Screening*, 2006, **9**, 213–228.

31. S. Weaver and M. P. Gleeson, *J. Mol. Graphics Modell.*, 2008, **26**, 1315–1326.

32. T. Hastie, R. Tibshirani and J. Friedman, *The Elements of Statistical Learning: Data Mining, Inference, and Prediction*, Springer Series in Statistics, Springer, New York, 2nd edn, 2013.

33. N. Busschaert, S. J. Bradberry, M. Wenzel, C. J. E. Haynes, J. R. Hiscock, I. L. Kirby, L. E. Karagiannidis, S. J. Moore, N. J. Wells, J. Herniman, G. J. Langley, P. N. Horton, M. E. Light, I. Marques, P. J. Costa, V. Félix, J. G. Frey and P. A. Gale, *Chem. Sci.*, 2013, **4**, 3036–3045.

34. J. Verma, V. M. Khedkar and E. C. Coutinho, *Curr. Top. Med. Chem.*, 2010, **10**, 95–115.

35. C. C. Melo-Filho, R. C. Braga and C. H. Andrade, *Curr. Comput.-Aided Drug Des.*, 2014, **10**, 148–159.

36. X. Zhao, M. Chen, B. Huang, H. Ji and M. Yuan, *Int. J. Mol. Sci.*, 2011, **12**, 7022–7037.

37. A. K. Gupta and A. K. Saxena, *Med. Chem. Res.*, 2011, **20**, 1455–1464.

38. C. Chang, S. Ekins, P. Bahadduri and P. W. Swaan, *Adv. Drug Delivery Rev.*, 2006, **58**, 1431–1450.

39. H. Xie, L. Chen, J. Zhang, X. Xie, K. Qiu and J. Fu, *Int. J. Mol. Sci.*, 2015, **16**, 12307–12323.

40. J. S. Mason, I. Morize, P. R. Menard, D. L. Cheney, C. Hulme and R. F. Labaudiniere, *J. Med. Chem.*, 1999, **42**, 3251–3264.

41. M. A. Demel, R. Schwaha, O. Krämer, P. Ettmayer, E. E. Haaksma and G. F. Ecker, *Expert Opin. Drug Metab. Toxicol.*, 2008, **4**, 1167–1180.

42. S. Ekins, J. S. Johnston, P. Bahadduri, V. M. D'Souza, A. Ray, C. Chang and P. W. Swaan, *Pharm. Res.*, 2005, **22**, 512–517, DOI: 10.1007/s11095-005-2505-y.

43. I. J. Enyedy, S. Sakamuri, W. A. Zaman, K. M. Johnson and S. Wang, *Bioorg. Med. Chem. Lett.*, 2003, **13**, 513–517, DOI: S0960894 × 02009435 [pii].

44. H. Murer and R. Kinne, *J. Membr. Biol.*, 1980, **55**, 81–95.

45. A. Aslamkhan, Y. H. Han, R. Walden, D. H. Sweet and J. B. Pritchard, *Am. J. Physiol. Renal Physiol.*, 2003, **285**, F775–F783, DOI: 10.1152/ajprenal.00140.2003.

46. N. R. Koopen, H. Wolters, M. Muller, I. J. Schippers, R. Havinga, H. Roelofsen, R. J. Vonk, B. Stieger, P. J. Meier and F. Kuipers, *J. Hepatol.*, 1997, **27**, 699–706, DOI: S0168-8278(97)80087-6 [pii].

47. V. Ganapathy and F. H. Leibach, *J. Biol. Chem.*, 1983, **258**, 14189–14192.

48. C. J. van Staden, R. E. Morgan, B. Ramachandran, Y. Chen, P. H. Lee and H. K. Hamadeh, in *Current Protocols in Toxicology*, John Wiley & Sons, Inc., 2012, ch. 23, unit 23.5.

49. K. L. Brouwer, D. Keppler, K. A. Hoffmaster, D. A. Bow, Y. Cheng, Y. Lai, J. E. Palm, B. Stieger, R. Evers and International Transporter Consortium, *Clin. Pharmacol. Ther.*, 2013, **94**, 95–112.

50. M. Jani and P. Krajcsi, *Drug Discov. Today Technol.*, 2014, **12**, e105–e112.

51. P. J. Meier and J. L. Boyer, *Methods Enzymol.*, 1990, **192**, 534–545.

52. S. L. Weinberg, G. Burckhardt and F. A. Wilson, *J. Clin. Invest.*, 1986, **78**, 44–50.

53. B. Hagenbuch, G. Stange and H. Murer, *Pflugers Arch.*, 1985, **405**, 202–208.

54. B. Stieger, K. Fattinger, J. Madon, G. A. Kullak-Ublick and P. J. Meier, *Gastroenterology*, 2000, **118**, 422–430.

55. M. M. van Oers, G. P. Pijlman and J. M. Vlak, *J. Gen. Virol.*, 2015, **96**, 6–23.

56. D. L. Jarvis, in *Methods in Enzymology*, ed. R. R. Burgess and M. P. Deutscher, Academic Press, 2009, pp. 191–222.

57. J. E. Karlsson, C. Heddle, A. Rozkov, J. Rotticci-Mulder, O. Tuvesson, C. Hilgendorf and T. B. Andersson, *Drug Metab. Dispos.*, 2010, **38**, 705–714.

58. E. Kis, E. Ioja, T. Nagy, L. Szente, K. Heredi-Szabo and P. Krajcsi, *Drug Metab. Dispos.*, 2009, **37**, 1878–1886.
59. C. Guyot, L. Hofstetter and B. Stieger, *Mol. Pharmacol.*, 2014, **85**, 909–920.
60. R. Elsby, V. Smith, L. Fox, D. Stresser, C. Butters, P. Sharma and D. D. Surry, *Xenobiotica*, 2011, **41**, 764–783.
61. B. Sarkadi, E. M. Price, R. C. Boucher, U. A. Germann and G. A. Scarborough, *J. Biol. Chem.*, 1992, **267**, 4854–4858.
62. R. R. Vethanayagam, H. Wang, A. Gupta, Y. Zhang, F. Lewis, J. D. Unadkat and Q. Mao, *Drug Metab. Dispos.*, 2005, **33**, 697–705.
63. Q. Mao, G. Conseil, A. Gupta, S. P. Cole and J. D. Unadkat, *Biochem. Biophys. Res. Commun.*, 2004, **320**, 730–737.
64. Y. Adachi, H. Suzuki and Y. Sugiyama, *Pharm. Res.*, 2001, **18**, 1660–1668.
65. C. Ozvegy, A. Varadi and B. Sarkadi, *J. Biol. Chem.*, 2002, **277**, 47980–47990.
66. P. Szeremy, A. Pal, D. Mehn, B. Toth, F. Fulop, P. Krajcsi and K. Heredi-Szabo, *J. Biomol. Screening*, 2011, **16**, 112–119.
67. A. Dreiseitel, B. Oosterhuis, K. V. Vukman, P. Schreier, A. Oehme, S. Locher, G. Hajak and P. G. Sand, *Br. J. Pharmacol.*, 2009, **158**, 1942–1950.
68. S. A. Bustin and R. Mueller, *Clin. Sci.*, 2005, **109**, 365–379.
69. N. Yosef and E. E. Ubogu, *Cell. Mol. Neurobiol.*, 2013, **33**, 175–186.
70. J. J. Castorino, S. Deborde, A. Deora, R. Schreiner, S. M. Gallagher-Colombo, E. Rodriguez-Boulan and N. J. Philp, *Traffic*, 2011, **12**, 483–498.
71. C. Hilgendorf, G. Ahlin, A. Seithel, P. Artursson, A. L. Ungell and J. Karlsson, *Drug Metab. Dispos.*, 2007, **35**, 1333–1340.
72. P. Sharma, V. E. Holmes, R. Elsby, C. Lambert and D. Surry, *Xenobiotica*, 2010, **40**, 24–37.
73. *US FDA Draft guidance 2012: Drug Interaction Studies—Study Design, Data Analysis, Implications for Dosing, and Labeling Recommendations.*
74. *EMA Guidance 2013: Guideline on the Investigation of Drug Interactions.*
75. Pharmaceuticals and Medical Devices Agency, *Drug interaction guideline for drug development and labeling recommendations*, PMDA, Japan, 2014.
76. F. L. Graham, J. Smiley, W. C. Russell and R. Nairn, *J. Gen. Virol.*, 1977, **36**, 59–74.
77. J. R. Masters, *Nat. Rev. Cancer.*, 2002, **2**, 315–319.
78. L. Zhang, M. E. Schaner and K. M. Giacomini, *J. Pharmacol. Exp. Ther.*, 1998, **286**, 354–361.
79. M. Sasaki, H. Suzuki, K. Ito, T. Abe and Y. Sugiyama, *J. Biol. Chem.*, 2002, **277**, 6497–6503.
80. Y. Cui, J. Konig and D. Keppler, *Mol. Pharmacol.*, 2001, **60**, 934–943.
81. K. Kopplow, K. Letschert, J. Konig, B. Walter and D. Keppler, *Mol. Pharmacol.*, 2005, **68**, 1031–1038.

82. A. Guo, W. Marinaro, P. Hu and P. J. Sinko, *Drug Metab. Dispos.*, 2002, **30**, 457–463.

83. D. Gartzke and G. Fricker, *J. Pharm. Sci.*, 2014, **103**, 1298–1304.

84. Q. Wang, J. D. Rager, K. Weinstein, P. S. Kardos, G. L. Dobson, J. Li and I. J. Hidalgo, *Int. J. Pharm.*, 2005, **288**, 349–359.

85. J. Noe, R. Portmann, M. E. Brun and C. Funk, *Drug Metab. Dispos.*, 2007, **35**, 1308–1314.

86. A. Poirier, S. Belli, C. Funk, M. B. Otteneder, R. Portmann, K. Heinig, E. Prinssen, S. E. Lazic, C. R. Rayner, G. Hoffmann, T. Singer, D. E. Smith and F. Schuler, *Drug Metab. Dispos.*, 2012, **40**, 1556–1565.

87. E. Sigel, *J. Membr. Biol.*, 1990, **117**, 201–221.

88. T. Nakanishi, L. A. Doyle, B. Hassel, Y. Wei, K. S. Bauer, S. Wu, D. W. Pumplin, H. B. Fang and D. D. Ross, *Mol. Pharmacol.*, 2003, **64**, 1452–1462.

89. Y. Uwai, R. Taniguchi, H. Motohashi, H. Saito, M. Okuda and K. Inui, *Drug Metab. Pharmacokinet.*, 2004, **19**, 369–374.

90. B. L. Shneider and M. S. Moyer, *J. Biol. Chem.*, 1993, **268**, 6985–6988.

91. A. Schroeder, U. Eckhardt, B. Stieger, R. Tynes, C. D. Schteingart, A. F. Hofmann, P. J. Meier and B. Hagenbuch, *Am. J. Physiol.*, 1998, **274**, G370–G375.

92. B. Hagenbuch, B. Stieger, M. Foguet, H. Lubbert and P. J. Meier, *Proc. Natl. Acad. Sci. U. S. A.*, 1991, **88**, 10629–10633.

93. A. Ungell and J. Karlsson, in *Drug Bioavailability. Estimation of Solubility, Permeability, Absorption and Bioavailability*, ed. H. van de Waterbeemd, H. Lennernas and P. Artursson, Wiley-VCH Verlag GmbH & Co KGaA, Weinheim, 2003, pp. 90–131.

94. P. Anderle, Y. Huang and W. Sadee, *Eur. J. Pharm. Sci.*, 2004, **21**, 17–24.

95. A. Seithel, J. Karlsson, C. Hilgendorf, A. Bjorquist and A. L. Ungell, *Eur. J. Pharm. Sci.*, 2006, **28**, 291–299.

96. R. Hayeshi, C. Hilgendorf, P. Artursson, P. Augustijns, B. Brodin, P. Dehertogh, K. Fisher, L. Fossati, E. Hovenkamp, T. Korjamo, C. Masungi, N. Maubon, R. Mols, A. Mullertz, J. Monkkonen, C. O'Driscoll, H. M. Oppers-Tiemissen, E. G. Ragnarsson, M. Rooseboom and A. L. Ungell, *Eur. J. Pharm. Sci.*, 2008, **35**, 383–396.

97. A. H. Eriksson, P. L. Elm, M. Begtrup, B. Brodin, R. Nielsen and B. Steffansen, *Eur. J. Pharm. Sci.*, 2005, **25**, 145–154.

98. R. Elsby, D. D. Surry, V. N. Smith and A. J. Gray, *Xenobiotica*, 2008, **38**, 1140–1164.

99. P. Prieto, S. Hoffmann, V. Tirelli, F. Tancredi, I. Gonzalez, M. Bermejo and I. De Angelis, *Altern. Lab. Anim.*, 2010, **38**, 367–386.

100. K. E. Sampson, A. Brinker, J. Pratt, N. Venkatraman, Y. Xiao, J. Blasberg, T. Steiner, M. Bourner and D. C. Thompson, *Drug Metab. Dispos.*, 2015, **43**, 199–207.

101. T. Terasaki and K. Hosoya, *Biol. Pharm. Bull.*, 2001, **24**, 111–118.

102. F. Roux and P. O. Couraud, *Cell. Mol. Neurobiol.*, 2005, **25**, 41–58.

103. N. J. Abbott, A. A. Patabendige, D. E. Dolman, S. R. Yusof and D. J. Begley, *Neurobiol. Dis.*, 2010, **37**, 13–25.

104. N. J. Abbott, D. E. Dolman and A. K. Patabendige, *Curr. Drug Metab.*, 2008, **9**, 901–910.

105. A. Patabendige, R. A. Skinner and N. J. Abbott, *Brain Res.*, 2013, **1521**, 1–15.

106. M. J. Wilmer, M. A. Saleem, R. Masereeuw, L. Ni, T. J. van der Velden, F. G. Russel, P. W. Mathieson, L. A. Monnens, L. P. van den Heuvel and E. N. Levtchenko, *Cell Tissue Res.*, 2010, **339**, 449–457.

107. C. M. Schophuizen, M. J. Wilmer, J. Jansen, L. Gustavsson, C. Hilgendorf, J. G. Hoenderop, L. P. van den Heuvel and R. Masereeuw, *Pflugers Arch.*, 2013, **465**, 1701–1714.

108. L. Aschauer, G. Carta, N. Vogelsang, E. Schlatter and P. Jennings, *Toxicol. In Vitro*, 2014.

109. A. Guillouzo, A. Corlu, C. Aninat, D. Glaise, F. Morel and C. Guguen-Guillouzo, *Chem. Biol. Interact.*, 2007, **168**, 66–73.

110. M. Le Vee, G. Noel, E. Jouan, B. Stieger and O. Fardel, *Toxicol. In Vitro*, 2013, **27**, 1979–1986.

111. L. Badolo, M. M. Trancart, L. Gustavsson and C. Chesne, *Chem. Biol. Interact.*, 2011, **190**, 165–170.

112. P. Lundquist, J. Loof, A. K. Sohlenius-Sternbeck, E. Floby, J. Johansson, J. Bylund, J. Hoogstraate, L. Afzelius and T. B. Andersson, *Drug Metab. Dispos.*, 2014, **42**, 469–480.

113. T. R. Grandjean, M. J. Chappell, A. M. Lench, J. W. Yates and C. J. O'Donnell, *Xenobiotica*, 2014, **44**, 961–974.

114. M. G. Soars, P. J. Webborn and R. J. Riley, *Mol. Pharm.*, 2009, **6**, 1662–1677.

115. Y. Yabe, A. Galetin and J. B. Houston, *Drug Metab. Dispos.*, 2011, **39**, 1808–1814.

116. P. Nordell, S. Winiwarter and C. Hilgendorf, *Mol. Pharm.*, 2013, **10**, 4443–4451.

117. A. S. Ray, T. Cihlar, K. L. Robinson, L. Tong, J. E. Vela, M. D. Fuller, L. M. Wieman, E. J. Eisenberg and G. R. Rhodes, *Antimicrob. Agents Chemother.*, 2006, **50**, 3297–3304.

118. E. I. Lepist, X. Zhang, J. Hao, J. Huang, A. Kosaka, G. Birkus, B. P. Murray, R. Bannister, T. Cihlar, Y. Huang and A. S. Ray, *Kidney Int.*, 2014, **86**, 350–357.

119. G. Cramb and J. W. Dow, *Biochem. Pharmacol.*, 1983, **32**, 227–231.

120. M. G. Soars, K. Grime, J. L. Sproston, P. J. Webborn and R. J. Riley, *Drug Metab. Dispos.*, 2007, **35**, 859–865.

121. K. Yang, K. Kock and K. L. R. Brouwer, in *Transporters in Drug Development: Discovery, Optimization, Clinical Study and Regulation*, ed. Y. Sugiyama and B. Steffansen, Springer, New York, 2013, pp. 201–240.

122. U. Baccarani, A. Sanna, A. Cariani, M. Sainz, G. L. Adani, D. Lorenzin, D. Montanaro, M. Scalamogna, G. Piccolo, A. Risaliti, F. Bresadola and A. Donini, *Transplant. Proc.*, 2005, **37**, 256–259.

123. P. Godoy, N. J. Hewitt, U. Albrecht, M. E. Andersen, N. Ansari, S. Bhattacharya, J. G. Bode, J. Bolleyn, C. Borner, J. Bottger, A. Braeuning, R. A. Budinsky, B. Burkhardt, N. R. Cameron, G. Camussi, C. S. Cho, Y. J. Choi, J. Craig Rowlands, U. Dahmen, G. Damm, O. Dirsch, M. T. Donato, J. Dong, S. Dooley, D. Drasdo, R. Eakins, K. S. Ferreira, V. Fonsato, J. Fraczek, R. Gebhardt, A. Gibson, M. Glanemann, C. E. Goldring, M. J. Gomez-Lechon, G. M. Groothuis, L. Gustavsson, C. Guyot, D. Hallifax, S. Hammad, A. Hayward, D. Haussinger, C. Hellerbrand, P. Hewitt, S. Hoehme, H. G. Holzhutter, J. B. Houston, J. Hrach, K. Ito, H. Jaeschke, V. Keitel, J. M. Kelm, B. Kevin Park, C. Kordes, G. A. Kullak-Ublick, E. L. LeCluyse, P. Lu, J. Luebke-Wheeler, A. Lutz, D. J. Maltman, M. Matz-Soja, P. McMullen, I. Merfort, S. Messner, C. Meyer, J. Mwinyi, D. J. Naisbitt, A. K. Nussler, P. Olinga, F. Pampaloni, J. Pi, L. Pluta, S. A. Przyborski, A. Ramachandran, V. Rogiers, C. Rowe, C. Schelcher, K. Schmich, M. Schwarz, B. Singh, E. H. Stelzer, B. Stieger, R. Stober, Y. Sugiyama, C. Tetta, W. E. Thasler, T. Vanhaecke, M. Vinken, T. S. Weiss, A. Widera, C. G. Woods, J. J. Xu, K. M. Yarborough and J. G. Hengstler, *Arch. Toxicol.*, 2013, **87**, 1315–1530.

124. B. J. Blaauboer and A. J. Paine, *Biochem. Biophys. Res. Commun.*, 1979, **90**, 368–374.

125. U. Hahn, A. Stallmach, E. G. Hahn and E. O. Riecken, *Gastroenterology*, 1990, **98**, 322–335.

126. Y. Chen, P. P. Wong, L. Sjeklocha, C. J. Steer and M. B. Sahin, *Hepatology*, 2012, **55**, 563–574.

127. N. J. Liptrott, M. Penny, P. G. Bray, J. Sathish, S. H. Khoo, D. J. Back and A. Owen, *Br. J. Pharmacol.*, 2009, **156**, 497–508.

128. P. Lundquist, J. Loof, U. Fagerholm, I. Sjogren, J. Johansson, S. Briem, J. Hoogstraate, L. Afzelius and T. B. Andersson, *Drug Metab. Dispos.*, 2014, **42**, 459–468.

129. T. De Bruyn, S. Chatterjee, S. Fattah, J. Keemink, J. Nicolai, P. Augustijns and P. Annaert, *Expert Opin. Drug Metab. Toxicol.*, 2013, **9**, 589–616.

130. G. Elaut, T. Henkens, P. Papeleu, S. Snykers, M. Vinken, T. Vanhaecke and V. Rogiers, *Curr. Drug Metab.*, 2006, **7**, 629–660.

131. C. D. Brown, R. Sayer, A. S. Windass, I. S. Haslam, M. E. De Broe, P. C. D'Haese and A. Verhulst, *Toxicol. Appl. Pharmacol.*, 2008, **233**, 428–438.

132. T. Nakanishi, A. Fukushi, M. Sato, M. Yoshifuji, T. Gose, Y. Shirasaka, K. Ohe, M. Kobayashi, K. Kawai and I. Tamai, *Mol. Pharm.*, 2011, **8**, 2142–2150.

133. F. Tasnim and D. Zink, *Am. J. Physiol. Renal Physiol.*, 2012, **302**, F1055–F1062.

134. P. Fisel, O. Renner, A. T. Nies, M. Schwab and E. Schaeffeler, *Expert Opin. Drug Metab. Toxicol.*, 2014, **10**, 395–408.

135. D. Beghin, J. L. Delongeas, N. Claude, R. Farinotti, F. Forestier and S. Gil, *Toxicol. In Vitro*, 2010, **24**, 630–637.

136. D. G. Hemmings, B. Lowen, R. Sherburne, G. Sawicki and L. J. Guilbert, *Placenta*, 2001, **22**, 70–79.

137. P. Berveiller, S. A. Degrelle, N. Segond, H. Cohen, D. Evain-Brion and S. Gil, *Placenta*, 2015, **36**, 93–96.

138. D. A. Evseenko, J. W. Paxton and J. A. Keelan, *Am. J. Physiol. Regul. Integr. Comp. Physiol.*, 2006, **290**, R1357–R1365.

139. D. Beghin, J. L. Delongeas, N. Claude, F. Forestier, R. Farinotti and S. Gil, *Toxicol. In Vitro*, 2009, **23**, 141–147.

140. E. Gabriel, S. Schievenbusch, E. Kolossov, J. G. Hengstler, T. Rotshteyn, H. Bohlen, D. Nierhoff, J. Hescheler and I. Drobinskaya, *PLoS One*, 2012, **7**, e44912.

141. M. Ulvestad, P. Nordell, A. Asplund, M. Rehnstrom, S. Jacobsson, G. Holmgren, L. Davidson, G. Brolen, J. Edsbagge, P. Bjorquist, B. Kuppers-Munther and T. B. Andersson, *Biochem. Pharmacol.*, 2013, **86**, 691–702.

142. K. Narayanan, K. M. Schumacher, F. Tasnim, K. Kandasamy, A. Schumacher, M. Ni, S. Gao, B. Gopalan, D. Zink and J. Y. Ying, *Kidney Int.*, 2013, **83**, 593–603.

143. Y. Li, K. Kandasamy, J. K. Chuah, Y. N. Lam, W. S. Toh, Z. Y. Oo and D. Zink, *Mol. Pharm.*, 2014, **11**, 1982–1990, DOI: 10.1021/mp400637s.

144. A. L. Kauffman, A. V. Gyurdieva, J. R. Mabus, C. Ferguson, Z. Yan and P. J. Hornby, *Front. Pharmacol.*, 2013, **4**, 79.

145. R. Cecchelli, S. Aday, E. Sevin, C. Almeida, M. Culot, L. Dehouck, C. Coisne, B. Engelhardt, M. P. Dehouck and L. Ferreira, *PLoS One*, 2014, **9**, e99733.

146. E. S. Lippmann, S. M. Azarin, J. E. Kay, R. A. Nessler, H. K. Wilson, A. Al-Ahmad, S. P. Palecek and E. V. Shusta, *Nat. Biotechnol.*, 2012, **30**, 783–791.

147. B. Swift, N. D. Pfeifer and K. L. Brouwer, *Drug Metab. Rev.*, 2010, **42**, 446–471.

148. X. Liu, J. P. Chism, E. L. LeCluyse, K. R. Brouwer and K. L. Brouwer, *Drug Metab. Dispos.*, 1999, **27**, 637–644.

149. G. Ghibellini, L. S. Vasist, E. M. Leslie, W. D. Heizer, R. J. Kowalsky, B. F. Calvo and K. L. Brouwer, *Clin. Pharmacol. Ther.*, 2007, **81**, 406–413.

150. G. Ghibellini, A. S. Bridges, C. N. Generaux and K. L. Brouwer, *Drug Metab. Dispos.*, 2007, **35**, 345–349.

151. N. D. Pfeifer, K. Yang and K. L. Brouwer, *J. Pharmacol. Exp. Ther.*, 2013, **347**, 727–736.

152. K. Yang, N. D. Pfeifer, R. N. Hardwick, W. Yue, P. W. Stewart and K. L. Brouwer, *Mol. Pharm.*, 2014, **11**, 766–775.

153. D. Huh, G. A. Hamilton and D. E. Ingber, *Trends Cell Biol.*, 2011, **21**, 745–754.

154. T. M. Desrochers, E. Palma and D. L. Kaplan, *Adv. Drug Delivery Rev.*, 2014, **69–70**, 67–80.
155. M. R. Ebrahimkhani, J. A. Neiman, M. S. Raredon, D. J. Hughes and L. G. Griffith, *Adv. Drug Deliv. Rev.*, 2014, **69–70**, 132–157.
156. O. Ukairo, C. Kanchagar, A. Moore, J. Shi, J. Gaffney, S. Aoyama, K. Rose, S. Krzyzewski, J. McGeehan, M. E. Andersen, S. R. Khetani and E. L. Lecluyse, *J. Biochem. Mol. Toxicol.*, 2013, **27**, 204–212.
157. N. Li, D. Wang, Z. Sui, X. Qi, L. Ji, X. Wang and L. Yang, *Tissue Eng. Part C*, 2013, **19**, 708–719.
158. A. Dash, M. B. Simmers, T. G. Deering, D. J. Berry, R. E. Feaver, N. E. Hastings, T. L. Pruett, E. L. LeCluyse, B. R. Blackman and B. R. Wamhoff, *Am. J. Physiol. Cell Physiol.*, 2013, **304**, C1053–C1063.
159. S. Messner, I. Agarkova, W. Moritz and J. M. Kelm, *Arch. Toxicol.*, 2013, **87**, 209–213.
160. P. Gunness, D. Mueller, V. Shevchenko, E. Heinzle, M. Ingelman-Sundberg and F. Noor, *Toxicol. Sci.*, 2013, **133**, 67–78.
161. K. J. Jang, A. P. Mehr, G. A. Hamilton, L. A. McPartlin, S. Chung, K. Y. Suh and D. E. Ingber, *Integr. Biol.*, 2013, **5**, 1119–1129.
162. M. Ulvestad, M. Darnell, E. Molden, E. Ellis, A. Asberg and T. B. Andersson, *J. Pharmacol. Exp. Ther.*, 2012, **343**, 145–156.
163. R. de Kanter, A. Tuin, E. van de Kerkhof, M. Martignoni, A. L. Draaisma, M. H. de Jager, I. A. de Graaf, D. K. Meijer and G. M. Groothuis, *J. Pharmacol. Toxicol. Methods*, 2005, **51**, 65–72.
164. M. J. Sanderson, *Pulm. Pharmacol. Ther.*, 2011, **24**, 452–465.
165. M. G. Elferink, P. Olinga, A. L. Draaisma, M. T. Merema, K. N. Faber, M. J. Slooff, D. K. Meijer and G. M. Groothuis, *Am. J. Physiol. Gastrointest. Liver Physiol.*, 2004, **287**, G1008–G1016.
166. D. H. Sweet, D. S. Miller, J. B. Pritchard, Y. Fujiwara, D. R. Beier and S. K. Nigam, *J. Biol. Chem.*, 2002, **277**, 26934–26943.
167. M. B. Thompson, D. G. Davis and R. W. Morris, *J. Lipid Res.*, 1993, **34**, 553–561.
168. J. W. Lohr, G. R. Willsky and M. A. Acara, *Pharmacol. Rev.*, 1998, **50**, 107–141.
169. N. Ishiguro, A. Saito, K. Yokoyama, M. Morikawa, T. Igarashi and I. Tamai, *Drug Metab. Dispos.*, 2005, **33**, 495–499.
170. M. Bebawy and M. Chetty, *Curr. Drug Metab.*, 2009, **10**, 322–328.
171. R. Chambrey, I. Kurth, J. Peti-Peterdi, P. Houillier, J. M. Purkerson, F. Leviel, M. Hentschke, A. A. Zdebik, G. J. Schwartz, C. A. Hubner and D. Eladari, *Proc. Natl. Acad. Sci. U. S. A.*, 2013, **110**, 7928–7933.
172. D. S. Miller, *Am. J. Physiol.*, 1995, **269**, R370–R379.
173. S. Notenboom, D. S. Miller, P. Smits, F. G. Russel and R. Masereeuw, *Am. J. Physiol. Renal Physiol.*, 2004, **287**, F33–F38.
174. H. Gutmann, D. S. Miller, A. Droulle, J. Drewe, A. Fahr and G. Fricker, *Br. J. Pharmacol.*, 2000, **129**, 251–256.
175. S. A. Terlouw, C. Graeff, P. H. Smeets, G. Fricker, F. G. Russel, R. Masereeuw and D. S. Miller, *J. Pharmacol. Exp. Ther.*, 2002, **301**, 578–585.

176. A. Lungkaphin, V. Chatsudthipong, K. K. Evans, C. E. Groves, S. H. Wright and W. H. Dantzler, *Am. J. Physiol. Renal Physiol.*, 2004, **286**, F68–F76.

177. R. Kovacs, C. Raue, S. Gabriel and U. Heinemann, *J. Neurosci. Methods*, 2011, **200**, 164–172.

178. M. Friden, F. Ducrozet, B. Middleton, M. Antonsson, U. Bredberg and M. Hammarlund-Udenaes, *Drug Metab. Dispos.*, 2009, **37**, 1226–1233.

179. I. Loryan, M. Friden and M. Hammarlund-Udenaes, *Fluids Barriers CNS*, 2013, **10**, 6-8118-10-6.

180. K. S. Pang, W. F. Lee, W. F. Cherry, V. Yuen, J. Accaputo, S. Fayz, A. J. Schwab and C. A. Goresky, *J. Pharmacokinet. Biopharm.*, 1988, **16**, 595–632.

181. A. Blom, A. H. Scaf and D. K. Meijer, *Biochem. Pharmacol.*, 1982, **31**, 1553–1565.

182. H. M. Nijssen, T. Pijning, D. K. Meijer and G. M. Groothuis, *Hepatology*, 1992, **15**, 302–309.

183. A. M. Evans, Z. Hussein and M. Rowland, *J. Pharm. Sci.*, 1993, **82**, 421–428.

184. N. J. Patel, M. J. Zamek-Gliszczynski, P. Zhang, Y. H. Han, P. L. Jansen, P. J. Meier, B. Stieger and K. L. Brouwer, *Mol. Pharmacol.*, 2003, **64**, 154–159.

185. H. Xiong, H. Suzuki, Y. Sugiyama, P. J. Meier, G. M. Pollack and K. L. Brouwer, *Drug Metab. Dispos.*, 2002, **30**, 962–969.

186. C. L. Booth, K. R. Brouwer and K. L. Brouwer, *Cancer Res.*, 1998, **58**, 3641–3648.

187. Y. Y. Lau, C. Y. Wu, H. Okochi and L. Z. Benet, *J. Pharmacol. Exp. Ther.*, 2004, **308**, 1040–1045.

188. C. Weiss, H. Passow and A. Rothstein, *Am. J. Physiol.*, 1959, **196**, 1115–1118.

189. I. Bekersky, *Drug Metab. Rev.*, 1983, **14**, 931–960.

190. H. Chang, B. Choong, A. Phillips and K. Loomes, *Exp. Anim.*, 2013, **62**, 19–23.

191. R. Masereeuw, M. M. Moons and F. G. Russel, *Eur. J. Pharmacol.*, 1997, **321**, 315–323.

192. J. Wang, R. L. Nation, A. M. Evans and S. Cox, *J. Pharmacol. Toxicol. Methods*, 2004, **49**, 105–113.

193. M. A. Alam, F. I. Al-Jenoobi and A. M. Al-Mohizea, *J. Pharm. Pharmacol.*, 2012, **64**, 326–336.

194. A. Sjoberg, M. Lutz, C. Tannergren, C. Wingolf, A. Borde and A. L. Ungell, *Eur. J. Pharm. Sci.*, 2013, **48**, 166–180.

195. A. D. Mottino, T. Hoffman, L. Jennes and M. Vore, *J. Pharmacol. Exp. Ther.*, 2000, **293**, 717–723.

196. R. A. Van Aubel, A. Hartog, R. J. Bindels, C. H. Van Os and F. G. Russel, *Eur. J. Pharmacol.*, 2000, **400**, 195–198.

197. I. S. Haslam, J. A. Wright, D. A. O'Reilly, D. J. Sherlock, T. Coleman and N. L. Simmons, *Drug Metab. Dispos.*, 2011, **39**, 2321–2328.

198. T. Knutson, P. Fridblom, H. Ahlstrom, A. Magnusson, C. Tannergren and H. Lennernas, *Mol. Pharm.*, 2009, **6**, 2–10, DOI: 10.1021/mp800145r.

199. R. Sandstrom, A. Karlsson, L. Knutson and H. Lennernas, *Pharm. Res.*, 1998, **15**, 856–862.

200. T. De Bruyn, G. J. van Westen, A. P. Ijzerman, B. Stieger, P. de Witte, P. F. Augustijns and P. P. Annaert, *Mol. Pharmacol.*, 2013, **83**, 1257–1267.

201. M. Karlgren, G. Ahlin, C. A. Bergstrom, R. Svensson, J. Palm and P. Artursson, *Pharm. Res.*, 2012, **29**, 411–426.

202. C. Gui, A. Obaidat, R. Chaguturu and B. Hagenbuch, *Curr. Chem. Genomics*, 2010, **4**, 1–8.

203. H. Tang, D. R. Shen, Y. H. Han, Y. Kong, P. Balimane, A. Marino, M. Gao, S. Wu, D. Xie, M. G. Soars, J. C. O'Connell, A. D. Rodrigues, L. Zhang and M. E. Cvijic, *J. Biomol. Screen.*, 2013, **18**, 1072–1083.

204. M. G. Soars, P. Barton, L. L. Elkin, K. W. Mosure, J. L. Sproston and R. J. Riley, *Xenobiotica*, 2014, **44**, 657–665.

205. J. M. Pedersen, P. Matsson, C. A. Bergstrom, J. Hoogstraate, A. Noren, E. L. LeCluyse and P. Artursson, *Toxicol. Sci.*, 2013, **136**, 328–343.

206. M. G. Soars, P. Barton, M. Ismair, R. Jupp and R. J. Riley, *Drug Metab. Dispos.*, 2012, **40**, 1641–1648.

207. J. Rautio, J. E. Humphreys, L. O. Webster, A. Balakrishnan, J. P. Keogh, J. R. Kunta, C. J. Serabjit-Singh and J. W. Polli, *Drug Metab. Dispos.*, 2006, **34**, 786–792.

208. D. A. Volpe, *Future Med. Chem.*, 2011, **3**, 2063–2077.

209. J. Bentz, M. P. O'Connor, D. Bednarczyk, J. Coleman, C. Lee, J. Palm, Y. A. Pak, E. S. Perloff, E. Reyner, P. Balimane, M. Brannstrom, X. Chu, C. Funk, A. Guo, I. Hanna, K. Heredi-Szabo, K. Hillgren, L. Li, E. Hollnack-Pusch, M. Jamei, X. Lin, A. K. Mason, S. Neuhoff, A. Patel, L. Podila, E. Plise, G. Rajaraman, L. Salphati, E. Sands, M. E. Taub, J. S. Taur, D. Weitz, H. M. Wortelboer, C. Q. Xia, G. Xiao, J. Yabut, T. Yamagata, L. Zhang and H. Ellens, *Drug Metab. Dispos.*, 2013, **41**, 1347–1366.

210. International Transporter Consortium, K. M. Giacomini, S. M. Huang, D. J. Tweedie, L. Z. Benet, K. L. Brouwer, X. Chu, A. Dahlin, R. Evers, V. Fischer, K. M. Hillgren, K. A. Hoffmaster, T. Ishikawa, D. Keppler, R. B. Kim, C. A. Lee, M. Niemi, J. W. Polli, Y. Sugiyama, P. W. Swaan, J. A. Ware, S. H. Wright, S. W. Yee, M. J. Zamek-Gliszczynski and L. Zhang, *Nat. Rev. Drug Discovery*, 2010, **9**, 215–236.

211. C. Funk, C. Ponelle, G. Scheuermann and M. Pantze, *Mol. Pharmacol.*, 2001, **59**, 627–635.

212. L. Fossati, R. Dechaume, E. Hardillier, D. Chevillon, C. Prevost, S. Bolze and N. Maubon, *Int. J. Pharm.*, 2008, **360**, 148–155.

213. R. Elsby, L. Fox, D. Stresser, M. Layton, C. Butters, P. Sharma, V. Smith and D. Surry, *Eur. J. Pharm. Sci.*, 2011, **43**, 41–49.

214. S. Dawson, S. Stahl, N. Paul, J. Barber and J. G. Kenna, *Drug Metab. Dispos.*, 2012, **40**, 130–138.

215. M. J. Zamek-Gliszczynski, C. A. Lee, A. Poirier, J. Bentz, X. Chu, H. Ellens, T. Ishikawa, M. Jamei, J. C. Kalvass, S. Nagar, K. S. Pang, K. Korzekwa, P. W. Swaan, M. E. Taub, P. Zhao, A. Galetin and International Transporter Consortium, *Clin. Pharmacol. Ther.*, 2013, **94**, 64–79.

216. S. Agarwal, V. Arya and L. Zhang, *J. Clin. Pharmacol.*, 2013, **53**, 228–233.

217. R. Z. Cer, U. Mudunuri, R. Stephens and F. J. Lebeda, *Nucleic Acids Res.*, 2009, **37**, W441–W445.

218. *US FDA Draft guidance 2006: Drug Interaction Studies—Study Design, Data Analysis, Implications for Dosing, and Labeling Recommendations*, US FDA.

219. R. Amundsen, H. Christensen, B. Zabihyan and A. Asberg, *Drug Metab. Dispos.*, 2010, **38**, 1499–1504.

220. R. Elsby, M. Gillen, C. Butters, G. Imisson, P. Sharma, V. Smith and D. D. Surry, *Drug Metab. Dispos.*, 2011, **39**, 275–282.

221. J. Hunter, M. A. Jepson, T. Tsuruo, N. L. Simmons and B. H. Hirst, *J. Biol. Chem.*, 1993, **268**, 14991–14997.

222. J. C. Kalvass and G. M. Pollack, *Pharm. Res.*, 2007, **24**, 265–276.

223. K. Menochet, K. E. Kenworthy, J. B. Houston and A. Galetin, *J. Pharmacol. Exp. Ther.*, 2012, **341**, 2–15.

224. H. M. Jones, H. A. Barton, Y. Lai, Y. A. Bi, E. Kimoto, S. Kempshall, S. C. Tate, A. El-Kattan, J. B. Houston, A. Galetin and K. S. Fenner, *Drug Metab. Dispos.*, 2012, **40**, 1007–1017.

225. P. Nordell, P. Svanberg, J. Bird and K. Grime, *Drug Metab. Dispos.*, 2013, **41**, 836–843.

226. R. Elsby, C. Hilgendorf and K. Fenner, *Clin. Pharmacol. Ther.*, 2012, **92**, 584–598.

227. P. Sharma, C. J. Butters, V. Smith, R. Elsby and D. Surry, *Eur. J. Pharm. Sci.*, 2012.

228. M. J. Zamek-Gliszczynski, J. C. Kalvass, G. M. Pollack and K. L. Brouwer, *Drug Metab. Dispos.*, 2009, **37**, 386–390.

229. K. Korzekwa and S. Nagar, *Pharm. Res.*, 2014, **31**, 335–346.

230. Y. Cheng and W. H. Prusoff, *Biochem. Pharmacol.*, 1973, **22**, 3099–3108.

Knockout and Humanised Animal Models to Study Membrane Transporters in Drug Development

NICO SCHEER,*[a] XIAOYAN CHU,[b] LAURENT SALPHATI[c] AND MACIEJ J. ZAMEK-GLISZCZYNSKI[d]

[a] Independent Consultant, Cologne, Germany; [b] Merck & Co., Rahway, New Jersey, USA; [c] Genentech, Inc., 1 DNA Way, South San Francisco, CA 94080, USA; [d] GlaxoSmithKline, 5 Moore Drive, Research Triangle Park, NC 27709, USA
*Email: nicoscheer3@gmail.com

8.1 Introduction

Membrane transporters are increasingly recognised as major determinants of the pharmacokinetics, tissue distribution, efficacy, and safety of drugs and metabolites.[1–3] Of the more than 400 human membrane transporters from the two major families, solute carrier (SLC) uptake transporters and ATP binding cassette (ABC) efflux pumps, only a small number are currently considered important to drug pharmacokinetics and drug–drug interactions (DDIs; Figure 8.1). Based on their clinical relevance, the International Transporter Consortium recommended the following subset of SLCs and ABC transporters from Figure 8.1 for evaluation of substrate and inhibitor interactions during drug development (see also Chapter 1): organic

RSC Drug Discovery Series No. 54
Drug Transporters: Volume 1: Role and Importance in ADME and Drug Development
Edited by Glynis Nicholls and Kuresh Youdim
© The Royal Society of Chemistry 2016
Published by the Royal Society of Chemistry, www.rsc.org

anion-transporting polypeptide (OATP) 1B1 (*SLCO1B1*), OATP1B3 (*SLCO1B3*), organic anion transporter (OAT) 1 (*SLC22A6*) and OAT3 (*SLC22A8*), organic cation transporter (OCT) 2 (*SLC22A2*), multidrug and toxin extrusion (MATE) 1 (*SLC47A1*) and MATE2-K (*SLC47A2*), P-glycoprotein (P-gp)/multidrug resistance protein (MDR) 1 (*ABCB1*), breast cancer resistance protein (BCRP; *ABCG2*); as well as bile salt export pump (BSEP; *ABCB11*) and multidrug resistance protein (MRP) 2 (*ABCC2*) inhibition for mechanistic understanding of cholestasis and conjugated hyperbilirubinaemia, respectively.[1,3]

Various tools have been developed to understand the impact of these transporters on the disposition of compounds in drug development. *In vitro* models, such as expression systems (*e.g.* immortalised cell lines, oocytes, and vesicles) and whole cells (*e.g.* hepatocytes and derived cell lines) are commonly used to assess the potential for interactions with uptake and efflux transporters and determine basic kinetic parameters (*e.g.* Michaelis constant (K_m), half maximal inhibitory concentration (IC_{50})). These *in vitro* assays have been successful in identifying many substrates and inhibitors of transporters. Physiologically-based pharmacokinetic (PBPK) modelling approaches have also been increasingly used to predict complex DDIs, and in combination with static *in vitro* to *in vivo* extrapolation (IVIVE) techniques, have facilitated predictions of DDIs stemming from transporter inhibition in humans.[4] However, for transporter substrates, direct extrapolation of *in vitro* results to complex transporter mediated drug disposition in humans is limited by, for example, differential transporter function and coordination with other processes in various organs,[5] such that static IVIVE and dynamic PBPK approaches may not accurately predicting the impact of transporters on clinical pharmacokinetics from *in vitro* data alone. Drug-independent empirical scaling factors have been proposed as a means to bridge this gap; however, in practice these correction factors are drug-dependent and require determination on a case-by-case basis based on detailed knowledge of clinical pharmacokinetics (Figure 8.2[6,7]). This greatly limits the ability to predict the effects of transporters on pharmacokinetics and the victim DDI potential from *in vitro* data alone, highlighting the need for *in vivo* transporter models to assist with translation to the clinic.

For transporter substrates, preclinical *in vivo* studies can provide useful complementary information towards understanding transporter-limited or transporter-mediated drug absorption, distribution, and excretion by contextualising the relevance of *in vitro* substrate findings in terms of *in vivo* pharmacokinetics. As further discussed in Sections 8.3 and 8.4, additional supporting information may be required for extrapolation of results from animal studies to humans depending on the specific transporter-related processes under investigation. Transporter gene knockout models (Figure 8.3A and Table 8.1) are particularly useful in determining the *in vivo* impact of transporters on the fraction absorbed orally, the fraction excreted through various elimination pathways, as well as on the distribution to tissues of interest, such as the liver or brain.[5] Furthermore, knockout models

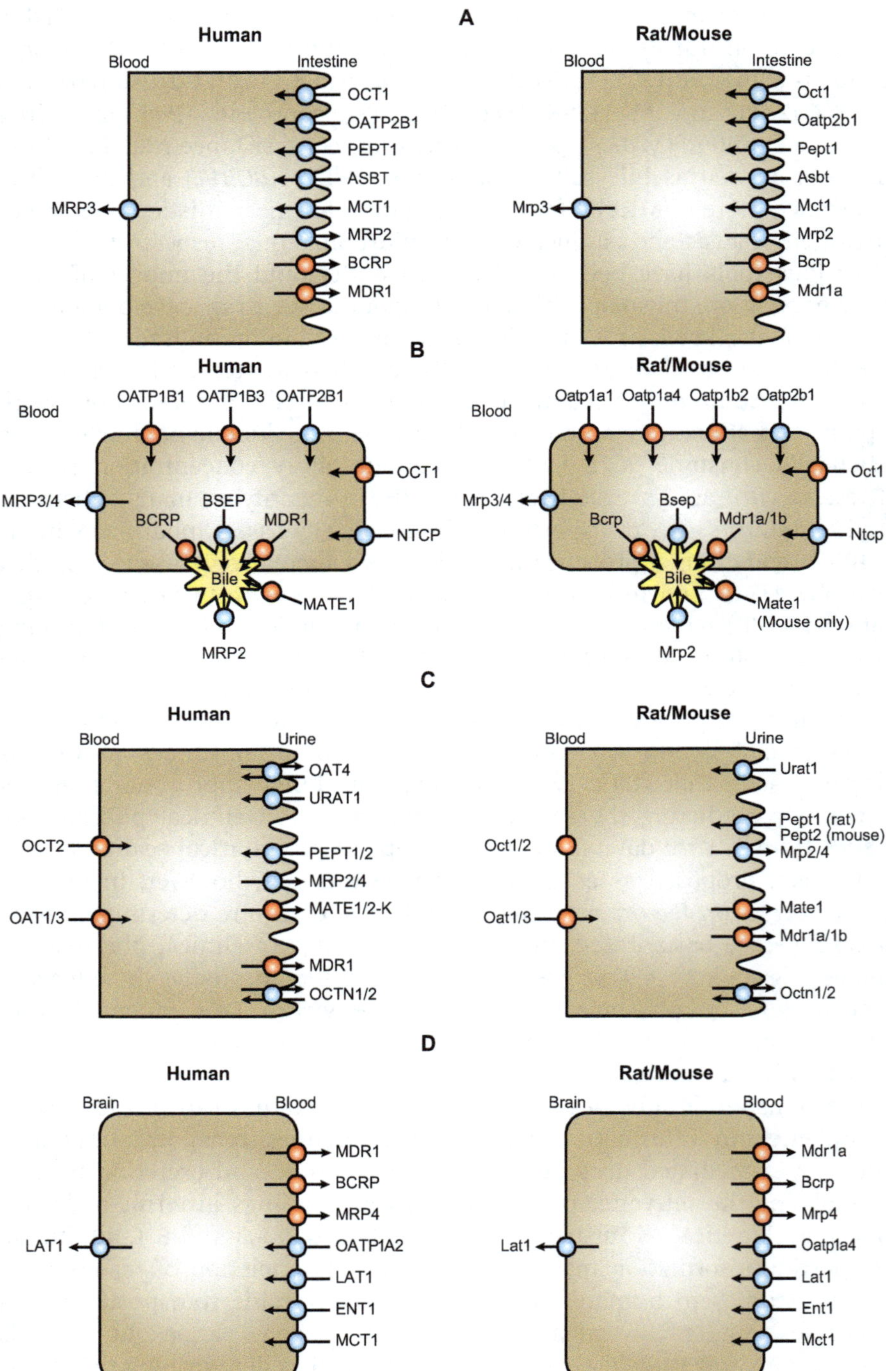
A
Human
Blood
Intestine
OCT1
OATP2B1
PEPT1
ASBT
MCT1
MRP2
BCRP
MDR1
MRP3
Rat/Mouse
Blood
Intestine
Oct1
Oatp2b1
Pept1
Asbt
Mct1
Mrp2
Bcrp
Mdr1a
Mrp3
B
Human
Blood
OATP1B1
OATP1B3
OATP2B1
OCT1
NTCP
MRP3/4
BCRP
BSEP
MDR1
Bile
MATE1
MRP2
Rat/Mouse
Blood
Oatp1a1
Oatp1a4
Oatp1b2
Oatp2b1
Oct1
Ntcp
Mrp3/4
Bcrp
Bsep
Mdr1a/1b
Bile
Mate1
(Mouse only)
Mrp2
C
Human
Blood
Urine
OAT4
URAT1
PEPT1/2
MRP2/4
MATE1/2-K
MDR1
OCTN1/2
OCT2
OAT1/3
Rat/Mouse
Blood
Urine
Urat1
Pept1 (rat)
Pept2 (mouse)
Mrp2/4
Mate1
Mdr1a/1b
Octn1/2
Oct1/2
Oat1/3
D
Human
Brain
Blood
MDR1
BCRP
MRP4
OATP1A2
LAT1
ENT1
MCT1
LAT1
Rat/Mouse
Brain
Blood
Mdr1a
Bcrp
Mrp4
Oatp1a4
Lat1
Ent1
Mct1
Lat1

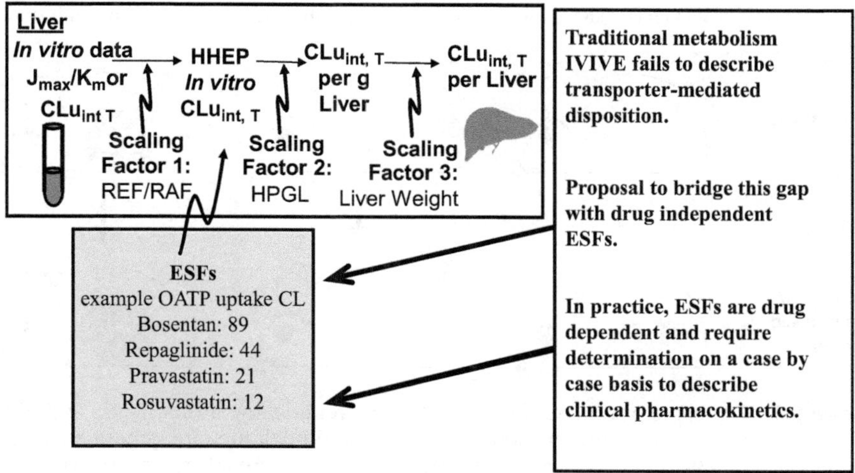

Figure 8.2   IVIVE limitations for transporter-mediated or transporter-limited absorption, distribution, and excretion. Static IVIVE and dynamic PBPK approaches are unable to accurately predict the clinical impact of transporters on pharmacokinetics from *in vitro* data alone. Drug independent empirical scaling factors (ESFs) were proposed as a means to bridge this gap; however, in practice these correction factors turned out to be drug-dependent and require determination on a case by case basis from clinical pharmacokinetics. This greatly limits the potential to predict pharmacokinetic profiles and victim DDIs from *in vitro* data alone, and highlights the need for *in vivo* transporter models to assist with the clinical translation. CL: clearance; CLuint, T: transport unbound intrinsic clearance; HHEP: human hepatocytes; HPGL: hepatocytes per gram of liver; J_{max}: Maximum number of compounds translocated across an area per time by a given transporter; RAF: relative activity factors; REF: relative transporter expression factors. Adapted from Jones *et al.*[6] and Varma *et al.*[7]

have also proven useful in improving mechanistic understanding of the disposition of drug metabolites, which are usually more polar and less permeable and thus more reliant on transporters for excretion from their sites of formation and any subsequent disposition.[2] Initially, targeted transporter knockouts were only available in mice, but recent progress with

---

Figure 8.1   Membrane transporters with demonstrated functional relevance in humans and rodents. Diagrams show the localisation of relevant transporters in major drug disposition organs: (A) intestine, (B) liver, (C) kidney, and (D) brain. Most notable species differences include: (1) no direct orthologues between human and rodent hepatic OATP isoforms, which despite these fundamental differences are functionally comparable at a collective level;[11,129] (2) differences in renal OCT/MATE secretion (human: OCT2, MATE1/2-K; rodent: Oct1/2, Mate1), which ultimately do not result in differences at a functional level in active tubular secretion;[5] and (3) single *MDR1* gene encoding P-gp in humans *versus* two in rodents (*Mdr1a* and *Mdr1b*), which nonetheless does not result in functional differences for P-gp.[73,80]

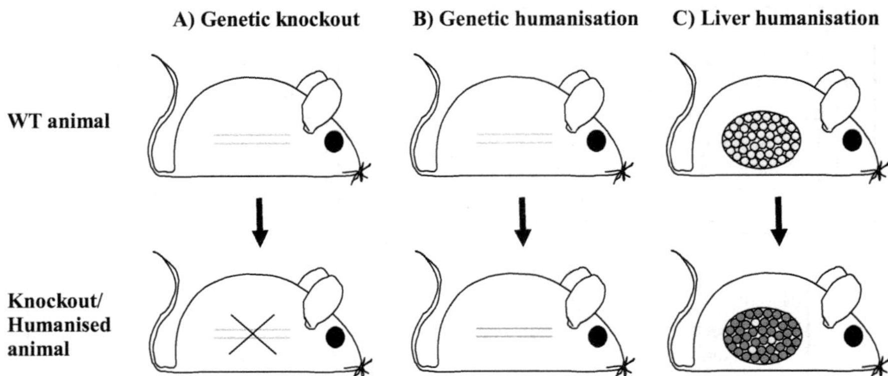

Figure 8.3   Concepts for generating knockout, genetically humanised, and liver humanised animal models. (A) Genetic knockouts are generated by the deletion of a particular animal gene (*e.g.* a transporter gene). The light grey lines represent the two gene alleles, which are mutated in the knocked out model. (B) The animal gene (light grey lines), is replaced by the corresponding human gene (dark grey lines) in genetically humanised animal models. Genetic humanisation can be achieved by replacing the two orthologous mouse genes with their human counterparts within the corresponding mouse locus, as shown here, or by combining a knockout of the animal gene with a transgene expressing the human protein (*e.g.* a transporter) inserted at a different site in the animal genome. The latter can be obtained by breeding a mouse model carrying the human transgene with a knockout model for the corresponding mouse gene. (C) Animal hepatocytes (light grey circles) are replaced with human hepatocytes (dark grey circles) in chimeric liver humanised animal models. The remaining light grey circles in the liver humanised model indicate the presence of residual mouse hepatocytes, because only partial (up to 95%) replacement with human hepatocytes has been achieved in the models described to date.[14] WT: wild-type.

gene editing technologies has now enabled gene knockouts for selected transporters to also be generated in rats.[8,9]

An emerging type of *in vivo* model for studying membrane transporters is the so called genetically humanised transporter animal, carrying a genetic replacement of a specific animal transporter gene with its human counterpart (Figure 8.3B and Table 8.2). This field of research is relatively new and only a few models have so far been described in mice.[10] Such models can be used, for example, to assess the contribution of the corresponding human transporter to drug disposition (by comparison with the relevant rodent knockout control).[11] These genetically humanised transporter models can help to overcome limitations stemming from species differences in transporter expression and substrate specificity. Hepatic OATPs, which have no direct orthologues between rodents and humans and yet collectively perform very similar functions between species, are a prime example.[12] An alternative approach to generate humanised animal models is by transplantation of human cells, rather than genetic exchange of particular genes. Such models

Table 8.1   Selected knockout animal models of key transporters described in the literature.[a]

Transporter gene(s) [synonyms]	Species	Ref.
Abcb1a [*Mdr1a*, *P-gp*]	Mouse[b]	33
	Rat[c]	8, 32
Abcb1a/b [*Mdr1a/b*]	Mouse[b]	34
Abcb1a/b/Abcc2/Abcg2 [*Mdr1a/b/Mrp2/Bcrp*]	Mouse	53
Abcb1a/b/Abcg2 [*Mdr1a/b/Bcrp*]	Mouse[b]	52
Abcb11 [*Bsep*]	Mouse	35
Abcc1 [*Mrp1*]	Mouse[b]	39
Abcc2 [*Mrp2*]	Mouse[b]	22, 38
	Rat[c]	9
Abcc2/3/Abcg2 [*Mrp2/3/Bcrp*]	Mouse	54
Abcc3 [*Mrp3*]	Mouse	36
Abcc4 [*Mrp4*]	Mouse	37
Abcg2 [*Bcrp*]	Mouse[b]	40
	Rat[c]	9
Slc1b2 [*Oatp1b2*]	Mouse	41–43
Slc1a1/4/5/6/1b2 [*Oatp1a1/4/5/6/1b2*]	Mouse[b]	44
Slc15a1 [*Pept1*]	Mouse	45
Slc15a2 [*Pept2*]	Mouse	46
Slc22a1 [*Oct1*]	Mouse	47
Slc22a1/2 [*Oct1/2*]	Mouse[b]	48
Slc22a2 [*Oct2*]	Mouse	48
Slc22a6 [*Oat1*]	Mouse	49
Slc22a8 [*Oat3*]	Mouse	50
Slc47a1 [*Mate1*]	Mouse	51

[a]Additional unpublished knockout models that are commercially offered: *Oat1/2* mice (Taconic) and *Bsep*, *Oat1*, *Oat3*, *Oct1*, and *Oct2* knockout rats (SAGE). Models not supplied by Taconic or SAGE might be obtained from various academic institutions.
[b]Commercially available from Taconic Biosciences (Hudson, NY, USA).
[c]Commercially available from SAGE Labs (St Louis, MO, USA).

Table 8.2   Genetically humanised transporter mice.[a]

Humanised transporter	Type of genetic modification	Promoter driving human gene expression	Mouse gene(s) deleted	Ref.
MRP2	TR[b]	Mouse Mrp2	*Abcc2*	59
OATP1B1	RT[c]	Human ApoE	None	55
	KO[d] + RT[c]	Human ApoE	*Slc1a1, 1a4, 1a5, 1a6* and *1b2*	57
OATP1B3	KO[d] + RT[c]	Human ApoE	*Slc1a1, 1a4, 1a5, 1a6* and *1b2*	57
OATP1A2	KO[d] + RT[c]	Human ApoE	*Slc1a1, 1a4, 1a5, 1a6* and *1b2*	57
OATP1B1–1B3	KO[d] + RT[c]	Human ApoE	*Slc1a1, 1a4, 1a5, 1a6* and *1b2*	58
PEPT1	KO[d] + RT[c]	Human PEPT1	*Pept1*	60

[a]All models, except OATP1A2 humanised mice are commercially available from Taconic Biosciences (Hudson, NY, USA).
[b]TR: targeted replacement.
[c]RT: random transgenesis.
[d]KO: knockout.

Table 8.3   Liver humanised mice.[a]

Model	Genetic cause of mouse hepatocyte ablation	Control of mouse hepatocyte ablation	Immune deficient background	Ref.
uPA-SCID	Uroplasminogen activator (uPA)	None	SCID (adaptive)	65, 68
FRG	Fumarylacetoacetate hydrolase deficiency	$\pm$ NTBC $\pm$ Low tyrosine diet	Il2rg$^{-/-}$ (innate) Rag2$^{-/-}$ (adaptive)	64
TK-NOG	Herpes simplex virus thymidine kinase	$\pm$ Ganciclovir	Il2rg$^{-/-}$ (innate) SCID (adaptive)	69
AFC8	FK508–caspase 8 fusion	$\pm$ AP20187	Il2rg$^{-/-}$ (innate) Rag2$^{-/-}$ (adaptive)	67

[a]uPA-SCID mice are commercially available from PhoenixBio (Hiroshima, Japan) and FRG mice are commercially available from Yecuris Incorporation (Portland, OR, USA).

are of potential interest for studying membrane transporters, if tissues with relevant transporter activities are humanised. This has been achieved in chimeric liver humanised mouse models with extensive replacement of murine with human hepatocytes (Figure 8.3C and Table 8.3), which already have shown utility in studies related to drug metabolising enzymes.[13–15]

This chapter, in addition to briefly summarising the methods and technologies that are used to generate transporter knockout and humanised rodent models, highlights the corresponding *in vivo* models described to date and discusses their utility, rational study design, data interpretation, key case studies, and future directions.

8.2   Methods for Generating Transporter Knockout and Humanised Animal Models for Use in Drug Development

A description of all technical details and intricacies of generating knockout and humanised animal models is beyond the scope of this chapter, and the interested reader is referred to more comprehensive reviews or specific research articles.[14,16,17] The aim here is to give a general overview of such technologies. Furthermore, a comprehensive, though not exhaustive, overview of the currently available *in vivo* transporter models generated over the last two decades by using these approaches is provided. The use of these models in drug development will be discussed further in Section 8.3.

8.2.1   Transporter Knockout Animals

While mutant animal models for particular transporters can arise spontaneously, for example, in the case of Mrp2-deficient rats,[18,19] relying on such unforeseeable events does not provide a sustainable strategy. Furthermore, the transport deficient (TR$^-$) rat with a naturally occurring mutation in Mrp2 has been used in research for some time and is heavily inbred. This results in health problems and exaggerated compensatory differences, such as marked upregulation of hepatic Mrp3 and down-regulation of hepatic Bcrp,[20,21] whereas these changes are far more modest or non-existent in *Mrp2*$^{-/-}$ knockout mice or rats generated by a targeted deletion of this gene.[22,23] Accordingly, researchers have made use of such gene targeting approaches to inactivate selected transporters (Figure 8.3A). Initially, this was only possible in the mouse, due to the isolation and establishment of mouse embryonic stem (ES) cell lines in the early 1980s,[24,25] and the subsequent discovery and application of gene targeting to ES cells.[26] The pluripotent mouse ES cells can be maintained in culture and retain the capacity to contribute to all cell lineages of the mouse. Gene targeting utilises the natural cellular repair mechanism of homologous recombination, which exchanges nucleotide sequences between two related DNA molecules. Due to the relatively low frequency of altering a gene of interest by homologous recombination, this approach initially relied on appropriate screening strategies to identify these rare events in millions of mouse ES cells and on selecting correctly modified ES cells for subsequent generation of a genetically engineered mouse model. Dozens of knockout mouse models for most of the major transporters involved in drug disposition have been generated over the last two decades (Table 8.1).

More recently, so-called gene editing technologies, which induce targeted, site-specific double-strand DNA breaks in the genes of interest that can be repaired by error-prone non-homologous end joining (NHEJ), have been employed by different researchers for the generation of knockout animals. For example, knockout rats and mice have been generated by pronuclear or intra-cytoplasmic injection of engineered zinc-finger nucleases (ZFNs), transcription activator-like effector nuclease (TALEN) or clustered regularly interspaced short palindromic repeats (CRISPR)/CRISPR associated (Cas) into rat or mouse embryos.[27–31] Since the efficiency of induced mutations by these approaches is much higher than by the aforementioned gene targeting by homologous recombination, it was possible to generate these knockout rats and mice *in vivo*, *i.e.* from injected embryos, at an acceptable frequency, without the need to screen and select correctly modified ES cells *in vitro*. Many transporter knockout rats have also become available recently through the use of gene editing technologies[8,9,32] (see below), which in principle should also enable the generation of gene knockouts in other species.

Knockout mouse models of most membrane transporters with known relevance to clinical pharmacokinetics and/or drug interactions (Figure 8.1) have been described in the literature (Table 8.1). Since the generation of *Mdr1a*$^{-/-}$ mice in 1994,[33] knockout mouse models have also been published for *Mdr1a/1b*,[34] *Bsep*,[35] *Mrp1-4*,[22,36–39] *Bcrp*,[40] *Oatp1b2*,[41–43] *Oatp1a/1b* gene cluster,[44] *Pept1* and *2*,[45,46] *Oct1* and *2*,[47,48] *Oct1/2*,[48] *Oat1* and *3*,[49,50] and *Mate1*.[51] Furthermore, complex, multiple transporter knockout models have been generated by breeding the corresponding single knockout mice, which can be used to study the combined and overlapping roles of transporters in drug disposition, as exemplified by the *Mdr1a/b/Bcrp*, *Mdr1a/b/Mrp2/Bcrp*, or *Mrp2/3/Bcrp* knockout mice.[52–54] More recently *Mdr1a*, *Mrp2*, and *Bcrp* knockout rats have also been described (Table 8.1).[8,9,32] Importantly, most of these mice and rats, as well as unpublished models, such as *Bsep, Oat1, Oat3, Oct1*, and *Oct2* knockout rats, are commercially available from companies such as Taconic Biosciences and SAGE Labs (Table 8.1). Commercial availability is advantageous not only in terms of broad access to the models, but also because it standardises knockout studies across groups by circumventing issues related to, for example, differences in housing and dietary conditions, as well as genetic drift that may occur within small in-bred colonies over time.[5]

8.2.2   Genetically Humanised Transporter Models

Genetic humanisation describes a method by which one or more animal genes are replaced by the corresponding human gene(s) (Figure 8.3B). As of 2014, genetically humanised transporter models have only been described in mice. A large variety of methods to generate genetically humanised mouse models have been used and are described elsewhere.[17] In summary, genetically humanised mouse models can be generated by: (1) random transgenesis, such that the DNA construct expressing the human gene is inserted into the mouse genome at an undefined position; (2) integrating it at a predefined position by a targeted approach using homologous recombination (see above); or (3) introducing a freely segregating mouse or human artificial chromosome (MAC or HAC) expressing the human gene of interest, which is propagated during cell division like a normal mouse chromosome. As the corresponding orthologous mouse gene is not deleted by random transgenesis or introduction of a freely segregating extra chromosome, and its expression is usually undesirable, it needs to be deleted separately in these cases. In contrast, inserting the human transgene into the mouse genome by targeted integration offers the possibility of directly replacing the mouse gene with its human counterpart. Furthermore, different promoters can be employed to express the human transgene, depending on its desired pattern of expression. For example, a heterologous promoter can be used to express the human gene in particular organs, such as the liver, intestine, or kidney. Alternatively, the corresponding mouse promoter or the cognate human promoter can be selected, if physiological regulation of human gene expression is preferable. Finally, the human gene can be encoded by a cDNA, a genomic construct, or a mixture of both. Due to their small sizes, cDNAs can facilitate cloning of the expression vectors, while larger genomic constructs allow for the expression of various splice variants of the human gene and often confer more robust expression levels. A more detailed discussion of the advantages and disadvantages of these different approaches can be found in the literature.[17]

Compared with transporter gene knockouts, the field of transporter humanised animal models is still very new and only a few such models have so far been described in mice (Table 8.2).[10] The first transporter humanised mouse model was generated in 2009 with liver specific expression of human

OATP1B1.[55] The human transporter in this model was expressed under control of the liver specific ApoE promoter and the construct was inserted by random transgenesis into the mouse genome. While no mouse genes were deleted in this initial model, it was subsequently further crossed to an *Oatp1a/1b*$^{-/-}$ gene cluster knockout mouse line in order to express the human OATP1B1 transporter in the absence of the major murine hepatic Oatps.[56,57] In a similar manner, *OATP1B3* and *1A2* humanised/*Oatp1a/1b*$^{-/-}$ mice were generated by the same group.[57] Double humanised *OATP1B1/1B3* mice on the *Oatp1a/1b*$^{-/-}$ knockout background were subsequently generated by cross-breeding the single humanised mice.[58] As the expression of OATP1B1 and 1B3 is essentially restricted to the human liver, the use of the ApoE promoter to express the human transporters in these mice mimics the natural situation relatively well. On the other hand, it should be noted that the physiological regulation of OATP1B1 and 1B3 expression cannot be studied using humanised mice. Therefore, it is not possible to predict DDIs mediated by the induction of OATP1B1 or 1B3 expression by a compound when given in combination with a substrate of these transporters. It should be noted, however, that hepatic OATP induction has not been identified as a major cause for changes in clinical pharmacokinetics. A different approach was chosen in a *MRP2* humanised mouse in which the coding region of the mouse gene was replaced by the corresponding human sequence, such that the human transporter is expressed under control of the murine *Mrp2* promoter.[59] In this model, the expression of the human transporter is physiologically regulated and reflects that of murine Mrp2 in wild-type animals. Furthermore, a *PEPT1* humanised model was recently generated by inserting a bacterial artificial chromosome carrying the human transporter gene and its regulatory elements into the genome of a *Pept1*$^{-/-}$ knockout mouse.[60] It can be anticipated that additional humanised mouse models for other clinically-relevant transporters, such as P-gp, BCRP, OAT1/3 or OCT2, will become available in the near future.

In addition to the use of gene editing technologies to generate knockout animals as described in Section 8.2.1, these methods allow for the introduction of sequence-specific modifications by increasing the efficiency of homologous recombination in rats and mice.[28,31,61,62] While no genetically humanised animal models have been generated by this approach to date, and there might be limitations with regard to the DNA size that can be efficiently integrated, embryo injection of ZFNs, TALENs or CRISPR/Cas might be used to generate transporter humanised animal models in species other than the mouse, *e.g.* the rat, which would offer benefits in terms of sample sizes for bioanalysis and the preference for rats compared with mice for certain applications.[63]

A general limitation of genetically humanised mouse models is the restriction of humanisation to one or a few mouse genes. Accordingly, these models should not be used generically to predict human outcomes, but studies should be designed carefully to assess the role of a particular transporter in drug development (see also Sections 8.3.6 and 8.4).

8.2.3   Liver Humanised Animal Models

In order to provide *in vivo* models humanised for the major organ involved in drug metabolism and disposition, chimeric mice with humanised livers have been generated by different groups[64–67] through transplantation of human hepatocytes into mice (Figure 8.3C). These mice have to carry two types of genetic alterations in order to efficiently repopulate the mouse liver with human hepatocytes: (1) mutations in genes coding for certain proteins of the immune system, so that the mice are immunocompromised and do not reject the human cells; and (2) a genetic modification, either a knockout or a transgene, that is toxic to the mouse hepatocytes, and therefore, mediates their ablation. Different genetic alterations have been employed by different groups resulting in up to 95% replacement of mouse with human hepatocytes.[13–15] So far, extensive repopulation with human hepatocytes has only been achieved in mice, but it can be anticipated that this will also become possible in other species, such as rats, in the future.

Liver humanised mice theoretically provide an alternative approach for studying hepatic drug transport in the human liver *in vivo*, although their utility in this regard requires further study. Several liver humanised mouse models have been generated by different groups, the major differences between these models being the immunocompromised background and the method used for ablation of the mouse hepatocytes (Table 8.3).[13–15] The first such model was derived by constitutively expressing a urokinase-type plasminogen activator (uPA) in the liver of a transgenic mouse line, causing injury to mouse hepatocytes.[65,66] When expressed on a severe combined immunodeficiency (SCID) background, the uPA-SCID model could be efficiently repopulated with human hepatocytes.[68] A triple knockout mouse for fumarylacetoacetate hydrolase (Fah), recombination activating gene 2 (Rag2), and common γ-chain of the interleukin 2 receptor (Il2rg), called FRG (Fah/Rag2/γ-chain of Il2rg), was described in 2007.[64] Liver injury in this model is caused by a deficiency in the Fah gene coding for an enzyme in the tyrosine catabolic pathway, resulting in the intracellular hepatic accumulation of the toxic fumarylacetoacetate metabolite. In contrast to uPA-SCID mice, the onset and severity of hepatocellular injury in FRG mice is controllable through the administration and withdrawal of the small molecule 2-(2-nitro-4-trifluoro-methylbenzoyl)-1,3-cyclohexanedione (NTBC), which blocks an upstream enzyme in tyrosine catabolism and thereby prevents the accumulation of the toxic metabolite. Accordingly, liver injury is circumvented as long as the FRG mice are maintained on NTBC, but it is induced when the compound is withdrawn. After combination of the Fah mutation with deletions of alleles resulting in immune deficiencies, namely in the common γ-chain of Il2rg and rag2, extensive liver humanisation has been achieved.[64] Two additional liver humanised mouse models have been described more recently, where the mouse hepatocellular injury was controlled by small molecules. In the TK-NOG model, a herpes simplex virus type 1 thymidine kinase (TK) transgene was constitutively expressed within the liver of a highly immunodeficient mouse strain (NOG). While the TK transgene itself is not harmful, it can convert the antiviral synthetic 2′-deoxyguanosine analogue ganciclovir into a toxic product. Accordingly, the mouse liver cells can be efficiently ablated by a brief exposure to ganciclovir in TK-NOG mice, allowing for subsequent repopulation with human hepatocytes.[69] Finally, in so called AFC8 mice, a fusion protein of the FK506 binding protein and caspase 8, under the control of the liver specific albumin promoter, was expressed on an Il2rg/rag2-deficient background. The FK506 binding protein dimerises in the presence of AP20187, leading to activation of caspase 8 activity and resulting in hepatocyte apoptosis in this model. These AFC8 mice were also successfully repopulated with human hepatocytes.[67]

While tissue humanisation in mice is currently restricted to the liver and has not been reported for other absorption, distribution, metabolism, and excretion (ADME) relevant organs, this might also become possible for other organs such as the intestine or kidney in the future. Additionally, the first attempts at generating chimeric liver humanised rats have recently been made.[70–72] While further optimisation will be required, it can be expected that robust and efficient repopulation of rat liver with human hepatocytes will be achieved in the near future, which would be beneficial for the same reasons elucidated in Section 8.2.2.

Chimeric liver humanised mice, similar to genetically humanised mouse models, have their limitations as well. On a single cell level, they outcompete the genetically humanised mouse models, because in a given human hepatocyte all genes and proteins are of human origin. However, all organs other than the liver are mouse, limiting the use of the chimeric mice to studying hepatic transporter processes. Furthermore, the incomplete repopulation of the mouse liver with up to 95% human hepatocytes can complicate the interpretation of data when mouse and human hepatocytes contribute with varying degrees to the metabolism and disposition of a compound.

8.3 Knockout and Humanised Animal Models in the Study of Transporter-mediated Drug Disposition

Results obtained from *in vivo* studies in transporter knockout animals, when appropriately contextualised with *in vitro* and clinical pharmacokinetic data, can provide useful information towards understanding transporter-limited absorption and distribution, as well as transporter-mediated drug absorption, distribution, and excretion.[1] As discussed below, they can assist in determining the impact of transporters on the fraction absorbed after oral administration, the fraction excreted through specific routes, or the distribution to tissues of interest, such as the brain or liver.[11,73,74] Genetically humanised transporter models are promising tools that could improve preclinical to clinical translation in cases where species differences are known to exist (*e.g.* no direct rodent–human hepatic OATP orthologues).

8.3.1 Use of Transporter Knockout Animals to Study Efflux Transporter-limited Absorption

Transporter knockout animals have proven useful in assessing efflux transporter-limited drug absorption.[73,75] An important question in this regard is whether P-gp and/or BCRP efflux observed *in vitro* indicates an *in vivo* impediment to intestinal absorption following oral drug administration, and if so, the extent of *in vivo* attenuation. Especially for moderate to high permeability drugs, it is difficult to determine whether active efflux detected *in vitro* in systems over-expressing transporters will in fact limit *in vivo* drug absorption (Figure 8.4A). Therefore, data obtained from *P-gp*[−/−] and/or *Bcrp*[−/−] knockout animals can elucidate the *in vivo* oral absorption relevance of *in vitro* results and help to predict the potential for these transporters to impact drug absorption in humans. This information ultimately helps to determine whether clinical drug interaction studies may be warranted.[1] In fact, a clinical DDI study related to intestinal efflux might not be required for a P-gp and/or BCRP substrate drug if nearly complete absorption with minimal *in vivo* impact of efflux can be demonstrated in a preclinical species with sufficient data supporting the clinical relevance (*e.g.* fraction absorbed, dose, formulation) of this observation.[5] Other uptake and efflux transporters are expressed in the intestinal epithelia (Figure 8.1); however, their impact on oral drug absorption is less evident.

Numerous case studies highlight the utility of *P-gp*[−/−] and/or *Bcrp*[−/−] knockout mice in determining whether oral absorption of substrate drugs is limited by active efflux *in vivo*. For example, the oral bioavailability of paclitaxel and docetaxel is increased up to nine-fold by genetic knockout or chemical inhibition of intestinal P-gp in mice, and clinically, co-administration of paclitaxel or docetaxel with the oral P-gp inhibitor cyclosporine A increased oral bioavailability by up to 12-fold.[73] For the BCRP substrate sulfasalazine, oral bioavailability was 9–21-fold increased in *Bcrp*[−/−] knockout mice and rats, as well as 5–13-fold increased by chemical inhibition in mice.[9,75,76] In humans homozygous for the low activity BCRP variant (*ABCG2* 421C>A) and following co-administration with the BCRP inhibitor curcumin, sulfasalazine oral exposure was increased 3.2–3.5-fold.[77,78] These studies highlight the utility of *P-gp*[−/−] and/or *Bcrp*[−/−] knockout models in predicting the extent to which these efflux transporters attenuate drug absorption in humans, as well as the worst case victim DDI potential caused by efflux-limited oral drug exposure. While chemical inhibition of intestinal P-gp and/or BCRP in wild-type animals provides an alternative to using knockout models, this approach is generally limited by the lack of specificity of the currently available inhibitors.

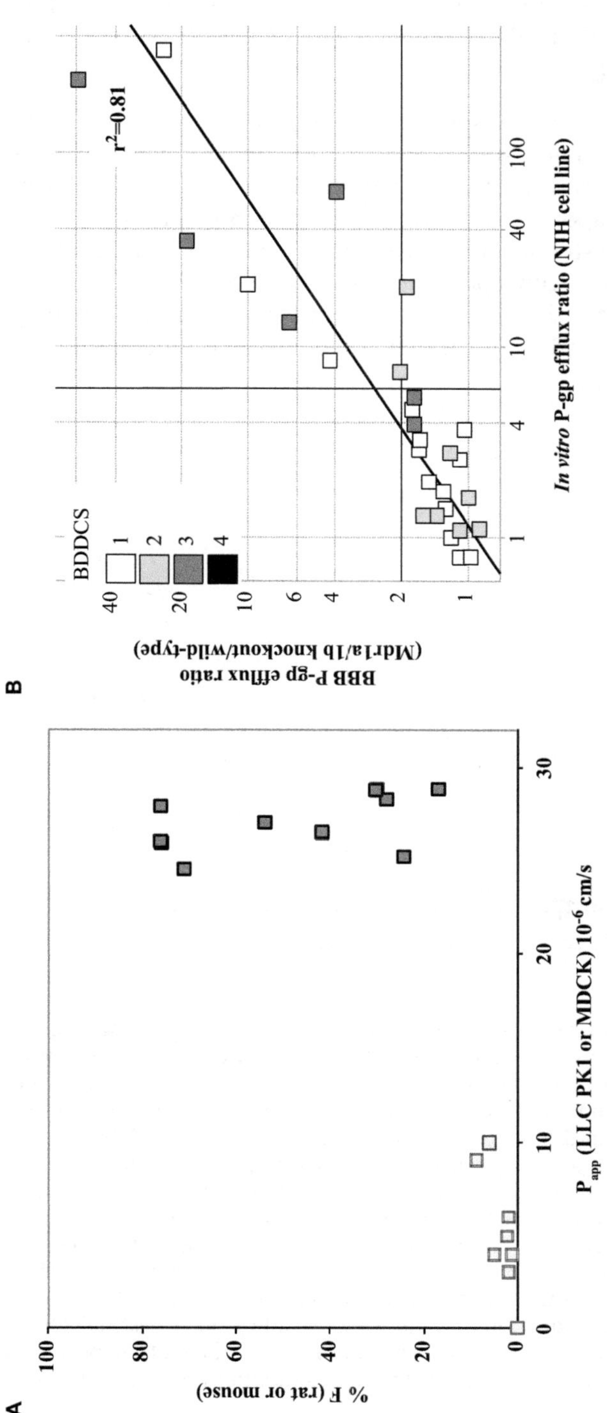

Figure 8.4 Differential influence of permeability on intestinal absorption and brain distribution of efflux substrates. (A) Oral bioavailability (F) of P-gp and/or BCRP substrates. Note that high permeability can overcome P-gp and/or BCRP efflux in the intestine due to a uniquely high luminal to blood concentration gradient following oral drug administration. (B) In contrast, attenuation of brain distribution by P-gp is present regardless of permeability [denoted by Biopharmaceutical Drug Disposition and Classification System (BDDCS) division of compounds into four classes based on their permeability and solubility class]. P_{app}: apparent permeability; NIH: National Institutes of Health. Reprinted with permission from ref. 81.

8.3.2 Fraction Transported Determination using Transporter Knockout Animals: Insight into Transporter-mediated DDI Potential

Drug exposure changes in transporter knockout animals enable determination of the *in vivo* fraction transported (f_t; Figure 8.5), which is a fundamentally important pharmacokinetic parameter analogous to the fraction metabolised for metabolic clearance.[74,79] This parameter can be used to describe changes in systemic and tissue exposure, as well as drug recovery in excreta. This theoretical concept, which is described in more detail within the literature, indicates that a fold change in exposure is governed by the relationship $1/(1-f_t)$ and is illustrated in detailed case studies in the following three sections.

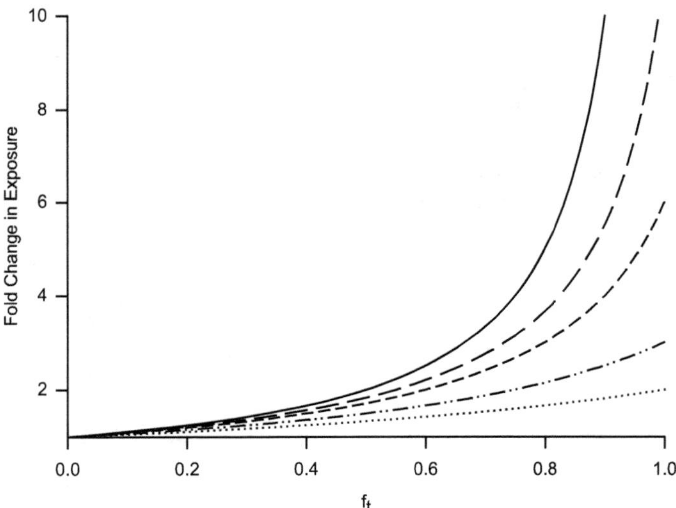

Figure 8.5 Relationship between change in exposure and fraction transported. The different lines show the relationship between changes in exposure and the fraction transported (f_t) by a particular transporter in a genetic knockout animal ($I/K_i = \infty$) or at different inhibitor concentrations. $I/K_i = 1$ (dotted line), 2 (dot-dash line), 5 (short-dashed line), 10 (long-dashed line), and ∞ (equivalent to genetic ablation; solid line). Exposure refers to systemic or tissue exposure, as well recovery in any particular excreta relative to untreated wild-type controls. Note that the curve becomes steeper as the $I:K_i$ ratio increases. As such, the magnitude of change in a knockout animal ($I/K_i = \infty$) always exceeds partial chemical inhibition, and the difference can be particularly large when comparing with clinical transporter DDIs where, aside from intestinal efflux and hepatic OATP inhibitors, concentrations of marketed transporter inhibitors generally do not achieve unbound systemic concentrations far above the K_i.[1,80]

8.3.3 Use of Transporter Knockout Animals to Study Brain Distribution

P-gp and, to a much lesser extent, BCRP and MRP4 are currently considered the major blood–brain barrier (BBB) efflux transporters limiting the CNS penetration of substrate drugs.[80] In contrast to its influence on intestinal absorption, P-gp limits brain penetration and CNS distribution regardless of permeability (Figure 8.4B[81]) due to the tightness of the endothelial barrier and the absence of high concentration gradients (such as those between the intestine and blood following oral drug administration).

The functional importance of BBB P-gp was first demonstrated in $Mdr1a^{-/-}$ knockout mice with ivermectin, an anti-parasitic drug transported by P-gp.[33] Ivermectin exhibited 87-fold greater CNS exposure in $P\text{-}gp^{-/-}$ knockout mice, resulting in serious CNS toxicity and deaths at dose levels that are safe in wild-type animals [$\sim$100-fold decrease in lethal dose, 50% (LD_{50})]. Whereas P-gp effectively restricts brain exposure to its substrates, the impact of BCRP by itself is usually less pronounced.[81,82] Protein quantification by liquid chromatography–mass spectrometry (LC-MS) has demonstrated that, relative to mice, humans express half of the P-gp and up to twice the BCRP protein at the BBB.[83] These modest transporter expression differences have been used to propose that studies in $P\text{-}gp^{-/-}$ knockout mice may exaggerate the importance of P-gp when used to predict CNS distribution in humans; however, it is not true for drugs extensively effluxed at the BBB by P-gp (Figure 8.6[80]). From a kinetic perspective, expression changes $\leq$2-fold in magnitude have little impact on predictions of whether active efflux extensively limits brain penetration. As such, the differences between mouse and human BBB P-gp and BCRP expressions are considered too small to fundamentally alter the conclusion whether efflux limits drug brain exposure when mice are used as a model for humans in this regard.

BCRP ablation in mice already deficient in P-gp ($Mdr1a/b/Bcrp^{-/-}$ knockout) can significantly increase the CNS penetration of dual P-gp/BCRP substrates.[84–87] As it is difficult to predict from *in vitro* data the contribution of P-gp and BCRP to restricting brain penetration of a dual substrate compound, due to the poor *in vitro–in vivo* correlation for BCRP, $Mdr1a^{-/-}$ single knockout, $Mdr1a/1b^{-/-}$ double knockout, $Bcrp^{-/-}$ single knockout, and $Mdr1a/1b/Bcrp^{-/-}$ triple knockout animals are widely used to dissect the role of these two efflux transporters in the overall process (Figure 8.7[80]).

Lapatinib, a dual P-gp/BCRP substrate, exemplifies the utility of knockout mice in delineating the contribution of the individual efflux transporters to limiting brain penetration. CNS exposure of lapatinib was increased approximately 1.3- and 3–4-fold in $Bcrp^{-/-}$ and $P\text{-}gp^{-/-}$ knockout mice, respectively, but was increased 26-fold in the absence of both transporters, approximately an order of magnitude higher than the sum of the increase in the individual knockouts.[85] However, in interpreting such results, it should be noted that brain exposure increases exponentially as a function of the fraction transported (Figure 8.7[80]). As such, increases in exposure are not directly additive

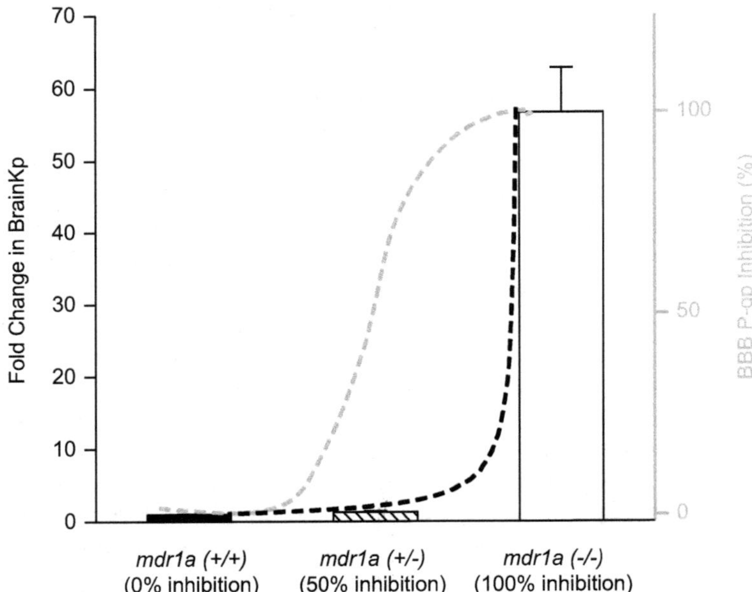

Figure 8.6 CNS distribution of loperamide in wild-type ($mdr1a^{+/+}$) mice, heterozygous ($mdr1a^{+/-}$) mice expressing 50% of wild-type mouse P-gp, and P-gp knockout mice. These three types of mice represent 0%, 50%, and 100% BBB P-gp inhibition, respectively. Note that loperamide's brain K_p is ~60-fold increased when efflux is 100% impaired; however, a 50% decrease in BBB P-gp function does not elicit half of the maximal increase in K_p (*i.e.* ~30), but a <2-fold increase relative to wild-type mice. The increase in CNS distribution (dashed black line) is not proportional to the extent of BBB efflux inhibition (dashed grey line). Thus, at most, a two-fold decrease in P-gp mediated attenuation of CNS distribution is expected in humans relative to mice based on the less than two-fold greater P-gp expression at the murine BBB relative to humans.[83]
Reprinted with permission from ref. 80.

but are instead an exponential function of the fraction transported. Therefore, brain exposure data generated from $Mdr1a/1b/Bcrp^{-/-}$ knockout mice and individual P-$gp^{-/-}$ and $Bcrp^{-/-}$ knockouts should be interpreted using the fraction transported approach, and not by direct addition of the observed increase in CNS exposure. Finally, it is worth noting that other proposed explanations for these findings, including biochemical synergism between P-gp and BCRP, and compensatory upregulation, have been ruled out.[79,80,88]

8.3.4　Use of Transporter Knockout Animals to Study Hepatic Uptake

Human OATP1B1 and 1B3 are primarily expressed in the liver and are of key importance for the hepatic uptake of many classes of drugs, such as

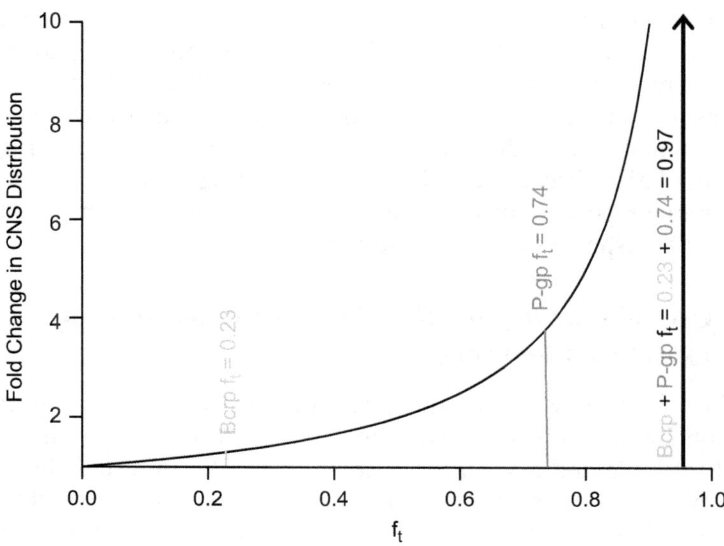

Figure 8.7 Relationship between the change in CNS distribution in a knockout animal *versus* wild-type control and the fraction transported (f_t) by the knocked out efflux transporter. Using lapatinib as an example, a 1.3-fold increase in CNS exposure in *Bcrp*$^{-/-}$ knockout mice (253 *versus* 194 ng ml^{-1}) means that the f_t is 0.23, while the 3.8-fold increase in the absence of P-gp (732 *versus* 194 ng ml^{-1}) means that the f_t is 0.74.[74,85] The fraction transported by both Bcrp and P-gp is $0.23 + 0.74 = 0.97$, which predicts an increase in CNS exposure comparable to the experimental observation of 26-fold (4938 *versus* 194 ng ml^{-1}).[74,85] Brain concentrations were used directly in this example, because lapatinib was infused to comparable steady-state systemic levels between wild-type and knockout mice. Note that directly adding the magnitude of increase in CNS exposure in individual P-gp and Bcrp knockouts to predict exposure in the absence of both efflux pumps would under-predict CNS exposure by an order of magnitude (4.1-fold $\ll$ 26-fold). Raw increase in exposure is not additive, but is a nonlinear function of the f_t (which itself is an additive parameter).
Reprinted with permission from ref. 80.

3-hydroxyl-3-methylglutaryl-CoA reductase inhibitors (statins), angiotensin II receptor antagonists (sartans), and several anticancer drugs.[89] Due to the presence of multiple hepatic OATP isoforms with overlapping substrate specificity and the lack of *bonafide* orthologues between humans and rodents, the use of single gene knockout models to study hepatic OATP function was initially limited.[41–43] To overcome this limitation, an *Oatp1a/1b*$^{-/-}$ knockout mouse model with deletion of all five established mouse *Slco1a* and *-1b* genes, including the three major hepatic Oatp isoforms, has recently been generated.[44] This model has now been used to study the role of liver OATPs in hepatic uptake,[11,44,90–93] and to reproduce and provide further insight into the human Rotor syndrome, which is caused by mutations resulting in non-functional OATP1B1 and 1B3.[56] Of the other uptake transporters with functional expression in the liver, knockout mice have only been described for Oct1. The known functional relevance of hepatic OCT1 is presently limited to metformin distribution to the liver, which is the most important site of its pharmacological activity.[94] *Oct1*$^{-/-}$ knockout models have also been valuable in understanding metformin liver exposure during OCT/MATE modulation, because distribution of this drug is not readily predicted by its systemic exposure due to its lack of passive membrane permeability and reliance on transporters to cross lipid bilayers.[95]

8.3.5   Use of Transporter Knockout Animals to Study Excretory Clearance

The kidney is the most important organ involved in direct excretory clearance of unchanged drugs.[96,97] Drug metabolites formed in the liver can be excreted into bile and also into sinusoidal blood for eventual clearance by the kidneys. While other organs, such as the lungs, can contribute to drug elimination, this section will focus on renal and biliary clearance due to their relative importance and the established role of membrane transporters in these processes.

MRP2, BCRP, and P-gp are key efflux transporters that determine the biliary excretion of many drugs and metabolites. The contribution of these apical ABC transporters to drug excretion can be effectively studied in the corresponding knockout animal models.[98] It should be noted, however, that extensive biliary clearance in animal bile duct cannulation studies may be an artifact of disrupted enterohepatic recycling, and such studies should be accompanied by studies in intact animals to ensure that the systemic pharmacokinetics are not affected by bile collection. For example, 81% of the pravastatin dose was shown to be excreted as the parent compound in bile *via* Mrp2[99] using cannulated animals, but clinical translation of these data was poor due to disruption of enterohepatic cycling, where several passes through the liver would normally result in more extensive recovery of the dose as metabolites in excreta. However, biliary excretion is a more common route of elimination for drug metabolites than for parent drugs.[97,100,101] As discussed for drug brain distribution above, the fold change in recovery of polar metabolites in excreta of knockout animals is also described by the relationship $1/(1 - f_t)$.[74]

The utility of knockout animals in the study of metabolite disposition was recently illustrated with LY2090314, an intravenous oncolytic agent in clinical studies for the treatment of solid tumours and leukaemia.[102] Parent drug is cleared by extensive hepatic metabolism, and the numerous metabolites are extensively and irreversibly eliminated into faeces *via* biliary excretion. Note that the absence of metabolite absorption from the gut is not an artifact of disrupted enterohepatic cycling during bile collection, as the metabolites were not absorbed in intact rats either. Biliary excretion of metabolites was conceptually consistent with an active process and was therefore investigated further in transporter knockout rats. The tested hypothesis was that in single gene knockout rats, lacking one of the three key hepatic canalicular transporters (Mdr1a P-gp, Bcrp, or Mrp2), biliary excretion of LY2090314 metabolites may be impaired, resulting in excretion of these metabolites across the sinusoidal membrane into the circulation.[102] Unexpectedly, in all three single gene knockout rats, metabolites did not appear in the circulation, and their urinary recovery was not enhanced. Metabolite biliary excretion was therefore either maintained by the other liver efflux transporters (*e.g.* Mdr1a, P-gp and Bcrp in the Mrp2 knockout) or the metabolites were metabolised further into more polar species, which were then excreted into the bile by the remaining transporters. Thus, even complete inhibition of individual canalicular efflux transporters did not significantly alter LY2090314 metabolite disposition, *i.e.* shunt excretion into the circulation or accumulation in the liver. As such, these preclinical studies dismissed the risk of victim drug interactions at the level of biliary excretion of LY2090314 metabolites. Specifically, it was reasoned that co-administration with drugs that inhibit canalicular transporters would not present a risk of systemic exposure to or hepatic accumulation of LY2090314 metabolites.

The kidney proximal tubules contain a variety of basolateral uptake transporters and apical (luminal) uptake and efflux transporters (Figure 8.1C) whose role in urinary excretion can be studied using knockout rodent models. The renal uptake transporters OAT1/3 and OCT2, and the efflux transporters MATE1 and MATE2-K have demonstrated clinical relevance.[1,5,103] OCT2 mediates the basolateral entry step in tubular secretion of typically cationic substrates, which is most commonly followed by MATE secretion into urine. MATEs present an interesting case in which the secondary excretion step can be the rate determining step in overall renal clearance due to the unique OCT/MATE driving force energetic properties.[3,51,104] OAT1/3 mediate uptake of typically anionic substrates, which may be followed by secretion *via* MRP4, MRP2, and/or OAT4. However, unlike MATEs, MRP4, MRP2, and OAT4 are not usually rate-determining, and so their relevance is not readily apparent in terms of systemic drug pharmacokinetics. However, their inhibition may have important implications in kidney drug exposure.[105]

The role of proximal tubule uptake transporters in urinary excretion has been studied extensively in knockout mice.[48–50,95,106] Likewise, *Mate1*$^{-/-}$ knockout mice have been used to study kidney and systemic exposure of metformin following ablation of excretion from renal proximal tubule cells into urine,[51] while *Mrp4*$^{-/-}$ knockout mice have been used to study the impact of Mrp4 on systemic and kidney exposure of adefovir, tenofovir, and cidofovir.[105] However, species differences in the relative contribution of active tubular secretion to overall renal clearance can be significant and more pronounced than in other drug disposition processes influenced by active transport.[1,99,107] Accordingly, results from preclinical studies should not be extrapolated directly to humans until contextualised with both *in vitro* and clinical renal clearance data. As such, it is critical to confirm clinical translation of excretory clearance on a pharmacokinetic level. Specifically, for renal clearance, it should be demonstrated that the ratio of renal clearance to unbound glomerular filtration supports active tubular secretion in both humans and the species studied, that renal clearance is substantial in both species (>25% of total clearance), and ideally that a similar *in vitro* substrate profile for renal transporters with regard to species differences in renal expression is present (*e.g.* OCT2 *versus* both Oct1 and Oct2 in rodents).

8.3.6 Utility of Genetically Humanised Mouse Models

Humanised mouse models can provide additional assistance with pre-clinical to clinical translation of the impact of transporters on pharmacokinetics.[11] In principle, genetically humanised mouse models can be used either instead of or alongside wild-type animals, utilising transporter knockout models as controls. Using this approach, the contribution of the human rather than the mouse transporter to drug disposition can be determined *in vivo*. However, it is important to note that since at most a few human genes are expressed in these mouse models, complex disposition pathways that may involve several transporters (and enzymes) cannot be directly extrapolated to humans. Rather, the role of the transporter(s) in the handling of xenobiotics or endogenous substrates should be understood in order to carefully design studies that will address specific question(s). Theoretically, such models also enable investigation of the *in vivo* DDI potential of an inhibitor, as well as the potential for interference with the transport of endogenous substrates. Examples for the latter would be to study drug induced hyperbilirubinaemia *via* the inhibition of OATP1B1, OATP1B3, and/or MRP2, or to determine whether BSEP inhibition elicits cholestasis *in vivo*,[108] but it should be noted that proof of concept data for such applications are still missing. The field of genetically humanised transporter models is emerging, and to date, humanised mice have only been described for OATP1B1, 1B3, and 1A2,[56,57] as well as MRP2.[59] A prerequisite for their routine use in translational research is thorough characterisation of these models, as exemplified by studies conducted with OATP and MRP2 humanised mice.[11,57–59,109]

Van de Steeg *et al.* first described the generation of mice with liver-specific expression of OATP1B1 and the impact of this human transporter on the pharmacokinetics of methotrexate.[55] Although the human OATP was expressed without deletion of the corresponding mouse Oatp transporters, these studies showed a 32% reduction in methotrexate systemic exposure, with an approximately two-fold increase in liver concentrations of methotrexate relative to wild-type controls, suggesting that OATP1B1 played a role in the hepatic uptake of methotrexate. However, as demonstrated with methotrexate, adding a human OATP to a full complement of murine hepatic Oatps is likely to elicit relatively small pharmacokinetic changes. To address this shortcoming, the role of OATP1B1 and 1B3 in methotrexate disposition was recently confirmed in humanised mice expressing the human transporters on an *Oatp1a/1b*$^{-/-}$ knockout background, which effectively lacks murine hepatic Oatp function.[57]

The characterisation and evaluation of transgenic mice expressing human OATP1B1, 1B3, or 1A2,[57] and protein quantification studies using targeted quantitative proteomic LC-MS/MS analysis have confirmed that human OATP1B1 and 1B3 proteins are expressed at levels within a few fold of those in the human liver (2.2-fold lower OATP1B1 and 3.3-fold higher OATP1B3 expression in humanised mouse *versus* human livers[11]). OATP1B1 and 1B3 could reverse, almost completely, the markedly reduced uptake of bilirubin in *Oatp1a/1b*$^{-/-}$ knockout mice, and return plasma (and urine) levels of bilirubin (total, unconjugated, and conjugated) to those observed in wild-type animals.[56] These mice also confirmed the involvement of human OATPs in the uptake of methotrexate and paclitaxel.[57] Hepatic expression of OATPs resulted in substantial although incomplete, rescue of the liver uptake of the OATP substrate methotrexate when compared with the *Oatp1a/1b*$^{-/-}$ knockout animals, with plasma levels 2-fold lower than in *Oatp*$^{-/-}$ knockout animals, but still twice as high as in the wild-type mice. Liver concentrations were up to nine-fold higher than in the *Oatp*-deficient mice. The fact that methotrexate liver uptake was only partially restored compared with wild-type controls might be explained by the replacement of five functional murine Oatp1a/1b genes with only OATP1B1 or 1B3 in the respective humanised model. Increases in paclitaxel plasma levels were mild in the *Oatp1a/1b*$^{-/-}$ knockout *versus* wild-type animals, suggesting plasma levels were not the most sensitive measure of the loss of Oatp function. In contrast, liver concentrations, which were markedly reduced in these knockout mice, were restored to wild-type levels by human OATP1B3, but not by OATP1B1. More recent studies have confirmed the role of human OATPs in the hepatic uptake of pravastatin, atorvastatin, and simvastatin in transgenic mice.[11,58] Despite some differences in pharmacokinetic results, which may be related to differences in the studied dose levels, overall these reports confirmed that the humanised OATPs were active, able to transport some statins tested, and could at least partially recover the liver uptake activity suppressed in *Oatp1a/1b*$^{-/-}$ mice. In addition, based on the difference in the liver partition coefficient between the humanised mice and the *Oatp* knockout animals, Higgins *et al.* proposed an approach for estimating the contribution of OATP1B1 to total hepatic uptake,[11] and translated it to humans by correcting for the difference in protein expression between human and transgenic mouse livers. It should be noted, however, that the expression of OATP1B1 and 1B3 in the humanised models did not rescue liver or plasma concentrations of rosuvastatin and pitavastatin compared with *Oatp1a/1b*$^{-/-}$ knockout controls under the conditions employed in a recent study.[58] The reason for this finding is currently unknown and requires further investigation to better define the use and limitations of these models in the prediction of drug disposition and DDIs in humans.

Humanised OATP1B1 and 1B3 mice have helped to confirm the involvement of these transporters in the hepatic uptake of the tyrosine kinase inhibitor sorafenib and its glucuronic acid metabolite.[109] Many tyrosine kinase inhibitors undergo extensive hepatic metabolism, but the mechanisms of their hepatocellular uptake remain poorly understood. While the plasma exposure of the parent compound was not markedly affected by the ablation of Oatp1a/1b, exposure of sorafenib glucuronide increased by up to 30-fold in the $Oatp1a/1b^{-/-}$ knockout mice, indicating that it is a substrate of OATPs. Furthermore, the expression of OATP1B1 or 1B3 provided partial restoration of the uptake function, with sorafenib glucuronide exposure decreasing approximately two-fold relative to knockout mice. The partial recovery of the uptake function in the humanised mice with sorafenib glucuronide was consistent with the results obtained with statins;[11,58] plasma exposure of sorafenib glucuronide as well as statins in the humanised mice generally showed intermediate levels, between those of the wild-type animals and the $Oatp1a/1b^{-/-}$ knockout mice. In addition, the results obtained with sorafenib and its glucuronide in the OATP knockout and humanised mice are in agreement with the physiological role of OATP in bilirubin elimination hypothesised by van de Steeg *et al.*, based on findings in Rotor syndrome and studies in knockout mice.[56] The transgenic mice expressing human OATP1B1 or 1B3 supported the proposed role of OATPs in bilirubin handling and the hypothesised cycle of efflux–reuptake of conjugated bilirubin through the basolateral efflux transporter (MRP3) and uptake by OATPs ("hepatocyte hopping" hypothesis).

These results, obtained with both prototypical OATP drug and endogenous substrates, suggest that humanised OATP mice could be valuable in studying the functions of OATPs and potentially predicting the occurrence and magnitude of DDIs in humans.

Humanised MRP2 mice have also been generated, and the function of human MRP2 was confirmed by the reversion of hyperbilirubinaemia observed in $Mrp2^{-/-}$ knockout animals.[59] In these transgenic mice, bilirubin levels in blood were similar to those in wild-type animals, confirming that the human transporters were fulfilling their expected physiological function. Furthermore, it was recently demonstrated that while $Mrp2^{-/-}$ knockout mice are more susceptible to cisplatin-mediated nephrotoxicity compared with wild-type controls, the expression of the human transporter protected the MRP2 humanised mice against cisplatin toxicity.[110]

While further studies will be necessary to assess the potential for humanised mice to explain or predict clinical outcomes, the limited results obtained so far suggest that they would be valuable additions to the models currently available for transporter investigations.

8.3.7   Utility of Liver Humanised Mouse Models

While in principle the genetically humanised mouse models can be used to study transporter-related processes in any organ, this is presently restricted to hepatic function in the case of the chimeric liver humanised animals. An advantage of this system over the genetically humanised models is the nearly complete humanisation of the liver, which therefore enables investigations of complex disposition pathways involving both drug metabolism and drug transport within the human liver. However, the maintenance of expression of key transporters in the transplanted hepatocytes needs to be ascertained prior to the use of these in hepatobiliary transport studies. In this regard, 21 transporters in human hepatocytes of chimeric uPA-SCID mice were shown to have protein levels comparable to the human liver.[111–113] Likewise, the basal gene expression levels of BSEP and MRP2 in cultures of isolated hepatocytes derived from chimeric FRG mice were equivalent to those found in cultured human hepatocytes[64] and, in livers obtained from chimeric TK-NOG mice, the mRNAs of five SLC and four ABC transporters were expressed at levels comparable to hepatocytes from the original donors.[69] Chimeric liver humanised mouse models therefore have potential utility in studying the role of human hepatic transporters in drug liver distribution, biliary excretion, DDIs, and liver toxicity. However, reports of using chimeric liver humanised mice to study drug transporters are still limited. While a few examples indicate the utility of such models to study transporter-related hepatotoxicity[114] and processes in liver biology,[115] the utility of these mice in studying biliary excretion has not been evaluated.

Troglitazone, the first of a class of thiazolidinedione drugs registered for the treatment of type 2 diabetes, was withdrawn in the United States in 2000 because of cases of serious idiosyncratic liver injury.[116,117] Troglitazone decreased the expression of BSEP and MRP2 in humans and humanised chimeric uPA-SCID mice, but not normal mice.[114] Furthermore, troglitazone treatment was associated with a decrease in the amount of neutral lipid in the humanised mouse livers. It was speculated that, due to the cholestatic component of troglitazone-induced hepatotoxicity, and the involvement of BSEP and MRP2 in bile acid transport, the observed changes in the humanised mouse livers might partially explain the liver injury seen in humans.[114] However, this hypothesis remains to be tested.

In order to analyse the role of MDR3 in cholestatic liver disease, the effect of the peroxisome proliferator activating receptor agonist bezafibrate (a fibrate drug used for the treatment of hyperlipidaemia) on the expression level of MDR3 was investigated using a humanised liver mouse model. After oral administration, bezafibrate caused an increase in the amount of MDR3 protein and its redistribution to the bile canaliculi in the chimeric liver humanised mice, but not to the same extent in wild-type mice.[115] The authors proposed that these results supported the beneficial role of bezafibrate in reducing cholestasis and/or cholangitis associated with defective MDR3 expression and function in various types of cholestatic liver diseases.

8.4   Study Design and Data Interpretation

Due to the species differences of transporters and other elimination pathways between animals and humans, and the complexity of transporters regarding *in vivo* drug disposition, rational design and interpretation of transporter animal studies will be critical to understand preclinical to clinical translation. When designing and interpreting pharmacokinetic studies in transporter knockout/humanised models, it is helpful to consider the following points.

- It is critical to confirm that the compound exhibits similar behaviour regarding the transporter-mediated pathways and other ADME properties between humans and the knockout/humanised species. For instance, in both mice and humans, metformin is entirely cleared by urinary excretion of the parent, which is mediated by one-third glomerular filtration and two-thirds active tubular secretion *via* OCT/MATE transport (OCT2 and MATE1/2-K in humans; Oct1/2 and Mate1 in mice). Thus, in this case, *Oct1/2* and *Mate1* knockout mice are relevant models in understanding the roles of these transporters for *in vivo* disposition of metformin in humans.[51,95] In contrast, pemetrexed is also cleared by urinary excretion of the parent in humans, with OAT3/4-mediated active secretion contributing to 80% of this process. However, in mice, clearance of pemetrexed is predominantly metabolic, with only 32% of the intravenous dose excreted as parent in urine *via* glomerular filtration.[107] This discrepancy may in part be explained by the absence of Oat4 in mice (Figure 8.1). Therefore, pemetrexed is an example where renal clearance in mice is not representative of human clearance, and for which murine knockout models would provide no value in understanding the clinical mechanism(s) of active tubular secretion and DDI potential.
- The selection of dose and the route of administration is also an important aspect to consider when designing *in vivo* studies in knockout/humanised animals. Within the analytical detection limit and acceptable safety dose range, it is important to ensure that clinically-relevant systemic exposure is achieved in the animals being used. Pilot dose-ranging studies in wild-type animals may be needed, if pharmacokinetic data are not available in the knockout species. The studies following oral and intravenous administration will provide additional insights to differentiate the impact of the transporter on pre-systemic and systemic elimination and tissue distribution.
- Species differences may exist in the localisation, expression, affinity, and sensitivity to inhibitors between rodent and human transporters (Figure 8.1).[12] With recent advances in LC-MS/MS protein quantification, data on the expression levels of key transporters in different tissues of rodents and humans have become available.[12] However, further studies are needed to understand if a difference at the level of the transporter protein expression is correlated with functional activity and can be translated to a change in systemic/tissue exposure *in vivo*. *In vitro* studies to determine the species differences in transporter activity, and inhibition potential, should also be conducted to support/verify the translation of the observations from animals to humans.
- Another general concern in the use of transporter knockout/humanised animals for pharmacokinetic studies is the potential up- or down-regulation of other transporters and drug metabolising enzymes, which

may complicate the interpretation of data obtained with these models. As discussed previously, Mrp2 transporter-deficient rat (TR⁻) and Eisai hyperbilirubinaemic rat (EHBR) strains show a strong induction of Mrp3 in hepatocytes under conditions in which the function of Mrp2 is impaired.[20,118] Interpretation of data obtained with TR⁻ and EHBR in some cases is further complicated by the induction of UGT1a, and significant differences in cytochrome (CYP) P450 enzyme levels between wild-type and mutant strains.[119–121] Interestingly, such compensatory changes in ADME-related genes have not been observed, or have been at most modest, in $Mrp2^{-/-}$ gene knockout mice and rats.[9,22,23,122] Likewise, other transporter knockout/humanised animal models published thus far have not exhibited marked compensatory changes in ADME-relevant genes that would preclude their use in pharmacokinetic studies. Therefore, the compensatory changes may not be as much of a concern as initially suggested in inbred naturally-occurring mutants. Nevertheless, determining the compensatory changes of major ADME-related genes in newly established transporter knockout/humanised animal models is necessary for their rational use in pharmacokinetic studies.

- The fraction transported approach is useful to interpret the pharmacokinetic data generated from knockout models. Where parallel transport pathways are involved, the f_t can be used as an additive parameter in predicting the contribution of multiple transporters to the pharmacokinetics of overlapping substrates. Using this kinetic concept, CNS exposure of dual P-gp/Bcrp substrates, such as lapatinib, in P-gp/Bcrp knockout mice can be explained, as illustrated in Section 8.3.3. Therefore, adding together the exposure changes in single knockouts of different transporters directly, but not f_t values, should never be used to predict modulation of multiple transport pathways.

- Transporter knockout mice can be used as a reference for complete inhibition of a specific transporter, representing the 'worst case' DDI scenario. However, the data in knockout mice cannot be directly translated to humans for predicting the magnitude of DDIs, as complete inhibition of transporters at clinically-relevant exposures of inhibitors is not likely. This is illustrated by the nearly eight-fold increase in brain penetration of verapamil in the $Mdr1a/b^{-/-}$ knockout mice compared with control mice, whereas an increase of less than two-fold in the area under the curve $(AUC)_{brain} : AUC_{blood}$ ratio of ^{11}C-verapamil in humans was reported in the presence of a high intravenous dose of the P-gp inhibitor cyclosporine.[123,124] These data illustrate the point that in most clinical DDI situations, inhibition will not be as pronounced as that observed in a knockout model representing 100% inhibition of the transporter. However, using the knockout to assess the fraction transported (f_t), the magnitude of the DDI can be estimated at a given $I : K_i$ ratio,[5,74] assuming that the f_t values are similar in humans and mice.

- Caution is needed in the interpretation of the extent of biliary excretion in bile cannulation studies, which is not the true *in vivo* clearance of the parent drug, since bile is removed during collection and is therefore not made available for intestinal reabsorption, as would normally occur.[98] The prominence of the biliary route may also be different between humans and preclinical species, and studies in knockout animals, while providing mechanistic insight, cannot be quantitatively extrapolated to humans.

- Finally, in most cases, it is not possible to predict human pharmacokinetics based solely on data obtained from rodent studies, and preclinical-to-clinical translation must be established on a case-by-case basis with supporting *in vitro* and human pharmacokinetic data. For example, the effects of CYP3A and P-gp on the pharmacokinetics of docetaxel were investigated in $Mdr1a/1b^{-/-}$, $Cyp3a^{-/-}$, and $Cyp3a/Mdr1a/1b^{-/-}$ mice. Based on these data, a mouse pharmacokinetic model was developed using nonlinear mixed effect modelling. With the integration of human pharmacokinetic parameters following intravenous administration of docetaxel, this combined model predicted the exposure of orally administered docetaxel in combination with ritonavir in humans,[125] thus illustrating the importance of incorporating clinical data to validate preclinical to clinical translation.

8.5 Conclusions and Perspectives

Knockout and humanised transporter animal models have and will continue to play an important role in understanding the involvement of transporters in drug pharmacokinetics. Clinical transporter inhibition risk can be adequately predicted from *in vitro* data and prediction of clinical DDIs with prototypical substrate drugs is relatively well established.[1,126–128] In contrast, the transporter-mediated pharmacokinetics, and associated victim DDI potential, are considerably more complex, and IVIVE has fundamentally failed to predict clinical pharmacokinetics in these cases.[6,7] *In vitro* data are useful for qualitatively identifying transporter substrate interactions, but animal models are critical to understanding the relevance of these findings for drug pharmacokinetics *in vivo*, in combination with clinical pharmacokinetic data.

The perception of "poor clinical translation" from animal models arose from the unrealistic expectation that pharmacokinetic changes following complete genetic ablation of a transporter (*i.e.* 100% inhibition) would be comparable to drug interactions in humans, where transporter inhibitors often circulate at unbound concentrations that are not far in excess of the inhibitory potency ($I_u \approx K_i = 50\%$ inhibition).[5,80] As demonstrated by the f_t data analysis approach (Figures 8.5–8.7), change in exposure (systemic, tissue, or recovery in excreta) is an exponential function of the f_t, with the curve becoming steeper as the I_u/K_i increases. Exposure changes in knockout models ($I_u/K_i = \infty$) can be used to estimate the f_t, which can then be used to predict the magnitude of the clinical DDIs at the appropriate $I_u : K_i$ ratio of the inhibitor drug in humans.[1,5,74,80] However, change in exposure in knockout models ($I_u/K_i = \infty$) will typically exceed the magnitude of the clinical DDI at a far lower $I_u : K_i$ ratio. This concept was recently reviewed in detail by the International Transporter Consortium to reconcile the disconnect between CNS exposure changes of P-gp substrate drugs in P-gp knockout animals and clinical co-administration of P-gp inhibitors.[80]

Compensatory changes confounding data interpretation in knockout and humanised animal drug transporter models are a concern that emerged after a multitude of unexpected changes in naturally occurring *Mrp2*-deficient rats that have been inbred within small populations for nearly three decades.[20,21] Notably, these compensatory changes are minimal in healthier *Mrp2*$^{-/-}$ gene knockout rats and mice.[9,22,23,122] Generally, genetic knockouts of efflux transporters have been independently verified to be healthy and to exhibit no major compensatory changes in expression of ADME-relevant genes that would preclude their use in pharmacokinetic studies.[9,22,23,88,122] Likewise, *Oatp1a/1b*$^{-/-}$ and *Oct1/2*$^{-/-}$ knockout mice have not exhibited phenotypic, gene expression, or pathological compensatory changes that would preclude their use in the study of drug disposition.[44,95] Also, attempts have been made to explain the BBB P-gp/Bcrp "synergy" phenomenon through compensatory upregulation of *P-gp* in the *Bcrp*$^{-/-}$ knockout mouse and *vice versa*. However, subsequent studies demonstrated that no such compensatory changes exist,[88] and that the "synergism" could be simply explained through basic pharmacokinetic f_t principles (Figure 8.3).

Humanised animal drug transporter models are an important emerging area in translational science for studying the impact of transporters on drug pharmacokinetics. OATP1B1 humanised mice generated on the *Oatp1a/1b* knockout background have proven powerful in determining the fraction of hepatic uptake mediated by human OATP1B1.[11] Otherwise, at present, it is premature to speculate about the translational utility of these animals, because clinical translation has not been established for other humanised transporter models. The clinical translation potential for hepatobiliary drug transport in the humanised liver mouse model appears promising, but at present remains to be evaluated.

In conclusion, animal drug transporter models have an important role in investigating transporter-mediated drug disposition. These models have to be used thoughtfully with due regard for pharmacokinetic principles (*i.e.* partial clinical inhibition *versus* complete inhibition due to genetic ablation), and the results must be placed in proper context with supporting *in vitro* and clinical pharmacokinetic data.

Declaration of Interest

The authors declare no financial conflicts of interest beyond employment noted in the affiliation section. The authors alone are responsible for the content and writing of this paper.

References

1. K. M. Giacomini, S. M. Huang, D. J. Tweedie, L. Z. Benet, K. L. Brouwer, X. Chu, A. Dahlin, R. Evers, V. Fischer, K. M. Hillgren, K. A. Hoffmaster,

T. Ishikawa, D. Keppler, R. B. Kim, C. A. Lee, M. Niemi, J. W. Polli, Y. Sugiyama, P. W. Swaan, J. A. Ware, S. H. Wright, S. W. Yee, M. J. Zamek-Gliszczynski and L. Zhang, *Nat. Rev. Drug Discovery*, 2010, **9**, 215–236.

2. M. J. Zamek-Gliszczynski, X. Chu, J. W. Polli, M. F. Paine and A. Galetin, *Drug Metab. Dispos.*, 2014, **42**, 650–664.

3. M. J. Zamek-Gliszczynski, K. A. Hoffmaster, D. J. Tweedie, K. M. Giacomini and K. M. Hillgren, *Clin. Pharmacol. Ther.*, 2012, **92**, 553–556.

4. A. Rostami-Hodjegan, *Clin. Pharmacol. Ther.*, 2012, **92**, 50–61.

5. M. J. Zamek-Gliszczynski, C. A. Lee, A. Poirier, J. Bentz, X. Chu, H. Ellens, T. Ishikawa, M. Jamei, J. C. Kalvass, S. Nagar, K. S. Pang, K. Korzekwa, P. W. Swaan, M. E. Taub, P. Zhao and A. Galetin, *Clin. Pharmacol. Ther.*, 2013, **94**, 64–79.

6. H. M. Jones, H. A. Barton, Y. Lai, Y. A. Bi, E. Kimoto, S. Kempshall, S. C. Tate, A. El-Kattan, J. B. Houston, A. Galetin and K. S. Fenner, *Drug Metab. Dispos.*, 2012, **40**, 1007–1017.

7. M. V. Varma, Y. Lai, B. Feng, J. Litchfield, T. C. Goosen and A. Bergman, *Pharm. Res.*, 2012, **29**, 2860–2873.

8. X. Chu, Z. Zhang, J. Yabut, S. Horwitz, J. Levorse, X. Q. Li, L. Zhu, H. Lederman, R. Ortiga, J. Strauss, X. Li, K. A. Owens, J. Dragovic, T. Vogt, R. Evers and M. K. Shin, *Mol. Pharmacol.*, 2011, **81**, 220–227.

9. M. J. Zamek-Gliszczynski, D. W. Bedwell, J. Q. Bao and J. W. Higgins, *Drug Metab. Dispos.*, 2012, **40**, 1825–1833.

10. N. Scheer and C. R. Wolf, *Xenobiotica*, 2014, **44**, 96–108.

11. J. W. Higgins, J. Q. Bao, A. B. Ke, J. R. Manro, J. K. Fallon, P. C. Smith and M. J. Zamek-Gliszczynski, *Drug Metab. Dispos.*, 2014, **42**, 182–192.

12. X. Chu, K. Bleasby and R. Evers, *Expert Opin. Drug Metab. Toxicol.*, 2013, **9**, 237–252.

13. J. R. Foster, G. Lund, S. Sapelnikova, D. L. Tyrrell and N. M. Kneteman, *Xenobiotica*, 2014, **44**, 109–122.

14. M. Grompe and S. Strom, *Gastroenterology*, 2013, **145**, 1209–1214.

15. G. Peltz, *Trends Pharmacol. Sci.*, 2013, **34**, 255–260.

16. B. Hall, A. Limaye and A. B. Kulkarni, *Curr. Protoc. Cell Biol.*, 2009, **Chapter 19**, Unit 19 12 19 12 11–17.

17. N. Scheer, M. Snaith, C. R. Wolf and J. Seibler, *Drug Discovery Today*, 2013, **18**, 1200–1211.

18. S. Hosokawa, O. Tagaya, T. Mikami, Y. Nozaki, A. Kawaguchi, K. Yamatsu and M. Shamoto, *Lab Anim. Sci.*, 1992, **42**, 27–34.

19. P. L. Jansen, W. H. Peters and W. H. Lamers, *Hepatology*, 1985, **5**, 573–579.

20. T. Hirohashi, H. Suzuki, K. Ito, K. Ogawa, K. Kume, T. Shimizu and Y. Sugiyama, *Mol. Pharmacol.*, 1998, **53**, 1068–1075.

21. W. Yue, J. K. Lee, K. Abe, Y. Sugiyama and K. L. Brouwer, *Drug Metab. Dispos.*, 2011, **39**, 441–447.

22. X. Y. Chu, J. R. Strauss, M. A. Mariano, J. Li, D. J. Newton, X. Cai, R. W. Wang, J. Yabut, D. P. Hartley, D. C. Evans and R. Evers, *J. Pharmacol. Exp. Ther.*, 2006, **317**, 579–589.

23. M. J. Zamek-Gliszczynski, K. M. Goldstein, A. Paulman, T. K. Baker and T. P. Ryan, *Drug Metab. Dispos.*, 2013, **41**, 1174–1178.
24. M. J. Evans and M. H. Kaufman, *Nature*, 1981, **292**, 154–156.
25. G. R. Martin, *Proc. Natl. Acad. Sci. U. S. A.*, 1981, **78**, 7634–7638.
26. K. R. Thomas and M. R. Capecchi, *Nature*, 1986, **324**, 34–38.
27. A. M. Geurts, G. J. Cost, Y. Freyvert, B. Zeitler, J. C. Miller, V. M. Choi, S. S. Jenkins, A. Wood, X. Cui, X. Meng, A. Vincent, S. Lam, M. Michalkiewicz, R. Schilling, J. Foeckler, S. Kalloway, H. Weiler, S. Menoret, I. Anegon, G. D. Davis, L. Zhang, E. J. Rebar, P. D. Gregory, F. D. Urnov, H. J. Jacob and R. Buelow, *Science*, 2009, **325**, 433.
28. M. Meyer, M. H. de Angelis, W. Wurst and R. Kuhn, *Proc. Natl. Acad. Sci. U. S. A.*, 2010, **107**, 15022–15026.
29. B. Shen, J. Zhang, H. Wu, J. Wang, K. Ma, Z. Li, X. Zhang, P. Zhang and X. Huang, *Cell Res.*, 2013, **23**, 720–723.
30. Y. H. Sung, I. J. Baek, D. H. Kim, J. Jeon, J. Lee, K. Lee, D. Jeong, J. S. Kim and H. W. Lee, *Nat. Biotechnol.*, 2013, **31**, 23–24.
31. H. Wang, H. Yang, C. S. Shivalila, M. M. Dawlaty, A. W. Cheng, F. Zhang and R. Jaenisch, *Cell*, 2013, **153**, 910–918.
32. C. Bundgaard, C. J. Jensen and M. Garmer, *Drug Metab. Dispos.*, 2012, **40**, 461–466.
33. A. H. Schinkel, J. J. Smit, O. van Tellingen, J. H. Beijnen, E. Wagenaar, L. van Deemter, C. A. Mol, M. A. van der Valk, E. C. Robanus-Maandag, H. P. te Riele *et al.*, *Cell*, 1994, **77**, 491–502.
34. A. H. Schinkel, U. Mayer, E. Wagenaar, C. A. Mol, L. van Deemter, J. J. Smit, M. A. van der Valk, A. C. Voordouw, H. Spits, O. van Tellingen, J. M. Zijlmans, W. E. Fibbe and P. Borst, *Proc. Natl. Acad. Sci. U. S. A.*, 1997, **94**, 4028–4033.
35. R. Wang, M. Salem, I. M. Yousef, B. Tuchweber, P. Lam, S. J. Childs, C. D. Helgason, C. Ackerley, M. J. Phillips and V. Ling, *Proc. Natl. Acad. Sci. U. S. A.*, 2001, **98**, 2011–2016.
36. M. G. Belinsky, P. A. Dawson, I. Shchaveleva, L. J. Bain, R. Wang, V. Ling, Z. S. Chen, A. Grinberg, H. Westphal, A. Klein-Szanto, A. Lerro and G. D. Kruh, *Mol. Pharmacol.*, 2005, **68**, 160–168.
37. M. Leggas, M. Adachi, G. L. Scheffer, D. Sun, P. Wielinga, G. Du, K. E. Mercer, Y. Zhuang, J. C. Panetta, B. Johnston, R. J. Scheper, C. F. Stewart and J. D. Schuetz, *Mol. Cell Biol.*, 2004, **24**, 7612–7621.
38. M. L. Vlaming, K. Mohrmann, E. Wagenaar, D. R. de Waart, R. P. Elferink, J. S. Lagas, O. van Tellingen, L. D. Vainchtein, H. Rosing, J. H. Beijnen, J. H. Schellens and A. H. Schinkel, *J. Pharmacol. Exp. Ther.*, 2006, **318**, 319–327.
39. J. Wijnholds, R. Evers, M. R. van Leusden, C. A. Mol, G. J. Zaman, U. Mayer, J. H. Beijnen, M. van der Valk, P. Krimpenfort and P. Borst, *Nat. Med.*, 1997, **3**, 1275–1279.
40. J. W. Jonker, M. Buitelaar, E. Wagenaar, M. A. Van Der Valk, G. L. Scheffer, R. J. Scheper, T. Plosch, F. Kuipers, R. P. Elferink, H. Rosing, J. H. Beijnen and A. H. Schinkel, *Proc. Natl. Acad. Sci. U. S. A.*, 2002, **99**, 15649–15654.

41. C. Chen, J. L. Stock, X. Liu, J. Shi, J. W. Van Deusen, D. A. DiMattia, R. G. Dullea and S. M. de Morais, *Drug Metab. Dispos.*, 2008, **36**, 1840–1845.

42. H. Lu, S. Choudhuri, K. Ogura, I. L. Csanaky, X. Lei, X. Cheng, P. Z. Song and C. D. Klaassen, *Toxicol. Sci.*, 2008, **103**, 35–45.

43. H. Zaher, H. E. Meyer zu Schwabedissen, R. G. Tirona, M. L. Cox, L. A. Obert, N. Agrawal, J. Palandra, J. L. Stock, R. B. Kim and J. A. Ware, *Mol. Pharmacol.*, 2008, **74**, 320–329.

44. E. van de Steeg, E. Wagenaar, C. M. van der Kruijssen, J. E. Burggraaff, D. R. de Waart, R. P. Elferink, K. E. Kenworthy and A. H. Schinkel, *J. Clin. Invest.*, 2010, **120**, 2942–2952.

45. Y. Hu, D. E. Smith, K. Ma, D. Jappar, W. Thomas and K. M. Hillgren, *Mol. Pharm.*, 2008, **5**, 1122–1130.

46. I. Rubio-Aliaga, I. Frey, M. Boll, D. A. Groneberg, H. M. Eichinger, R. Balling and H. Daniel, *Mol. Cell Biol.*, 2003, **23**, 3247–3252.

47. J. W. Jonker, E. Wagenaar, C. A. Mol, M. Buitelaar, H. Koepsell, J. W. Smit and A. H. Schinkel, *Mol. Cell Biol.*, 2001, **21**, 5471–5477.

48. J. W. Jonker, E. Wagenaar, S. Van Eijl and A. H. Schinkel, *Mol. Cell Biol.*, 2003, **23**, 7902–7908.

49. S. A. Eraly, V. Vallon, D. A. Vaughn, J. A. Gangoiti, K. Richter, M. Nagle, J. C. Monte, T. Rieg, D. M. Truong, J. M. Long, B. A. Barshop, G. Kaler and S. K. Nigam, *J. Biol. Chem.*, 2006, **281**, 5072–5083.

50. D. H. Sweet, D. S. Miller, J. B. Pritchard, Y. Fujiwara, D. R. Beier and S. K. Nigam, *J. Biol. Chem.*, 2002, **277**, 26934–26943.

51. M. Tsuda, T. Terada, T. Mizuno, T. Katsura, J. Shimakura and K. Inui, *Mol. Pharmacol.*, 2009, **75**, 1280–1286.

52. J. W. Jonker, J. Freeman, E. Bolscher, S. Musters, A. J. Alvi, I. Titley, A. H. Schinkel and T. C. Dale, *Stem Cells*, 2005, **23**, 1059–1065.

53. M. L. Vlaming, S. F. Teunissen, E. van de Steeg, A. van Esch, E. Wagenaar, L. Brunsveld, T. F. de Greef, H. Rosing, J. H. Schellens, J. H. Beijnen and A. H. Schinkel, *Mol. Pharmacol.*, 2014, **85**, 520–530.

54. M. L. Vlaming, A. van Esch, Z. Pala, E. Wagenaar, K. van de Wetering, O. van Tellingen and A. H. Schinkel, *Mol. Cancer Ther.*, 2009, **8**, 3350–3359.

55. E. van de Steeg, C. M. van der Kruijssen, E. Wagenaar, J. E. Burggraaff, E. Mesman, K. E. Kenworthy and A. H. Schinkel, *Drug Metab. Dispos.*, 2009, **37**, 277–281.

56. E. van de Steeg, V. Stranecky, H. Hartmannova, L. Noskova, M. Hrebicek, E. Wagenaar, A. van Esch, D. R. de Waart, R. P. Oude Elferink, K. E. Kenworthy, E. Sticova, M. al-Edreesi, A. S. Knisely, S. Kmoch, M. Jirsa and A. H. Schinkel, *J. Clin. Invest.*, 2012, **122**, 519–528.

57. E. van de Steeg, A. van Esch, E. Wagenaar, K. E. Kenworthy and A. H. Schinkel, *Clin. Cancer Res.*, 2013, **19**, 821–832.

58. L. Salphati, X. Chu, L. Chen, B. Prasad, S. Dallas, R. Evers, D. Mamaril-Fishman, E. G. Geier, J. Kehler, J. Kunta, M. Mezler, L. Laplanche,

J. Pang, A. Rode, M. G. Soars, J. D. Unadkat, R. A. van Waterschoot, J. Yabut, A. H. Schinkel and N. Scheer, *Drug Metab. Dispos.*, 2014, **42**, 1301–1313.

59. N. Scheer, P. Balimane, M. D. Hayward, S. Buechel, G. Kauselmann and C. R. Wolf, *Drug Metab. Dispos.*, 2012, **40**, 2212–2218.

60. Y. Hu, Y. Xie, Y. Wang, X. Chen and D. E. Smith, *Mol. Pharm.*, 2014, **11**, 3737–3746.

61. X. Cui, D. Ji, D. A. Fisher, Y. Wu, D. M. Briner and E. J. Weinstein, *Nat. Biotechnol.*, 2011, **29**, 64–67.

62. B. Wefers, M. Meyer, O. Ortiz, M. Hrabe de Angelis, J. Hansen, W. Wurst and R. Kuhn, *Proc. Natl. Acad. Sci. U. S. A.*, 2013, **110**, 3782–3787.

63. P. M. Iannaccone and H. J. Jacob, *Dis. Models & Mech.*, 2009, **2**, 206–210.

64. H. Azuma, N. Paulk, A. Ranade, C. Dorrell, M. Al-Dhalimy, E. Ellis, S. Strom, M. A. Kay, M. Finegold and M. Grompe, *Nat. Biotechnol.*, 2007, **25**, 903–910.

65. J. A. Rhim, E. P. Sandgren, J. L. Degen, R. D. Palmiter and R. L. Brinster, *Science*, 1994, **263**, 1149–1152.

66. E. P. Sandgren, R. D. Palmiter, J. L. Heckel, C. C. Daugherty, R. L. Brinster and J. L. Degen, *Cell*, 1991, **66**, 245–256.

67. M. L. Washburn, M. T. Bility, L. Zhang, G. I. Kovalev, A. Buntzman, J. A. Frelinger, W. Barry, A. Ploss, C. M. Rice and L. Su, *Gastroenterology*, 2011, **140**, 1334–1344.

68. C. Tateno, Y. Yoshizane, N. Saito, M. Kataoka, R. Utoh, C. Yamasaki, A. Tachibana, Y. Soeno, K. Asahina, H. Hino, T. Asahara, T. Yokoi, T. Furukawa and K. Yoshizato, *Am. J. Pathol.*, 2004, **165**, 901–912.

69. M. Hasegawa, K. Kawai, T. Mitsui, K. Taniguchi, M. Monnai, M. Wakui, M. Ito, M. Suematsu, G. Peltz, M. Nakamura and H. Suemizu, *Biochem. Biophys. Res. Commun.*, 2011, **405**, 405–410.

70. Y. Igarashi, C. Tateno, Y. Tanaka, A. Tachibana, R. Utoh, M. Kataoka, H. Ohdan, T. Asahara and K. Yoshizato, *Xenotransplantation*, 2008, **15**, 235–245.

71. A. Tachibana, C. Tateno and K. Yoshizato, *Xenotransplantation*, 2013, **20**, 227–238.

72. C. H. Wu, E. C. Ouyang, C. M. Walton and G. Y. Wu, *J. Viral Hepatitis*, 2001, **8**, 111–119.

73. P. Breedveld, J. H. Beijnen and J. H. Schellens, *Trends Pharmacol. Sci.*, 2006, **27**, 17–24.

74. M. J. Zamek-Gliszczynski, J. C. Kalvass, G. M. Pollack and K. L. Brouwer, *Drug Metab. Dispos.*, 2009, **37**, 386–390.

75. H. Zaher, A. A. Khan, J. Palandra, T. G. Brayman, L. Yu and J. A. Ware, *Mol. Pharm.*, 2006, **3**, 55–61.

76. S. Shukla, H. Zaher, A. Hartz, B. Bauer, J. A. Ware and S. V. Ambudkar, *Pharm. Res.*, 2009, **26**, 480–487.

77. H. Kusuhara, H. Furuie, A. Inano, A. Sunagawa, S. Yamada, C. Wu, S. Fukizawa, N. Morimoto, I. Ieiri, M. Morishita, K. Sumita, H. Mayahara, T. Fujita, K. Maeda and Y. Sugiyama, *Br. J. Pharmacol.*, 2012, **166**, 1793–1803.
78. Y. Yamasaki, I. Ieiri, H. Kusuhara, T. Sasaki, M. Kimura, H. Tabuchi, Y. Ando, S. Irie, J. Ware, Y. Nakai, S. Higuchi and Y. Sugiyama, *Clin. Pharmacol. Ther.*, 2008, **84**, 95–103.
79. H. Kodaira, H. Kusuhara, J. Ushiki, E. Fuse and Y. Sugiyama, *J. Pharmacol. Exp. Ther.*, 2010, **333**, 788–796.
80. J. C. Kalvass, J. W. Polli, D. L. Bourdet, B. Feng, S. M. Huang, X. Liu, Q. R. Smith, L. K. Zhang and M. J. Zamek-Gliszczynski, *Clin. Pharmacol. Ther.*, 2013, **94**, 80–94.
81. R. Kikuchi, S. M. de Morais and J. C. Kalvass, *Drug Metab. Dispos.*, 2013, **41**, 2012–2017.
82. R. Zhao, T. J. Raub, G. A. Sawada, S. C. Kasper, J. A. Bacon, A. S. Bridges and G. M. Pollack, *Drug Metab. Dispos.*, 2009, **37**, 1251–1258.
83. Y. Uchida, S. Ohtsuki, Y. Katsukura, C. Ikeda, T. Suzuki, J. Kamiie and T. Terasaki, *J. Neurochem.*, 2011, **117**, 333–345.
84. N. A. de Vries, J. Zhao, E. Kroon, T. Buckle, J. H. Beijnen and O. van Tellingen, *Clin. Cancer Res.*, 2007, **13**, 6440–6449.
85. J. W. Polli, K. L. Olson, J. P. Chism, L. S. John-Williams, R. L. Yeager, S. M. Woodard, V. Otto, S. Castellino and V. E. Demby, *Drug Metab. Dispos.*, 2009, **37**, 439–442.
86. L. Salphati, L. B. Lee, J. Pang, E. G. Plise and X. Zhang, *Drug Metab. Dispos.*, 2010, **38**, 1422–1426.
87. L. Zhou, K. Schmidt, F. R. Nelson, V. Zelesky, M. D. Troutman and B. Feng, *Drug Metab. Dispos.*, 2009, **37**, 946–955.
88. S. Agarwal, Y. Uchida, R. K. Mittapalli, R. Sane, T. Terasaki and W. F. Elmquist, *Drug Metab. Dispos.*, 2012, **40**, 1164–1169.
89. I. Ieiri, S. Higuchi and Y. Sugiyama, *Expert Opin. Drug Metab. Toxicol.*, 2009, **5**, 703–729.
90. D. Iusuf, M. Ludwig, A. Elbatsh, A. van Esch, E. van de Steeg, E. Wagenaar, M. van der Valk, F. Lin, O. van Tellingen and A. H. Schinkel, *Mol. Cancer Ther.*, 2014, **13**, 492–503.
91. D. Iusuf, R. W. Sparidans, A. van Esch, M. Hobbs, K. E. Kenworthy, E. van de Steeg, E. Wagenaar, J. H. Beijnen and A. H. Schinkel, *Mol. Pharm.*, 2012, **9**, 2497–2504.
92. D. Iusuf, E. van de Steeg and A. H. Schinkel, *Trends Pharmacol. Sci.*, 2012, **33**, 100–108.
93. D. Iusuf, A. van Esch, M. Hobbs, M. Taylor, K. E. Kenworthy, E. van de Steeg, E. Wagenaar and A. H. Schinkel, *Mol. Pharmacol.*, 2013, **83**, 919–929.
94. M. J. Zamek-Gliszczynski, J. Q. Bao, J. S. Day and J. W. Higgins, *Drug Metab. Dispos.*, 2013, **41**, 1967–1971.
95. J. W. Higgins, D. W. Bedwell and M. J. Zamek-Gliszczynski, *Drug Metab. Dispos.*, 2012, **40**, 1170–1177.

96. D. C. Brater, *Br. J. Clin. Pharmacol.*, 2002, **54**, 87–95.
97. J. A. Williams, R. Hyland, B. C. Jones, D. A. Smith, S. Hurst, T. C. Goosen, V. Peterkin, J. R. Koup and S. E. Ball, *Drug Metab. Dispos.*, 2004, **32**, 1201–1208.
98. X. Tian, J. Li, M. J. Zamek-Gliszczynski, A. S. Bridges, P. Zhang, N. J. Patel, T. J. Raub, G. M. Pollack and K. L. Brouwer, *Antimicrob. Agents Chemother.*, 2007, **51**, 3230–3234.
99. K. T. Kivisto, O. Grisk, U. Hofmann, K. Meissner, K. U. Moritz, C. Ritter, K. A. Arnold, D. Lutjoohann, K. von Bergmann, I. Kloting, M. Eichelbaum and H. K. Kroemer, *Drug Metab. Dispos.*, 2005, **33**, 1593–1596.
100. K. M. Morrissey, S. L. Stocker, M. B. Wittwer, L. Xu and K. M. Giacomini, *Annu. Rev. Pharmacol. Toxicol.*, 2013, **53**, 503–529.
101. M. V. Varma, B. Feng, R. S. Obach, M. D. Troutman, J. Chupka, H. R. Miller and A. El-Kattan, *J. Med. Chem.*, 2009, **52**, 4844–4852.
102. M. J. Zamek-Gliszczynski, T. L. Abraham, J. J. Alberts, P. Kulanthaivel, K. A. Jackson, K. H. Chow, D. J. McCann, H. Hu, S. Anderson, N. A. Furr, R. J. Barbuch and K. C. Cassidy, *Drug Metab. Dispos.*, 2013, **41**, 714–726.
103. K. M. Hillgren, D. Keppler, A. A. Zur, K. M. Giacomini, B. Stieger, C. E. Cass and L. Zhang, *Clin. Pharmacol. Ther.*, 2013, **94**, 52–63.
104. S. L. Stocker, K. M. Morrissey, S. W. Yee, R. A. Castro, L. Xu, A. Dahlin, A. H. Ramirez, D. M. Roden, R. A. Wilke, C. A. McCarty, R. L. Davis, C. M. Brett and K. M. Giacomini, *Clin. Pharmacol. Ther.*, 2013, **93**, 186–194.
105. T. Imaoka, H. Kusuhara, M. Adachi, J. D. Schuetz, K. Takeuchi and Y. Sugiyama, *Mol. Pharmacol.*, 2007, **71**, 619–627.
106. A. M. Torres, A. V. Dnyanmote, K. T. Bush, W. Wu and S. K. Nigam, *J. Biol. Chem.*, 2011, **286**, 26391–26395.
107. J. M. Woodland, C. J. Barnett, D. E. Dorman, J. M. Gruber, C. Shih, L. A. Spangle, T. M. Wilson and W. J. Ehlhardt, *Drug Metab. Dispos.*, 1997, **25**, 693–700.
108. R. E. Morgan, M. Trauner, C. J. van Staden, P. H. Lee, B. Ramachandran, M. Eschenberg, C. A. Afshari, C. W. Qualls, Jr., R. Lightfoot-Dunn and H. K. Hamadeh, *Toxicol. Sci.*, 2010, **118**, 485–500.
109. E. I. Zimmerman, S. Hu, J. L. Roberts, A. A. Gibson, S. J. Orwick, L. Li, A. Sparreboom and S. D. Baker, *Clin. Cancer Res.*, 2013, **19**, 1458–1466.
110. X. Wen, B. Buckley, E. McCandlish, M. J. Goedken, S. Syed, R. Pelis, J. E. Manautou and L. M. Aleksunes, *Am. J. Pathol.*, 2014, **184**, 1299–1308.
111. M. Nishimura, H. Yoshitsugu, T. Yokoi, C. Tateno, M. Kataoka, T. Horie, K. Yoshizato and S. Naito, *Xenobiotica*, 2005, **35**, 877–890.
112. S. Ohtsuki, H. Kawakami, T. Inoue, K. Nakamura, C. Tateno, Y. Katsukura, W. Obuchi, Y. Uchida, J. Kamiie, T. Horie and T. Terasaki, *Drug Metab. Dispos.*, 2014, **42**, 1039–1043.
113. Y. Sato, H. Yamada, K. Iwasaki, C. Tateno, T. Yokoi, K. Yoshizato and I. Horii, *Toxicol. Pathol.*, 2008, **36**, 581–591.

114. J. R. Foster, M. Jacobsen, G. Kenna, T. Schulz-Utermoehl, Y. Morikawa, J. Salmu and I. D. Wilson, *Toxicol. Pathol.*, 2012, **40**, 1106–1116.
115. J. Shoda, K. Okada, Y. Inada, H. Kusama, H. Utsunomiya, K. Oda, T. Yokoi, K. Yoshizato and H. Suzuki, *Hepatol. Res.*, 2007, **37**, 548–556.
116. J. Kohlroser, J. Mathai, J. Reichheld, B. F. Banner and H. L. Bonkovsky, *Am. J. Gastroenterol.*, 2000, **95**, 272–276.
117. P. B. Watkins and R. W. Whitcomb, *N. Engl. J. Med.*, 1998, **338**, 916–917.
118. J. Konig, D. Rost, Y. Cui and D. Keppler, *Hepatology*, 1999, **29**, 1156–1163.
119. W. Jager, M. Sartori, W. Herzog and T. Thalhammer, *Res. Commun. Mol. Pathol. Pharmacol.*, 1998, **100**, 105–116.
120. B. M. Johnson, P. Zhang, J. D. Schuetz and K. L. Brouwer, *Drug Metab. Dispos.*, 2006, **34**, 556–562.
121. D. J. Newton, R. W. Wang and D. C. Evans, *Life Sci.*, 2005, **77**, 1106–1115.
122. K. Nezasa, X. Tian, M. J. Zamek-Gliszczynski, N. J. Patel, T. J. Raub and K. L. Brouwer, *Drug Metab. Dispos.*, 2006, **34**, 718–723.
123. P. Hsiao and J. D. Unadkat, *Mol. Pharm.*, 2012, **9**, 629–633.
124. L. Sasongko, J. M. Link, M. Muzi, D. A. Mankoff, X. Yang, A. C. Collier, S. C. Shoner and J. D. Unadkat, *Clin. Pharmacol. Ther.*, 2005, **77**, 503–514.
125. S. L. Koolen, R. A. van Waterschoot, O. van Tellingen, A. H. Schinkel, J. H. Beijnen, J. H. Schellens and A. D. Huitema, *J. Clin. Pharmacol.*, 2012, **52**, 370–380.
126. C. f. H. M. P. C. European Medicine Agency (EMA), 2012.
127. J. P. Pharmaceutical and Medical Devices Agency, www.pmda.go.jp/english, 2014.
128. F. a. D. A. U.S. Department of Health and Human Services, Center for Drug Evaluation and Research (CDER). http://www.fda.gov/downloads/Drugs/GuidanceComplianceRegulatoryInformation/Guidances/ucm292362.pdf, 2012.
129. L. Badolo, L. M. Rasmussen, H. R. Hansen and C. Sveigaard, *Eur. J. Pharm. Sci.*, 2010, **40**, 282–288.

Mechanistic Modelling to Predict Transporter-mediated Drug Disposition and Drug–Drug Interactions

RUI LI,[a] KRISHNA K. MACHAVARAM,[b] SIMON THOMAS[c] AND MANTHENA V. VARMA*[d]

[a] Systems Modeling and Simulation, Pharmacokinetics, Dynamics and Metabolism, Worldwide Research and Development, Pfizer Inc., 610 Main Street, Cambridge, MA 02139, USA; [b] Simcyp Limited (a Certara Company), Blades Enterprise Centre, John Street, Sheffield, S2 4SU, UK; [c] Scientific Computing Group, Cyprotex Discovery Ltd, 15 Beech Lane, Macclesfield, SK10 2DR, UK; [d] Pharmacokinetics, Dynamics and Metabolism, Worldwide Research and Development, Pfizer Inc., Eastern Point Road, Groton, CT 06340, USA
*Email: manthena.v.varma@pfizer.com

9.1 Introduction

Transporters of the solute carrier (SLC) and ATP binding cassette (ABC) superfamilies are expressed in a variety of organs, including the intestine, liver, kidney and brain, and can play a key role in drug disposition.[1–7] Additionally, transporters may govern exposure at the target site and determine adverse reactions and therapeutic efficacy. Understanding the role of drug transporters is particularly important in modern drug discovery, as the

RSC Drug Discovery Series No. 54
Drug Transporters: Volume 1: Role and Importance in ADME and Drug Development
Edited by Glynis Nicholls and Kuresh Youdim
© The Royal Society of Chemistry 2016
Published by the Royal Society of Chemistry, www.rsc.org

chemistry efforts to reduce cytochrome P450 (CYP)-mediated clearance and lipophilicity driving off-target pharmacology have led to a chemical space with an increased prevalence of transporter involvement in drug disposition.[8] The International Transporter Consortium (ITC) give an overview of transporters involved in drug absorption, distribution, metabolism and excretion (ADME), summarising the evidence for clinically-relevant drug–drug interactions (DDIs) and providing decision trees for the *in vitro* evaluation of drugs as substrates or inhibitors of seven transporters, including P-glycoprotein (P-gp), breast cancer resistance protein (BCRP), organic anion transporting polypeptides (OATPs) and organic anion and cation transporters (OATs and OCTs).[1]

It is well known that many transporter proteins are expressed in the human intestine.[4,9] Evidence suggests that influx transporters such as peptide transporter 1 (PEPT1), OATP2B1 and monocarboxylate transporter 1 (MCT1) facilitate, while efflux pumps P-gp and BCRP limit, the oral absorption of drugs. The human liver also contains a variety of transporters, including drug transporting proteins belonging to the gene families of *SLCO1B, SLCO2B, SLC22A, SLC10A, ABCB, ABCC* and *ABCG.* The OATP isoforms OATP1B1 (*SLCO1B1*), OATP1B3 (*SLCO1B3*) and OATP2B1 (*SLCO2B1*), together with OCT1 (*SLC22A1*), sodium taurocholate co-transporting polypeptide (NTCP; *SLC10A1*) and OAT2 (*SLC22A7*) are expressed on the sinusoidal membrane and are involved in the hepatic uptake of bile acids, hormones, steroid conjugates and numerous drugs (Figure 9.1).[2,10–15] Given the demonstrated clinical relevance of transporters such as OATP1B1 and OATP1B3, the European Medicines Agency (EMA), the Japanese Pharmaceuticals and Medical Devices Agency (PMDA) and the United States Food and Drug Administration (US FDA) regulatory guidances now recommend appropriate assessment of investigational drugs for potential transporter DDIs.[16–18] HMG-CoA reductase inhibitors (statins) represent an important therapeutic class for which OATPs may play a key role in determining their pharmacokinetics and the target site exposure that influences pharmacodynamic activity.[19–22] Polymorphisms in *SLCO1B1* (encoding OATP1B1) have been demonstrated to alter transporter activity leading to changes in systemic exposure for some statins and other drugs.[23–25] Therefore, it is relevant within drug discovery to assess OATP1B1 activity for new molecular entities (NMEs), *i.e.*, to identify substrates and inhibitors, quantitate transport kinetics and assess DDI liability. A wide variety of drugs are recognised by the canalicular efflux transporter proteins, including multidrug resistance protein 2 (MRP2; *ABCC2*), P-gp (*ABCB1*), bile salt export pump (BSEP; *ABCB11*), multidrug and toxin extrusion 1 (MATE1; *SLC47A1*) and BCRP (*ABCG2*) (Figure 9.1).[26,27] These efflux pumps can play an important role in the biliary clearance of drugs, regulating hepatocellular drug concentrations. In addition, the basolateral efflux transporters MRP3 (*ABCC3*), MRP4 (*ABCC4*) and MRP6 (*ABCC6*) may contribute to the hepatic disposition of substrate drugs, particularly for conjugated drug metabolites.[26,28,29] In the kidney, transporters expressed on the basolateral and apical sides of renal proximal tubule cells mediate the active renal secretion and reabsorption of

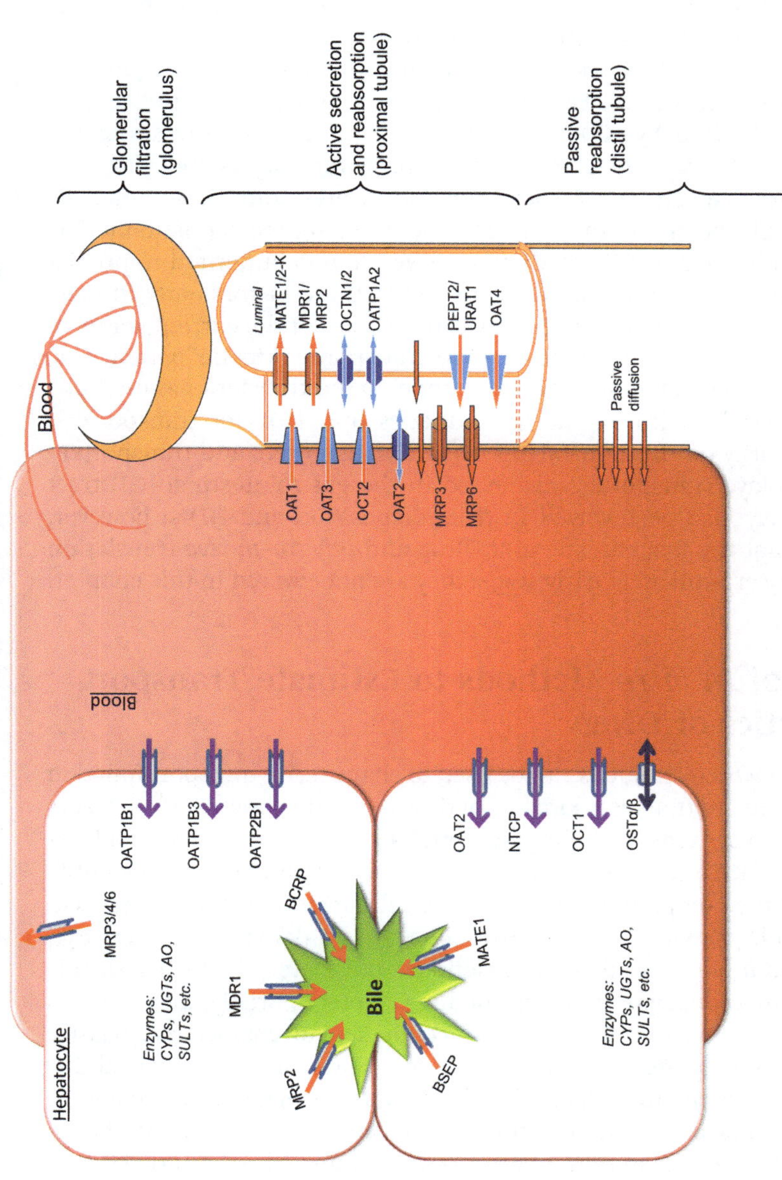

Figure 9.1 Membrane transporters known to be involved in the hepatobiliary and renal clearance of drugs. Transporter–enzyme interplay in hepatic clearance can be mechanistically described by the extended clearance concept, eqn (9.4), where the rate determining process for hepatic clearance is defined by metabolic/biliary clearance or uptake clearance, or a combination of both. Renal clearance is a function of glomerular filtration, active secretion and tubular reabsorption; and transporters play a key role in the active secretion process. AO: aldehyde oxidase; SULF: sulfotransferase.

several endogenous compounds and xenobiotics (Figure 9.1).[5,6] OCT2 (*SLC22A2*) and OAT1/3 (*SLC22A6/8*) are major transporters on the basolateral membrane involved in active renal secretion, whereas PEPT2 (*SLC15A2*) and urate anion transporter 1 (URAT1, *SLC22A12*) are more involved in active reabsorption of compounds. MATE1/2-K, P-gp and MRP2 are involved in the apical efflux of several drugs, driving renal secretion.

Transporters may also work in conjunction with metabolic enzymes in determining drug clearance. For instance, the OATP substrate atorvastatin is largely metabolised by CYP3A4, while repaglinide and cerivastatin are metabolised by CYP2C8 and CYP3A4. The transporter–enzyme interplay may determine the disposition of such dual substrates and these multiple processes should be simultaneously considered using mechanistic models for pharmacokinetic and DDI predictions. Overall, it is important to predict the pharmacokinetic variability associated with these factors early in drug development. In this chapter, we present an overview of the current status of the approaches used to estimate *in vitro* transporter kinetic parameters, together with the current and emerging model-based translational methods used to predict the clinical pharmacokinetics and DDIs of transporters, focusing primarily on those involving hepatic transporters and transporter–enzyme interplay. Transporters expressed in the gastrointestinal (GI) tract and kidney may also be involved in drug disposition and DDIs; however, considering that the mechanistic modelling and *in vitro–in vivo* translation of their activity are still at an early stage, they are not covered in this chapter.

9.2 Use of *In vitro* Methods to Estimate Transport Kinetics of Drugs

A number of *in vitro* techniques can be used to characterise transport function and further estimate transport kinetics of drugs and metabolites.[1,30,31] Assays using expression systems, including immortalised animal or human cell lines (*e.g.*, CHO, MDCKII and HEK), oocytes and vesicles, are often used to identify the affinity of the compound (substrate or inhibitor) to the transporters. On the other hand, primary cells (*e.g.*, hepatocytes) and derived cell lines (*e.g.*, Caco-2) are commonly used for translational purposes.[1,31] Further details about *in vitro* methods have been provided in Chapter 7. Briefly, there are two kinds of experiments that are routinely conducted to characterise passive diffusion and transporter-mediated processes. These include permeability assessments across cell monolayers in the apical to basolateral (A→B) and/or basolateral to apical (B→A) directions, and cell or vesicle accumulation measurements. Kinetic studies may also be conducted at multiple drug concentrations to determine *in vitro* parameters such as the K_m (Michaelis constant) or K_I (inhibition constant). Several approaches can be used to estimate transporter kinetics from these *in vitro* experiments. However, these estimates are usually based on the assumption that transporter kinetics obey enzyme kinetic principles, which may not always be true.[32]

9.2.1 Basic (Static) Approaches to Estimate Active and Passive Transport

The conventional two-step approach is used to evaluate transporter kinetics of drugs from studies using primary or cultured cells or expression systems.[33] Primarily, this approach assumes that the initial uptake rate is a function of the media concentration, considering the saturable active uptake (*i.e.*, Michaelis–Menten kinetics) and uni-directional passive diffusion (*i.e.*, non-saturable uptake) from media to cells. Here, the accumulation of a compound in cells is measured and plotted against time to obtain initial uptake rates. The obtained rates are then plotted against initial (nominal) medium concentrations to estimate maximal uptake rates (V_{max}), K_m and passive diffusion ($CL_{passive}$).[33] The cellular uptake rate can be measured at 37 and 4 °C. $CL_{passive}$ is obtained from the uptake rate at 4 °C using eqn (9.1), and is then introduced to eqn (9.2) as a constant to estimate the active uptake kinetic parameters.[34]

$$v = CL_{passive} \times S \tag{9.1}$$

$$v = \frac{V_{max} \times S}{K_m + S} + CL_{passive} \times S \tag{9.2}$$

where v is the uptake rate, V_{max} the maximum uptake rate, K_m the Michaelis constant and S the substrate concentration.

However, the conventional approach has a number of limitations, which include: (1) nominal (initial) concentrations in the media are used to estimate transporter kinetics; (2) the unbound drug concentration at the transporter binding site(s) and time variation of passive and active processes are not considered; (3) it assumes that the accumulation of intracellular compound is within the linear phase, which may not always be true; (4) it does not account for the experimental system/conditions used (*i.e.*, volumes of media, cells and membrane compartments are important in governing the drug concentration as well as driving the transport processes); and (5) this is a static model that considers uptake to be an independent process and does not account for the simultaneous impact of bidirectional passive diffusion, intracellular binding, metabolism or active efflux. In addition, if altered cellular membrane fluidity at 4 °C is used, temperature dependent passive diffusion could be a limitation and could potentially impact estimating the transporter kinetics.[33] To overcome this practical problem, conducting transport studies at 37 °C in the presence of specific active transport inhibitors is useful. However, the lack of specific inhibitors[35,36] and potential substrate-dependent inhibition are still limiting factors of this approach.[37] In assays using expression systems, saturable active uptake by the overexpressed transporter can be calculated by subtracting uptake into vector control cells from the transfected cells. Utilising a conventional approach, where time variation is not considered, the estimation of kinetic parameters could be inaccurate. If unreliable *in vitro* estimates were used as input parameters in an *in vitro–in vivo* extrapolation (IVIVE) linked physiologically-based pharmacokinetic (PBPK) model, the predicted outcomes would be deemed to be inaccurate and variable. To overcome some of the issues mentioned above, mechanistic compartmental models have, therefore, been proposed.[33,34,38]

9.2.2 Mechanistic (Dynamic) Approaches to Delineate Hepatic Uptake, Efflux and Metabolism

Unlike the conventional approaches, mechanistic models allow for a dynamic assessment of bidirectional passive diffusion between the cellular and media compartments.[34] Additionally compartmental approaches are commonly utilised to characterise uptake[39,40] and efflux transport,[41,42] metabolism,[34,43] and combinations of these processes. A schematic representation of the simplified mathematical models is shown in Figure 9.2.

In addition, mechanistic models can be useful for the model-driven design of cellular uptake experiments. For example, the duration of incubation required to estimate uptake and metabolism simultaneously could be predicted successfully from uptake kinetic parameters obtained from the two-compartment model and the metabolic clearance determined in the

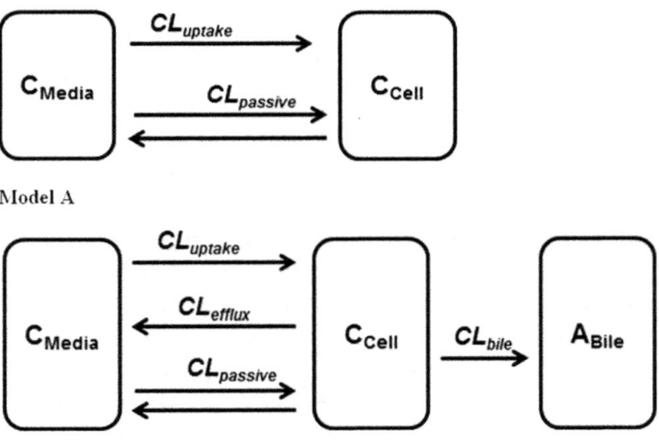

Figure 9.2 Schematic representation of simplified compartmental models describing hepatic uptake, efflux and passive diffusion. Model A: two-compartment model describing hepatic uptake and passive diffusion. Model B: three-compartment model describing cellular uptake, passive diffusion and biliary elimination. A_{Bile}: amount of substrate in bile; C_{Cell}: intracellular concentration of substrate; CL_{bile}: biliary clearance; CL_{efflux}: active basolateral efflux clearance; $CL_{passive}$: bidirectional passive diffusion clearance; CL_{uptake}: active uptake clearance; C_{Media}: substrate concentration in the media.

substrate depletion assays.[34] In the case of drugs undergoing both uptake and metabolism, a systematic investigation (in the presence or absence of the enzyme inhibitor 1-aminobenzotriazole (ABT)) can be used to differentiate phase I and II metabolic pathways and evaluate their impact on the estimation of uptake parameters. However, in the case of phase II enzymes (*e.g.*, for uridine 5-diphospho-glucuronosyltransferases (UGT)), selective enzyme inhibitors are not currently available. These mechanistic compartmental modelling approaches have also been applied to characterising kinetics in sandwich-cultured hepatocyte systems.[29,38,39,44] It should be noted that increasing the complexity of the data analysis could be useful for describing *in vitro* parameters more mechanistically, but this may demand more data inputs and may still result in parameter identifiability issues.

9.2.3 Permeability Models for Assessing Cellular Efflux and Transport

Monolayer systems have been used to identify efflux transporter substrates and to determine transport kinetic parameters. Direct application of enzyme kinetics to monolayer flux data can lead to inconsistencies in the estimation of efflux transport parameters. For example, P-gp K_m estimates varied up to 27-fold when estimated by applying Michaelis–Menten kinetics directly (using the basic static approach) to flux data from five different mono-layers,[39] but were within $\sim$2-fold when these data were analysed using a compartmental mechanistic model.[43] Schematic representations of these mathematical models are described in Figure 9.3. A number of reports have used three-compartment models to represent Caco-2 or MDCK cell monolayers.[41,43,45] Three-compartment models accounting for passive permeability, active uptake, efflux transport and metabolism in Caco-2 cells have been applied for parameter estimations, wherein the time-varying amount transported across the monolayer in absorptive and secretory directions is modelled.[43] Such mechanistic modelling of drug permeability has highlighted the importance of actual intracellular concentrations (at the binding site) and correct interpretation of drug transport studies for efflux transporters such as P-gp and BCRP. Using verapamil, quinidine and vinblastine as substrates, K_m values that were determined based on intracellular concentrations calculated using a three-compartment model were consistent for different cell systems, whereas apparent K_m ($K_{m,app}$), estimated based on extracellular substrate concentrations, varied greatly with transporter expression.[43] Use of calculated intracellular concentrations (instead of apical or basolateral media concentrations) resulted in more consistent kinetic parameters. Recently, Nagar *et al.*[46] compared a number of mechanistic compartment models (*i.e.*, three, five, six and nine compartments) to determine P-gp kinetics of six drugs using transwell monolayer assay systems. These mechanistic models were evaluated by considering the experimental variability and the impact of other factors such

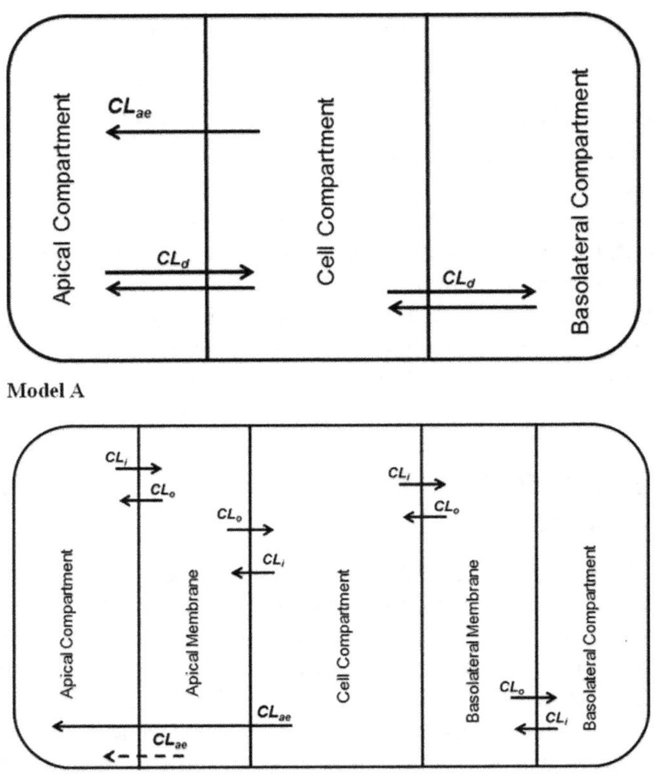

Figure 9.3 Schematic representation of compartmental models describing drug permeability and efflux transport in monolayer cell line assays. A three-compartment model (Model A) and a five-compartment model (Model B) describing drug efflux and passive diffusion. CL_{ae}: apical efflux clearance, modeled either out of the cell (solid arrow) or out of the apical membrane (dashed arrow); CL_d: diffusional clearance; CL_i and CL_o: diffusional clearances in and out of the membrane, respectively.
Figure adapted from ref. 47.

as membrane configuration, lipid content and apical surface area on model performance. Based on their analysis, a three-compartment model was indicated to be sufficient to predict the impact of transporters that efflux substrates directly from the cell (*i.e.*, the drug binds to the transporter in the cytosol and the intracellular concentration drives the active efflux process) and a five-compartment model with an explicit membrane compartment may be useful for predicting intracellular concentrations when efflux of substrates occurs directly from the membrane (*i.e.*, the drug binds to the transporter in the membrane and the membrane concentration drives the active efflux process). This study suggested that further complexity in models may not be required to determine efflux kinetics.[47] Overall, mechanistic compartmental models may provide a path forward to better characterise *in vitro* transporter data for *in vivo* extrapolation.

9.3 Pharmacokinetic Models for Hepatic Transporter Substrates

Mathematical modelling is frequently used within drug development, integrating information from physiology, physicochemistry and biochemistry to predict pharmacokinetics, and analysing data generated from preclinical or clinical studies. Various models, discussed in the following sections, along with their pros and cons, have been proposed.[31,48] For example, the whole body PBPK model has the potential to predict concentration–time profiles of a drug; however, it requires comprehensive understanding of the ADME of the drug, as well as various input information.

9.3.1 Static Model

The static model represents the simplest model for describing the pharmacokinetics of compounds. Generally, the static models for liver transporter substrates only focus on clearance. Clearance can be described based on the well-stirred liver model:[49]

$$\mathrm{CL_H} = \frac{Q_\mathrm{H} \times f_\mathrm{u,p} \times \mathrm{CL_{int,H}}}{Q_\mathrm{H} + f_\mathrm{u,p} \times \mathrm{CL_{int,H}}/R_\mathrm{B/P}} \tag{9.3}$$

where $\mathrm{CL_H}$, Q_H, $\mathrm{CL_{int,H}}$, $R_\mathrm{B/P}$ and $f_\mathrm{u,p}$ represent plasma hepatic clearance, hepatic blood flow, apparent hepatic 'intrinsic' clearance, blood to plasma ratio and fraction unbound in the plasma, respectively. $\mathrm{CL_{int,H}}$ can be described by an extended clearance term.[2,50–53]

$$\mathrm{CL_{int,H}} = \frac{\mathrm{CL_{passive}} + \mathrm{CL_{uptake}}}{\mathrm{CL_{passive}} + \mathrm{CL_{efflux}} + \mathrm{CL_{met+bile}}} \times \mathrm{CL_{met+bile}} \tag{9.4}$$

where $\mathrm{CL_{uptake}}$ and $\mathrm{CL_{efflux}}$ represent active hepatic uptake and efflux clearances across the sinusoidal membrane, respectively, and $\mathrm{CL_{met+bile}}$ represents intrinsic hepatic metabolic and biliary clearance. When a compound is subjected to active transport, particularly hepatic uptake mediated by OATPs, it is important to consider the extended clearance term. On the other hand, for highly permeable compounds with no substrate affinity to transporters, $\mathrm{CL_{int,H}}$ can be approximated by $\mathrm{CL_{met+bile}}$ and drug entry into hepatocytes is limited by Q_H. Historically, $\mathrm{CL_H}$ has also been predicted using allometric scaling methods;[54] however, such methods may suffer from species differences in the transporter and/or enzyme expression and activity.[55] A recent study for four Pfizer compounds indicated that allometric scaling yielded a less accurate prediction compared with whole body PBPK modelling incorporating *in vitro* data using human reagents.[38]

9.3.2 Empirical Compartment Model and Reduced PBPK Model

The empirical pharmacokinetics model typically has a central compartment representing plasma linked to one or two peripheral compartments *via* constants that quantify the rate of compound transfer between them. Since all the parameters in these models are estimated by fitting the observed pharmacokinetics of individual drugs, such models are useful in offering a concise and standard presentation of data, but not in extrapolating pharmacokinetics to similar drugs or to different physiological conditions.[56] To increase the predictive power of empirical compartment models, reduced PBPK (minimal PBPK or semi-mechanistic pharmacokinetics) models can be created by having a few compartments with physiological meaning that mechanistically capture the disposition, while grouping other tissues into central and peripheral compartments.[57] For example, by having additional compartments for liver blood and liver tissue, a reduced PBPK model has been applied to investigate the effect of OATP1B1 genotype on repaglinide pharmacokinetics and DDI risks.[58]

9.3.3 Whole Body PBPK Model

The whole body PBPK model is different to the reduced PBPK model because it has compartments for all major tissues.[59] To date, most published PBPK models for liver transporter substrates are full models.[52] In a generic PBPK model, tissues including the liver are connected by blood circulation and modelled assuming rapid equilibrium between blood and tissue compartments.[52,56] For example, assuming a well-stirred compartment, the liver is modelled as:

$$
\begin{aligned}
V_{\text{liver}} \frac{\mathrm{d}C_{\text{liver}}}{\mathrm{d}t} &= Q_{\text{H}} \times C_{\text{input}} - Q_{\text{H}} \times C_{\text{liver}} \times \frac{R_{\text{B/P}}}{\text{Kp}} \\
&\quad - \text{CL}_{\text{int,H}} \times C_{\text{liver}} \times \frac{R_{\text{B/P}}}{\text{Kp}} \times f_{\text{u,p}}
\end{aligned}
\tag{9.5}
$$

where V_{liver}, C_{liver}, C_{input} and Q_{H} represent the volume of the liver, drug concentration in the liver, concentration of mixed venous return from splanchnic organs and arterial blood, and liver blood flow, respectively. $f_{\text{u,b}}$ represents the fraction unbound in the blood, $R_{\text{B/P}}$ is the blood to plasma concentration ratio and Kp is the tissue–plasma partition coefficient for the liver. Non-eliminating tissues are modelled in a similar fashion but without the clearance term. The well-stirred liver model incorporating *in vitro* CL_{met} can reasonably predict the pharmacokinetics of structurally diverse molecules cleared *via* non-permeability-limited CYP-mediated clearance; however, it cannot adequately describe the pharmacokinetics of compounds actively transported into the liver due to permeability-limited hepatic disposition.[52] As such, a PBPK model incorporating permeability-limited distribution is proposed for hepatic transporter substrates, where the liver is split into extracellular (*i.e.*, liver blood and liver extracellular tissue) and intracellular (*i.e.*, liver intracellular tissue) compartments.[38,60,61]

$$V_{EC}\frac{dC_{EC}}{dt} = Q_H \times C_{input} - Q_H \times C_{EC}$$
$$- CL_{passive,u} \times [C_{EC} \times f_{u,p} - C_{IC} \times f_{u,cell}] - CL_{uptake,u} \times C_{EC} \times f_{u,b}$$

$$V_{IC}\frac{dC_{IC}}{dt} = CL_{passive,u} \times [C_{EC} \times f_{u,p} - C_{IC} \times f_{u,cell}] + CL_{uptake,u} \times C_{EC} \times f_{u,b}$$
$$- CL_{met,u} \times C_{IC} \times f_{u,cell} - CL_{bile,u} \times C_{IC} \times f_{u,cell}$$

$$(9.6)$$

where, V_{EC} and V_{IC} represent the volume of the extracellular and intracellular liver compartments, C_{EC} and C_{IC} represent extracellular and intracellular concentration, $f_{u,cell}$ is the fraction unbound in the hepatocytes, and the model assumes no sinusoidal efflux. To date, only a few studies have been done to incorporate non-hepatic transporter activity into PBPK models,[62,63] potentially due to the challenges in the *in vitro* to *in vivo* translation.

9.3.4 Characteristics and Applications of Different Pharmacokinetic Models for Transporter Substrates

Although all of the models mentioned above can describe pharmacokinetic time course data or characteristics (*e.g.*, total plasma clearance (CL), plasma area under the curve (AUC) and volume of distribution (V_d)), they have different strengths and weaknesses, and hence should be applied based on the question to be addressed. Due to their simpler structures, empirical compartment models are ideal for pharmacokinetic data characterisation, particularly for population data. For example, parameter values derived from different individuals can be compared, and values derived can be used to simulate the systemic concentration to aid in dosing regimen design. Empirical compartment models are also used to simulate systemic concentrations for pharmacodynamic modelling. However, because the general assumption that liver and plasma concentrations reach equilibrium instantaneously does not apply to liver transporter substrates, empirical pharmacokinetic models should not be used in conjugation with pharmacodynamic models for these compounds.[64] In addition, because the compartments of such models do not have physiological meaning, and all of the parameter values are estimated by fitting observed pharmacokinetics, these models in general cannot be used for forward pharmacokinetic prediction.[56]

Static models have an even simpler model structure than empirical compartment models, and are also generally used to generate data characteristics. For example:

$$CL = \frac{\text{Dose}}{\text{AUC}} \tag{9.7}$$

$$V_d = CL \times \frac{\text{AUMC}}{\text{AUC}} \tag{9.8}$$

where AUMC is the area under the first-order moment curve of the plasma concentration–time profile. Under certain assumptions, parameters with physiological meaning (*e.g.*, CL and V_d) can be potentially predicted from *in silico* approaches, *in vitro* assay data or allometric scaling.[52,65] The static model approach has also been used in DDI prediction if the change in plasma AUC is the only concern. However, by definition, the static model ignores the dynamic nature of the system; hence, description of the concentration–time course of the drugs or the time-varying interplay between the inhibitor and victim drug across different tissues cannot be achieved.

The whole body and reduced PBPK models both have compartments with physiological meaning, and have the ability to mechanistically incorporate *in vitro* estimated parameters such as CL_{uptake}, CL_{efflux}, $CL_{passive}$ and $CL_{met+bile}$. To model compound distribution into non-liver tissues, the reduced models only use the observed volumes of the central and peripheral compartments, while the full models use multiple tissue volumes and Kp for the same purpose. It should be noted that the two volumes in the reduced models are fitted,[58] and tissue volumes and Kp values in the full models are usually predicted by *in silico* approaches.[66] The structure of whole body PBPK models is more complicated, and hence they require more parameters than reduced models, but this does not necessarily mean that the full models have more uncertainty associated with their parameters, because the values of these additional parameters are usually fixed. Overall, in comparison to whole body PBPK models, reduced models have limited power to prospectively predict drug pharmacokinetics, due to the use of compound specific empirical parameters.

9.3.5 Pharmacokinetic Prediction and IVIVE of Transporter Activity

The transporter-related parameters (CL_{uptake}, $CL_{passive}$, CL_{met} and CL_{bile}) required by PBPK models for pharmacokinetic prediction can be estimated from multiple *in vitro* assays, as discussed in Section 9.2.[31] Similar to the requirements previously described for predicting metabolic clearance for compounds that are not transporter substrates,[67] PBPK models for transporter substrates require empirical scaling factors (SFs; additional to the physiological scaling) to accurately recapitulate human pharmacokinetics.[38,60,61,68] One contributing factor of SFs is the differing transporter abundance between *in vitro* and *in vivo* systems, leading to the concept of the relative expression factor (REF; *i.e.*, the ratio between *in vivo* and *in vitro* protein expression).[69] After accounting for the experimentally determined abundance difference, Bosgra *et al.* showed reasonable agreement between HEK cell based rosuvastatin pharmacokinetic prediction and clinical observation.[70] However, only minimal differences in protein expression are noted between the *in vitro* systems and liver tissue (based on the proteomics of OATPs),[71–73] implying that REF cannot always explain the large SFs for active uptake needed to recover the *in vivo* clearance of OATP substrate drugs.

Using sandwich-cultured human hepatocyte (SCHH) data, the empirical SFs for the active transport rate of OATP substrates can range from 12 to 161, which is much higher than the abundance difference reported in several recent protein expression studies on OATPs,[52] further suggesting that the abundance difference is not the only source of the SF in these studies. As such, in these studies, the SF is estimated by fitting models to observed animal or clinical data.[38,60,61,63,74] However, the introduction of these 'top-down' fitted empirical SFs belies the extrapolative power commonly attributed to PBPK. Furthermore, fitting the pharmacokinetics of individual compounds leads to compound specific SFs, raising concerns about the feasibility and reliability of prospective pharmacokinetic predictions for novel compounds. To increase the confidence in using SFs estimated by fitting observed human pharmacokinetics, Jones *et al.* estimated SFs for seven compounds individually and proposed to use the geometric mean value of these SFs for novel compounds.[38] However, when three SFs (*i.e.*, SFs for CL_{uptake}, $CL_{passive}$, and CL_{met} or CL_{bile}) are estimated simultaneously by fitting the observed plasma concentration–time profiles of individual compounds, the model can be over-parameterised and results in many seemingly acceptable SF combinations.[38,52] As such, when multiple SFs are simultaneously estimated by fitting systemic plasma data, it is necessary to perform statistical analyses to understand if these SFs can be estimated with enough confidence. In previous work, in order to decrease the uncertainty in parameter identification, the SF of CL_{uptake} was estimated, while SFs of $CL_{passive}$ and CL_{met} were assumed to be 1,[38] although this assumption still needs necessary validation.[52] For example, assuming SF for $CL_{passive}$ of 1 requires evidence that the *in vitro* surface area between the hepatocytes and assay buffer (or blood) is equal to the *in vivo* value. In a follow-up study, a non-numerical global optimisation method (*i.e.*, the 'brute force' grid search method) that simultaneously leverages data for seven compounds when searching for a unique set of four SFs was applied, and determined that such SFs could be identified within relatively narrow ranges by visualising the SF space and associated total residual sum of squares between the observation and simulation.[75] The globally fitted SFs also have an overall better fit on the pharmacokinetics of the seven compounds compared with the geometric mean of individually fitted SFs. In the same study, statistical methods were proposed to construct confidence intervals for globally optimised SFs, as well as prediction intervals for the simulated plasma concentration–time courses.[75] However, considering that the fraction of each transporter contributing to the total active uptake may be different for different compounds, this approach essentially assumes that all transporters share a similar SF, which requires further validation. In order to estimate transporter-specific SFs, one may need to know the contribution each transporter made to the total active uptake, which has been previously approximated by simultaneously using data from overexpression systems and hepatocytes.[69] In addition, one may need data for many more training compounds (potentially including transporter-selective substrates) in simultaneous fitting to estimate IVIVE SFs for individual transporters. It should be noted that empirical SFs are not only specific to the model structure, but also the laboratory specific implementation of *in vitro* experimental systems.

9.3.6 Determining Values of Other Key Parameters in PBPK Models

9.3.6.1 *Partition Coefficients (*K_p*)*

K_p is defined as the partition coefficient between tissue and plasma. K_p values in PBPK models are usually estimated using *in silico* methods based on tissue composition,[38,60,61] due to the lack of methods to acquire data in human tissue. Early *in silico* methods focused on water–lipid partitioning in tissues.[76] Drug interactions with extracellular proteins were incorporated into later methods,[77] leading to an improved prediction accuracy for rodent K_p and V_d.[78] For acids, these methods may still mispredict K_p because there are other possible mechanisms missing, including specific or nonspecific binding to intracellular protein, potentially unknown active process(es) or compound exclusion due to membrane potential.[52,79] K_p values have also been experimentally determined using rodents or predicted using an empirical regression model based on rodent V_d and/or other physiological properties of the compound observed in rodents.[80] However, these methods ignore potential species differences, which may result in misprediction in humans.

It is challenging to validate K_p values predicted from the above methods, again because of a lack of techniques to monitor drug concentration in human tissues. In a few cases, predicted K_p values result in V_d values deviating from the observed value to such an extent that the model cannot describe the data even when hepatic clearances or SFs are fitted.[66] However, in other cases, to compensate for under- or over-predicted V_d resulting from incorrect K_p estimates, hepatic uptake clearance or SFs can be over- or under-estimated when fitting plasma concentration–time courses. In a recently developed method to validate K_p values using positron emission tomography (PET) imaging, using telmisartan as an example, the authors converted the radioactivity detected in the human tissues into a corresponding telmisartan

K_p value.[79] In the example, compared with *in silico* predicted and rodent K_p, the PET derived K_p led to a larger V_d and the prediction better matched the data, particularly during the distribution phase.[79] However, since this example only used a single compound, it is still unclear how well the current *in silico* methods predict actual human K_p values. In addition, although in the telmisartan PET study only a very limited amount of metabolite was detected in systemic plasma, for other compounds disposed through metabolism, the PET signal will be a combination of both parent and metabolite(s).[81]

9.3.6.2   Plasma Unbound Fraction (f_{u,p}), Blood to Plasma Ratio (R_{B/P}) and Liver Unbound Fraction (f_{u,liver})

Among the compound-specific input parameters in PBPK models, simulated plasma concentration–time profiles are also sensitive to $f_{u,p}$ and $R_{B/P}$.[38,79] Different assays performed by different laboratories often lead to very different values. For example, the $f_{u,p}$ of valsartan has been reported as 0.04[82] and as 0.001.[38] These differences in estimated values would mean different SFs to recover the observed clearance. The $R_{B/P}$ of acids is, however, relatively consistent considering that most acids cannot penetrate the red blood cell membrane, hence $R_{B/P}$ is essentially the ratio of volumetric fractions between plasma and blood. Simulated concentrations in the liver, but not in plasma, are usually sensitive to $f_{u,liver}$, although the parameter does have some impact on the elimination phase of the plasma pharmacokinetics by affecting hepatic metabolism and biliary excretion.[75,79] As such, for most compounds, active uptake and passive diffusion SFs estimated by fitting plasma data are not likely to be confounded by this parameter. However, one important role of PBPK is to simulate liver concentration for DDI and pharmacodynamic predictions, which can consequently be seriously affected by the value of $f_{u,liver}$. Unfortunately, to date, there is no ideal method to determine this value. Assuming the albumin concentration in the liver is half of that in the plasma, $f_{u,liver}$ has been calculated based on $f_{u,p}$ values.[38] It has also been experimentally determined using a dialysis assay with homogenised human liver tissue.[75] When such $f_{u,liver}$ values are used in their PBPK models, the authors assume that the intact membrane has limited impact on the equilibrium concentration in the liver. By using the liver model structure initially proposed by Rodgers and Rowland,[77] Jamei *et al.* bypassed the problem of determining overall $f_{u,liver}$.[63] However, the Rodgers and Rowland model assumes that intracellular binding solely depends on binding to neutral phospholipids, neutral lipids and acidic phospholipids, an assumption that may be limiting, as discussed above for Kp prediction methods.

9.3.6.3   Physiological Parameters

PBPK simulations are in general also sensitive to physiological parameters that affect hepatic clearance. These parameters include liver size, number of hepatocytes (or weight of microsome) per unit of liver weight, liver blood flow (as a proportion of cardiac output), as well as cardiac output itself. The values of these parameters are usually obtained from the published mean values,[65] which may not necessarily be the same as the values of the small group of individuals from whom the pharmacokinetic data are collected. Such heterogeneity involved in physiological parameters is likely and not necessarily readily resolvable. Because these parameters are fixed in the PBPK model, they contribute to the variable SF values and the residual errors in the predictions.

9.4   Transporter-mediated DDIs

DDIs refer to the phenomenon whereby the administration of one drug alters the pharmacokinetics of another drug *in vivo*. The drug causing the effect can be referred to as the 'perpetrator' and the drug that is affected as the 'victim'. Pharmacokinetic interactions can occur through a number of mechanisms, including direct inhibitory mechanisms of the perpetrator at enzymes or transporters, or induction of proteins that influence the pharmacokinetics of the victim.[83] Thus, if one compound (perpetrator) inhibits the active transport into the liver of a second (victim), the initial effects are that the concentration of the victim in the plasma will increase, relative to the concentrations in the absence of the perpetrator. These changes could thus alter the pharmacological and toxicological effects. Interactions are usually strongly influenced by the relative times of administration and the amount of each drug administered. In general, the closer in time the two drugs are administered the greater the chance of an interaction between them. At one extreme, in cases when interaction is solely due to direct inhibition of enzymes or transporters, the perpetrator must still be present in the body. In cases involving protein induction or time-varying inhibition mechanisms, it is possible for the interaction effect to be present even after the perpetrator has been cleared from the body. Finally, it should be remembered that two compounds can be both perpetrator and victim with respect to one another.[84]

DDIs involving direct inhibition of transport by the perpetrator are the most frequently studied. Nevertheless, other mechanisms, including induction of transporter activity by means of drug-induced transcriptional activation are known.[85] The following sub-sections contain descriptions of two major modelling approaches to predict the magnitude of DDIs occurring *via* inhibition of hepatic transport. In 'static' approaches, the concentration of the inhibitor is considered at a fixed value. The determination of an appropriate inhibitor concentration is crucial to acceptable prediction by static methods. Conversely, 'dynamic' approaches explicitly allow for the variation of inhibitor concentration over time, usually by means of using PBPK models. Comprehensive reviews[48,52] and recommendations[86,87] regarding these approaches have been published recently.

9.4.1   Static Approaches

Changes in the pharmacokinetics of a victim drug can be expressed in terms of changes in pharmacokinetic parameters such as $C_{\max}$ or AUC. For analytical purposes, AUC has the advantage of being related to clearance, as mentioned above in eqn (9.7). Consequently, an increase in the AUC in the presence of a clearance inhibitor can be related to the causative decrease in clearance:

$$R\text{-value} = \frac{\text{AUC}'}{\text{AUC}} = \frac{\text{CL}}{\text{CL}'} \tag{9.9}$$

where AUC' and CL' are, respectively, the area under the curve and clearance in the presence of the inhibitor, and the *R-value* is defined as the ratio of AUCs in the absence and presence of the inhibitor. For a compound whose clearance is mediated by active transport into the liver, the rate can be defined in terms of standard enzyme kinetics:

$$v = \frac{V_{\max} \times S}{K_m + S} = \frac{V_{\max}/K_m \times S}{1 + S/K_m} \tag{9.10}$$

where v is the rate of uptake and $V_{\max}$ its maximum rate, K_m the Michaelis constant, and S the substrate concentration. Making an assumption of linearity with respect to substrate concentration (*i.e.*, $S \ll K_m$), and defining active uptake clearance ($\text{CL}_{\text{uptake}}$) as V_m/K_m:

$$v \approx \text{CL}_{\text{uptake}} \times S \tag{9.11}$$

The effect of a competitive inhibitor can be easily incorporated into this framework:

$$v' = \frac{V_{\max} \times S}{K_m \times \left(1 + \dfrac{I}{K_I}\right) + S} = \frac{\dfrac{V_{\max}}{K_m} \times S}{1 + \dfrac{I}{K_I} + \dfrac{I}{K_m}} \tag{9.12}$$

where v' is the uptake rate in the presence of the inhibitor, I is the inhibitor concentration and K_I its inhibition constant with respect to the uptake of the victim compound. Making the same assumption of linearity with respect to substrate concentration:

$$v' \approx \frac{\text{CL}_{\text{uptake}} \times S}{1 + \dfrac{I}{K_I}} \tag{9.13}$$

From the above, the uptake clearance in the presence of the inhibitor ($\text{CL}_{\text{uptake}}'$) is given by:

$$\text{CL}_{\text{uptake}}' = \frac{\text{CL}_{\text{uptake}}}{1 + \dfrac{I}{K_I}} \tag{9.14}$$

If total clearance is equated to uptake clearance, then:

$$\frac{\text{AUC}'}{\text{AUC}} = \frac{\text{CL}_{\text{uptake}}}{\text{CL}_{\text{uptake}}'} = 1 + \frac{I}{K_{\text{I}}} \tag{9.15}$$

This can be generalised to multiple interacting inhibitors:

$$\frac{\text{AUC}'}{\text{AUC}} = \frac{\text{CL}_{\text{uptake}}}{\text{CL}_{\text{uptake}}'} = 1 + \sum_{j=1}^{n} \frac{I}{K_{\text{I},j}} \tag{9.16}$$

where $K_{\text{I},j}$ is the inhibition constant of the jth inhibitor. Under these assumptions, the presence of an inhibitor at a constant concentration equal to the K_{I} will result in a doubling of the AUC of the victim compound. For operational purposes, this equality provides a convenient way of defining significant interactions (R-*value* > 2), as well as false positive ($I/K_{\text{I}} > 1$, but R-*value* < 2) and false negative ($I/K_{\text{I}} < 1$, but R-*value* > 2) predictions. *In vivo*, the inhibitor concentration, I, is a time-varying quantity. Consequently, a static approach to predict DDIs is a conservative approximation to avoid clinical risk.

Based on DDIs involving hepatic metabolic clearance, Ito *et al.* showed that the estimated maximum total plasma concentration at the inlet to the liver (I_{in}) provided improved predictions with the fewest false negative predictions.[88]

$$I_{\text{in}} = I_{\text{av}} + \frac{k_{\text{a}} \times F_{\text{a}} \times D}{Q_{\text{h}}} \tag{9.17}$$

where k_{a} is the rate constant for absorption from the GI tract, F_{a} the fraction absorbed from the GI tract, Q_{h} the hepatic blood flow, D the dose and I_{av} the average systemic plasma concentration after repeated oral administration:

$$I_{\text{av}} = \frac{D/\tau}{\text{CL}/F} \tag{9.18}$$

Here, τ is the dosing interval, CL the clearance and F the bioavailability of the inhibitor.

Another study[89] considered the case of multiple uptake routes, where only one route is inhibited:

$$\frac{\text{AUC}'}{\text{AUC}} = \frac{1}{\dfrac{f_{\text{t}}}{1 + \sum\limits_{j=1}^{n} \dfrac{I}{K_{\text{I},j}}} + 1 - f_{\text{t}}} \tag{9.19}$$

where $K_{\text{I},j}$ is the inhibition constant of the jth inhibitor, f_{t} is the fraction of victim compound transported, in the absence of inhibitor, by a specific transporter, and I is the average plasma concentration, I_{av}, given by eqn (9.18). In the limit of $f_{\text{t}} = 1$, eqn (9.19) reduces to eqn (9.16), as expected.

On the other hand, when considering the simultaneous inhibition of hepatic uptake and either intrahepatic metabolism and/or hepatic efflux, the following model has been proposed for the overall increase in AUC, which is referred to as an *R-value* product:[87]

$$\frac{\text{AUC}'}{\text{AUC}} = \frac{\text{CL}}{\text{CL}'} = \left(1 + \sum_{j=1}^{n} \frac{I}{K_{\text{Iu},j}}\right) \times \left(1 + \sum_{j=1}^{n} \frac{I}{K_{\text{Ie},j}}\right) \tag{9.20}$$

where $K_{\text{Iu},j}$ and $K_{\text{Ie},j}$ are, respectively, the inhibition constants for hepatic uptake and elimination (whether by efflux or by metabolism) of the jth inhibitor. The *R-value* product is an extension of eqn (9.16), expressing the concept that the decreases in clearance for eliminating processes that act in series are multiplicative.

Generally, *R-value* and *R-value* product models provide an oversimplification of the transporter and transporter/enzyme-mediated DDI risks, and a predicted AUC ratio >2 suggests that more sophisticated modelling approaches with more refined static models or whole body PBPK models are needed for quantitative DDI predictions.[1,68,74,90–92] For DDI predictions involving transporters and enzymes, the 'extended net effect' model, eqn (9.21), which is based on the extended clearance term, eqn (9.4), could be considered.[92] In the presence of a perpetrator, the expected net effect of reversible inhibition of uptake or biliary efflux, and reversible inhibition, time-dependent inhibition and induction of CYPs can be assessed with this model:

$$\text{CL}'_{\text{int,H}} = \left(\frac{\text{CL}_{\text{uptake}}}{\text{RI}_{\text{H}}} + \text{CL}_{\text{passive}}\right) \times \frac{\left(\sum \dfrac{\text{CL}_{\text{CYP}}}{\text{RI}_{\text{H}} \times \text{TDI}_{\text{H}} \times \text{IND}_{\text{H}}} + \dfrac{\text{CL}_{\text{bile}}}{\text{RI}_{\text{H}}}\right)}{\left(\text{CL}_{\text{passive}} + \sum \dfrac{\text{CL}_{\text{CYP}}}{\text{RI}_{\text{H}} \times \text{TDI}_{\text{H}} \times \text{IND}_{\text{H}}} + \dfrac{\text{CL}_{\text{bile}}}{\text{RI}_{\text{H}}}\right)} \tag{9.21}$$

where RI_{H} is the competitive inhibition term, TDI_{H} the time-dependent inhibition term and IND_{H} the hepatic induction term.[92–94] While comprehensive experimental data on the individual intrinsic clearances are required for applying the extended net effect model, it incorporates the mechanistic aspects of transporter–enzyme interplay and it has been proposed that this method will be capable of producing quantitative predictions.[92–94]

9.4.2 Dynamic Approaches

DDI prediction using a dynamic model that captures the time variable inhibitor concentration should, in theory, be more reliable than the static approaches. PBPK models can more readily incorporate fractional transport or metabolism along with considerations of non-hepatic disposition processes. Also, PBPK models can help to consider extremes in population variability by incorporating variability into the *in vivo* drug disposition and polymorphic clearance pathways, or by simulating drug disposition in disease states or special populations.[95]

Limited examples exist demonstrating the utility of PBPK modelling for transporter-mediated DDIs.[58,63,66,68,74,96,97] An early study on the effect of cyclosporine, gemfibrozil and rifampicin on the pharmacokinetics of pravastatin, mediated largely *via* inhibition of OATP1B1, illustrated some of the challenges in predicting transporter-mediated pharmacokinetics and DDIs.[68] Passive and active uptake into hepatocytes was measured using SCHH, and the inhibition potency of inhibitors against OATP1B1 was obtained from the literature. Simulation of intravenous pravastatin pharmacokinetics in the control group yielded under-prediction of hepatic clearance, requiring a SF ($\sim$31) for hepatic uptake derived by optimisation of the uptake rate against the observed plasma concentrations. As demonstrated with the pravastatin PBPK approach, application of empirical SFs for uptake clearance alone was proven to successfully recover a variety of DDIs for transporter and enzyme substrates.[63,66,68,74,93,98] For example, repaglinide, a substrate of OATP1B1 that is mainly metabolised by CYP2C8 and CYP3A4,[99–103] shows a wide range of DDIs with various perpetrator drugs. In particular, gemfibrozil causes about a 5–8-fold increase in the repaglinide AUC. Based on *in vitro* studies, the major circulating metabolite of gemfibrozil (gemfibrozil 1-O-β-glucuronide) showed time-dependent inhibition of CYP2C8,[104] with both parent and metabolite inhibiting OATP1B1.[105,106] The hepatic transport of repaglinide was characterised using the SCHH model, and a PBPK model was developed to predict its pharmacokinetics and DDIs.[74,94] While an empirical SF (16.9) for hepatic active uptake was necessary to recover the pharmacokinetics of repaglinide, the optimised PBPK model was able to closely predict repaglinide DDIs with gemfibrozil and other perpetrator drugs. Reduced PBPK models using a small system of equations have also been used to explore transporter-mediated DDIs. For example, a model was used to predict the impact of OATP1B1 inhibition by cyclosporine on the pharmacokinetics of repaglinide, although no *in vivo* DDI data were provided for comparison with the predictions.[58]

The expected advantage of dynamic over static approaches derives from the ability to provide more realistic estimates of the relevant inhibitor concentration, I_j. This advantage stems, in turn, from two sources. First, predicting the time-dependence of the inhibitor concentration obviates the necessity to select a suitable static concentration that is intended to approximate the actual varying concentration. Second, such models enable the simulation of the inhibitor concentration at the active site of the transporter, removing the need to make approximations of that concentration. This, in turn, places the onus on dynamic methods to reliably predict inhibitor concentrations. For many perpetrator drugs with pharmacokinetic data available from clinical studies, this should be relatively straightforward, as models can be optimised against clinical data.[38] For potential perpetrator compounds in preclinical stages, the challenge is to predict the concentration from existing *in vitro* ADME and animal *in vivo* pharmacokinetics data.[107,108]

Despite the expectations regarding dynamic approaches, there is limited evidence for their significant superiority of prediction accuracy over static models. For example, a study comparing the predictions of 35 metabolism-related interactions by both PBPK and static approaches[109] found that, of the 35 interactions, the static approach predicted 77% of the AUC increase within two-fold of the observed values, compared with about 71% for the PBPK approach. For predictions within 1.5-fold of the observed values, the static method (60% within 1.5-fold) outperformed the PBPK approach (43% within 1.5-fold) by a greater margin. A study of repaglinide–rifampicin interactions involving both inhibition of OATP1B1 and induction of CYP3A4 by rifampicin showed similar performance on AUC prediction by both the static and PBPK methods.[92] Nevertheless, dynamic methods offer significantly greater scope and flexibility than static approaches. In particular, static approaches are predominantly expressed in terms of linearised victim pharmacokinetics, changes in inhibitor-mediated clearance and consequent changes in plasma AUC. Dynamic (particularly PBPK) approaches offer the scope of predicting changes in multiple metrics, including predicted victim concentration at any time in any compartment included within the model. By this means they offer the potential of providing information on DDI induced changes in victim drug concentrations at target and non-target sites.

9.4.3 Limitations of Current Approaches for DDI Predictions

As previously discussed, the prediction of DDIs mediated *via* transporters is still limited compared with the study of metabolism-mediated DDIs,[88,110,111] but substantial progress has been made in developing both static and dynamic approaches in recent years.[58,63,66,68,74,86,87,89,92,93,112] Some of the areas that need further resolution are:

- The reliability of the inhibition constant, K_I, and its relevance to the *in vivo* situation, which impacts both static and dynamic approaches.
- The mismatch between uptake rates as measured using *in vitro* systems and the rates required for PBPK model predictions to match *in vivo* plasma concentrations. This phenomenon, which has been noted using SCHH and other systems, prevents the reliable prediction of the plasma concentration of the victim compound in the absence of measured *in vivo* data against which the actual uptake rate can be optimised.
- In general, both static and dynamic approaches have drawn conclusions from relatively small numbers of compounds. In particular, the necessity to scale transporter-mediated hepatic uptake for the PBPK approach has made the availability of *in vivo* plasma concentration data for the victim a requisite. These data will not be available for novel compounds that are potential victims, in which case validated SFs should be used in the early predictions with an emphasis on refining the models as the clinical data become available.

9.5 Summary

Hepatic transporters play an important role in the disposition of many drugs; hence, it is important to be able to predict their pharmacokinetics and DDIs based on transport kinetics and consideration of transporter–enzyme interplay. Recently, mechanism-based IVIVE modelling approaches have become more valued, with enormous progress being made in developing both assays and modelling methodologies. Although a mechanism based approach is commonly considered to be more powerful in extrapolation than static models, difficulties still exist in reliably and accurately translating *in vitro* transporter data for *in vivo* prediction. These difficulties are largely due to limited understanding of the functional activity of both *in vitro* and *in vivo* systems, and the lack of reliable protein abundance information. Data from experimental techniques that could better mimic *in vivo* systems, and new non-invasive techniques that could measure compound distribution in human tissues, may better inform model building for reliable, prospective predictions of transporter disposition and DDIs. Nonetheless, several reports discussed in this chapter provide increasing confidence regarding the model-based predictions of hepatic transporter-mediated disposition and DDIs for novel compounds. More efforts are required in incorporating non-hepatic (*e.g.*, GI tract and kidney) transporter activity into such predictions.

References

1. K. M. Giacomini, S. M. Huang, D. J. Tweedie, L. Z. Benet, K. L. Brouwer, X. Chu, A. Dahlin, R. Evers, V. Fischer, K. M. Hillgren, K. A. Hoffmaster, T. Ishikawa, D. Keppler, R. B. Kim, C. A. Lee, M. Niemi, J. W. Polli, Y. Sugiyama, P. W. Swaan, J. A. Ware, S. H. Wright, S. W. Yee, M. J. Zamek-Gliszczynski and L. Zhang, *Nat. Rev. Drug Discovery*, 2010, **9**, 215.
2. Y. Shitara, T. Horie and Y. Sugiyama, *Eur. J. Pharm. Sci.*, 2006, **27**, 425.
3. H. Koepsell, K. Lips and C. Volk, *Pharm. Res.*, 2007, **24**, 1227.
4. M. V. Varma, C. M. Ambler, M. Ullah, C. J. Rotter, H. Sun, J. Litchfield, K. S. Fenner and A. F. El-Kattan, *Curr. Drug Metab.*, 2010, **11**, 730.
5. B. Feng, J. L. LaPerle, G. Chang and M. V. Varma, *Expert. Opin. Drug Metab. Toxicol.*, 2010, **6**, 939.
6. W. Lee and R. B. Kim, *Annu. Rev. Pharmacol. Toxicol.*, 2004, **44**, 137.
7. D. J. Begley, *Curr. Pharm. Des.*, 2004, **10**, 1295.
8. J. D. Hughes, J. Blagg, D. A. Price, S. Bailey, G. A. Decrescenzo, R. V. Devraj, E. Ellsworth, Y. M. Fobian, M. E. Gibbs, R. W. Gilles, N. Greene, E. Huang, T. Krieger-Burke, J. Loesel, T. Wager, L. Whiteley and Y. Zhang, *Bioorg. Med. Chem. Lett.*, 2008, **18**, 4872.
9. J. R. Kunta and P. J. Sinko, *Curr. Drug Metab.*, 2004, **5**, 109.
10. C. Fahrmayr, M. F. Fromm and J. Konig, *Drug Metab. Rev.*, 2010, **42**, 380.

11. K. S. Fenner, H. M. Jones, M. Ullah, S. Kempshall, M. Dickins, Y. Lai, P. Morgan and H. A. Barton, *Xenobiotica*, 2011, **42**, 28.

12. P. Borst, R. Evers, M. Kool and J. Wijnholds, *J. Natl. Cancer Inst.*, 2000, **92**, 1295.

13. B. Hagenbuch and P. Meier, *Biochim. Biophys. Acta, Biomembr.*, 2003, **1609**, 1.

14. B. Hagenbuch and P. J. Meier, *Pflügers Archiv.*, 2004, **447**, 653.

15. I. Tamai, J. Nezu, H. Uchino, Y. Sai, A. Oku, M. Shimane and A. Tsuji, *Biochem. Biophys. Res. Commun.*, 2000, **273**, 251.

16. U.S. Food and Drug Adminstration. *Drug Interaction Studies— Study Design, Data Analysis, Implications for Dosing, and Labeling Recommendations FDA Guidance for Industry Center for Drug Evaluation and Research (CDER)* 2012, http://www.fda.gov/downloads/drugs/guidancecomplianceregulatoryinformation/guidances/ucm292362.pdf.

17. European Medicines Agency. *Guideline on the Investigation of Drug Interactions European Medicines Agency Guidline Committee for Human Medicinal Products (CHMP)* 2012, http://www.ema.europa.eu/docs/en_GB/document_library/Scientific_guideline/2012/07/WC500129606.pdf.

18. Y. Saito, K. Maekawa and Y. Ohno, *Regul. Sci. Med. Prod.*, 2014, **4**, 249.

19. Y. Shitara and Y. Sugiyama, *Pharmacol. Ther.*, 2006, **112**, 71.

20. S. C. Group, E. Link, S. Parish, J. Armitage, L. Bowman, S. Heath, F. Matsuda, I. Gut, M. Lathrop and R. Collins, *N. Engl. J. Med.*, 2008, **359**, 789.

21. A. Kalliokoski and M. Niemi, *Br. J. Pharmacol.*, 2009, **158**, 693.

22. M. Niemi, M. K. Pasanen and P. J. Neuvonen, *Pharmacol. Rev.*, 2011, **63**, 157.

23. Y. Lai, M. Varma, B. Feng, J. C. Stephens, E. Kimoto, A. El-Kattan, K. Ichikawa, H. Kikkawa, C. Ono and A. Suzuki, *Expert Opin. Drug Metab. Toxicol.*, 2012, **8**, 723.

24. M. Ingelman-Sundberg, *Pharmacogenomics J.*, 2004, **5**, 6.

25. C. International Warfarin Pharmacogenetics, T. E. Klein, R. B. Altman, N. Eriksson, B. F. Gage, S. E. Kimmel, M. T. Lee, N. A. Limdi, D. Page, D. M. Roden, M. J. Wagner, M. D. Caldwell and J. A. Johnson, *N. Engl. J. Med.*, 2009, **360**, 753.

26. K. Kock and K. L. Brouwer, *Clin. Pharmacol. Ther.*, 2012, **92**, 599.

27. C. D. Klaassen and L. M. Aleksunes, *Pharmacol. Rev.*, 2010, **62**, 1.

28. N. D. Pfeifer, R. N. Hardwick and K. L. Brouwer, *Annu. Rev. Pharmacol. Toxicol.*, 2014, **54**, 509.

29. N. D. Pfeifer, K. Yang and K. L. Brouwer, *J. Pharmacol. Exp. Ther.*, 2013, **347**, 727.

30. K. L. Brouwer, D. Keppler, K. A. Hoffmaster, D. A. Bow, Y. Cheng, Y. Lai, J. E. Palm, B. Stieger and R. Evers, *Clin. Pharmacol. Ther.*, 2013, **94**, 95.

31. M. J. Zamek-Gliszczynski, C. A. Lee, A. Poirier, J. Bentz, X. Chu, H. Ellens, T. Ishikawa, M. Jamei, J. C. Kalvass, S. Nagar, K. S. Pang, K. Korzekwa, P. W. Swaan, M. E. Taub, P. Zhao and A. Galetin, *Clin. Pharmacol. Ther.*, 2013, **94**, 64.

32. T. Tachibana, S. Kitamura, M. Kato, T. Mitsui, Y. Shirasaka, S. Yamashita and Y. Sugiyama, *Pharm. Res.*, 2010, **27**, 442.
33. A. Poirier, T. Lave, R. Portmann, M. E. Brun, F. Senner, M. Kansy, H. P. Grimm and C. Funk, *Drug Metab. Dispos.*, 2008, **36**, 2434.
34. K. Menochet, K. E. Kenworthy, J. B. Houston and A. Galetin, *J. Pharmacol. Exp. Ther.*, 2012, **341**, 2.
35. R. Amundsen, H. Christensen, B. Zabihyan and A. Asberg, *Drug Metab. Dispos.*, 2010, **38**, 1499.
36. R. G. Tirona, B. F. Leake, A. W. Wolkoff and R. B. Kim, *J. Pharmacol. Exp. Ther.*, 2003, **304**, 223.
37. J. Noe, R. Portmann, M. E. Brun and C. Funk, *Drug Metab. Dispos.*, 2007, **35**, 1308.
38. H. M. Jones, H. A. Barton, Y. Lai, Y. A. Bi, E. Kimoto, S. Kempshall, S. C. Tate, A. El-Kattan, J. B. Houston, A. Galetin and K. S. Fenner, *Drug Metab. Dispos.*, 2012, **40**, 1007.
39. Y. Shirasaka, T. Sakane and S. Yamashita, *J. Pharm. Sci.*, 2008, **97**, 553.
40. K. Menochet, K. E. Kenworthy, J. B. Houston and A. Galetin, *Drug Metab. Dispos.*, 2012, **40**, 1744.
41. K. R. Korzekwa, S. Nagar, J. Tucker, E. A. Weiskircher, S. Bhoopathy and I. J. Hidalgo, *Drug Metab. Dispos.*, 2012, **40**, 865.
42. D. Agnani, P. Acharya, E. Martinez, T. T. Tran, F. Abraham, F. Tobin, H. Ellens and J. Bentz, *PLoS One*, 2011, **6**, e25086.
43. H. Sun and K. S. Pang, *Drug Metab. Dispos.*, 2008, **36**, 102.
44. N. D. Pfeifer, K. B. Harris, G. Z. Yan and K. L. Brouwer, *Drug Metab. Dispos.*, 2013, **41**, 1949.
45. J. Bentz, T. T. Tran, J. W. Polli, A. Ayrton and H. Ellens, *Pharm. Res.*, 2005, **22**, 1667.
46. S. Nagar, J. Tucker, E. A. Weiskircher, S. Bhoopathy, I. J. Hidalgo and K. Korzekwa, *Pharm. Res.*, 2014, **31**, 347.
47. K. Korzekwa and S. Nagar, *Pharm. Res.*, 2014, **31**, 335.
48. K. Yoshida, K. Maeda and Y. Sugiyama, *Annu. Rev. Pharmacol. Toxicol.*, 2013, **53**, 581.
49. K. S. Pang and M. Rowland, *J. Pharmacokinet. Biopharm.*, 1977, **5**, 625.
50. L. Liu and K. S. Pang, *Drug Metab. Dispos.*, 2005, **33**, 1.
51. G. L. Sirianni and K. S. Pang, *J. Pharmacokinet. Biopharm.*, 1997, **25**, 449.
52. R. Li, H. A. Barton and M. V. Varma, *Clin. Pharmacokinet.*, 2014, **53**, 659.
53. Y. Shitara, K. Maeda, K. Ikejiri, K. Yoshida, T. Horie and Y. Sugiyama, *Biopharm. Drug Dispos.*, 2013, **34**, 45.
54. T. Lave, P. Coassolo, G. Ubeaud, R. Brandt, C. Schmitt, S. Dupin, D. Jaeck and R. C. Chou, *Pharm. Res.*, 1996, **13**, 97.
55. K. Grime and S. W. Paine, *Drug Metab. Dispos.*, 2013, **41**, 372.
56. H. Jones and K. Rowland-Yeo, *CPT: Pharmacometrics Syst. Pharmacol.*, 2013, **2**, e63.
57. Y. Cao and W. J. Jusko, *J. Pharmacokinet. Pharmacodyn.*, 2012, **39**, 711.
58. M. Gertz, N. Tsamandouras, C. Sall, J. B. Houston and A. Galetin, *Pharm. Res.*, 2014, **31**, 2367.

59. I. Nestorov, *Clin. Pharmacokinet.*, 2003, **42**, 883.
60. A. Poirier, A. C. Cascais, C. Funk and T. Lave, *J. Pharmacokinet. Pharmacodyn.*, 2009, **36**, 585.
61. T. Watanabe, H. Kusuhara, K. Maeda, Y. Shitara and Y. Sugiyama, *J. Pharmacol. Exp. Ther.*, 2009, **328**, 652.
62. S. Neuhoff, K. R. Yeo, Z. Barter, M. Jamei, D. B. Turner and A. Rostami-Hodjegan, *J. Pharm. Sci.*, 2013, **102**, 3145.
63. M. Jamei, F. Bajot, S. Neuhoff, Z. Barter, J. Yang, A. Rostami-Hodjegan and K. Rowland-Yeo, *Clin. Pharmacokinet.*, 2014, **53**, 73.
64. R. H. Rose, S. Neuhoff, K. Abduljalil, M. Chetty, A. Rostami-Hodjegan and M. Jamei, *CPT: Pharmacometrics Syst. Pharmacol.*, 2014, **3**, e124.
65. H. M. Jones, N. Parrott, K. Jorga and T. Lave, *Clin. Pharmacokinet.*, 2006, **45**, 511.
66. M. V. Varma, R. J. Scialis, J. Lin, Y. A. Bi, C. J. Rotter, T. C. Goosen and X. Yang, *AAPS J.*, 2014, **16**, 736.
67. K. Ito and J. B. Houston, *Pharm. Res.*, 2005, **22**, 103.
68. M. V. Varma, Y. Lai, B. Feng, J. Litchfield, T. C. Goosen and A. Bergman, *Pharm. Res.*, 2012, **29**, 2860.
69. M. Hirano, K. Maeda, Y. Shitara and Y. Sugiyama, *J. Pharmacol. Exp. Ther.*, 2004, **311**, 139.
70. S. Bosgra, E. V. Steeg, M. L. Vlaming, K. C. Verhoeckx, M. T. Huisman, M. Verwei and H. M. Wortelboer, *Eur. J. Pharm. Sci.*, 2014, **65**, 156.
71. J. Badee, B. Achour, A. Rostami-Hodjegan and A. Galetin, *Drug Metab. Dispos.*, 2015, **43**, 424.
72. E. Kimoto, K. Yoshida, L. M. Balogh, Y. A. Bi, K. Maeda, A. El-Kattan, Y. Sugiyama and Y. Lai, *Mol. Pharm.*, 2012, **9**, 3535.
73. B. Prasad, R. Evers, A. Gupta, C. E. Hop, L. Salphati, S. Shukla, S. V. Ambudkar and J. D. Unadkat, *Drug Metab. Dispos.*, 2014, **42**, 78.
74. M. V. Varma, Y. Lai, E. Kimoto, T. C. Goosen, A. F. El-Kattan and V. Kumar, *Pharm. Res.*, 2013, **30**, 1188.
75. R. Li, H. A. Barton, P. D. Yates, A. Ghosh, A. C. Wolford, K. A. Riccardi and T. S. Maurer, *J. Pharmacokinet. Pharmacodyn.*, 2014, **41**, 197.
76. P. Poulin and F. P. Theil, *J. Pharm. Sci.*, 2000, **89**, 16.
77. T. Rodgers and M. Rowland, *J. Pharm. Sci.*, 2006, **95**, 1238.
78. H. Graham, M. Walker, O. Jones, J. Yates, A. Galetin and L. Aarons, *J. Pharm. Pharmacol.*, 2012, **64**, 383.
79. R. Li, A. Ghosh, T. S. Maurer, E. Kimoto and H. A. Barton, *Drug Metab. Dispos.*, 2014, **42**, 1646.
80. Y. E. Yun and A. N. Edginton, *Xenobiotica*, 2013, **43**, 839.
81. J. Stangier, J. Schmid, D. Turck, H. Switek, A. Verhagen, P. A. Peeters, S. P. van Marle, W. J. Tamminga, F. A. Sollie and J. H. Jonkman, *J. Clin. Pharmacol.*, 2000, **40**, 1312.
82. D. M. Colussi, C. Parisot, M. L. Rossolino, L. A. Brunner and G. Y. Lefevre, *J. Clin. Pharmacol.*, 1997, **37**, 214.
83. L. Zhang, Y. D. Zhang, P. Zhao and S. M. Huang, *AAPS J.*, 2009, **11**, 300.
84. S. Haddad, P. Poulin and C. Funk, *J. Pharm. Sci.*, 2010.

85. E. S. Tien and M. Negishi, *Xenobiotica*, 2006, **36**, 1152.
86. M. Karlgren, G. Ahlin, C. A. Bergstrom, R. Svensson, J. Palm and P. Artursson, *Pharm. Res.*, 2012, **29**, 411.
87. K. Yoshida, K. Maeda and Y. Sugiyama, *Clin. Pharmacol. Ther.*, 2012, **91**, 1053.
88. K. Ito, H. S. Brown and J. B. Houston, *Br. J. Clin. Pharmacol.*, 2004, **57**, 473.
89. L. K. Hinton, A. Galetin and J. B. Houston, *Pharm. Res.*, 2008, **25**, 1063.
90. T. Watanabe, H. Kusuhara and Y. Sugiyama, *J. Pharmacokinet. Pharmacodyn.*, 2010, **37**, 575.
91. M. Jamei, S. Marciniak, K. Feng, A. Barnett, G. Tucker and A. Rostami-Hodjegan, *Expert. Opin. Drug Metab. Toxicol.*, 2009, **5**, 211.
92. M. V. Varma, J. Lin, Y. A. Bi, C. J. Rotter, O. A. Fahmi, J. L. Lam, A. F. El-Kattan, T. C. Goosen and Y. Lai, *Drug Metab. Dispos.*, 2013, **41**, 966.
93. M. V. Varma, Y.-a. Bi, E. Kimoto and J. Lin, *J. Pharmacol. Exp. Ther.*, 2014, **351**, 214.
94. M. V. Varma, J. Lin, Y. A. Bi, E. Kimoto and A. D. Rodrigues, *Drug Metab. Dispos.*, 2015, **43**, 1108.
95. M. Rowland, C. Peck and G. Tucker, *Annu. Rev. Pharmacol. Toxicol.*, 2011, **51**, 45.
96. T. Kudo, A. Hisaka, Y. Sugiyama and K. Ito, *Drug Metab. Dispos.*, 2013, **41**, 362.
97. N. D. Pfeifer, S. L. Goss, B. Swift, G. Ghibellini, M. Ivanovic, W. D. Heizer, L. M. Gangarosa and K. L. Brouwer, *CPT: Pharmacometrics Syst. Pharmacol.*, 2013, **2**, e20.
98. M. V. Varma, R. Scialis, J. Lin, Y. A. Bi, C. J. Rotter, T. C. Goosen and X. Yang, *AAPS J.*, 2014, **16**, 736.
99. T. B. Bidstrup, I. Bjornsdottir, U. G. Sidelmann, M. S. Thomsen and K. T. Hansen, *Br. J. Clin. Pharmacol.*, 2003, **56**, 305.
100. L. I. Kajosaari, J. Laitila, P. J. Neuvonen and J. T. Backman, *Basic Clin. Pharmacol. Toxicol.*, 2005, **97**, 249.
101. C. Sall, J. B. Houston and A. Galetin, *Drug Metab. Dispos.*, 2012, **40**, 1279.
102. L. I. Kajosaari, M. Niemi, M. Neuvonen, J. Laitila, P. J. Neuvonen and J. T. Backman, *Clin. Pharmacol. Ther.*, 2005, **78**, 388.
103. M. Niemi, J. T. Backman, L. I. Kajosaari, J. B. Leathart, M. Neuvonen, A. K. Daly, M. Eichelbaum, K. T. Kivisto and P. J. Neuvonen, *Clin. Pharmacol. Ther.*, 2005, **77**, 468.
104. B. W. Ogilvie, D. Zhang, W. Li, A. D. Rodrigues, A. E. Gipson, J. Holsapple, P. Toren and A. Parkinson, *Drug Metab. Dispos.*, 2006, **34**, 191.
105. Y. Shitara, M. Hirano, H. Sato and Y. Sugiyama, *J. Pharmacol. Exp. Ther.*, 2004, **311**, 228.
106. H. Fujino, S. Shimada, I. Yamada, M. Hirano, Y. Tsunenari and J. Kojima, *Arzneimittelforschung*, 2003, **53**, 701.

107. S. A. Peters, A. L. Ungell and H. Dolgos, *Curr. Opin. Drug Discovery Dev.*, 2009, **12**, 509.
108. P. Zhao, L. Zhang, J. A. Grillo, Q. Liu, J. M. Bullock, Y. J. Moon, P. Song, S. S. Brar, R. Madabushi, T. C. Wu, B. P. Booth, N. A. Rahman, K. S. Reynolds, E. Gil Berglund, L. J. Lesko and S. M. Huang, *Clin. Pharmacol. Ther.*, 2011, **89**, 259.
109. E. J. Guest, K. Rowland-Yeo, A. Rostami-Hodjegan, G. T. Tucker, J. B. Houston and A. Galetin, *Br. J. Clin. Pharmacol.*, 2011, **71**, 72.
110. R. J. Dimelow, P. D. Metcalfe and S. Thomas, In silico models of drug metabolism and drug interactions, in *Encyclopedia of Drug Metabolism and Interactions*, ed. A. V. Lyubimov, John Wiley & Sons, Inc., 2012, p. 1.
111. R. S. Obach, R. L. Walsky, K. Venkatakrishnan, J. B. Houston and L. M. Tremaine, *Clin. Pharmacol. Ther.*, 2005, **78**, 582.
112. M. Hirano, K. Maeda, Y. Shitara and Y. Sugiyama, *Drug Metab. Dispos.*, 2006, **34**, 1229.

Section III: Importance and Clinical Impact of Transporter-mediated Drug–Drug Interactions

Transporter Drug–Drug Interactions: A Pharmaceutical Industry Perspective

SILKE SIMON,*[a] MOHAMMED ULLAH,[a]
ROBERT VAN WATERSCHOOT,[a] DIETMAR SCHWAB,[b]
SUSAN GRANGE,[b] CAROLINE A. LEE[c] AND CHRISTOPH FUNK[a]

[a] Pharmaceutical Sciences, Pharma Research & Early Development, Roche Innovation Center Basel, F. Hoffmann-La Roche Ltd, Basel, Switzerland;
[b] Clinical Pharmacology, Pharma Research & Early Development, Roche Innovation Center Basel, F. Hoffmann-La Roche Ltd, Basel, Switzerland;
[c] Ardea Biosciences, Inc., San Diego, CA, USA
*Email: silke.simon@roche.com

10.1 Introduction and Overview of Clinical Drug–Drug Interactions

Drug–drug interactions (DDIs), in general, describe a situation where two or more co-administered pharmacologically active substances impact each other in their pharmacological effect, which can be synergistic (activity enhancement), antagonistic (activity mitigation) or additive in unrelated effects. Generally, interactions can occur between co-administered drugs, but also with certain food ingredients (drug–food interactions) or herbal substances (drug–plant interactions). Furthermore, one distinguishes between "pharmacodynamic" (PD) interactions, when a receptor, enzyme or a biochemical process in general is affected by the two compounds, and

RSC Drug Discovery Series No. 54
Drug Transporters: Volume 1: Role and Importance in ADME and Drug Development
Edited by Glynis Nicholls and Kuresh Youdim

Published by the Royal Society of Chemistry, www.rsc.org

"pharmacokinetic" (PK) interactions, which arise from changes in compound concentrations in body tissues or within the systemic circulation over time. Since in most reported cases transport proteins impact the disposition of drugs within the body, this chapter will only discuss PK-related DDIs.

DDIs involving drug transporters can occur in various tissues within the body during the phases of absorption, distribution or excretion, and can result in altered drug exposure in blood (plasma) or local tissue compartments (such as the brain or liver), which may lead to unwanted pharmacological or toxicological effects. The severity of these consequences is dependent on the efficacy and safety margins of the "victim" drug (transporter substrate). Thus, if the PK is altered by the co-medicated "perpetrator" drug (transporter inhibitor), this may result in non-significant changes without any health risk to the patient for some drugs with a wide therapeutic index, but multi-fold changes in victim drug exposure could render the drug either inefficacious or toxic.

Figure 10.1 illustrates the potential consequence of a DDI on the *in vivo* plasma exposure of a victim drug, when a transporter, important for its removal from the circulation, is inhibited by a co-medicated perpetrator.

In drug development, it is necessary to assess the risk for potential transporter-mediated DDIs that might impact drug efficacy or safety prior to clinical studies. However, identification of the key transporter(s) that drive the disposition of a compound, and hence might be responsible for a victim DDI, is not straightforward, since many transporters exhibit a high degree of substrate overlap and multiplicity in tissue expression. An example is the

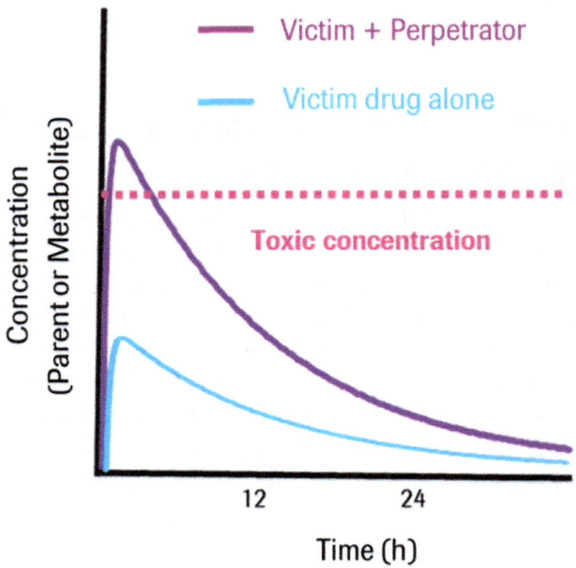

Figure 10.1   Plasma exposure of a "victim" drug (transporter substrate) over time in the absence and presence of a "perpetrator" drug (transporter inhibitor).

ATP binding cassette (ABC) family of efflux transporters, where drugs that specifically interact with only P-glycoprotein (P-gp) or breast cancer resistance protein (BCRP)—to name only two—are very rare. As a consequence, victim DDI assessments can be conducted and interpreted most reliably, when human PK data are available to guide the identification of liable elimination pathways (*e.g.*, active renal or biliary secretion).

By contrast, perpetrator DDI assessments are more straightforward, since the transporter(s) most frequently involved in, for example, hepatic or renal uptake are already known based on previous clinical findings (*e.g.*, organic anion transporting polypeptide (OATP) transporter OATP1B1 and statin interactions, discussed in more detail below). It is, therefore, usually routine for a candidate drug to be tested for inhibition against these clinically relevant transporters during the early drug development phase.

This chapter provides an overview of clinical transporter-mediated DDIs and the approaches used to assess transporter DDIs during early and late drug development. Moreover, it comments on the current gaps and future challenges in this emerging field of science.

An overview of clinical transporter-mediated DDIs at the main barrier tissues—the gut, liver, blood–brain barrier (BBB) and kidney—is given in Table 10.1. Table 10.1 lists known victim–perpetrator pairs together with the respective PK effects caused by this specific interaction and the implicated therapeutic consequences of co-administration. In indicated cases, the DDI is also associated with drug metabolising enzymes. Examples of particular interest are discussed subsequently to familiarise the reader with a broad range of DDIs in terms of their severity and clinical relevance.

10.1.1 DDIs in the Intestine

The intestinal epithelium (enterocyte monolayer) is lined with uptake and efflux transporters, some of which may be important determinants of drug absorption and bioavailability. The significance of intestinal transport is dependent on the physicochemical characteristics of the drug; principally its permeability, solubility and local exposure along the gut lumen, and whether uptake or efflux transporters are involved after an oral dose. However, for highly soluble and permeable drugs, which are also likely to be metabolised [*i.e.*, Biopharmaceutics Drug Disposition Classification System (BDDCS) Class I drugs; see Figure 10.2A and Chapter 1], transporter effects will be minimal. Typical oral doses (1–500 mg) result in high drug concentrations in the gut lumen, where transporters are likely to be saturated and therefore passive diffusion dominates. For BDDCS Class 1 drugs, intestinal absorption is therefore linear and bioavailability is dose proportional. The role of intestinal efflux transport is most pronounced for drug substrates with low bioavailability, low solubility and/or low permeability (BDDCS Classes 3 and 4). Such drugs are particularly susceptible to DDIs that can impact oral absorption in terms of rate, extent or both. While changes in the rate of absorption are in most cases not clinically relevant, except for acutely

Table 10.1 Overview of clinical DDIs, classified by drug class of the victim, site of interaction, fold change in PK parameters (maximal reported effect in the literature) and concluding clinical implications for co-administration.[a,b]

Victim drug class (alphabetical order)	Victim drug	Victim drug ADME properties (F, CL, CYP enzyme and transporter substrate properties, etc.)	Perpetrator drug	Implicated site(s) of interaction	Implicated transporter(s)	Observed fold effect on clinical PK, PD or reported adverse effect	Clinical Implications	Ref.
Analgesics	**Morphine**	$F \sim 100\%$; CL_{ren} and CL_{hep} n/a; P-gp	**Elacridar** **Quinidine**	BBB BBB	P-gp, BCRP P-gp	None/minimal None	No clinical relevance None	11, 106 11, 106
	Methadone	F 36–100%; CL_{ren} and CL_{hep} n/a; CYP3A4, 2B6, 2C19, 2D6; P-gp		BBB	P-gp	None	None related to transporter DDIs[c]	11, 107, 108
	Fentanyl	$F \sim 54\%$; $CL_{ren} < 7\%$ unchanged, CL_{hep} >90% → metabolites mainly renally excreted; in feces 1% unchanged; CYP3A4; P-gp	**Quinidine**	BBB	P-gp	Altered CNS activity, but probably due to C_p ↑	None	11, 109
	Procainamide	$F \sim 85\%$; CL_{ren} 50% unchanged (30% NAPA); CYP2D6; OCT2, MATEs (parent and active metabolite NAPA)	**Cimetidine**	Kidney	OCT2, MATEs	AUC ↑ 1.44; CL_{ren} ↓ 0.57–0.60[d]	Monitoring of serum procainamide and NAPA levels, electrocardiogram changes, and hemodynamic status. Famotidine and nizatidine to be considered as alternatives	72, 110–116
Antibiotics	**Ciprofloxacin**	$F \sim 70\%$; CL_{ren} 40–50% unchanged (including 5% active tubular secretion), in feces 20–35%; CYP1A2; OAT1/3	**Probenecid**	Kidney	OAT1/3	AUC ↑ 1.75; CL_{ren} ↓ 0.36	Careful monitoring of ciprofloxacin serum levels during co-medication	117–121

	Drug	Properties	Interacting drug	Organ	Transporter	Effect	Clinical consequence	Ref.
	Ticarcillin	$F \sim 0\%$ → i.v. or intramuscularly administered; CL_{ren} 60–70% unchanged (including active tubular secretion); OAT1/3		Kidney	OAT1/3	$C_p \uparrow$; $t_{1/2} \uparrow$; $CL_{renal} \downarrow 0.78$	Potential for toxicity → caution when high penicillin G dosages are administered i.v.	117, 122–126
	Azlocillin	F n/a; $CL_{ren} \sim 60\%$ unchanged, CL_{hep} 40–50%, CL_{bil} NS		Kidney	OAT1/3	AUC $\uparrow$ 2.23; $CL_{ren} \downarrow 0.53$; $t_{1/2} \uparrow$		117, 127, 128
	Temocillin	F n/a; $CL_{ren} \sim 100\%$ (including active tubular secretion); OAT1/3		Kidney	OAT1/3	$CL_{ren} \downarrow 0.66$		117, 129
Anti-coagulants	**Dabigatran** *(given as etexilate prodrug)*	F 3–7%; CL_{ren} 80%; not a CYP3A substrate; etexilate prodrug is a P-gp substrate	**Clarithromycin**	Intestine	P-gp	AUC $\uparrow$ 1.49; $C_{max} \uparrow$ 1.6; $F \uparrow$ from 6.5% to 10.1%	Dose adjustment (reduction) and monitoring of coagulation parameters if co-administering with P-gp inhibitors. Combination with P-gp inhibitors is contraindicated for severely renally impaired patients.	130
			Verapamil	Intestine	P-gp	AUC $\uparrow$ 1.43; $C_{max} \uparrow$ 1.79; $F \uparrow \sim$2-fold		131
Antidiabetics	**Metformin**	F 50–60%; CL_{ren} 100% unchanged (tubular secretion); OCT2/1; MATE1/2-K	**Cimetidine**	Kidney	OCT2, MATEs	AUC $\uparrow$ 1.5; $CL_{ren} \downarrow 0.72$	Risk of lactic acidosis → particularly slow and cautious titration of metformin dosage; maximum dose of metformin might be reduced; close monitoring of blood glucose	115, 132

Table 10.1 (*Continued*)

Victim drug class (*alphabetical order*)	Victim drug	Victim drug ADME properties (*F*, CL, CYP enzyme and transporter substrate properties, *etc.*)	Perpetrator drug	Implicated site(s) of interaction	Implicated transporter(s)	Observed fold effect on clinical PK, PD or reported adverse effect	Clinical Implications	Ref.
			Pyrimethamine	Kidney	OCT2, MATEs	AUC $\uparrow$ 1.39–1.42; $CL_{ren} \downarrow$ 0.65–0.77	Careful monitoring of patients and doses of metformin and perpetrator	21
	Repaglinide	$F \sim 56\%$; CYP2C8, 3A4; OATP1B1	**Cyclosporine**	Intestine, liver	OATP1B1	AUC $\uparrow$ 2.4; $C_{max} \uparrow$ 1.75	Contraindicated with cyclosporine and gemfibrozil	133
			Gemfibrozil	Intestine, liver	OATP1B1	AUC $\uparrow$ 8.2; $C_{max} \uparrow$ 2.6		134, 135
Antidiarrheals	**Loperamide**	$F \sim 0.3\%$. CL_{ren} 1%, CL_{hep} 99%; CYP3A4; P-gp	**Quinidine**	BBB	P-gp	Altered CNS activity possible, but probably due to $C_p \uparrow$	Risk of opiate CNS side effects (but probably due to $C_p \uparrow$) $\rightarrow$ monitoring of side effects and respiratory depression; dose reductions may be required	11, 136
			Ritonavir	BBB	P-gp	None	No serious adverse events, no changes in PD effects of loperamide $\rightarrow$ no intervention needed	11
			Cyclosporine	BBB	P-gp	CNS exposure $\uparrow$ 2.0 (PET studies)	None	11
Antihistamines	**Cetirizine**	$F \sim 100\%$; mainly CL_{ren} 70%, CL_{hep} 10%; P-gp	**Diverse**	BBB	P-gp	None reported so far	No CNS side effects reported $\rightarrow$ not considered clinically relevant	11

	Fexofenadine	F 30–41%; CL_{ren} 11–12%, CL_{hep} 5%, $\sim$80% unchanged parent in feces; P-gp, OATP1B1/1B3/1A2/2B1	Verapamil	Intestine	P-gp	AUC ↑ 2.48; C_{max} ↑ 2.92; CL_{ren} ↔	No adverse effects → no dose adjustment necessary	3
			Itraconazole	Intestine	P-gp	AUC ↑ 2.73; C_{max} ↑ 1.93; CL_{ren} ↔		
			Ketoconazole	Intestine	P-gp	AUC ↑2.64; C_{max} ↑ 2.35		
			Diverse	BBB	P-gp	None reported so far	No CNS side effects reported → not considered clinically relevant	11, 137–139
	Loratadine	$F\sim$100%; CL_{ren} 40% (conjugated metabolite), in feces 40%; CYP2D6, 3A4; P-gp		BBB	P-gp	None reported so far		11
Antihyper-tensives	Bosentan	$F\sim$50%, CL_{ren} 3%, mainly excreted into bile as metabolites; CYP2C9, 3A4; OATP1B1	Sildenafil citrate	Liver	OATP1B1	AUC ↑ 1.5; C_{max} ↑ 1.42	Contraindicated	140
			Cyclosporine	Liver	OATP1B1 (+CYP2C9, 3A4)	AUC ↑ 2		141
Antiparasitics	Ivermectin	F moderate; CL_{ren} <1%, CL_{hep} 99%; CYP3A4; P-gp, BCRP	Diverse	BBB	P-gp	None reported so far	No CNS side effects reported → not considered clinically relevant	11
Anti-rheumatics	Sulfasalazine	$F\sim$3–12%, low permeability; substrate of gut bacterial azo-reductases; CL_{ren} 37%; OATP2B1	Pantoprazole	Intestine	BCRP	NS	None	142
			Curcumin	Intestine	BCRP (+ azo-reductases)	AUC ↑ 3.2; C_{max} ↑ 3.8; CL_{total} ↓ 3.2		143
Antiulcer	Ranitidine	F 90–100%; CL_{ren} 70% unchanged *via* cationic renal secretion	Cimetidine	Kidney	OCT2, MATEs	AUC ↑ 1.21; CL_{ren} ↓ 0.75	None	144, 145

Table 10.1   (*Continued*)

Victim drug class (*alphabetical order*)	Victim drug	Victim drug ADME properties (*F*, CL, CYP enzyme and transporter substrate properties, *etc.*)	Perpetrator drug	Implicated site(s) of interaction	Implicated transporter(s)	Observed fold effect on clinical PK, PD or reported adverse effect	Clinical Implications	Ref.
Antivirals	**Nelfinavir**	*F* uncertain; in feces 87% (unchanged 22%; metabolite 78%), CL_{ren} 1–2%; CYP3A, 2C19; P-gp	**Cyclosporine**	BBB	P-gp	None reported so far, but predicted CNS exposure ↑ ~2.4	None	146
	Ganciclovir	*F* ~ 5%; CL_{ren} >90% unchanged (including active tubular secretion); OAT1/3	**Probenecid**	Kidney	OAT1/3	AUC ↑ 1.53; CL_{ren} ↓ 0.81	Caution, if ganciclovir or its prodrug, valganciclovir, are used concomitantly with probenecid → close monitoring for toxicities associated with ganciclovir	117, 147, 148
	Zidovudine	*F* ~ 100%; CL_{ren} 14% unchanged, CL_{hep} 74%		Kidney	OAT1/3	AUC ↑ 2.06–2.14; CL_{ren} ↓ 0.54–0.75	Greater risk of zidovudine toxicity → close monitoring for signs and possible need for zidovudine dose reduction, or increase of the dosing interval	117, 149–152
			Cimetidine	Kidney	OCT2, MATEs	AUC ↔; CL_{ren} ↓ 0.44	Unlikely to be clinically significant (except in patients with hepatic impairment)	114, 153
	Oseltamivir	*F* ~ 75% as carboxylate; CL_{ren} >99% (carboxylate) including tubular secretion; esterases; OAT1/3	**Probenecid**	Kidney	OAT1/3	AUC ↑ 2.52; CL_{ren} ↓ 0.48	Due to the safety margin of oseltamivir carboxylate no dose adjustments are required	117, 154, 155

	Cidofovir	F n/a; CL_{ren} 80–100% unchanged (including active tubular secretion); OAT1/3		Kidney	OAT1/3	AUC ↑ 1.29–1.44; CL_{ren} ↓ 0.68	If both drugs co-dosed with zidovudine, zidovudine needs to be either temporarily discontinued or decreased	117, 156, 157
	Zalcitabine	F >80%; CL_{ren} 80% unchanged (including active tubular secretion); OAT1/3		Kidney	OAT1/3	AUC ↑ 1.54; CL_{ren} ↓ 0.58	Close clinical monitoring for evidence of zalcitabine toxicity is recommended; a dose reduction of zalcitabine may be warranted	117, 158, 159
Cardiacs	Digoxin	F 60–80%; CL_{ren} 84%, CL_{hep} 16%; CYP3A4 (minor); P-gp	Verapamil	Intestine, kidney	P-gp	AUC ↑ 1.5; C_{max} ↑ 1.7	Close monitoring of serum digoxin levels and pharmacologic effects: dose adjustment with all P-gp inhibitors	3
			Valspodar	Intestine, kidney	P-gp	AUC ↑ 3.05; C_{max} ↑ 2.44; CL_{ren} ↓ 0.25		
			Quinidine	Intestine, kidney	P-gp	AUC ↑ 1.76; C_{max} ↑ 1.75		
				Kidney	P-gp	C_p ↑ 2.0. (*Due to intestinal and/or renal P-gp interactions → effect was assessed after i.v. administration → Δ* CL_{ren})		160–162
			Itraconazole	Intestine, kidney	P-gp	AUC ↑ 2.53; C_{max} ↑ 2.05; CL_{ren} ↓ 1.04		3
	Talinolol	$F \sim$ 50%, CL_{ren} 55% (parent); P-gp	Verapamil	Intestine	P-gp	AUC ↑ 0.76; C_{max} ↑ 0.64; CL_{ren} ↓ 1.01	Dose adjustment (reduction) with P-gp inhibitors	3

Table 10.1 (*Continued*)

Victim drug class (*alphabetical order*)	Victim drug	Victim drug ADME properties (*F*, CL, CYP enzyme and transporter substrate properties, *etc.*)	Perpetrator drug	Implicated site(s) of interaction	Implicated transporter(s)	Observed fold effect on clinical PK, PD or reported adverse effect	Clinical Implications	Ref.
			Simvastatin	Intestine	P-gp	AUC ↑ 1.07; C_{max} ↑ 1.13; CL_{ren} ↓ 1.13		
			Erythromycin	Intestine	P-gp	AUC ↑1.5; C_{max} ↑ 1.26; CL_{ren} ↔		
Verapamil		F 20–35%; CL_{ren} 70% (metabolite), 3–4% (unchanged), in feces ∼16%; CYP3A4 (1A2, 2C8, 2C9, 2C18)	**Cyclosporine**	BBB	P-gp	CNS exposure ↑ 3.3–5.8 (1.6–2.0 in PET studies)	None related to transporter DDIs[e]	11, 163–165
Pindolol		F > 95%; CL_{ren} 35–40% unchanged, CL_{hep} 60–65%	**Cimetidine**	Kidney	OCT2, MATEs	AUC ↑ 1.38–1.4; CL_{ren} ↓ 0.61–0.70	Clinical monitoring of patient response and tolerance is recommended, but low clinical relevance	115, 166–168
			Trimethoprim	Kidney	OCT2, MATEs	AUC ↑ 1.1-1.37; CL_{ren} ↓ 0.63-0.75	Consider slight change in the beta blocking activity of pindolol, if trimethoprim is added to or discontinued from the patient's therapy	169
Dofetilide		F >90%; CL_{ren} 80% unchanged by glomerular filtration and active tubular secretion, CL_{hep} 20%; OCT2, MATEs	**Cimetidine**	Kidney	OCT2, MATEs	AUC ↑ 1.48; CL_{ren} ↓ 0.67	Strong increase in systemic exposure of dofetilide → all renal cationic transport inhibitors should be contraindicated	115, 170, 171

Chemo-therapeutics	Topotecan	$F \sim 40\%$, minimal metabolism; P-gp	Elacridar	Intestine	BCRP, P-gp	AUC $\uparrow$ 1.43; $F \uparrow$ from 40 to 100%	Patient monitoring when co-administered with BCRP and P-gp inhibitors	46
	Lapatinib	F variable and incomplete; $CL_{ren} < 2\%$, $CL_{hep} > 90\%$, in feces $\sim 27\%$ unchanged; CYP3A4, 3A5 (CYP2C19/2C8); BCRP, P-gp	Cyclosporine	BBB	BCRP, P-gp	Combined effect: CNS exposure $\uparrow$ 1.9 (maximum)	None related to transporter DDIs[f]	11, 172
			Efavirenz	BBB	BCRP, P-gp			
	Cisplatin	F 100%; CL_{ren} 35–51%, 13–17% unchanged (including active tubular secretion); OCT2	Vandetanib	Kidney	OCT2, MATEs	AUC $\uparrow$ 1.33	Use with caution and closely monitor for toxicities	173, 174
CNS agents	Gabapentin enacarbil	$F \sim 75\%$, CL_{ren} 94% (including active tubular secretion); MCT1, OCT2	Cimetidine	Kidney	OCT2, MATEs	AUC $\uparrow$ 1.24; CL_{ren} $\downarrow$ 0.83	Not considered clinically relevant	115, 175, 176
Diuretics	Furosemide	F 60–64%; CL_{ren} 80–90% unchanged; OAT1/3	Probenecid	Kidney	OAT1/3	CL_{ren} $\downarrow$ 0.28	Only high dose treatment may result in elevated serum levels and may potentiate furosemide toxicity $\rightarrow$ no particular intervention necessary, but potential for interaction should be considered	119, 177–179

Table 10.1 (*Continued*)

Victim drug class (*alphabetical order*)	Victim drug	Victim drug ADME properties (*F*, CL, CYP enzyme and transporter substrate properties, *etc.*)	Perpetrator drug	Implicated site(s) of interaction	Implicated transporter(s)	Observed fold effect on clinical PK, PD or reported adverse effect	Clinical Implications	Ref.
	Triamterene	F 30–70%, CL_{ren} <50%, 21% unchanged; OAT1/3	Diflunisal	Kidney	OAT1/3	AUC ↑ 4.6;[g] CL_{ren} ↓ 0.2[g]	None	180, 181
			Cimetidine	Kidney	OCT2, MATEs	AUC ↑ 1.22; CL_{ren} ↓ 0.38[h]	Monitor side effects of triamterene during co-administration (but considered as a minor interaction)	182
HMG CoA reductase inhibitors (statins)	**Rosuvastatin**	F ~ 50%, CL_{ren} 30%, CL_{hep} 72%; OATP1B1/B3, NTCP, BCRP	Cyclosporine	Intestine	BCRP, OATP1B1/1B3	AUC ↑ 7.1; C_{max} ↑ 11; T_{max} ↓ 0.33	Dose adjustment required with inhibitors of OATP and BCRP	46, 96
			Eltrombopag	Intestine, liver	BCRP, OATP1B1/1B3	AUC ↑ 1.55; C_{max} ↑ 2		183
			Gemfibrozil	Intestine, liver	OATP1B1/1B3	AUC ↑ 1.9; C_{max} ↑ 2.2		184
			Lopinavir/ritonavir	Intestine, liver	OATP1B1/1B3	AUC ↑ 2.1; C_{max} ↑ 4.7		185
			Atazanavir/ritonavir	Intestine, liver	BCRP, OATP1B1/1B3	AUC ↑ 3.1; C_{max} ↑ 7; T_{max} ↓ 0.4		186
	Simvastatin acid (administered as lactone)	Conversion by carboxyesterases; F ~ 5%, CL_{ren} ~ 13%, CL_{feces} 60%, CL_{bil} ~ 25%; CYP3A4, 2C8; OATP1B1	Cyclosporine	Liver	OATP1B1 (+CYP3A4)	AUC ↑ 8.0; C_{max} ↑ 7.6	Contraindicated with cyclosporine and gemfibrozil (and other strong CYP3A4 and OATP1B1 inhibitors)	187
			Gemfibrozil	Intestine, liver	OATP1B1 (+CYP3A4)	AUC ↑ 2.85; C_{max} ↑ 2.18		188
	Atorvastatin	F ~ 14%, CL_{ren} <2%, $CL_{hep,bil}$ >90%; CYP3A4, UGT1A1,	Cyclosporine	Liver	OATP1B1	AUC ↑ 8.7; C_{max} ↑ 10.7; T_{max} ↑ 1.9	Dose adjustment required: limited to 10 or 20 mg if taken	63, 189

		UGT1A3; OATP1B1, BCRP	Lopinavir/ ritonavir	Liver	OATP1B1	AUC ↑ 5.9; C_{max} ↑ 4.7	with cyclosporine or HIV protease	190
			Clarithromycin	Liver	OATP1B1	AUC ↑ 4.4; C_{max} ↑ 5.4	inhibitors, respectively	93
	Cerivastatin	$F \sim 60\%$, CL_{ren} 30%, CL_{bil} 70%; CYP2C8, 4A4; UGT1A (metabolism major); OATP1B1/1B3, P-gp, BCRP, MRP2	Cyclosporine	Intestine, liver	OATP1B3 (+BCRP, CYPs)	AUC ↑ 3.8; C_{max} ↑ 5	Withdrawn from the market: fatalities associated with severe	191
			Gemfibrozil	Liver	OATP1B1 (+CYP2C8)	AUC ↑ 4.4; C_{max} ↑ 4.8	rhabdomyolysis	65, 99, 192
	Fluvastatin	$F \sim 30\%$; CYP2C9 (metabolism major), 3A4; BCRP, OATP1B1/ 1B3/2B1	Cyclosporine	Liver	OATP1B1	AUC ↑ 1.9; C_{max} ↑ 1.3	Dose adjustment required: limit to	193, 194
			Fluconazole	Intestine, liver	OATP1B3	AUC ↑ 1.84; C_{max} ↑ 1.44; $t_{1/2}$ ↑ 0.8	20 mg when co-administering with OATP inhibitors	195
	Pitavastatin	$F \sim 80\%$; CYP2C9 (metabolism minor); OATP1B1/1B3, BCRP	Cyclosporine	Intestine, liver	OATP1B1	AUC ↑ 4.55; C_{max} ↑ 6.58	Dose adjustment when co-administered with	97, 196, 197
			Erythromycin	Liver	OATP1B1	AUC ↑ 2.8; C_{max} ↑ 3.6	erythromycin, rifampicin or fibrates	198
			Gemfibrozil	Liver	OATP1B1	AUC ↑ 1.45; C_{max} ↑ 1.31	(1–2 mg limit); contraindicated with lopinavir/ritonavir	199
	Pravastatin	$F \sim 17\%$, CL_{ren} 47% (≥80% active), CL_{hep} 53%; MRP2, OAT1B1/ 1B3, OAT3	Cyclosporine	Intestine, liver	OATP1B1	AUC ↑ 3.82; C_{max} ↑ 4.27	Contraindicated with gemfibrozil; limit	196, 200
			Clarithromycin	Intestine, liver	OATP1B1	AUC ↑ 2.1; C_{max} ↑ 2.28	dose with other OATP inhibitors such as	93
			Gemfibrozil	Intestine, liver	OATP1B1	AUC ↑ 2.0; C_{max} ↑ 1.81	cyclosporine (20 mg) or clarithromycin (40 mg)	201
Immuno-suppressants	Methotrexate	F 60%; CL_{ren} 80–90% unchanged (filtration and active tubular secretion), $CL_{hep\ (bil)}$ <10%; OAT1/3	Amoxicillin	Kidney	OAT1/3	Cp ↑ 13-30; CL_{ren} ↓ 0.44	Increased Mtx serum concentrations with concomitant hematologic and gastrointestinal toxicity have been observed for high and	202, 203

Table 10.1   (*Continued*)

Victim drug class (*alphabetical order*)	Victim drug	Victim drug ADME properties (F, CL, CYP enzyme and transporter substrate properties, *etc.*)	Perpetrator drug	Implicated site(s) of interaction	Implicated transporter(s)	Observed fold effect on clinical PK, PD or reported adverse effect	Clinical Implications	Ref.
			Naproxen (and other NSAIDS)	Kidney	OAT1/3	$CL_{ren} \downarrow 0.82$	low dose Mtx → co-therapy with penicillins should be carefully monitored Potential for enhanced Mtx toxicity → caution and careful monitoring, when NSAIDs and salicylates are administered concomitantly with lower doses of Mtx	119, 204
			Probenecid	Kidney	OAT1/3	$Css \uparrow 4.40$; $CL_{ren} \downarrow 0.44$	Careful monitoring of this drug combination	119, 205

[a]DDIs are attributed to the implicated transport proteins and, when indicated, also to drug metabolising enzymes. In some cases, the applied perpetrator doses exceeded the therapeutic range. Clinical DDI information was mainly retrieved from the University of Washington Drug Interaction Database Program (http://www.druginteractioninfo.org/), from the Drugs Interactions Checker provided by Cerner Multum, Inc. (http://www.drugs.com/) and from prescribing information.
[b]↔: non-significant change (change of PK parameters <20%); ↑: statistically significant increase (x-fold change in the presence of perpetrator); ↓: statistically significant decrease (x-fold change in the presence of perpetrator); n/a: not available; NAPA: *N*-acetyl-procainamide; NS: non-significant.
[c]Co-administration may result in additive effects and increased risk of ventricular arrhythmias including torsade de pointes and sudden death.
[d]Effect of maximal oral perpetrator dose.
[e]Increased risk of cyclosporine nephrotoxicity → monitoring of cyclosporine plasma levels and renal function, dosage adjustments as needed.
[f]Risk of adverse side effects (diarrhea, nausea, vomiting, dyspepsia and fatigue) due to CYP3A4 inhibition → monitor pharmacologic response to lapatinib and adjust the lapatinib dosage as necessary.
[g]Refers to the PK changes of the active metabolite *p*-hydroxytriamterene sulfate.
[h]CL reduction due to inhibition of hepatic CYP P450 metabolism and renal active secretion.

administered medication (*e.g.*, sedative hypnotics or analgesics), a change in the extent will have an impact on the amount of systemic exposure (*i.e.*, area under the curve (AUC) and maximal plasma concentration (C_{max})), which becomes particularly significant in the case of a drug with a narrow therapeutic index (such as digoxin—discussed further below).

DDIs involving intestinal uptake transporters are very limited in the number of cases reported, with most such cases being food–drug interactions. For instance, the concomitant intake of fruit juices can result in a decreased bioavailability of numerous drugs, including the antihistamine fexofenadine, several β-blockers such as talinolol, celiprolol and atenolol, as well as the renin inhibitor, aliskiren.[1] In the case of fexofenadine, it was shown in healthy volunteers that a concomitant intake of fruit juice (grapefruit, orange or apple) leads to an approximately 60% reduction in the fexofenadine plasma AUC and C_{max}. Although fexofenadine is a substrate of P-gp (Table 10.1), and thus inhibition of P-gp should lead to an increase in the fexofenadine plasma AUC and C_{max}, it is also a substrate of OATP1A2 and OATP2B1 (Table 10.1), both of which are localised in the luminal membrane of intestinal epithelia. Although it is unclear which OATP isoform is predominantly involved, it is likely that the overall mechanism of fruit–fexofenadine interactions is predominantly due to inhibition of OATP1A2 and/or OATP2B1 uptake carriers at the level of the small intestine.

P-gp is expressed in the apical membrane of enterocytes, where it exports substrates back into the lumen, limiting their oral bioavailability or, if dosed intravenously (i.v.), enhancing systemic clearance. P-gp also shares a substantial substrate (and inhibitor) overlap with the cytochrome P450 enzyme (CYP) 3A4, and they can act together in the intestine to reduce oral drug bioavailability. Due to this overlap, many intestinal DDIs often involve an effect on both P-gp and CYP3A4.[2] Accordingly, P-gp substrates that are not metabolised in humans to any great extent can be useful as sensitive probes for clinical P-gp DDI studies.[3] Some of these probe substrates are discussed below.

Digoxin, a cardiac glycoside, with a relatively high bioavailability of $\sim$70%, is mainly eliminated by glomerular filtration and P-gp-mediated tubular secretion ($\sim$60% of absorbed dose). It is an exception to most pharmaceutical compounds in that it has a narrow therapeutic index, where only a small increase in digoxin plasma exposure ($\sim$25%) can lead to toxicity. Several P-gp inhibitors have been reported to increase the plasma levels of digoxin above its upper limit of safety.[4] The biggest effects of P-gp inhibitors on digoxin plasma exposure in humans have been observed for valspodar, followed by quinidine, cyclosporine, itraconazole and clarithromycin (Table 10.1). Such interactions resulted in two- or three-fold increases in digoxin serum concentrations, and were frequently accompanied by adverse drug reactions (see Section 10.4.1).

The anti-coagulant prodrug dabigatran etexilate is also a P-gp substrate (but not the parent drug), and its low oral bioavailability is increased by about 1.5-fold when co-administered with clarithromycin or verapamil, resulting in the need for dose adjustment and monitoring (Table 10.1). Other drugs such as talinolol and fexofenadine are P-gp substrates and show increased bioavailability and plasma exposure when co-administered with inhibitor drugs such as verapamil (Table 10.1). However, their use as potential clinical P-gp probes is restricted due to the need to exclude confounding factors (intestinal OATP inhibition in the case of fexofenadine) or approval restrictions (talinolol is only approved in Germany).

Induction of intestinal P-gp has also been described in a small number of cases. P-gp induction in the intestine would result in enhanced pre-systemic extraction and a consequent reduction in oral bioavailability. For example, pre-treatment with the known P-gp inducer rifampicin decreased the oral bioavailability of the P-gp substrates digoxin[5] and talinolol[6] by around 30% and 20%, respectively.

The efflux transporters BCRP and multidrug resistance associated protein 2 (MRP2) are also expressed in the luminal membrane of enterocytes. In contrast to P-gp-mediated clinical DDIs, there are only a few reported clinical DDIs associated with these additional transporters, and in the majority of cases, they are confounded by interactions with other transport proteins and metabolising enzymes. This is exemplified by BCRP, with substrates that include methotrexate (Mtx), sulfasalazine, rosuvastatin and topotecan (Table 10.1). In clinical studies, while there were no significant changes in the PK of sulfasalazine after co-administration with pantoprazole, a known inhibitor of BCRP *in vitro*, a 3.2-fold increase in the AUC was observed when co-administered with high doses of the BCRP inhibitor curcumin. For the other BCRP substrates mentioned above, while clinically significant changes in PK (or adverse effects) are caused by known inhibitors of BCRP (*e.g.*, omeprazole, pantoprazole, eltrombopag, ritonavir and cyclosporine), the substrate and inhibitor overlap with other pathways confound the understanding of the role of BCRP in such cases (*e.g.*, where there is an inhibitor overlap, OATPs are a more important component than BCRP in DDI cases involving rosuvastatin; Table 10.1).

10.1.2 DDIs in the Liver

The liver plays a central role in drug disposition: it is the major component in first-pass elimination of orally administered drugs, as well as for the plasma clearance of systemically distributed drugs. The majority of the blood supplied to the liver is *via* the portal vein (about 70–80%), while the remainder is supplied by the hepatic artery. Both influxes lead to the hepatic sinusoid, where drugs are taken up into hepatocytes, metabolised and subsequently excreted. Transporters are important modulators of this process: they facilitate drug entry into hepatocytes, as well as driving the export of drugs and their metabolites into the bile or back into the blood. Overall, hepatic drug disposition is a balance between the rates of uptake (active and passive), metabolism and efflux. The rate-determining component is the critical step, as it will impact the overall rate of clearance and is the most susceptible site for DDIs.

A transporter-mediated DDI could be caused at each membrane barrier due to (1) basolateral uptake, (2) basolateral efflux, *i.e.*, back into the blood stream, and (3) biliary (apical) excretion.[7] Transporters can modulate intracellular drug concentrations, which in turn affect intrinsic metabolic clearance.[8] This interplay between hepatobiliary transport and metabolism is the basis for the high complexity in the assessment of the relevance of hepatic transporters in clinical DDIs. Hepatic sinusoidal uptake transporters (such as OATP1B1/OATP1B3, organic cation transporter 1 (OCT1) and the sodium/bile acid co-transporter (NTCP)) transport substrates from a range of different drug classes. There are many *in vitro* and *in vivo* examples of clinically relevant interactions (Table 10.1). Such interactions are most notable for the OATP transporters and 3-hydroxy-3-methyl-glutaryl-CoA (HMG CoA) reductase inhibitors (statins); concomitant administration with OATP inhibitors such as cyclosporine can lead to clinically substantial changes in the AUC or C_{max} of substrate drugs such as rosuvastatin (by up to seven-fold; see Table 10.1). For the statins, such interactions raise concerns for adverse effects, in particular for rhabdomyolysis—often necessitating either a contraindication or dose adjustment (see Section 10.4.2). The PK of other OATP victim drugs, such as the anti-diabetic drug repaglinide, as well as the anti-hypertensive bosentan, are also affected by concomitantly administered OATP inhibitors (Table 10.1).

Several ATP-dependent efflux transporters, including P-gp, MRP2, bile salt export pump (BSEP) and BCRP, as well as the transmembrane gradient-driven multidrug and toxin extrusion (MATE) transporter MATE1, are localised in the canalicular membrane of hepatocytes. Multiple drugs from various therapeutic areas have been identified as substrates of these biliary efflux transporters *in vitro*, and *in vivo* there is evidence for their biliary secretion (*e.g.*, BCRP for rosuvastatin; Table 10.1). In addition, clinical studies on digoxin using a duodenal catheter have observed a reduction in digoxin biliary clearance by 45% following concomitant administration of quinidine or verapamil.[9] This interaction is likely due to inhibition of canalicular P-gp. Cases of clinical DDIs involving the inhibition of biliary efflux are rare. This might be due to the fact that many drug substrates would have confounding effects from overlapping transporter and metabolism contributions. Adverse hepatic drug reactions can occur as a result of drugs inhibiting the biliary export of potentially hepatotoxic endogenous compounds (*e.g.*, BSEP inhibition and its association with cholestatic liver injury; see Chapter 2).

10.1.3   DDIs at the Blood–Brain Barrier

The blood–brain barrier (BBB) is one of the tightest cellular barriers in the human body. Its physiological role is to protect the brain and central nervous system (CNS) from exposure to harmful toxicants and xenobiotics. The BBB consists of endothelial cells held together by tight junctions, effectively restricting any paracellular transport, as well as efflux transporters, which pump substrates back into the blood. In addition, various influx transporters of the solute carrier SLC super family—glucose transporter (GLUT), peptide transporter (PEPT), OCTs, OATPs and equilibrative nucleoside transporter (ENT)—can be found, which ensure and control the supply of crucial nutrients, such as glucose, nucleosides, amino acids, *etc.*, to the brain. Some of these uptake carriers can interact with drugs as well, thereby enabling their uptake into the CNS. Prominent examples are the triptans, indicated for migraine treatment, which are described as being transported by OATP1A2.[10] However, in the context of DDIs at the BBB, the transporters of highest clinical relevance are the ABC efflux pumps P-gp and BCRP, with broad and overlapping affinities for structurally diverse drug substrates.[11]

The importance of P-gp at the BBB first became apparent when it was observed that P-gp knock-out (KO) mice were ~100-fold more sensitive to the neurotoxicity of the P-gp substrate ivermectin than wildtype (wt) mice.[12,13] Further analysis revealed that the brain exposure to ivermectin was almost 90-fold higher in P-gp KO mice.[12] This seminal work sparked a number of follow-up preclinical studies aimed at exploring further the significance of P-gp at the BBB. Importantly, large increases in CNS exposures to P-gp substrates that were observed in P-gp KO mice could also be reproduced in wt mice when co-administering a P-gp inhibitor. Verapamil brain exposure was elevated 11-fold when co-dosed with cyclosporine, and nelfinavir in combination with the dual BCRP–P-gp model inhibitor elacridar showed a 100-fold increase in brain exposure.[11]

Measuring the PK effects of efflux transporter-mediated interactions at the BBB, typically expressed as an increase in CNS exposure of the transporter substrate, is however challenging in humans; the unbound drug concentration in brain interstitial fluid (ISF) is considered the appropriate exposure parameter for a drug targeting the brain (in terms of PK/PD). To measure ISF, the common approach is to use cerebrospinal fluid (CSF), the fluid circulating in the subarachnoid space surrounding brain and spinal cord that is separated from the brain parenchyma by the blood–CSF barrier (BCSFB), which is also equipped with various efflux transporters (P-gp, BCRP and MRP4).[14] CSF is an easy to sample surrogate for unbound ISF, but only works well for compounds displaying a moderate to high passive permeability and not undergoing active transport at the BBB and BCSFB. For slowly permeating substances with substrate properties for efflux transporters in these barriers, CSF concentrations can be very misleading. The alternative is to use cost-intensive positron emission tomography (PET) studies or PD readouts.

Human PET studies with ^{11}C-verapamil demonstrated that clinical DDIs at the BBB can theoretically occur. For example, an 88% increase in the blood to plasma (B:P) ratio of ^{11}C-verapamil was observed when co-administered with supratherapeutic doses of cyclosporine.[15] Even though this is a very artificial scenario, it still remains a valid concern that any CNS restriction mediated by P-gp or BCRP-catalysed efflux might be impaired by co-dosed inhibitors, leading to elevated brain concentrations and subsequent adverse CNS side effects.

However, although it is well established that serious DDIs at the BBB are possible in preclinical species, as discussed above, these studies are not considered as predictive for clinically relevant interactions. The explanation for this mismatch between the preclinical and clinical situation lies in the fact that even though the elevation in CNS distribution of dual P-gp/BCRP substrates in *Mdr1a/1b*$^{-/-}$/*Bcrp*$^{-/-}$ KO mice can be very high, this effect requires nearly complete (>90%) blockade of these transporters. Such a state is, according to Kalvass *et al.*, almost never attainable with currently marketed drugs in humans, because most do not exhibit sufficient inhibitory potency at their unbound therapeutic plasma levels (unbound $C_p \ll IC_{50}$ (inhibitor concentration needed for 50% of maximally obtainable inhibition)).[11] Hence, for DDI risk assessment, it is not only important to consider the *in vitro* potency of the P-gp inhibitor, but also its unbound plasma concentration and, therefore, the ratio of $C_{p(unbound)}$ to K_i (inhibition constant) or IC_{50}.

Therefore, clinical concerns for such DDIs are low, with the exception of drugs such as loperamide, where the label warns of the risk of co-treatment with quinidine, ritonavir or saquinavir. However, the related DDI effect—which was observed in two thirds of the clinical DDI trials—is most probably attributable to an increase in plasma exposure and subsequently brain exposure (*i.e.*, a systemic effect rather than P-gp inhibition at the BBB).[11] For other drugs such as the antiparasitic drug ivermectin, second generation antihistamines (cetirizine, fexofenadine and loratadine), cardiac drugs (verapamil and digoxin), antivirals (nelfinavir), anticancer drugs (lapatinib) or opioids (morphine, methadone and fentanyl), which were administered in combination with therapeutic doses of potent marketed P-gp inhibitors, such as cyclosporine, quinidine and ritonavir, no CNS side effects were observed.

According to the International Transporter Consortium (ITC), only three approved drugs—quinidine, quinine and cyclosporine—exhibit significant systemic inhibition potential of P-gp at their therapeutic doses.[11] Nevertheless, the magnitude of clinical impact would be limited to a maximum two-fold increase in brain exposure. This range falls within the typical safety margin and inter-subject exposure variability of the vast majority of currently approved substrates under their clinical dosing regimens. Such interactions would only therefore be of concern for drugs with a narrow therapeutic index.[11] It is possible that future drugs will be approved that inhibit P-gp/BCRP with high potency at their clinical plasma concentrations. However, the approval of such inhibitors seems unlikely, or highly restricted to special MDR tumour-targeting applications, due to the risk of DDI liability at the BBB, as well as in other tissues.[11] In conclusion, the risk of clinically relevant DDIs due to efflux transport modulation at the BBB by currently-marketed drugs at clinical doses can be considered to be minimal.[11]

10.1.4 DDIs in the Kidney

Renal clearance represents the major elimination pathway for about one third of the top 200 marketed drugs. More than 90% of these drugs undergo so-called net secretion, meaning that the overall renal excretion exceeds the extent of glomerular filtration.[16] This active renal secretion takes place predominantly in the proximal tubule cells (PTC) and can be split into an anionic and a cationic pathway. Compounds falling into the first category are mainly transported by OAT1 and OAT3, while compounds of the cationic pathway are thought to be linked to uptake *via* OCT2 and apical efflux *via* MATE1 and MATE2-K.[17] For renally eliminated cationic compounds, the current regulatory guidance only suggests investigation of OCT2, although recent scientific evidence suggests that assessment of the interaction potential with MATEs is at least equally warranted.[18–20]

Renal DDIs can lead to various PK effects, such as elevations in C_{max}, plasma concentration at steady state (C_{ss}), AUC, half life ($t_{1/2}$) or a reduction in renal clearance (CL_{ren}; Table 10.1). Principally, those effects are more pronounced for drugs with extensive renal elimination, such as metformin, furosemide, digoxin, dofetilide, Mtx and penicillins. Nevertheless, renal DDIs can also occur for compounds (*e.g.*, triamterene, pindolol or zidovudine) cleared predominantly by hepatic pathways, and can become even more severe under hepatic impairment.

If measurable changes in one of the above mentioned PK parameters occur, a therapeutic intervention is not necessarily required. Examples are combinations of triamterene plus diflunisal (AUC change of >300%) and pindolol plus cimetidine (AUC change of >30%). One reason for this may be a sufficiently high safety margin of the victim drug that prevents the occurrence of severe adverse side effects, as in the case of the OAT substrate oseltamivir, where the AUC is increased by more than three-fold when co-dosed with the potent OAT inhibitor probenecid, but no dose adjustment is warranted.

There are currently only a few cases of clinically relevant DDIs mediated by renal transporters, as described in Table 10.1. One example for anionic transport involves the OAT1/OAT3 substrate Mtx. This drug should be cautiously combined with OAT inhibitors, such as beta-lactam antibiotics, non-steroidal anti-inflammatory drugs (NSAIDs), antivirals or the anti-gout medication probenecid. Even at low doses, patients should be monitored for elevated Mtx serum levels, which can be accompanied by pronounced haematologic and gut toxicity. For high dose treatment, severe cases of acute renal failure and even fatal events have been described. A co-therapy of the antibiotic ciprofloxacin (AUC increased by $\sim$75%) or antiviral ganciclovir (AUC increased by $\sim$50%) with probenecid should also be carefully monitored for abnormal elevations in drug exposure and victim-related toxicities.

For the cationic transporter DDIs, the most prominent victim is the frequently prescribed oral antidiabetic drug metformin. If elimination is impaired, supratherapeutic plasma levels can result in a highly severe state of lactic acidosis. Thus, since metformin is completely dependent on renal elimination (100% CL_{ren} for unchanged drug) *via* glomerular filtration and transporter-mediated secretion, the co-administration of OCT2/MATE inhibitors (cimetidine, pyrimethamine, *etc.*) should be closely monitored. Another important aspect is that these renal transporters are highly similar to the OCT1–MATE1 system in the liver, in terms of substrate and inhibitor interactions with drugs. Therefore, an overall multiplicative impact on the PK of metformin after co-administration of OCT/MATE inhibitors can be expected and—since its therapeutic activity is also localised in the hepatocytes—adverse events might occur.[21]

In the case of the cardiac drug dofetilide, AUC increases of ~50% already warrant a label stating the complete contraindication of cationic inhibitors. Also, interactions with P-gp are possible: the renal elimination of digoxin is strongly reduced in the presence of quinidine, implying the need for dose adjustment and serum level monitoring of the cardiac glycoside.

Transporters can relate directly to drug-induced nephrotoxicity, such as for cisplatin, where OCT2 accounts for its intra-PTC accumulation. This toxicity was shown to be partially reversed by OCT2 inhibitors (*e.g.*, the cationic anticancer agent imatinib)[22] or by exchanging with platinum agents, which due to their efficient efflux into the urine do not accumulate in the PTCs, *e.g.* oxaliplatin.[23,24]

Another victim for the renal transport system is the endogenous muscle breakdown product creatinine, which is used as a standard biomarker for renal performance, since it was thought to be eliminated solely by glomerular filtration.[25,26] However, according to current evidence, an immediate but moderate increase in serum creatinine upon drug administration, which is completely reversible after discontinuation of medication, might also be due to inhibition of OAT2, OCT2, OCT3 and/or MATEs by the administered compound.[27,28] Therefore, using creatinine as a biomarker for renal transport inhibition instead is being discussed, when renal impairment related effects on the GFR can be excluded.[29]

In summary, even though most renal transporter-mediated DDIs result in only modest PK effects, they can still lead to severe PD consequences, depending on the therapeutic window of the victim drug. Here, patients with renal impairment are at particular risk. Moreover, the change in plasma concentration does not necessarily reflect the change in intracellular concentration in the PTC, which accounts for nephrotoxicity or interactions with efflux transporters in the apical cell membrane. Finally, the kidney transporter systems are sometimes highly similar to other tissues, for example the renal OCT2–MATE1/2-K pathway and the hepatic OCT1–MATE1 system, leading to DDIs with an overall multiplicative impact on the PK of the victim drug.

10.1.5 DDIs in Other Tissues

Apart from the tissues discussed in detail in this section, transporter-mediated DDIs can, in principle, occur in every barrier tissue that is lined with epithelial/endothelial cells expressing drug transporters in their membranes.

To mention just a few:

- Alveolar epithelium of the lung[30]
- Barriers between the various compartments in the eye[10,31,32]

- Syncytiotrophoblast in the placenta[33]
- Mammary gland[33]
- Testis[34]
- Inner ear[35]

However, as these interactions are not necessarily relevant for each novel drug and only have to be considered for either specific forms of application or particular patient populations, they are not discussed in this chapter. Please refer to the literature citations provided to obtain more information on the individual topics.

10.2 Transporter Assessment Strategies

Having a good transporter strategy in place is important to address potential DDIs so that potential risks are well characterised in a timely manner. In this chapter, some general considerations for such a strategy are described. It is important to note that there will be notable differences among pharmaceutical companies at which stage they do certain studies, based on their internal insights as well as the regulatory requirements.

10.2.1 Which Transporters to Focus on for DDI Assessment: ITC Recommendations and Regulatory Requirements

In general, a DDI strategy within a pharmaceutical company will initially focus on those drug transporters for which there is clinical evidence that they are relevant for DDIs. Which transporters are clinically relevant has been addressed by the ITC, a group of drug transporter experts from academia, the pharmaceutical industry and regulatory agencies.[36] The ITC recommendations have also formed the basis for the current regulatory guidelines. Based on evolving science in this area, the ITC gives regular updates on several topics, including emerging transporters of clinical importance.[37,38]

The current draft US Food and Drug Administration (FDA) guidance on drug interaction studies (2012) recommends that all investigational drugs should be evaluated *in vitro* for substrate ("victim") interactions with P-gp and BCRP.[39] In addition, investigational drugs should be evaluated *in vitro* for substrate interactions with the hepatic uptake transporters OATP1B1 or OATP1B3, in cases where hepatic elimination is a significant pathway (>25%). Likewise, *in vitro* studies should be performed to check whether investigational drugs are substrates of the renal uptake carriers OAT1, OAT3 or OCT2 when renal active secretion is believed to be an important pathway (>25%). Beyond their evaluation as substrates, investigational drugs should also be evaluated *in vitro* as potential inhibitors ("perpetrators") of P-gp, BCRP, OATP1B1/OATP1B3, OAT1/OAT3 and OCT2. Importantly, the FDA guidance mentions that the recommendations are not necessarily restricted

Table 10.2 Summary of drug transporters to be studied, as recommended by the ITC, FDA, EMA and the PMDA (Japanese agency).

Transporter	ITC (2010)[36]	FDA (draft 2012)[39]	EMA (2012)[41]	PMDA (2014)[40]
P-gp	✓	✓	✓	✓
BCRP	✓	✓	✓	✓
OATP1B1	✓	✓	✓	✓
OATP1B3	✓	✓	✓	✓
OAT1	✓	✓	✓	✓
OAT3	✓	✓	✓	✓
OCT1			Consider	Consider
OCT2	✓	✓	✓	✓
BSEP		Consider	Consider	Consider
MRPs		Consider	Consider	Consider
MATE1/2-K		Consider	Consider	✓

to the characterisation of the transporters mentioned above, and other transporters such as MRPs, MATE1/2-K and BSEP should be considered when appropriate (Table 10.2).

The European and Japanese guidance essentially refer to the same panel of clinically relevant drug transporters (P-gp, BCRP, OATP1B1, OATP1B3, OAT1, OAT3 and OCT2) as the FDA for assessment as substrates, as well as inhibitors, but the Japanese authorities (Pharmaceuticals and Medical Devices Agency (PMDA)) also include the MATE1/2-K transporters as recommended (Table 10.2).[40] The European Medicines Agency (EMA) and PMDA both also, state that the inhibition of BSEP, OCT1 and MATE1/2-K could also be relevant to consider.[41]

Suitable *in vitro* methodologies, and their limitations, to investigate these interactions are discussed in Chapter 7.

Whether subsequent clinical DDI studies with an investigational drug are warranted for either victims or perpetrators is further specified in decision trees within the guidances (see Chapter 11). In general, this is dependent on an integrated assessment where not only the *in vitro* transporter data, but also the dose, the C_{max}, the relative importance of transporter-mediated pathway(s) for the PK, the therapeutic window, as well as the likelihood and properties of any co-medications interfering with major drug transporters are considered. These points will be highlighted in Section 10.3.

It is important to note that the regulatory guidelines also refer to characterisation of metabolites as potential substrates and/or inhibitors of drug transporters. Typically, this characterisation will be done during clinical development, once there is a good understanding regarding the *in vivo* human metabolites and their corresponding exposures. For a detailed discussion on the regulatory requirements for the study of transporter DDIs, please refer to Chapter 11. It is important to realise that it might occasionally be warranted to study additional transporters depending on the compound characteristics and potential co-medications.[39] For example, if a compound class is known to interact with the PEPT1 transporter, it may be important to investigate this transporter as well.

10.2.2   When to Investigate Risk for DDIs: Approaches in Transporter Assessment Strategies

It is important that both the need for and timing of transporter-related studies are carefully assessed to minimise the costs and time involved, and also to ensure that any risks are assessed prior to studies in humans. A general flow chart for the evaluation of transporter substrates and inhibitors within drug development is given in Figure 10.2.

Currently, the assessment of transporter interactions for investigational drugs typically starts in the discovery phase and continues until the post-marketing phase. It is important to have a good understanding of the

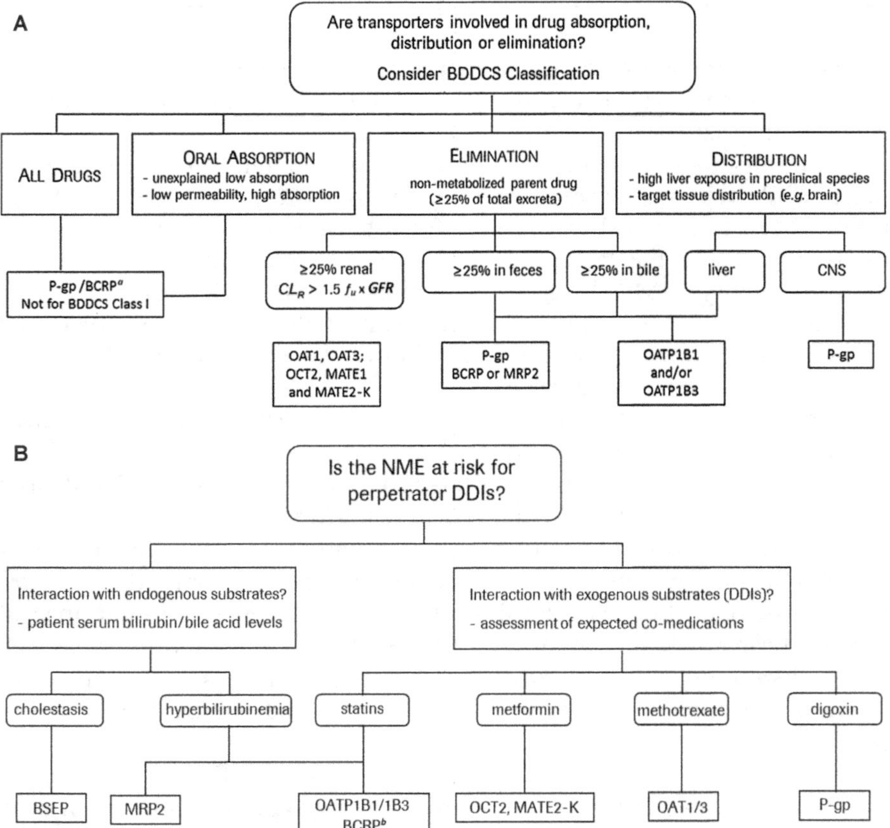

Figure 10.2   Typical decision trees for (A) substrate ("victim") and (B) inhibitor ("perpetrator") interaction investigations. [a]BCRP evaluation should not necessarily be routine: see text and Poirier *et al.*, 2014.[41] [b]BCRP inhibition assessment could be relevant when certain statins (*e.g.*, rosuvastatin[41]) are expected co-medications. Note, that depending on the treatment indication/patient population, other co-medications than the ones proposed in the figure might be relevant and will need to be investigated.

physicochemical properties of a compound (*e.g.*, passive permeability) and the major organ(s) of elimination, as these are important considerations for making an appropriate assessment of the potential *in vivo* impact of a transporter on the disposition of a drug.

Accordingly, for drugs that are highly permeable and highly soluble (*e.g.*, Biopharmaceutics Drug Disposition Classification System (BDDCS) Class 1 compounds), the *in vivo* impact of the transporter is considered to be limited. Typically, these drugs have a high oral bioavailability and there is low risk that drug transporters are an important determinant of their PK. Consequently, Class 1 compounds, in general, have a low risk for drug transporter-mediated DDIs as victims. In contrast, for drugs with low permeability, transporter interactions may be of greater consequence, for example, their oral absorption can be limited by transporters such as P-gp, posing a potential risk for a DDI.

In addition to the physicochemical and PK properties of the new molecular entity (NME), several general factors can influence the timing of when best to start investigating a particular drug transporter:

- An (anticipated) narrow therapeutic window of the NME can make the transporter DDI assessment more pressing, as even very small increases in exposure could have important safety implications. In contrast, for NMEs with a large therapeutic window, increases in exposures could be less critical, as they might not pose any safety risk.
- Knowledge of the likely co-medications can be important in prioritising the study of certain drug transporters. For example, for several therapeutic indications, there is a high likelihood that the NME will be co-administered with statins, which could justify an earlier assessment of OATP inhibition (see below).
- The intended route of administration: clearly, the risk of drug transporter-mediated DDIs at the level of the gut will be less of a concern for i.v. administered drugs than for orally dosed compounds.
- Whether the treatment duration will be chronic or short term could influence the DDI risk assessment. In the case of short term treatments (*e.g.*, antibiotics), temporarily suspending a chronic treatment (*e.g.*, statins) to avoid a DDI could be considered.
- Finally, the regulatory requirements also have to be considered in any transporter assessment strategy.

10.2.2.1　P-gp

10.2.2.1.1　Victim DDIs.　An *in vitro* P-gp substrate assessment is often done at an early stage of drug discovery. Especially for neuroscience projects, this is also an important screen to assess whether a compound will sufficiently penetrate into brain tissue. Typically, the early (high throughput) screens are followed by definitive studies that comply with regulatory recommendations. In cases where a NME appears to be a P-gp substrate, an assessment needs to be made regarding whether a clinical DDI study is warranted. Critical in this assessment is a good understanding of the relative contribution of P-gp to the overall absorption and clearance of the NME (see Section 10.3).

10.2.2.1.2 Perpetrator DDIs. In addition to a P-gp substrate assessment, a NME needs to be evaluated as an inhibitor of P-gp to allow an early assessment of whether a clinical (digoxin) DDI study is warranted. The IC_{50} (or K_i) value that is derived from the *in vitro* inhibition study is then—in conjunction with the information on exposure and dose—used to determine whether a clinical DDI study is indeed required. For example, the FDA guidance refers to using a static model for P-gp inhibition, where both the maximum plasma concentration, I_1, and the gut concentration, I_2 (defined as the maximal oral single dose per 250 ml), need to be considered. Accordingly, in cases where the $I_1 : IC_{50}$ ratio is higher than 0.1 or $I_2 : IC_{50}$ is higher than 10, a clinical DDI study is recommended by the FDA.[39]

It is important to note that there is substantial inter-laboratory variability in P-gp IC_{50} determinations.[42,43] For example, the inter-laboratory variability in IC_{50} determination for verapamil was more than 100-fold.[42] Clearly, such variability in IC_{50} determinations can have important implications and might lead to a completely different DDI risk assessment (see Section 10.4.1).

Another important consideration regarding P-gp perpetrator DDI assessment is substrate-specific inhibition.[44,45] Hence, in addition to evaluating the inhibition of digoxin transport by a NME, it is recommended to screen against at least one other substrate (*e.g.*, loperamide). When there is indeed evidence for substrate-specific inhibition, it might be necessary to evaluate the inhibition of a panel of P-gp substrates that are considered to be potential co-medications.

10.2.2.2   *BCRP Victim and Perpetrator DDIs*

Despite its functional similarity to P-gp, the clinical importance of BCRP is still under debate.[46] The need for and timing of BCRP substrate and inhibition assessment have recently been addressed by Poirier *et al.*,[47] with the authors suggesting that assessment of compounds as BCRP substrates during the preclinical development phase should not be routine. Rather, a case-by-case approach based on physicochemical and PK properties, as well as the therapeutic window of the NME, has been proposed.[47]

Essentially, similar considerations may also apply for the assessment of BCRP inhibition. Clearly, the *in vitro* assessment of BCRP may need to be completed to comply with current regulatory recommendations, but these studies can often be delayed until later stages of the development phase.

10.2.2.3   *OATP1B1/OATP1B3*

10.2.2.3.1 Victim DDIs. The necessity to evaluate a NME as an OATP substrate will be largely driven by evidence that hepatic or biliary secretion is a significant clearance pathway. In contrast to the OATP perpetrator assessment, OATP *in vitro* victim assessment will not necessarily be done during the preclinical phase, but could be addressed during clinical development. In cases where a compound has been identified as a substrate for OATP1B1 and/or OATP1B3, a clinical DDI study with rifampicin or cyclosporine could be considered. Also, investigating the PK of a NME in subjects with various OATP1B1 genotypes could be an option to further ascertain the importance of the OATP-mediated clearance pathway.

10.2.2.3.2 Perpetrator DDIs. Several clinically relevant DDIs have been attributed to interactions with OATPs (Table 10.1). It is therefore important to investigate the inhibition of these transporters, particularly OATP1B1 and OATP1B3, for NMEs that have a high likelihood of being co-administered with statins, a prominent drug class with a high likelihood of substrate properties for these transport proteins. A case study on the differing DDI potentials of various statins can be found in Section 10.4.2.

In vitro evaluation of whether compounds inhibit OATP1B1/OATP1B3 is typically completed during the preclinical phase. In some cases, OATP inhibition can even be an important parameter to guide lead optimisation, especially if the therapeutic area is associated with hyperlipidaemia. To predict a potential DDI, a static model based on the IC_{50} as well as the estimated maximum inhibitor concentration at the inlet to the liver ($I_{in,max}$) can be used.[39] This could subsequently trigger a clinical DDI study with a sensitive substrate such as rosuvastatin, pravastatin or pitavastatin.

Importantly, several limitations of the static model approach have been noted,[48] which are analogous to the issues described for P-gp/digoxin DDI extrapolations. Firstly, use of total C_{max} may be overly conservative, resulting in many false positive predictions. Secondly, drug interactions most often involve multiple mechanisms with other transporters and drug metabolising enzymes. Not properly taking these into account may over- or under-estimate the actual risk for DDIs or adverse effects (*e.g.*, in the case of statins, the role of OATP2B1 for peripheral distribution). Thirdly, the high inter-laboratory variability in experimental data would mean that the use of generic threshold criteria may result in incorrect predictions, depending on the source of the *in vitro* IC_{50}. Therefore, a more holistic approach to statin and also general DDI risk assessment is necessary, where the role of OATP inhibition—although a major component—can be put into context with other interactions involved in the overall DDI. For these reasons, physiologically-based PK (PBPK) modelling for predicting transporter DDIs has become more widespread (see Chapter 9).

10.2.2.4   OAT1/OAT3, OCT2 and MATE1/MATE2K

10.2.2.4.1 Victim DDIs. Whether the *in vitro* assessment of OAT1/OAT3 and OCT2 and MATE1/MATE2K substrate interactions is necessary is highly dependent on whether renal clearance (CL_{ren}) of the NME is an important contributor to the total clearance. Therefore, in cases where there is evidence for active secretion, defined as $CL_{ren} > 1.5$ fraction unbound (f_u) × glomerular filtration rate (GFR), an *in vitro* substrate assessment for OAT1/OAT3, OCT2 and MATE1/MATE2K should be conducted. The *in vitro* assessment may be started as soon as there is evidence for active renal secretion in preclinical species and when PBPK modelling also supports a significant renal component for humans. Often, however, these studies are delayed until the first human studies confirm significant renal clearance.

10.2.2.4.2 Perpetrator DDIs. It is recommended that an *in vitro* evaluation for inhibition of OAT1/OAT3, OCT2 and MATE1/2K is conducted for each NME. Whether subsequent clinical studies are required is dependent on the ratio of the unbound C_{max} to the IC_{50}. Typically, the assessment of these transporters is done during Phases I/II of clinical development. However, there can be reasons to frontload this investigation. For example, if it is known that metformin will be an important co-medication of the NME.

10.2.2.5 Overview of In vitro *Transporter DDI Assessments During Drug Development*

A generic flow chart for the timing of the various *in vitro* transporter studies during the different stages of drug development is given in Figure 10.3. Depending on the project, studies can be frontloaded (*e.g.*, OATP inhibition) or postponed (*e.g.*, BCRP substrate/inhibition). In addition, based on the characteristics of the NME, some studies might be completely waived (*e.g.*, OAT1/OAT3/OCT2 substrate assessment in cases where there is no significant renal clearance). Clearly, a transporter strategy will need to be regularly updated based on new scientific insights regarding the clinical relevance of certain transporters as well as changes in the regulatory requirements.

10.3 Clinical Interaction Studies

10.3.1 Introduction

The clinical DDI strategy aims for a quantitative mechanistic understanding of the drug metabolism and PK properties of the drug, based on which its propensity to act as a perpetrator or a victim in a DDI can be assessed. A clinical DDI study must be designed to reveal the quantitative contribution of the investigated processes and is guided by mechanistic information from nonclinical experiments. While the *in vitro* identification of a drug as a substrate of a transporter(s) is an important first step in the DDI assessment, the following data help to substantiate the assumption that transport mechanisms play an important role in the PK of a drug:

- The classification of the compound in the BDDCS system ≠ 1 (see Section 10.3.2)
- High tissue distribution is observed in animals (*vs.*/opposed to simulated PBPK prediction)

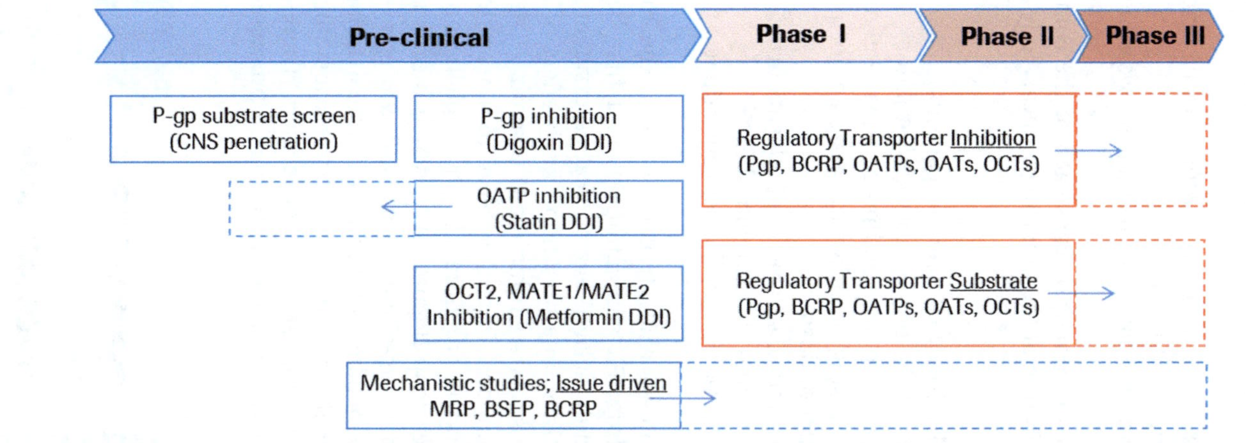

Figure 10.3 Generic overview of *in vitro* drug transporter studies in drug development. The testing for MATE1/MATE2K substrate and inhibition as of yet are not explicity mentioned by all regulatory authority guidances, but may warrant testing due to their clinical relevance in DDI.

- Low extent of metabolism and high recovery of parent drug (based on an appropriately designed study)
- Saturation of processes (*in vitro* and *in vivo*, the latter both non-clinically and clinically)
- Presence of active transport processes *in vitro* (*e.g.*, in isolated hepatocytes)
- *In vitro* transporter phenotyping (identification of involved transport proteins in an isolated manner)
- Correlation of transporter polymorphism(s) and systemic drug exposure

While in this chapter we cannot discuss each of the components extensively, thorough consideration of the following clinical study design features is essential:

- Study population (healthy volunteers or patients)
- Type and specificity of the probe substrate/inhibitor
- Dose selection of both the drug and probe substrate/inhibitor
- Single *vs.* multiple dosing of either the drug and/or probe substrate/inhibitor, and timing of dosing
- Sampling schedule and assessment of plasma/urine concentrations (amount) for the drug and substrate/inhibitor
- Quantification of plasma metabolite concentrations

Depending on the known characteristics of both the model substrate or inhibitor and the drug, PK simulations should first be performed to optimise the clinical study design. For clinical assessment of transporter-mediated DDIs and their interpretation, knowledge of the role and specificity of the probe substrate/inhibitor is essential. The translation of non-clinical data to humans is still subject to uncertainties, even for metabolic processes, and is accounted for by applying an iterative non-clinical–clinical feedback loop (Figure 10.4), which means that clinical results may lead to further non-clinical work and *vice versa* in an iterative process. For transporter-mediated processes, this translational step is still in the early stages, in particular because of the sparse availability of potent and selective probe inhibitors/substrates for transporters. Therefore, the identification and quantification of transporter-mediated processes in humans is

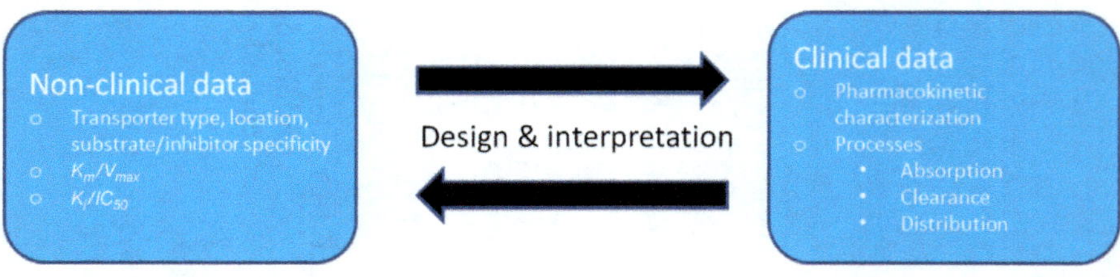

Figure 10.4   Interplay between non-clinical and clinical data.

a challenge and requires an overall thorough understanding of the PK properties of the drug and the prototype substrates/inhibitors.

Clinically, the PK of the drug can be described by the following different processes, based on well-established PK principles: absorption, distribution and elimination. However, inherently due to the nature of the role of transporters, distribution may play an important role in all of the different processes concerned, whether at a cellular or organ level. Clinical DDI examples at the barrier tissues of the intestine, BBB, liver and kidney can be found in Table 10.1.

There is a strong interplay between non-clinical and clinical data to be taken into consideration when assessing a transporter DDI (Figure 10.4). Non-clinical data provide information at a molecular level while clinical data characterise the PK processes.

What are the key questions to be answered when investigating the DDI potential of a drug? For the victim drug (as a NME or co-medication), it is important to know the quantitative contribution of the process. What are the changes in the PK of the victim drug, if the process of concern is being entirely blocked, as the worst case scenario, and what is the extent of change that can be expected if the interference is partial? Figure 10.5 gives an example where a process with a fraction metabolised (f_m) of 0.5 is inhibited. In PK terms, the contribution of an individual transporter to overall clearance, analogous to the f_m, is the parameter that needs to be quantified. As transporters contribute to f_m, this parameter may instead be described as fractional clearance, as the final question is the quantitative contribution of the transporter-mediated process to the overall process of concern.

For the perpetrator drug (as a NME or co-medication), it is important to know the quantitative interference with the process of concern, which is mainly dependent on the concentrations reached at the target (I) and the inhibition potential (K_i), expressed as the $I:K_i$ ratio. The clinical investigation of DDIs is dependent on both characteristics, the f_m of the process of concern and the $I:K_i$ ratio achieved (see example in Figure 10.5).

In this example, a maximum $I:K_i$ ratio of about 5 is achieved at the early time points, which translates into a minimum relative clearance of about 0.6, *i.e.*, does not achieve the theoretical value of 0.5. In this case, with a collection of samples taken up to 48 h post-dose, an AUC of 1.54 theoretically would be obtained in the presence of the inhibitor (real), while in cases of a full blockage (ideal), an AUC of 2 would be achievable. To achieve a higher theoretical value, a higher $I:K_i$ ratio would need to be sustained over a longer time period by adding multiple doses of the inhibitor.

Generally, in clinics, the extent of a DDI is described by the magnitude of changes in systemic exposure, *i.e.*, weak inhibition: $\geq$1.25- to <2-fold change in AUC; moderate inhibition: $\geq$2- to <5-fold change in AUC; and strong inhibition: $\geq$5-fold change in AUC. The assessment of the clinical implications of a change in exposure needs to take into account the risk/benefit for the specific drug of concern.

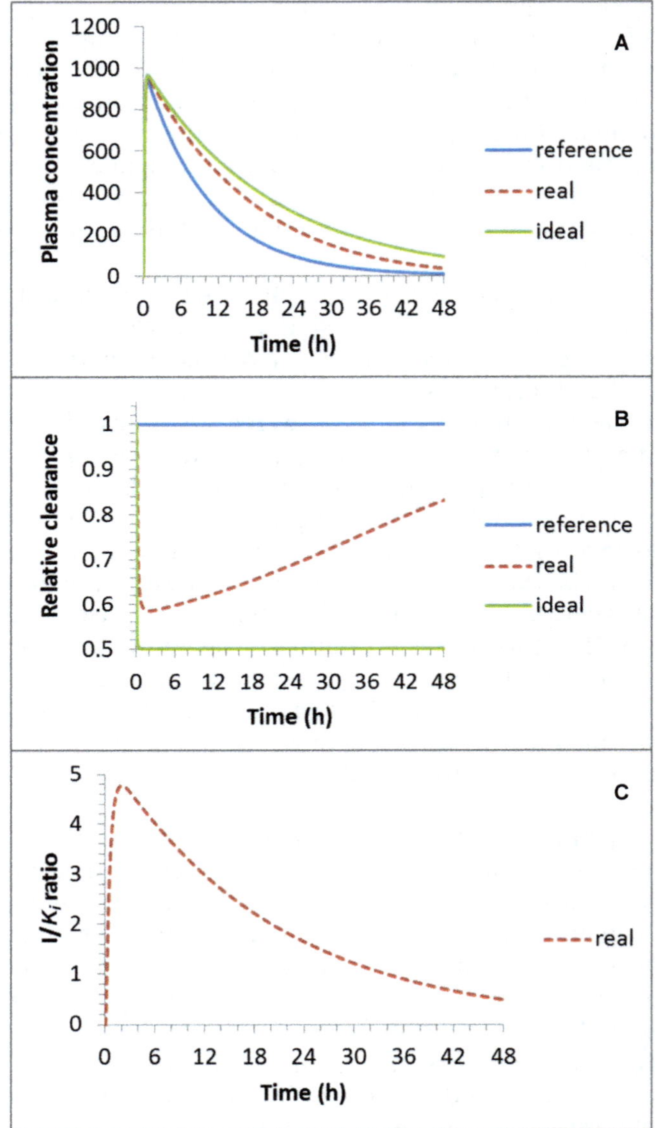

Figure 10.5 Simulation of plasma concentration time–profiles of a victim drug in the absence (reference) and presence (real) of an inhibitor, and in the case of complete blockage (ideal). (A) Plasma concentration–time profiles, (B) relative clearance–time profiles, and (C) I/K_i for a real example. The f_m was 0.5, $t_{1/2}$ was 7 h for the victim and 14 h for the perpetrator.

The clinical assessment of the DDI potential is not solely based on the results of one clinical DDI study, as all clinical data need to be taken into account. For example, absolute bioavailability (F) data are in many cases important to gain a full quantitative understanding of the relevance of a process, *i.e.*, *F* is not only important for assessment of absorption, but also clearance.[49] Furthermore, direct quantification of the amount of excreted parent drug and metabolites in a clinical mass balance study could in many cases be used to quantify f_m. As mentioned in Section 10.1.4, in particular for drugs that are mainly eliminated unmetabolised (such as metformin), transporter-mediated processes can play an inherently important role. In this case, a clinical mass balance study and *F* data are essential to quantify the role of transporters in the PK of the drug.

Correlation of transporter single nucleotide polymorphism (SNP) data with alterations in exposure of a NME is another way of quantitatively assessing the role of specific transporters in the PK of a drug.[50] Finally, integration of all available data by a PBPK model is highly recommended, in order to gain a full mechanistic and quantitative understanding of the pathways involved, and to extrapolate the results obtained to other co-medications or populations (*e.g.*, pediatrics). Even without a good quantitative understanding of the pathways involved, a model-based approach could be highly informative in cases where non-linear processes are observed[51] or to gain a mechanistic understanding, as applied in PBPK modelling.[52]

10.3.2 Absorption

The extent to which a drug is absorbed can be affected by many factors (*e.g.*, drug chelation in the gut, changes in gastrointestinal motility, gastro-intestinal pH, or intestinal CYP and transporter activity). The focus here is on how modulation of transporter activity in the gut influences the absorption process of a drug and its PK, and how best this can be assessed in clinics using non-clinical and clinical information. There are different PK methods available to separate the absorption process from systemic processes and to describe the absorption profile of a drug, such as compartmental modelling, deconvolution methods in PK data analysis or PBPK modelling. However, i.v. administration of the drug to determine *F* is required for a thorough characterisation of the absorption characteristics.

To evaluate the potential for a DDI at the absorption site, it is important to consider the following information:

- Physicochemical properties of the drug (BDDCS class), *e.g.*, lipophilicity and solubility
- Pharmaceutical properties of the drug, *e.g.*, dissolution rate and dosage form
- Absolute bioavailability of the drug
- Extent that the drug is excreted into faeces
- Assessment of dose proportionality:
  - Saturation of an efflux transporter may result in a more than dose-proportional increase in exposure
  - Saturation of an uptake transporter may result in a less than dose-proportional increase in exposure

Table 10.3 Modulation of P-gp activity in the intestine.[a]

Modulation of activity	Consequences
Inhibition of P-gp	Extrusion of P-gp substrate ↓
	Systemic availability of drug (P-gp substrate) ↑
Induction of P-gp	Extrusion of P-gp substrate ↑
	Systemic availability of drug (P-gp substrate) ↓

[a]Not all P-gp substrates will undergo reduced oral absorption in the presence of a P-gp inhibitor (see text).

- Non-clinical information (*e.g.*, K_i and IC_{50})
- History of transporter involvement at the absorption level within the same chemical class

To date, of all of the transporters known, the efflux pump P-gp is the best studied with a large amount of non-clinical and clinical data reported. In the context of P-gp DDIs and drug absorption, there is P-gp induction and inhibition (Table 10.3). The latter tends to be overestimated as a cause of clinically relevant DDIs (see below).

Many drugs that are P-gp substrates have a fairly good F (*e.g.*, digoxin, ritonavir, indinavir and verapamil).[53] This is due to saturation of intestinal P-gp at their applied clinical doses. Thus, these drugs are less sensitive to clinically relevant intestinal P-gp DDIs. Drug interactions arising from P-gp inhibition in the intestine are only likely to be clinically relevant when the drug that is a P-gp substrate:

- Is administered at small oral doses (<50 mg; *i.e.*, P-gp is not saturated)
- Shows slow dissolution
- Shows slow membrane diffusion rates

Exceptions to this are cyclosporine, paclitaxel and saquinavir hard gelatin capsules. These are all drugs that are administered at high doses (>50 mg), but the drug substance is either poorly soluble in water, large in size (>800 Da) or their formulations are slowly dissolving. Overall, P-gp inhibition in the intestine is rarely a source of clinically significant DDIs.[53,54]

Intestinal P-gp induction can lead to a reduction in F, which can be clinically relevant and needs to be assessed case by case. The study design that allows differentiation between the impact of the P-gp inducer on the absorption and the disposition of the P-gp substrate is to administer the P-gp substrate both i.v. and orally. A possible study design to study intestinal P-gp induction in clinics is first to give a single oral dose of the P-gp substrate together with a single i.v. radiolabelled microdose of the P-gp substrate, followed by a washout period in the case of a crossover design. Then the same subjects receive multiple doses of the P-gp inducer again followed by a single oral dose of the P-gp substrate together with a single i.v. radiolabelled microdose of the P-gp substrate, once maximum induction is reached. The concept of this study design can also be applied to investigate intestinal P-gp inhibition.

Even though it has a high F, most clinical studies have been performed, to date, with the model P-gp substrate digoxin, a narrow therapeutic index drug that is not metabolised by CYP enzymes (see Section 10.4.1). However, in the recent EMA DDI guidance, it is recommended to study *in vivo* inhibition of intestinal P-gp using a sufficiently specific P-gp substrate with low oral bioavailability, such as dabigatran etexilate or, if no OATP1B1 or OATP1B3 inhibition is expected, fexofenadine.[41] This recommendation is based on the lower F of dabigatran and fexofenadine, rendering these victim drugs more sensitive to intestinal P-gp inhibition than oral digoxin.

10.3.3   Tissue Distribution

Transporters can modulate intracellular drug concentrations. For the assessment of DDIs, there are two key criteria to consider, dependent on whether or not the tissue eliminates the drug.[8] Firstly, if transporters modulate intracellular concentrations in non-eliminating tissues, a DDI could result in specific organ safety or efficacy issues (*e.g.*, statin exposure in the muscle and associated muscle toxicity, or CNS penetration[55]). Secondly, if transporters modulate intracellular concentrations in eliminating tissues, a DDI at the transporter level could affect metabolic and uptake/elimination, rendering predictions based on *in vitro* data of higher uncertainty.

The clinical assessment of tissue distribution is challenging. Pharmacokinetically, a change in distribution could in principle be detected, but if specific tissues are affected with a low contribution to the overall volume of distribution at steady state (such as muscle or brain), in combination with a typical oral administration of a drug, changes are difficult to detect. Nevertheless, there are reports of pronounced changes in atorvastatin volume of distribution $(V_{ss})/F$, when the OATP1B inhibitor rifampicin is administered i.v.[56] Other approaches are indirect assessments by a PD or safety read-out in cases where the organ of concern is a PD target organ (*e.g.*, hepatic transporter polymorphism effects on efficacy, muscle toxicity of statins and respiratory depression in BBB P-gp inhibition).[57]

CSF sampling is used to assess CNS penetration of drugs clinically, but this method may not provide sufficient accuracy to assess inhibition at the BBB. Specific PET ligands can also be used clinically to assess tissue distribution and DDIs (Table 10.1, BBB DDIs).[58–60]

10.3.4   Hepatic Clearance

There are important design considerations to take into account for transporter-mediated processes in the liver. A typical clinical study to assess the potential of a drug to be susceptible to induction by rifampicin investigates the PK of the drug before and after multiple doses of rifampicin. Rifampicin is also an OATP inhibitor; thus, in cases where the drug is an

OATP substrate, its hepatic uptake could be inhibited by rifampicin. Inhibition of hepatic uptake by rifampicin may lead to increases in drug exposure in the plasma, which would potentially mask any induction effects by rifampicin that would otherwise reduce the plasma exposure.[61] A wash-out phase of rifampicin might therefore be required to avoid such interference, within a time frame that is not expected to alter the CYP expression levels to a relevant extent.

In order to discriminate between transporter and metabolic inhibition, the study should include an analysis of the metabolite to parent ratio. Inhibition of a metabolic pathway reduces the plasma concentration of the metabolite(s) formed by the metabolic pathway of concern, thus, decreasing the metabolite to parent ratio.[56] In cases where there is interference with a transporter-mediated pathway, there could be entirely different changes in metabolite to parent ratios.[62,63] While it is too complex to elaborate on this in more detail in this chapter, integration of these data in a PBPK model could provide valuable insights.[64]

In vitro data can be important in assessing whether an observed clinical interaction is caused by transporters or metabolism or both; a good example is the mechanistic evaluation of the DDI between gemfibrozil and cerivastatin.[65] Gemfibrozil and its glucuronide interfere with hepatic OATP1B1, but a quantitative assessment suggested that the observed interaction is most likely a result of a metabolic interaction.[65] Of note, this is also a good example to highlight that even the mechanisms of metabolic interactions could be subject to controversy as metabolites could also be perpetrators of DDIs.[66] More details on statin-mediated DDIs can be found in the respective case study in this chapter (Section 10.4.2).

Enterohepatic circulation is another important process in which transporters could potentially play an important role. Enterohepatic circulation of bile salts is a vital physiological process driven by specific bile salt transporters in the liver (basolateral NTCP and apical BSEP) and distal ileum (apical sodium dependent bile acid transporter (ASBT)/*SLC10A2*). The inhibition of the biliary bile salt transporter BSEP by bosentan and the subsequently increased risk for cholestasis due to intrahepatic accumulation of bile salts is a well-known example (see Chapter 2).[67] A drug itself could also be subject to enterohepatic circulation, indicated by a longer than expected half-life or a multiple peak phenomenon triggered by food intake. Clinically, interruption of enterohepatic circulation with oral administration of cholestyramine or charcoal, ideally following i.v. administration of the drug, is the method of choice to identify and quantify this process.[68,69]

10.3.5 Renal Elimination

The assessment of transporters involved in the CL_{ren} of drugs has a long-standing history, as active tubular secretion of drugs is a transporter-mediated process that can be specifically assessed. Clinical interaction studies use cimetidine and probenecid to inhibit organic cationic and anionic transport processes, respectively, and are currently characterised by the expected renal transporters involved: OCT2/MATEs and OAT1/OAT3.[70,71]

The presence of active tubular secretion is detected by a comparison of CL_{ren} and GFR multiplied by the f_u of the drug. For example, Somogyi *et al.* reported a CL_{ren} of 347 ml min^{-1} for procainamide[72] and a f_u in plasma of 0.82. Therefore, CL_{ren} was clearly greater than GFR$\times f_u$ (120 ml min$^{-1}\times$0.82), indicating the presence of net active tubular secretion for the PK of procainamide. Co-administration of cimetidine (400 mg 1 h pre-dose and 200 mg 4 and 12 h post-dose) decreased the CL_{ren} of procainamide from 347 to 196 ml min^{-1}, which was associated with a decrease in systemic clearance (CL) by 35%. The CL_{ren} with cimetidine co-administration (196 ml min^{-1}) was still higher than GFR $\times f_u$, indicating incomplete inhibition of the active tubular secretion process.

While CL_{ren} can be determined independently of the route of administration, for an assessment of the contribution of CL_{ren} and its individual processes to systemic CL, knowledge of F is required. This is important for the assessment of the DDI risk, as *e.g.*, a small contribution of CL_{ren} or its individual components to systemic CL translates even a large inhibition at the renal level into a small increase in systemic exposure (or small decrease in CL). This is further exemplified in Table 10.4 for two cases with all parameters being equal except for F. Depending on the magnitude of F, different predictions of the effect of changes in CL_{ren} on exposure are obtained.

With regards to study design, it is important to obtain a complete urine interval collection together with matched plasma sampling, in order to calculate the amount excreted into urine (A_e) and the plasma AUC for the same observation period.[73]

In patients with renal impairment, the GFR is decreased, which inherently will affect the contribution of CL_{ren} to CL. However, there is limited information available on the effect of renal impairment on active tubular secretion. As loss of functional nephrons is thought to be the main mechanism of renal impairment,[74] it is likely that active tubular secretion is diminished overall in patients with renal impairment.

Even though not directly related to the topic of this chapter, it is of note that renal impairment could also affect non-renal processes,[75] such as

Table 10.4 Example of how knowledge of F changes the prediction of the magnitude of a DDI where complete inhibition of active tubular secretion is observed.

Parameter	Case 1	Case 2
CL/F (mL min^{-1})	1000	1000
F	1	0.5
CL (mL min^{-1})	1000	500
CL_{ren} (mL min^{-1})	400	400
$f_u \times$ GFR (mL min^{-1})	100	100
ΔAUC, complete inhibition	1000/700 = 1.4	500/200 = 2.5

metabolism and/or active transport in other tissues. Therefore, drugs mainly cleared by non-renal processes may still be investigated in patients with renal impairment to assess how renal impairment may change the PK of a non-renally eliminated drug.

10.4 Case Studies

10.4.1 Digoxin DDIs

Drug interactions resulting in PK changes with digoxin have largely been attributed to P-gp efflux activity in the gut, liver and kidney.[76,77] Digoxin exhibits a bioavailability of approximately 70%, indicating that gut P-gp inhibition would result in about a 40% increase in exposure. Systemic elimination of digoxin is governed by transporters present in the liver and kidney, in which approximately 60% of the bioavailable dose is renally eliminated, the remaining being eliminated by hepatic processes.[78] Therefore, the dynamic window to observe a PK change with digoxin exposure theoretically is limited to an approximately 40% change, which is the difference to achieve 100% bioavailability.

Due to the safety concerns with digoxin, Fenner *et al.*[4] analysed 123 study reports that examined digoxin PK in the presence and absence of concomitant medications. Amongst these digoxin studies, 76% (93 out of 123 studies) reported changes in the AUC and maximum plasma concentration at steady state ($C_{\mathrm{max,ss}}$) in the presence and absence of co-medication ratios of <25%. The changes in the AUC and $C_{\mathrm{max,ss}}$ ratio were small, with only a handful of medicines (valspodar, quinidine, cyclosporine, amiodarone and dronedarone) exhibiting a two-fold change, despite many compounds being reported as P-gp inhibitors. A two-fold change is usually cited as the minimum change required in order to elicit an adverse effect. Besides digoxin, there is currently no other widely given medication that exhibits clinically significant P-gp related drug interactions. Moreover, from a safety perspective, as a medication with a narrow therapeutic window, digoxin requires close monitoring to safeguard exposures exceeding 25%, which may result in digitalis toxicities. In light of this, interactions related to the safety of digoxin are more frequently observed.

Digoxin was also believed to be a substrate for hepatic OATPs; however, four different laboratories recently refuted an earlier publication, stating that the disposition of digoxin is not mediated by human OATP1A2, OATP1B1, OATP1B3 or OATP2B1.[79,80] Digoxin has also been reported to be a substrate for a sodium-dependent uptake transporter identified in the HEK293 cell line[79] and a substrate for kidney OATP4C1.[81] Consequently, in addition to the kidney, it is possible that the sodium-dependent uptake transporter may be present in the gut and liver, facilitating the uptake of digoxin into various organs. Interestingly, several laboratories have also indicated the presence of an unidentified digoxin transporter on the basolateral membrane of polarised cell lines.[82,83] Therefore, *in vitro* studies conducted in polarised cell lines (Caco-2, MDCKII-MDR1 and LLCPK1-MDR1) to evaluate P-gp inhibition may also contain this sodium dependent uptake transporter. As such, inhibition studies using digoxin as the probe substrate likely reflect the overall inhibition, involving both uptake and efflux transporters, resulting in elevated drug exposures. Therefore, inhibition of digoxin should not necessarily be viewed as only a measure of P-gp, but may also be a measure of inhibition of uptake transporters. Thus, clinical digoxin interaction studies may also involve the inhibition of either or both uptake and efflux transporters.

The FDA,[39] EMA[41] and PMDA[40] provide a decision tree or guidelines as a means to facilitate studies to assess whether a new chemical entity can inhibit P-gp. As noted above, the disposition of digoxin is governed by several transporters, and therefore the strong safety risk viewed by the agencies should account for both uptake and efflux transporters rather than only P-gp. Hence, several groups (individual companies and a consortium) have published *in vitro* threshold criteria to identify potential digoxin DDIs that have both challenged and corroborated the thresholds published in the guidances (as discussed below). Internally generated IC_{50} values and a mix of internal and literature IC_{50} values determined with digoxin as the probe substrate in various *in vitro* cell systems were combined with *in vivo* maximum inhibitor concentration at steady state or intestinal concentration (dose per 250 mL) to construct inhibitory thresholds for I_1/IC_{50} and/or I_2/IC_{50},[3,84,85] to flag potential DDIs.[86]

As an alternative to generating individual thresholds, which is resource intensive, the variability in digoxin IC_{50} values was determined by 22 laboratories using the same set of P-gp inhibitors across three polarised cell systems and vesicles (N-methyl quinidine served as a surrogate for digoxin in the latter test system as there is high background with digoxin).[42] Substantial variability was observed for all co-medications; the lower end of the range was 20- and 24-fold for isradipine and sertraline, respectively, and the upper end was 406- and 796-fold for verapamil and telmisartan, respectively. This large variability was associated with inter-laboratory differences (35%), while variability due to the use of different cell lines was rather small ($\sim$14%), compared with the inter-laboratory variability.[43] One of the challenges with this large variability in IC_{50} determinations is the ability to adopt a universal criterion to determine the clinical risk of a digoxin–drug interaction, as provided in both the regulatory guidances.[39,41] The consortium group did, however, develop a universal threshold by incorporating the digoxin IC_{50} variability generated from the 22 laboratories and using "Receiver Operating Characteristics" (ROC), a statistical method to establish thresholds for I_1/IC_{50} and I_2/IC_{50} for digoxin DDIs that minimised false positive and false negative predictions. The ROC analysis incorporated the inter-laboratory variability observed in the 22 laboratories to determine the combined criteria for I_1/IC_{50} and I_2/IC_{50} of 0.03 and 45, respectively.[43]

As the disposition of digoxin is mediated by uptake and efflux transporters, inhibition of either or both transporters may result in increased clinical exposure. Similarly, the inhibition of digoxin flux using *in vitro* systems should not be interpreted as a direct assessment of P-gp inhibition, as the IC_{50} is likely to contain uptake components and therefore should be viewed as a "digoxin IC_{50}" value. Thus, the digoxin IC_{50} value should not be extrapolated to define "P-gp only" drug interactions. Moreover, the possible involvement of multiple transporters in digoxin DDIs highlights the importance of identifying the transporters involved in the disposition of a drug. It also underlines the importance of utilising an *in vitro* system that expresses the transporters of interest in order to use *in vitro* data to assess potential drug interactions. Therefore, *in vitro* evaluation of DDIs should involve the clinical victim drug as the probe substrate to assess the potential drug interactions of a new drug candidate, as multiple transporters are likely involved.

10.4.2 Role of OATP1B1 in Statin DDIs

Statins are a widely prescribed class of drugs indicated for the treatment of hyperlipidaemia. They are generally well tolerated and have favourable safety profiles, with the exception of the risk for skeletal muscle myopathy, which in rare cases has caused lethal rhabdomyolysis.[87] The causality of statin induced myopathy is directly associated with elevated plasma concentrations, which are influenced by the activity of transporters and CYP metabolising enzymes. The clinical recommendations vary on how such DDI liabilities should be managed. Options often include dose adjustments or switching to a non-interacting statin (*i.e.*, for chronic therapy), but a contraindication is the safest option. The cardiovascular risk of stopping statin therapy must also be considered, particularly for sufferers of acute myocardial infarction or ischemic stroke, where the risk of statin withdrawal is particularly high.[88]

The disposition of a statin is typically driven by active transport, CYP metabolism or by a complex interplay of both processes. The uptake of statins into the liver is of special importance, not only with regards to their PK but also their PD: inhibition of hepatocellular uptake may increase the risk of myopathy (due to increased plasma exposure) and impair therapeutic efficacy (since the liver is the target site of pharmacological activity). Mechanistically, the role of OATP1B1 is most significant for the hydrophilic statins with low passive permeability, such as pravastatin, rosuvastatin and pitavastatin.[89] For these statins, selective inhibitors of CYP3A4 do not result in significant interactions because CYP metabolism does not have an appreciable role in their disposition. They are eliminated mainly as unchanged parent drug. Pitavastatin, although quite lipophilic, is not metabolised to any great extent and is also excreted in the bile, mainly as unchanged parent. The lipophilic statins simvastatin, atorvastatin, cerivastatin, fluvastatin and lovastatin are subject to substantial CYP-mediated elimination.[90] Indeed, in the case of simvastatin, lovastatin and atorvastatin (CYP3A4 substrates), selective inhibitors of CYP3A4 can significantly increase their plasma concentrations.[91–93]

For pitavastatin, rosuvastatin and pravastatin, OATP1B1 plays the most substantial role in their uptake into the liver, proportionally accounting for >70% of active uptake.[89] Other carriers for their uptake include OATP1B3, OATP2B1 and NTCP, but these play a lesser role *in vivo* (Table 10.1). Interactions involving inhibition of OATP1B1 decrease the benefit:risk ratio of statins by interfering with their entry into hepatocytes. Cyclosporine, gemfibrozil and rifampicin are potent inhibitors of OATP1B1 *in vitro* (IC_{50} 0.1–10 μM) and can result in substantial interactions with OATP1B1 substrates. For example, cyclosporine causes between 4- and 20-fold mean AUC increases in organ transplant patients receiving fluvastatin, pitavastatin, pravastatin or rosuvastatin.[94–97] These interactions are clearly attributed to OATP1B1 inhibition, since they are not metabolised by CYP3A4 (cyclosporine is also a CYP3A4 inhibitor). Rifampicin, when dosed i.v. shortly before statin administration, increases the AUC of atorvastatin by around seven-fold.[98] Gemfibrozil caused a six-fold increase in the AUC of cerivastatin, which led to a number of fatalities from severe myotoxicity[99] and cerivastatin was subsequently withdrawn from the market in 2001. The latter example is mainly attributed to inhibition of CYP2C8, even though there is thought to be some involvement of OATP1B1. Gemfibrozil also increases the AUC by two- to three-fold of pravastatin, rosuvastatin and simvastatin, which are all substrates of OATP1B1, but not metabolised by CYP2C8.[89]

It is important to note the greater magnitude of interactions for OATP1B1-mediated statin DDIs compared with other transporters such as P-gp, where DDIs with for example digoxin, at most, result in a ∼two-fold increase in digoxin exposure. Furthermore, polymorphic variants of OATP1B1 that result in impaired transporter activity can affect the plasma concentrations and clinical outcomes of many statins, including those that are appreciably metabolised, such as simvastatin (see *Drug Transporters: Volume 2: Recent Advances and Emerging Technologies*, Chapter 5).[100] Furthermore, in cases where the statin is metabolised by CYP3A4 as well as transported by OATP1B1, the magnitude of interaction is most severe when the perpetrator drug inhibits both components. For instance, the combination ritonavir/saquinavir increases the AUC of simvastatin acid by 30-fold.[101]

These examples highlight the important role of OATP1B1 in statin disposition and DDIs. For drug development it is critical to understand whether a NME carries with it a clinical risk such as an OATP-related DDI. In order to avoid costly clinical DDI studies during development (and unwanted labeling should it make it to the market), an early assessment for statin DDI risks may, therefore, be necessary (see Section 10.2).

In summary, statin-related DDIs and associated adverse effects such as myopathy are of serious clinical concern, and OATP1B1 has been clearly identified as a key component in many of these cases. It is therefore necessary to routinely assess OATP inhibition during drug development in order to have safer drugs that are less prone to statin-related DDIs.

10.5 Conclusion and Outlook

Over the last few decades, membrane drug transporters have been the focus of basic and applied research, considerably enhancing our understanding of drug transporter (iso-)forms, their function and tissue expression. Growing numbers of examples have shown that drug transporters can substantially modulate drug ADME routes and that they are often involved in interactions with co-medications and endogenous products, thereby influencing drug efficacy and safety profiles. The effects are frequently interrelated with the well-established Phase I and II drug metabolic enzymes, due to overlapping substrate/inhibitor specificities and common regulation pathways. This results in complex, parallel elimination routes, which can be individually affected by inhibition, induction and polymorphic expression of drug transporters and metabolic enzymes. The respective processes are typically assessed in a tiered approach during drug research and development, and are part of the safety package of all NMEs. The wider availability of *in vitro* drug transporter tools enables laboratories to generate data on the propensity of a compound as a substrate or inhibitor of major drug transporters. However, one of the major challenges remains the assessment of the relevance of those diverse *in vitro* data for the *in vivo* situation.

10.5.1 Tailored, Step-wise Drug Transporter Testing Strategies in Drug Development

As outlined in sections of this chapter, substrate and inhibition data are requested by regulatory authorities for a number of major transporters. Based on current knowledge, it is evident that not all transporters are of equal importance with respect to their DDI potential with typical co-medications. Therefore, a custom-tailored, tiered testing strategy should, where possible, be developed for a given test compound.

For transporter inhibition, such a tiered testing strategy should primarily be driven by potential co-medications and the dosing regimen in the envisaged patient population. The list of transporters tested for inhibition by a test compound in the first tier likely includes MDR1 and OATP1B1/OATP1B3, based on the evidence for their clinically relevant DDIs with digoxin and statins, respectively. In the second tier, renal transporters, OAT1/OAT3 and OCT2/MATEs, might be tested based on anticipated co-medications (*e.g.*, metformin). Other transporters, such as BCRP, MRP2 and BSEP, may be tested on a case-by-case basis. The important question here is whether these transporters are likely to contribute to a DDI risk to a relevant extent. Moreover, this transporter panel can never be considered final, since novel clinically relevant drug transporters might be identified in the future that need to be taken into account as well.

Tests for the involvement of transporters in drug distribution and clearance should be based, above all, on the target organ, the clearance route and the therapeutic window of the test drug. The first tier should focus on transporters that may pose a safety liability (by altering levels of the test drug) for healthy volunteers and patients enrolled in the well-monitored early clinical trials (*e.g.*, OATPs when the drug undergoes active hepatic uptake). This information then guides the design of follow-up clinical DDI studies, which aid in the selection of the most promising compounds during early drug development (up to the end of Phase II). For example, identification of pathways contributing >50% of clearance rather than the >25% stipulated in health authority guidance may be, in most cases, a more appropriate criterion in the early phases of drug development. More quantitative information on compound clearance routes in humans is typically available only at later stages of compound development. A more focused approach can then be applied during late stage development (Phase III) when the drug label is being prepared.

10.5.2 Gaps Within *In vitro* Drug Transporter Tools

In spite of the considerable progress made in the field of drug transporter research, a number of major gaps remain in our currently available toolbox, summarised as follows:

- Large cross-laboratory variation in *in vitro* transporter assay results
- Emergence of new transporters that may change the planning of earlier assessments
- Poor extrapolation of *in vitro* to clinical data for many inhibitors, but even more so for substrates
- Multiple drug binding sites, which can limit interpretation of transporter data to the substrate/inhibitor combination used
- Lack of understanding of unbound inhibitor concentrations at the target site
- Lack of selective clinical substrates and inhibitors for transporters

Different *in vitro* systems and conditions are used across different laboratories. One possible way of dealing with individual assay performance is to "calibrate" an assay and evaluate the data relative to a set of model compounds/inhibitors to enable the translation to the *in vivo* situation,[3] in a similar manner to that done for digoxin/P-gp (see Section 10.4.1). However, a wider standardisation of assays and systems used would need to be a key requirement to allow cross-laboratory data comparison. Without this standardisation, public data repositories (*e.g.*, the Metabolism and Transport Drug Interaction Database; http://www.druginteractioninfo.org) are only of limited use. It is exactly this exchange and comparability of data on a global level that could broaden transporter knowledge and allow the implementation of more generic extrapolation approaches in translational applications.

Emerging transporters require the constant adaptation of tools and the reinterpretation of existing data.[38,102] Transporters such as PEPT1/2 and ENT are implicated in the disposition of some drugs and their importance may increase with the emergence of new chemical modalities, such as peptides and oligonucleotides. However, as long as clinical evidence of DDIs is sparse, a reasonable interpretation of experimental data obtained for these emerging transporters remains challenging. Also, additional functions of transport proteins, such as the binding of hepatitis B virus to NTCP as a hepatic receptor,[103] will put drug interactions with established transporters into a new context and open up new options.

10.5.3 The Challenge of Translating *In vitro* Drug Transporter Data to the Clinical Situation

Extrapolation approaches established for drug metabolising enzymes are only partly applicable to drug transporter data. It still remains very challenging to scale the frequently interrelated transport rates for multiple parallel transport routes of a compound, determined in different *in vitro* systems, to the *in vivo* situation. Firstly, the nature and concentration of substrate and inhibitor are important parameters for the translation of transporter inhibition data: model substrates, although widely utilised in *in vitro* studies, may be only of limited use if the transporter has multiple binding sites, and drug-like substrates (ideally the co-medications used in clinics *e.g.*, statins for OATP DDIs) should be considered for *in vitro* studies. Consequently, class statements that extrapolate to all substrates (within a certain affinity range) of a given transporter are currently of limited applicability for drug transporters. With a better understanding of the substrates common to one binding site of a given transporter, such class statements, as used for CYPs, might become possible in the future.

With regards to the effective inhibitor concentration, different approaches use an estimated gut concentration or the total or free plasma concentration as the input parameter for extrapolations. Such approaches might work best with data from one qualified assay with an established calibration. However, they are unlikely to work as generic approaches, with data collected from different assays, due to the inherent aforementioned variability. Standardisation and more systematic studies will help to improve generic extrapolation algorithms for *in vitro* transporter inhibition data.

The translation of transporter substrate data (*in vitro* transport rates) generated in different *in vitro* assays poses even more challenges. Frequently, the involvement of multiple transport processes in parallel limits the possibility of reliably estimating the individual fraction transported. Selective inhibitors would be essential, if primary cell models expressing multiple transporters are to be used. Here, new high-throughput transporter inhibition assays may offer the possibility to screen and counter-screen large compound libraries with the aim of identifying new specific and potent transporter inhibitors. Also, structure–activity relationships could be built from such large data sets. Finally, the information on the involvement of individual transporters in compound distribution and clearance is essential to qualify the impact of induction, inhibition, polymorphisms and ontogeny on the PK of a test compound.

PBPK models are often used for the scaling of *in vitro* transport and drug metabolism data to the *in vivo* situation. Depending on the experimental system used, different scaling factors are required for relating the functional activity of individual transporters in the *in vitro* assay to the *in vivo* functional activity. In addition to physiological scaling factors, which are established for primary cell systems (*e.g.*, hepatocytes), the application of experimentally-determined relative activity factors or differential protein expression levels determined by liquid chromatography-mass spectrometry (LC-MS) based quantitative proteomics has also been suggested,[104,105] although the multitude of input parameters, each with a certain degree of uncertainty, limits the accuracy of the final prediction. Nonetheless, this approach, once experimental limitations are solved, may finally enable better quantitative DDI predictions based on *in vitro* data to be obtained.

Acknowledgements

The authors want to acknowledge Edwige Ferrari for her extensive literature review on renal DDIs, Dr Agnès Poirier for scientific review of the chapter and her contribution to Figure 10.2, Dr Jacqueline Gillis for formal proof-reading of the manuscript, and Dr Franz Schuler, Dr Richard Peck and Dr Thierry Lavé for supporting the work on this project.

References

1. M. J. Dolton, B. D. Roufogalis and A. J. McLachlan, *Clin. Pharmacol. Ther.*, 2012, **92**, 622–630.
2. N. R. Srinivas, *Eur. J. Clin. Pharmacol.*, 2008, **64**, 1135–1136.
3. A. Poirier, A. C. Cascais, U. Bader, R. Portmann, M. E. Brun, I. Walter, A. Hillebrecht, M. Ullah and C. Funk, *Drug Metab. Dispos.*, 2014, **42**, 1411–1422.
4. K. S. Fenner, M. D. Troutman, S. Kempshall, J. A. Cook, J. A. Ware, D. A. Smith and C. A. Lee, *Clin. Pharmacol. Ther.*, 2009, **85**, 173–181.
5. B. Greiner, M. Eichelbaum, P. Fritz, H. P. Kreichgauer, O. von Richter, J. Zundler and H. K. Kroemer, *J. Clin. Invest.*, 1999, **104**, 147–153.
6. K. Westphal, A. Weinbrenner, M. Zschiesche, G. Franke, M. Knoke, R. Oertel, P. Fritz, O. von Richter, R. Warzok, T. Hachenberg, H. M. Kauffmann, D. Schrenk, B. Terhaag, H. K. Kroemer and W. Siegmund, *Clin. Pharmacol. Ther.*, 2000, **68**, 345–355.
7. F. Muller and M. F. Fromm, *Pharmacogenomics*, 2011, **12**, 1017–1037.
8. X. Chu, K. Korzekwa, R. Elsby, K. Fenner, A. Galetin, Y. Lai, P. Matsson, A. Moss, S. Nagar, G. R. Rosania, J. P. Bai, J. W. Polli, Y. Sugiyama and K. L. Brouwer, *Clin. Pharmacol. Ther.*, 2013, **94**, 126–141.
9. A. Hedman, B. Angelin, A. Arvidsson, O. Beck, R. Dahlqvist, B. Nilsson, M. Olsson and K. Schenck-Gustafsson, *Clin. Pharmacol. Ther.*, 1991, **49**, 256–262.
10. B. Stieger and B. Gao, *Clin. Pharmacokinet.*, 2015, **54**, 225–242.

11. J. C. Kalvass, J. W. Polli, D. L. Bourdet, B. Feng, S. M. Huang, X. Liu, Q. R. Smith, L. K. Zhang, M. J. Zamek-Gliszczynski and I. T. Consortium, *Clin. Pharmacol. Ther.*, 2013, **94**, 80–94.

12. A. H. Schinkel, J. J. M. Smit, O. Vantellingen, J. H. Beijnen, E. Wagenaar, L. Vandeemter, C. A. A. M. Mol, M. A. Vandervalk, E. C. Robanusmaandag, H. P. J. Teriele, A. J. M. Berns and P. Borst, *Cell*, 1994, **77**, 491–502.

13. P. Borst and A. H. Schinkel, *J. Clin. Invest.*, 2013, **123**, 4131–4133.

14. H. Kodaira, H. Kusuhara, T. Fujita, J. Ushiki, E. Fuse and Y. Sugiyama, *J. Pharmacol. Exp. Ther.*, 2011, **339**, 935–944.

15. L. Sasongko, J. M. Link, M. Muzi, D. A. Mankoff, X. D. Yang, A. C. Collier, S. C. Shoner and J. D. Unadkat, *Clin. Pharmacol. Ther.*, 2005, **77**, 503–514.

16. K. M. Morrissey, S. L. Stocker, M. B. Wittwer, L. Xu and K. M. Giacomini, *Annu. Rev. Pharmacol.*, 2013, **53**, 503.

17. Y. Shitara, H. Sato and Y. Sugiyama, *Annu. Rev.f Pharmacol. Toxicol.*, 2005, **45**, 689–723.

18. F. Staud, L. Cerveny, D. Ahmadimoghaddam and M. Ceckova, *Int. J. Biochem. Cell Biol.*, 2013, **45**, 2007–2011.

19. H. Motohashi and K. Inui, *Mol. Aspects Med.*, 2013, **34**, 661–668.

20. K. Damme, A. T. Nies, E. Schaeffeler and M. Schwab, *Drug Metab. Rev.*, 2011, **43**, 499–523.

21. H. Kusuhara, S. Ito, Y. Kumagai, M. Jiang, T. Shiroshita, Y. Moriyama, K. Inoue, H. Yuasa and Y. Sugiyama, *Clin. Pharmacol. Ther.*, 2011, **89**, 837–844.

22. Y. Tanihara, S. Masuda, T. Katsura and K. Inui, *Biochem. Pharmacol.*, 2009, **78**, 1263–1271.

23. S. Yokoo, A. Yonezawa, S. Masuda, A. Fukatsu, T. Katsura and K. Inui, *Biochem. Pharmacol.*, 2007, **74**, 477–487.

24. B. Feng, J. L. LaPerle, G. Chang and M. V. Varma, *Expert Opin Drug Metab. Toxicol.*, 2010, **6**, 939–952.

25. A. Reznichenko, S. J. Sinkeler, H. Snieder, J. van den Born, M. H. de Borst, J. Damman, M. C. R. F. van Dijk, H. van Goor, B. G. Hepkema, J. L. Hillebrands, H. G. D. Leuvenink, J. Niesing, S. J. L. Bakker, M. Seelen and G. Navis, *Physiol. Genomics*, 2013, **45**, 201–209.

26. Y. Urakami, N. Kimura, M. Okuda and K. Inui, *Pharm. Res.*, 2004, **21**, 976–981.

27. E. I. Lepist, X. X. Zhang, J. Hao, J. Huang, A. Kosaka, G. Birkus, B. P. Murray, R. Bannister, T. Cihlar, Y. Huang and A. S. Ray, *Kidney Int.*, 2014, **86**, 350–357.

28. H. Shen, T. T. Liu, B. L. Morse, Y. Zhao, Y. P. Zhang, X. Qiu, C. Chen, A. C. Lewin, X. T. Wang, G. W. Liu, L. J. Christopher, P. Marathe and Y. R. Lai, *Drug Metab. Dispos.*, 2015, **43**, 984–993.

29. P. D. Ward, D. La and J. E. McDuffie, in *New Insights into Toxicity and Drug Testing*, ed. S. Gowder, InTech, 2013, ch. 7.

30. J. J. Salomon and C. Ehrhardt, *Ther. Delivery*, 2012, **3**, 735–747.

31. P. Chen, H. Chen, X. J. Zang, M. Chen, H. R. Jiang, S. S. Han and X. G. Wu, *Drug Metab. Dispos.*, 2013, **41**, 1934–1948.
32. S. Dey, B. S. Anand, J. Patel and A. K. Mitra, *Expert Opin. Biol. Ther.*, 2003, **3**, 23–44.
33. A. Olagunju, A. Owen and T. R. Cressey, *Pharmacogenomics*, 2012, **13**, 1501–1522.
34. L. L. Su, D. D. Mruk, W. M. Lee and C. Y. Cheng, *J. Endocrinol.*, 2011, **209**, 337–351.
35. F. Lang, V. Vallon, M. Knipper and P. Wangemann, *Am. J. Physiol.: Cell Physiol.*, 2007, **293**, C1187–C1208.
36. K. M. Giacomini, S. M. Huang, D. J. Tweedie, L. Z. Benet, K. L. R. Brouwer, X. Y. Chu, A. Dahlin, R. Evers, V. Fischer, K. M. Hillgren, K. A. Hoffmaster, T. Ishikawa, D. Keppler, R. B. Kim, C. A. Lee, M. Niemi, J. W. Polli, Y. Sugiyama, P. W. Swaan, J. A. Ware, S. H. Wright, S. W. Yee, M. J. Zamek-Gliszczynski, L. Zhang and I. Transporter, *Nat. Rev. Drug Discovery*, 2010, **9**, 215–236.
37. M. J. Zamek-Gliszczynski, K. A. Hoffmaster, D. J. Tweedie, K. M. Giacomini and K. M. Hillgren, *Clin. Pharmacol. Ther.*, 2012, **92**, 553–556.
38. K. M. Hillgren, D. Keppler, A. A. Zur, K. M. Giacomini, B. Stieger, C. E. Cass and L. Zhang, *Clin. Pharmacol. Ther.*, 2013, **94**, 52–63.
39. *Guidance for Industry: Drug Interaction Studies – Study Design, Data Analysis, Implications for Dosing, and Labeling Recommendations*, Draft Guidance, FDA/CDER, 2006, 2012.
40. Y. Saito, K. Maekawa and Y. Ohno, *Regul. Sci. Med. Prod.*, 2014, **4**, 249–255.
41. *Guideline on the Investigation of Drug Interactions*, European Medicines Agency (EMA) / Committee for Human Medicinal Products (CHMP), 2012.
42. J. E. Bentz, M. O'Connor, D. Bednarczyk, J. Coleman, C. A. Lee, J. E. Palm, A. Pak, E. S. Perloff, E. L. Reyner, P. Balimane, M. Brannstrom, X. Chu, C. Funk, A. Guo, I. H. Hanna, K. Heredi-Szabo, K. M. Hillgren, L. Li, E. Hollnack-Pusch, M. Jamei, X. Lin, A. Mason, S. Neuhoff, A. Patel, L. Podila, E. Plise, G. Rajaraman, L. Salphati, E. Sands, M. E. Taub, J. Taur, D. Weitz, H. Wortelboer, C. Xia, G. Xiao, J. Yabut, T. Yamagata, L. K. Zhang and H. Ellens, *Drug Metab. Dispos.*, 2013, **41**, 1347–1366.
43. H. Ellens, S. B. Deng, J. Coleman, J. Bentz, M. E. Taub, I. Ragueneau-Majlessi, S. P. Chung, K. Heredi-Szabo, S. Neuhoff, J. Palm, P. Balimane, L. Zhang, M. Jamei, I. Hanna, M. O'Connor, D. Bednarczyk, M. Forsgard, X. Y. Chu, C. Funk, A. L. Guo, K. M. Hillgren, L. B. Li, A. Y. Pak, E. S. Perloff, G. Rajaraman, L. Salphati, J. S. Taur, D. Weitz, H. M. Wortelboer, C. Q. Xia, G. Q. Xiao, T. Yamagata and C. A. Lee, *Drug Metab. Dispos.*, 2013, **41**, 1367–1374.
44. J. K. Zolnerciks, C. L. Booth-Genthe, A. Gupta, J. Harris and J. D. Unadkat, *J. Pharm. Sci.*, 2011, **100**, 3055–3061.
45. D. Tweedie, J. W. Polli, E. G. Berglund, S. M. Huang, L. Zhang, A. Poirier, X. Chu, B. Feng and I. T. Consortium, *Clin. Pharmacol. Ther.*, 2013, **94**, 113–125.

46. R. Schnepf and O. Zolk, *Expert Opin Drug Metab. Toxicol.*, 2013, **9**, 287–306.

47. A. Poirier, R. Portmann, A. C. Cascais, U. Bader, I. Walter, M. Ullah and C. Funk, *Drug Metab. Dispos.*, 2014, **42**, 1466–1477.

48. M. J. Zamek-Gliszczynski, C. A. Lee, A. Poirier, J. Bentz, X. Chu, H. Ellens, T. Ishikawa, M. Jamei, J. C. Kavass, S. Nagar, K. S. Pang, K. Korzekwa, P. W. Swaan, M. E. Taub, P. Zhao, A. Galetin and I. T. Consortium, *Clin. Pharmacol. Ther.*, 2013, **94**, 64–79.

49. D. Schwab, A. Portron, Z. Backholer, B. Lausecker and K. Kawashima, *Clin. Pharmacokinet.*, 2013, **52**, 463–473.

50. I. Ieiri, S. Higuchi and Y. Sugiyama, *Expert Opin Drug Metab. Toxicol.*, 2009, **5**, 703–729.

51. T. Lehr, A. Staab, D. Trommeshauser, H. G. Schaefer and C. Kloft, *Clin. Pharmacokinet.*, 2010, **49**, 53–66.

52. N. Parrott, B. Davies, G. Hoffmann, A. Koerner, T. Lave, E. Prinssen, E. Theogaraj and T. Singer, *Clin. Pharmacokinet.*, 2011, **50**, 613–623.

53. J. H. Lin, *Adv. Drug Delivery Rev.*, 2003, **55**, 53–81.

54. J. H. Lin and M. Yamazaki, *Clin. Pharmacokinet.*, 2003, **42**, 59–98.

55. H. Sugimoto, H. Hirabayashi, N. Amano and T. Moriwaki, *Drug Metab. Dispos.*, 2013, **41**, 683–688.

56. Y. Y. Lau, Y. Huang, L. Frassetto and L. Z. Benet, *Clin. Pharmacol. Ther.*, 2007, **81**, 194–204.

57. A. J. Sadeque, C. Wandel, H. He, S. Shah and A. J. Wood, *Clin. Pharmacol. Ther.*, 2000, **68**, 231–237.

58. M. Bauer, R. Karch, M. Zeitlinger, J. Stanek, C. Philippe, W. Wadsak, M. Mitterhauser, W. Jager, H. Haslacher, M. Muller and O. Langer, *J. Nucl. Med.*, 2013, **54**, 1181–1187.

59. W. E. Hume, T. Shingaki, T. Takashima, Y. Hashizume, T. Okauchi, Y. Katayama, E. Hayashinaka, Y. Wada, H. Kusuhara, Y. Sugiyama and Y. Watanabe, *Bioorg. Med. Chem.*, 2013, **21**, 7584–7590.

60. B. Wulkersdorfer, T. Wanek, M. Bauer, M. Zeitlinger, M. Muller and O. Langer, *Clin. Pharmacol. Ther.*, 2014, **96**, 206–213.

61. J. L. Lam, S. B. Shugarts, H. Okochi and L. Z. Benet, *J. Pharmacol. Exp. Ther.*, 2006, **319**, 864–870.

62. J. L. Lam, S. B. Shugarts, H. Okochi and L. Z. Benet, *J. Pharmacol. Exp. Ther.*, 2006, **319**, 864–870.

63. Y. Y. Lau, Y. Huang, L. Frassetto and L. Z. Benet, *Clin. Pharmacol. Ther.*, 2007, **81**, 194–204.

64. M. V. S. Varma, Y. R. Lai, E. Kimoto, T. C. Goosen, A. F. El-Kattan and V. Kumar, *Pharm. Res.*, 2013, **30**, 1188–1199.

65. Y. Shitara, M. Hirano, H. Sato and Y. Sugiyama, *J. Pharmacol. Exp. Ther.*, 2004, **311**, 228–236.

66. T. Prueksaritanont, C. Y. Tang, Y. Qiu, L. Mu, R. Subramanian and J. H. Lin, *Drug Metab. Dispos.*, 2002, **30**, 1280–1287.

67. K. Fattinger, C. Funk, M. Pantze, C. Weber, J. Reichen, B. Stieger and P. J. Meier, *Clin. Pharmacol. Ther.*, 2001, **69**, 223–231.

68. N. M. Davies, J. K. Takemoto, D. R. Brocks and J. A. Yanez, *Clin. Pharmacokinet.*, 2010, **49**, 351–377.
69. T. Lehr, A. Staab, C. Tillmann, D. Trommeshauser, H. G. Schaefer and C. Kloft, *Clin. Pharmacokinet.*, 2009, **48**, 529–542.
70. F. Muller, J. Konig, H. Glaeser, I. Schmidt, O. Zolk, M. F. Fromm and R. Maas, *Antimicrob. Agents Chemother.*, 2011, **55**, 3091–3098.
71. V. Hsu, M. D. T. Vieira, P. Zhao, L. Zhang, J. H. Zheng, A. Nordmark, E. G. Berglund, K. M. Giacomini and S. M. Huang, *Clin. Pharmacokinet.*, 2014, **53**, 283–293.
72. A. Somogyi, A. McLean and B. Heinzow, *Eur. J. Clin. Pharmacol.*, 1983, **25**, 339–345.
73. G. T. Tucker, *Br. J. Clin. Pharmacol.*, 1981, **12**, 761–770.
74. H. W. Schnaper, *Pediatr. Nephrol.*, 2014, **29**, 193–202.
75. Y. Zhang, L. Zhang, S. Abraham, S. Apparaju, T. C. Wu, J. M. Strong, S. Xiao, A. J. Atkinson, Jr., K. E. Thummel, J. S. Leeder, C. Lee, G. J. Burckart, L. J. Lesko and S. M. Huang, *Clin. Pharmacol. Ther.*, 2009, **85**, 305–311.
76. M. E. Cavet, M. West and N. L. Simmons, *Br. J. Pharmacol.*, 1996, **118**, 1389–1396.
77. S. Ernest, S. Rajaraman, J. Megyesi and E. N. BelloReuss, *Nephron*, 1997, **77**, 284–289.
78. L. S. Goodman, A. Gilman, L. L. Brunton, J. S. Lazo and K. L. Parker, *Goodman & Gilman's the Pharmacological Basis of Therapeutics*, McGraw-Hill, New York, 11th edn, 2006.
79. M. E. Taub, K. Mease, R. S. Sane, C. A. Watson, L. F. Chen, H. Ellens, B. Hirakawa, E. L. Reyner, M. Jani and C. A. Lee, *Drug Metab. Dispos.*, 2011, **39**, 2093–2102.
80. G. A. Kullak-Ublick, M. G. Ismair, B. Stieger, L. Landmann, R. Huber, F. Pizzagalli, K. Fattinger, P. J. Meier and B. Hagenbuch, *Gastroenterology*, 2001, **120**, 525–533.
81. T. Mikkaichi, T. Suzuki, T. Onogawa, M. Tanemoto, H. Mizutamari, M. Okada, T. Chaki, S. Masuda, T. Tokui, N. Eto, M. Abe, F. Satoh, M. Unno, T. Hishinuma, K. Inui, S. Ito, J. Goto and T. Abe, *Proc. Natl. Acad. Sci. U. S. A.*, 2004, **101**, 3569–3574.
82. P. Acharya, M. P. O'Connor, J. W. Polli, A. Ayrton, H. Ellens and J. Bentz, *Drug Metab. Dispos.*, 2008, **36**, 452–460.
83. N. Ballatori, *Exp. Biol. Med.*, 2005, **230**, 689–698.
84. J. A. Cook, B. Feng, K. S. Fenner, S. Kempshall, R. Liu, C. Rotter, D. A. Smith, M. D. Troutman, M. Ullah and C. A. Lee, *Mol. Pharmaceutics*, 2010, **7**, 398–411.
85. R. Elsby, M. Gillen, C. Butters, G. Imisson, P. Sharma, V. Smith and D. D. Surry, *Drug Metab. Dispos.*, 2011, **39**, 275–282.
86. H. Sugimoto, S. Matsumoto, M. Tachibana, S. Niwa, H. Hirabayashi, N. Amano and T. Moriwaki, *J. Pharm. Sci.*, 2011, **100**, 4013–4023.
87. S. M. Jamal, M. J. Eisenberg and S. Christopoulos, *Am. Heart J.*, 2004, **147**, 956–965.

88. Y. H. G. Sandoval, M. V. Braganza and S. S. Daskalopoulou, *Curr. Pharm. Des.*, 2011, **17**, 3669–3689.

89. R. Elsby, C. Hilgendorf and K. Fenner, *Clin. Pharmacol. Ther.*, 2012, **92**, 584–598.

90. M. Schachter, *Fundam. Clin. Pharmacol.*, 2005, **19**, 117–125.

91. K. T. Kivisto, T. Kantola and P. J. Neuvonen, *Br. J. Clin. Pharmacol.*, 1998, **46**, 49–53.

92. A. L. Mazzu, K. C. Lasseter, E. C. Shamblen, V. Agarwal, J. Lettieri and P. Sundaresen, *Clin. Pharmacol. Ther.*, 2000, **68**, 391–400.

93. T. A. Jacobson, *Am. J. Cardiol.*, 2004, **94**, 1140–1146.

94. M. B. Regazzi, I. Iacona, C. Campana, V. Raddato, C. Lesi, G. Perani, A. Gavazzi and M. Vigano, *Transplant. Proc.*, 1993, **25**, 2732–2734.

95. J. W. Park, R. Siekmeier, P. Lattke, M. Merz, C. Mix, S. Schuler and W. Jaross, *J. Cardiovasc. Pharmacol. Ther.*, 2001, **6**, 351–361.

96. S. G. Simonson, A. Raza, P. D. Martin, P. D. Mitchell, J. A. Jarcho, C. D. A. Brown, A. S. Windass and D. W. Schneck, *Clin. Pharmacol. Ther.*, 2004, **76**, 167–177.

97. M. Hirano, K. Maeda, Y. Shitara and Y. Sugiyama, *Drug Metab. Dispos.*, 2006, **34**, 1229–1236.

98. J. T. Backman, H. Lulurila, M. Neuvonen and P. J. Neuvonen, *Clin. Pharmacol. Ther.*, 2005, **78**, 154–167.

99. J. A. Staffa, J. Chang and L. Green, *N. Engl. J. Med.*, 2002, **346**, 539–540.

100. J. E. Keskitalo, M. K. Pasanen, P. J. Neuvonen and M. Niemi, *Pharmacogenomics*, 2009, **10**, 1617–1624.

101. C. J. Fichtenbaum, J. G. Gerber, S. L. Rosenkranz, Y. Segal, J. A. Aberg, T. Blaschke, B. Alston, F. Fang, B. Kosel, F. Aweeka and N. A. C. T. Grp, *Aids*, 2002, **16**, 569–577.

102. K. M. Giacomini and S. M. Huang, *Clin. Pharmacol. Ther.*, 2013, **94**, 3–9.

103. H. Yan, G. Zhong, G. Xu, W. He, Z. Jing, Z. Gao, Y. Huang, Y. Qi, B. Peng, H. Wang, L. Fu, M. Song, P. Chen, W. Gao, B. Ren, Y. Sun, T. Cai, X. Feng, J. Sui and W. Li, *eLife*, 2012, **1**, e00049.

104. X. Qiu, H. Zhang and Y. R. Lai, *AAPS J.*, 2014, **16**, 714–726.

105. B. Prasad and J. D. Unadkat, *AAPS J.*, 2014, **16**, 634–648.

106. *FDA Approved Labeling Text for NDA 021264: APOKYNE (apomorphine hydrochloride injection)*, Tercica Inc., 2010.

107. E. O. Mokwe and O. Ositadinma, *Ann. Intern. Med.*, 2003, **139**, W64.

108. *Prescribing information: DISKETS Dispersible Tablets (methadone hydrochloride tablets for oral suspension)*, Boehringer Ingelheim, Roxane Laboratories, 2013.

109. *Prescribing information: ABSTRAL® (fentanyl) sublingual tablets*, Galena Biopharma, Inc., 2013.

110. M. Y. Lai, F. M. Jiang, C. H. Chung, H. C. Chen and P. D. Chao, *Int. J. Clin. Pharmacol. Ther. Toxicol.*, 1988, **26**, 118–121.

111. E. Mutschler, H. Spahn and W. Kirch, *Br. J. Clin. Pharmacol.*, 1984, **17**(Suppl 1), 51S–57S.

112. C. D. Christian, Jr., C. G. Meredith and K. V. Speeg, Jr., *Clin. Pharmacol. Ther.*, 1984, **36**, 221–227.

113. K. A. Rodvold, F. P. Paloucek, D. Jung and J. Gallastegui, *Ther. Drug Monit.*, 1987, **9**, 378–383.

114. T. Kosoglou, M. Salfi, J. M. Lim, V. K. Batra, M. N. Cayen and M. B. Affrime, *Br. J. Clin. Pharmacol.*, 2000, **50**, 581–589.

115. A. Somogyi, C. Stockley, J. Keal, P. Rolan and F. Bochner, *Br. J. Clin. Pharmacol.*, 1987, **23**, 545–551.

116. A. Somogyi and M. Muirhead, *Eur. J. Clin. Pharmacol.*, 1987, **33**, 109–110.

117. A. Selen, G. L. Amidon and P. G. Welling, *J. Pharm. Sci.*, 1982, **71**, 1238–1242.

118. U. Jaehde, F. Sorgel, A. Reiter, G. Sigl, K. G. Naber and W. Schunack, *Clin. Pharmacol. Ther.*, 1995, **58**, 532–541.

119. L. S. Goodman, J. G. Hardman, L. E. Limbird and A. G. Gilman, *Goodman & Gilman's the Pharmacological Basis of Therapeutics*, McGraw-Hill, New York, 10th edn, 2001.

120. C. B. Landersdorfer, C. M. Kirkpatrick, M. Kinzig, J. B. Bulitta, U. Holzgrabe, U. Jaehde, A. Reiter, K. G. Naber, M. Rodamer and F. Sorgel, *Br. J. Clin. Pharmacol.*, 2010, **69**, 167–178.

121. *Prescribing information: CIPRO*® *(ciprofloxacin hydrochloride) TABLETS, CIPRO*® *(ciprofloxacin*)* ORAL SUSPENSION, Bayer HealthCare Pharmaceuticals Inc., 2013.

122. B. E. Davies, M. J. Humphrey, P. F. Langley, L. Lees, B. Legg and G. A. Wadds, *Eur. J. Clin. Pharmacol.*, 1982, **23**, 167–172.

123. M. B. Allen, R. W. Fitzpatrick, A. Barratt and R. B. Cole, *Respir. Med.*, 1990, **84**, 143–146.

124. E. S. Waller, M. A. Sharanevych and G. J. Yakatan, *J. Clin. Pharmacol.*, 1982, **22**, 482–489.

125. R. Sutherland, E. A. Croydon and G. N. Rolinson, *Br. Med. J.*, 1972, **3**, 13–16.

126. *Prescribing information: TIMENTIN (ticarcillin disodium and clavulanate potassium) for Injection*, GSK group of companies, 2014.

127. A. Leroy, G. Humbert and J. P. Fillastre, *Scand. J. Infect. Dis.*, 1981, 49–54.

128. U. Gundert-Remy and E. Weber, *Eur. J. Clin. Pharmacol.*, 1982, **22**, 435–439.

129. D. Overbosch, C. van Gulpen and H. Mattie, *Drugs*, 1985, **29**(Suppl 5), 128–134.

130. X. Delavenne, E. Ollier, T. Basset, L. Bertoletti, S. Accassat, A. Garcin, S. Laporte, P. Zufferey and P. Mismetti, *Br. J. Clin. Pharmacol.*, 2013, **76**, 107–113.

131. S. Hartter, R. Sennewald, G. Nehmiz and P. Reilly, *Br. J. Clin. Pharmacol.*, 2013, **75**, 1053–1062.

132. *FDA NDA 20-357/S-031: GLUCOPHAGE*® *(metformin hydrochloride tablets)*, Bristol-Myers Squibb Company, 2008.

133. L. I. Kajosaari, M. Niemi, M. Neuvonen, J. Laitila, P. J. Neuvonen and J. T. Backman, *Clin. Pharmacol. Ther.*, 2005, **78**, 388–399.

134. A. Kalliokoski, J. T. Backman, K. J. Kurkinen, P. J. Neuvonen and M. Niemi, *Clin. Pharmacol. Ther.*, 2008, **84**, 488–496.

135. A. Tornio, M. Niemi, M. Neuvonen, J. Laitila, A. Kalliokoski, P. J. Neuvonen and J. T. Backman, *Clin. Pharmacol. Ther.*, 2008, **84**, 403–411.

136. *U.S. Patent 3, 714,159: IMODIUM® capsules (loperamide hydrochloride)*, Janssen Pharmaceutica Inc., 1998.

137. T. Nozawa, K. Imai, J. I. Nezu, A. Tsuji and I. Tamai, *J. Pharmacol. Exp. Ther.*, 2004, **308**, 438–445.

138. M. Cvetkovic, B. Leake, M. F. Fromm, G. R. Wilkinson and R. B. Kim, *Drug Metab. Dispos.*, 1999, **27**, 866–871.

139. M. Shimizu, K. Fuse, K. Okudaira, R. Nishigaki, K. Maeda, H. Kusuhara and Y. Sugiyama, *Drug Metab. Dispos.*, 2005, **33**, 1477–1481.

140. G. Burgess, H. Hoogkamer, L. Collings and J. Dingemanse, *Eur. J. Clin. Pharmacol.*, 2008, **64**, 43–50.

141. I. Binet, A. Wallnofer, C. Weber, R. Jones and G. Thiel, *Kidney Int.*, 2000, **57**, 224–231.

142. K. K. Adkison, S. S. Vaidya, D. Y. Lee, S. H. Koo, L. Li, A. A. Mehta, A. S. Gross, J. W. Polli, J. E. Humphreys, Y. Lou and E. J. Lee, *J. Pharm. Sci.*, 2010, **99**, 1046–1062.

143. H. Kusuhara, H. Furuie, A. Inano, A. Sunagawa, S. Yamada, C. Y. Wu, S. Fukizawa, N. Morimoto, I. Ieiri, M. Morishita, K. Sumita, H. Mayahara, T. Fujita, K. Maeda and Y. Sugiyama, *Br. J. Pharmacol.*, 2012, **166**, 1793–1803.

144. J. van Crugten, F. Bochner, J. Keal and A. Somogyi, *J. Pharmacol. Exp. Ther.*, 1986, **236**, 481–487.

145. *Prescribing information: ZANTAC® Injection (ranitidine hydrochlorid)*, GlaxoSmithKline, 2009.

146. P. Hsiao and J. D. Unadkat, *Mol. Pharmaceutics.*, 2014, **11**, 436–444.

147. P. J. Cimoch, J. Lavelle, R. Pollard, K. G. Griffy, R. Wong, T. L. Tarnowski, S. Casserella and D. Jung, *J. Acquired Immune Defic. Syndr.*, 1998, **17**, 227–234.

148. *Prescribing information: CYTOVENE-IV (ganciclovir sodium for injection)*, Roche Laboratories Inc., 2006.

149. P. de Miranda, S. S. Good, R. Yarchoan, R. V. Thomas, M. R. Blum, C. E. Myers and S. Broder, *Clin. Pharmacol. Ther.*, 1989, **46**, 494–500.

150. M. A. Hedaya, W. F. Elmquist and R. J. Sawchuk, *Pharm. Res.*, 1990, **7**, 411–417.

151. D. M. Kornhauser, B. G. Petty, C. W. Hendrix, A. S. Woods, L. J. Nerhood, J. G. Bartlett and P. S. Lietman, *Lancet*, 1989, **2**, 473–475.

152. *Prescribing information: RETROVIR (zidovudine) Tablets, Capsules, and Syrup*, ViiV Healthcare. GlaxoSmithKline, 2012.

153. C. V. Fletcher, W. K. Henry, S. E. Noormohamed, F. S. Rhame and H. H. Balfour, *Pharmacotherapy*, 1995, **15**, 701–708.

154. G. Hill, T. Cihlar, C. Oo, E. S. Ho, K. Prior, H. Wiltshire, J. Barrett, B. L. Liu and P. Ward, *Drug Metab. Dispos.*, 2002, **30**, 13–19.

155. *Prescribing information: TAMIFLU® (oseltamivir phosphate)* Genentech, Inc., 2014.

156. K. C. Cundy, *Clin. Pharmacokinet.*, 1999, **36**, 127–143.

157. *Prescribing information: VISTIDE® (cidofovir injection)*, Gilead Sciences, Inc., 2000.

158. J. W. Massarella, L. A. Nazareno, S. Passe and B. Min, *Pharm. Res.*, 1996, **13**, 449–452.

159. *Package insert: HIVID® (zalcitabine) tablets*, Roche Laboratories, Inc., 2001.

160. A. Mordel, H. Halkin, L. Zulty, S. Almog and D. Ezra, *Clin. Pharmacol. Ther.*, 1993, **53**, 457–462.

161. E. B. Leahey, Jr., J. T. Bigger, Jr., V. P. Butler, Jr., J. A. Reiffel, G. C. O'Connell, L. E. Scaffidi and J. N. Rottman, *Am. J. Cardiol.*, 1981, **48**, 1141–1146.

162. P. E. Fenster, W. D. Hager and M. M. Goodman, *Clin. Pharmacol. Ther.*, 1984, **36**, 70–73.

163. U. Jacobs, B. Klein and H. U. Klehr, *Transplant. Proc.*, 1994, **26**, 3122.

164. I. S. Sketris, M. E. Methot, D. Nicol, P. Belitsky and M. G. Knox, *Ann. Pharmacother.*, 1994, **28**, 1227–1231.

165. *Prescribing information: CALAN® (verapamil hydrochloride) tables*, G.D. Searle – Division of Pfizer Inc., 2013.

166. A. A. Somogyi, F. Bochner and B. C. Sallustio, *Clin. Pharmacol. Ther.*, 1992, **51**, 379–387.

167. J. Feely, G. R. Wilkinson and A. J. J. Wood, *N. Engl. J. Med.*, 1981, **304**, 692–695.

168. W. Kirch, I. Rose, I. Klingmann, J. Pabst and E. E. Ohnhaus, *Eur. J. Clin. Pharmacol.*, 1986, **31**, 59–62.

169. M. R. Ujhelyi, M. B. Bottorff, M. Schur, K. Roll, H. L. Zhang, J. Stewart and M. L. Markel, *Clin. Pharmacol. Ther.*, 1997, **62**, 117–128.

170. S. Abel, D. J. Nichols, C. J. Brearley and M. D. Eve, *Br. J. Clin. Pharmacol.*, 2000, **49**, 64–71.

171. *Prescribing information: TIKOSYN® (dofetilide) capsules*, Pfizer Labs – Division of Pfizer Inc., 2013.

172. *Prescribing information: TYKERB (lapatinib) tablets, for oral use*, GlaxoSmithKline group, 2013.

173. F. H. Blackhall, M. O'Brien, P. Schmid, M. Nicolson, P. Taylor, T. Milenkova, S. J. Kennedy and N. Thatcher, *J. Thorac. Oncol.*, 2010, **5**, 1285–1288.

174. *Prescribing information: PLATINOL® (cisplatin for injection, USP)*, Corden Pharma Latina S.p.A., 2011.

175. R. Lal, J. Sukbuntherng, W. Luo, V. Vicente, R. Blumenthal, J. Ho and K. C. Cundy, *Br. J. Clin. Pharmacol.*, 2010, **69**, 498–507.

176. *Prescribing information: HORIZANT (gabapentin enacarbil) Extended-Release Tablets for oral use*, GlaxoSmithKline, 2013.

177. P. Chennavasin, R. Seiwell, D. C. Brater and W. M. M. Liang, *Kidney Int.*, 1979, **16**, 187–195.
178. W. P. Leary and A. J. Reyes, *S. Afr. Med. J.*, 1984, **65**, 455–461.
179. *Prescribing information: Furosemide injection*, USP, American Regent, Inc., 2011.
180. S. S. Jacob, M. E. Franklin, R. G. Dickinson and W. D. Hooper, *Drug Metab. Drug Interact.*, 2000, **16**, 159–171.
181. B. Nuernberg, G. Koehler and K. Brune, *Clin. Pharmacokinet.*, 1991, **20**, 81–89.
182. M. R. Muirhead, A. A. Somogyi, P. E. Rolan and F. Bochner, *Clin. Pharmacol. Ther.*, 1986, **40**, 400–407.
183. A. J. Allred, C. J. Bowen, J. W. Park, B. Peng, D. D. Williams, M. B. Wire and E. Lee, *Br. J. Clin. Pharmacol.*, 2011, **72**, 321–329.
184. D. W. Schneck, B. K. Birmingham, J. A. Zalikowski, P. D. Mitchell, Y. Wang, P. D. Martin, K. C. Lasseter, C. D. A. Brown, A. S. Windass and A. Raza, *Clin. Pharmacol. Ther.*, 2004, **75**, 455–463.
185. J. J. Kiser, J. G. Gerber, J. A. Predhomme, P. Wolfe, D. M. Flynn and D. W. Hoody, *J. Acquir. Immune Defic. Syndr.*, 2008, **47**, 570–578.
186. A. J. Busti, A. M. Bain, R. G. Hall, R. G. Bedimo, R. D. Leff, C. Meek and R. Mehvar, *J. Cardiovasc. Pharmacol.*, 2008, **51**, 605–610.
187. N. Ichimaru, S. Takahara, Y. Kokado, J. D. Wang, M. Hatori, H. Kameoka, T. Inoue and A. Okuyama, *Atherosclerosis*, 2001, **158**, 417–423.
188. J. T. Backman, C. Kyrklund, K. T. Kivisto, J. S. Wang and P. J. Neuvonen, *Clin. Pharmacol. Ther.*, 2000, **68**, 122–129.
189. M. Hermann, A. Asberg, H. Christensen, H. Holdaas, A. Hartmann and J. L. E. Reubsaet, *Clin. Pharmacol. Ther.*, 2004, **76**, 388–391.
190. R. A. Carr, A. K. Andre, R. J. Bertz, W. Lam, M. Chang, P. Chen, L. Williams, B. Bernstein and E. Sun, presented in part at the 40th Interscience Conference on Antimicrobial Agents and Chemotherapy, Toronto, Canada, 2000 September 17–20, 2000.
191. W. Muck, I. Mai, L. Fritsche, K. Ochmann, G. Rohde, S. Unger, A. Johne, S. Bauer, K. Budde, I. Roots, H. H. Neumayer and J. Kuhlmann, *Clin. Pharmacol. Ther.*, 1999, **65**, 251–261.
192. J. T. Backman, C. Kyrklund, M. Neuvonen and P. J. Neuvonen, *Clin. Pharmacol. Ther.*, 2002, **72**, 685–691.
193. *Prescribing information: Lescol® (fluvastatin sodium) capsules/Lescol® XL (fluvastatin sodium) extended-release tablets for oral use*, Novartis Pharmaceuticals Corporation, 2012.
194. P. J. Neuvonen, M. Niemi and J. T. Backman, *Clin. Pharmacol. Ther.*, 2006, **80**, 565–581.
195. T. Kantola, J. T. Backman, M. Niemi, K. T. Kivisto and P. J. Neuvonen, *Eur. J. Clin. Pharmacol.*, 2000, **56**, 225–229.
196. X. F. Yin, Z. Q. Lin and J. Yang, *Yaoxue Xuebao*, 2011, **46**, 1108–1116.
197. N. M. Hasunuma, T. Yaji *et al.*, *J. Clin. Therap. Med.*, 2003, **19**, 381–389.

198. *Prescribing information: LIVALO (pitavastatin) Tablet, Film Coated for Oral Use*, Kowa Pharmaceuticals America, Inc., 2009.
199. P. Mathew, T. Cuddy, W. G. Tracewell and D. Salazar, *Clin. Pharmacol. Ther.*, 2004, **75**, P33.
200. *Prescribing information: PRAVACHOL® (pravastatin sodium) Tablets*, Bristol-Myers Squibb Company, 2011.
201. C. Kyrklund, J. T. Backman, M. Neuvonen and P. J. Neuvonen, *Clin. Pharmacol. Ther.*, 2003, **73**, 538–544.
202. C. L. Ronchera, T. Hernandez, J. E. Peris, F. Torres, L. Granero, N. V. Jimenez and J. M. Pla, *Ther. Drug Monit.*, 1993, **15**, 375–379.
203. NDA 011719/S-117: Methotrexate Injection, USP, Hospira, Inc., 2011.
204. T. S. Tracy, K. Krohn, D. R. Jones, J. D. Bradley, S. D. Hall and D. C. Brater, *Eur. J. Clin. Pharmacol.*, 1992, **42**, 121–125.
205. G. W. Aherne, E. Piall, V. Marks, G. Mould and W. F. White, *Br. Med. J.*, 1978, **1**, 1097–1099.

Transporter Drug–Drug Interactions: Regulatory Requirements and Drug Labelling

SUSAN M. COLE,*[a] GUSTAV AHLIN,[b] NAOMI NAGAI,[c] DAISUKE IWATA,[c] MASANOBU SATO[c] AND KENTA YOSHIDA[d]

[a] MHRA, 151 Buckingham Palace Road, London SW1W 9SZ, UK; [b] MPA, P.O. Box 26, SE-751 03 Uppsala, Sweden; [c] PMDA, Shin-Kasumigaseki Building, 3-3-2 Kasumigaseki, Chiyoda-ku, Tokyo 100-0013, Japan; [d] ORISE fellow, FDA/CDER/OTS/OCP, 10903 New Hampshire Avenue, Silver Spring, MD 20993, USA
*Email: susan.cole@mhra.gsi.gov.uk

11.1 Introduction

The role of transporters in the disposition and clearance of new drug entities is an important component in the regulatory dossier of any new drug. The regulatory assessment involves a careful consideration of the benefits *versus* the risks of the drug. This involves an understanding of the exposure of the population in general to the drug and consideration of factors that may alter drug exposure, for example, interactions with concomitant products or altered pharmacokinetics in special populations, including individuals with

RSC Drug Discovery Series No. 54
Drug Transporters: Volume 1: Role and Importance in ADME and Drug Development
Edited by Glynis Nicholls and Kuresh Youdim

Published by the Royal Society of Chemistry, www.rsc.org

genetic polymorphisms. An understanding of the transporters involved in the disposition and clearance of drugs is fundamental to support this assessment and gauge whether the appropriate studies have been done. Information is also required on any possible induction or inhibition of drug transporters by the drug so that the potential effects on other co-administered products can be predicted.

In recent years, regulatory agencies have issued guidance documents for the *in vitro* and *in vivo* drug interaction studies required to support drug development.[1–3] Contrary to past expectations, data are now required on a number of drug transporters. The US Food and Drug Administration (FDA) published a review of transporter information included in New Drug Applications (NDAs) in 2013.[4] Out of the 22 NDAs that contained drug–drug interaction (DDI) studies, 20 included some type of transporter study. Within those 20 NDAs, more than 120 *in vitro* transporter assays were described, against a total of 16 transporters. P-glycoprotein (P-gp) was the most represented transporter. The remaining transporters recommended in the FDA guideline were also well represented, along with: multidrug resistance associated protein 2 (MRP2), organic cation transporter 1 (OCT1), organic anion transporter polypeptide 2B1 (OATP2B1), bile salt export pump (BSEP), multidrug and toxin extrusion transporter 1 (MATE1), organic cation transporters novel, 1 and 2 (OCTN1 and 2), sodium-taurocholate co-transporting polypeptide (NTCP), and urate transporter.[5] In a more recent publication,[6] out of the 30 NDA approval packages released by the FDA in 2014, 22 (73%) contained *in vitro* transporter data, either substrate assays, inhibition assays or both. While this is a lower percentage than was seen in the previous year, the overall number of compounds (drugs and metabolites) tested against transporters increased. As a result of multiple combination drugs, there were 25 NMEs represented in the 22 NDA approval packages that contained transporter assays, and in addition, 17 individual metabolites were also evaluated; therefore, 42 new compounds were screened for *in vitro* transporter interactions in the approvals from 2014. To follow-up with the *in vitro* findings, nine drugs (a total of ten NMEs) were studied *in vivo* as substrates for P-gp, OATP1B1/1B3 and OAT3 using clinical inhibitors and/or inducers. A total of 23 clinical studies were performed, with 20 showing positive results [area under the curve (AUC) or maximum concentration (C_{max}) ratio $\geq$1.25 or $\leq$0.8, respectively). As perpetrators, ten drugs (a total of 13 NMEs) were evaluated clinically for the inhibition of P-gp, OATP1B1/1B3, OAT1/3, OCT2 and BCRP. Out of the 14 clinical studies conducted, half showed positive results.

In another publication, the trends (percentage of NDAs in Japan containing transporter-mediated *in vitro* drug interaction studies) were: 6% (7 out of 113 NDAs, 1997–2000), 12% (9 out of 77 NDAs, 2001–2004) and 19% (21 out of 110 NDAs, 2005–2008). P-gp was the primary target transporter, and information on P-gp was provided in either the Drug Interaction or Pharmacokinetics section in the package inserts (PIs) of 11 out of the 300 NMEs approved in Japan between 1997 and 2008.[7]

11.2 New Drug Applications

The purpose of the assessment of new drug applications is to ensure that an investigational drug is safe and effective for its intended uses, and that the labelling can appropriately communicate the key information for the correct use of the drugs to the healthcare providers. Results of preclinical and clinical investigations of an investigational drug are included in the application by drug sponsors to the regulatory agency.

New drug applications are usually submitted on a regional basis to the European Union (EU), the USA or Japan. The respective agencies are the European Medicines Agency (EMA) in Europe, the FDA in the USA and the Pharmaceuticals and Medical Devices Agency (PMDA) in Japan.

The content of the dossiers is similar for all of the agencies, but with some differences to the structure.

11.2.1 New Drug Applications to the EMA

Within Europe, the drug regulatory system currently covers 34 countries. The structure is complex, with the assessors of applications residing in the national agencies, while the EMA has the role of the overall co-ordinator of the application. An application for a new drug is made through the request for a Marketing Authorisation (MA). In many cases, this is through a centralised application, which would cover all member countries and for which two member countries are appointed Rapporteur and Co-Rapporteur by the EMA. The rapporteurs are responsible for assessing the application while other member countries provide comments on the assessment reports provided by the rapporteurs. A decision on the benefit/risk is made at the Committee for Human Medicinal Products (CHMP), which comprises delegates from all of the member states. At the close of the procedure, the CHMP makes a recommendation for approval to the European Commission, who make the final decision on approval. The majority of applications are received through this centralised procedure and there are some cases—for medicines for the treatment of certain conditions, biotech and advanced therapy medicines, and medicines used for rare human diseases (orphan products)—that require an application to be submitted for the centralised procedure.[8]

In other cases, an applicant can choose to take an application through the decentralised procedure. Here, an application is made to a National Competent Authority (known as the Reference Member State) and the applicant can choose which other countries (Concerned Member States) to include in the procedure.

For centralised and decentralised applications, the assessment process takes a total of up to 210 days with two to three additional 'clock stop' periods, where responses to any regulatory queries on the submission are prepared by the applicant. Post-approval, additional information may be provided based on additional clinical data or new *in vitro* data that allow an improved understanding of the drug (this application is classified as a variation). This would usually result in a change to the label and includes new dose forms, routes, doses or indications. In some cases, where there is a medical need, an accelerated assessment can be carried out, the assessment time in this case is 150 days.

11.2.2   New Drug Applications to the FDA

Within the US, NDA applications are submitted to the FDA, and the FDA website gives a good summary of NDA application processes.[9] After receiving a NDA, the FDA reviews the submitted data and decides whether the proposed indications are appropriate for sale and marketing in the US. Timelines for FDA review processes are currently determined under the Prescription Drug User Fee Act (PDUFA) V, which was signed into law on July 9, 2012, as part of the FDA Safety and Innovation Act (FDASIA). PDUFA V states that the FDA should review and act on standard applications within 10 months ($\sim$300 days) after NDA submissions are accepted. Priority Review status may be granted for new drugs whose approval would "provide significant improvements in the safety or effectiveness of the treatment, diagnosis, or prevention of serious conditions",[10] and the review periods for such drugs are 6 months ($\sim$180 days) after NDA submissions are accepted.

11.2.3   New Drug Applications to the MHLW/PMDA

The general process of new drug development and review in Japan is shown in Figure 11.1. The PMDA website also gives some useful information on the NDA processes in Japan.[11] The application dossier for the NDA is submitted to the Ministry of Health, Labour and Welfare (MHLW), then reviewed at the PMDA. The target total review time (median, 12 months) for standard review products was set in the Mid-Term Plan of the PMDA, authorised under the MHLW-Pharmaceutical and Food Safety Bureau (PFSB) No. 0331002, dated March 31, 2009. The target review time of 9 months was set for priority review products (orphan drugs and products designated by the MHLW based on their clinical usefulness and the seriousness of the diseases). Review teams evaluate the quality, pharmacology, pharmacokinetics, toxicology and clinical data of the NDA to ensure its quality, safety and efficacy in the light of current scientific and technological standards. The PMDA also conducts various inspections such as good laboratory practices (GLP) and good clinical practices (GCP) inspections *etc.*, to ensure the ethical and scientific background of the submitted data and to evaluate the manufacturing process. The PMDA then provides the MHLW with the review report and its judgement on approval. The Pharmaceutical Affairs and Food Sanitation Council of the MHLW then confirms the judgement, with final approval being determined by the MHLW based on the Pharmaceutical Affairs Law.

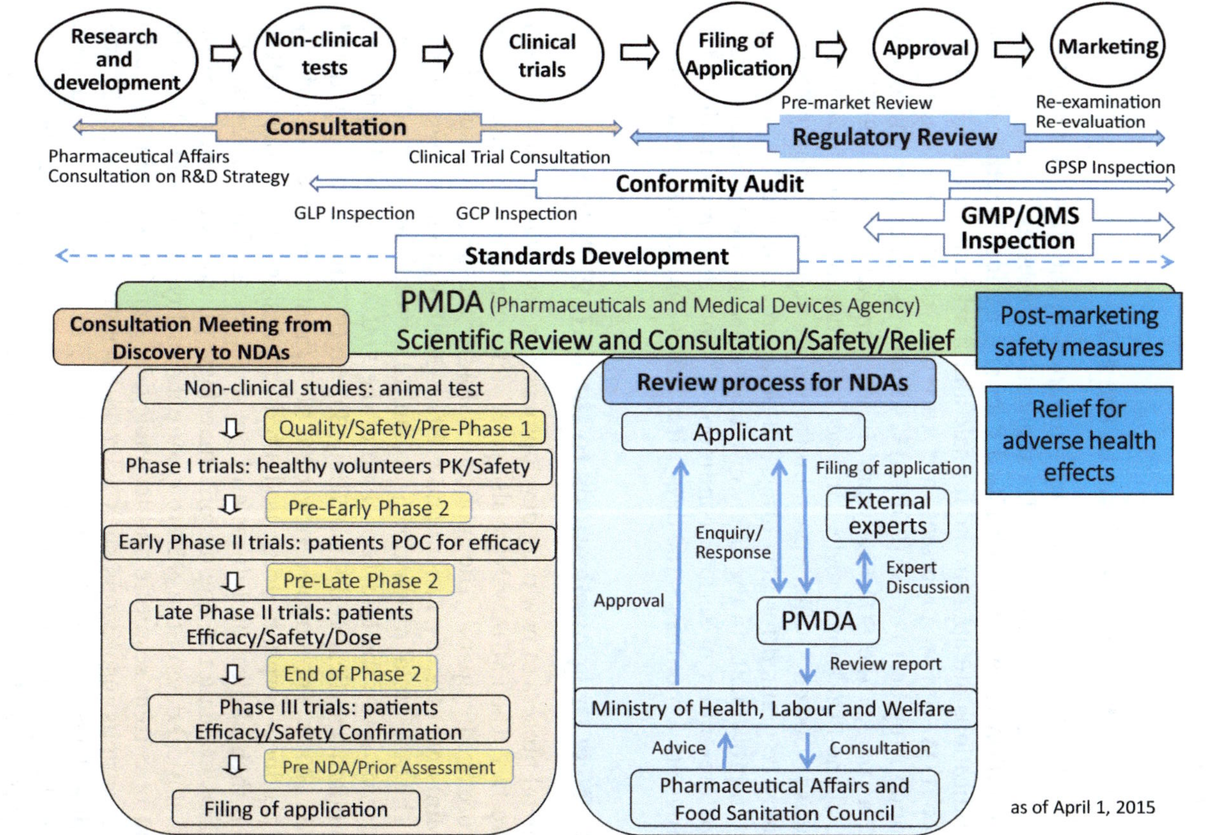

Figure 11.1 Overall process of new drug development and review in Japan. GMP: good manufacturing practices; GPSP: good post-marketing study practice; PK: pharmacokinetics; POC: proof of concept; QMS: quality management system; R&D: research and development.

After approval and marketing of a new drug submission, the drug is continuously evaluated through post-marketing surveillance (including post-marketing clinical studies) or clinical studies in all agencies.

11.2.4   Transporter Sections in New Drug Applications

Detailed information on the expected content of an application is available in guidelines on the respective agency websites.[1–3] For European submissions, information on transporters would be expected in the non-clinical pharmacokinetics section, describing the disposition of the drug in animals and *in vitro* studies to determine whether the drug is a substrate or an inhibitor of transporters. The *in vitro* inhibition data are also included in the clinical pharmacology section. Results of the *in vitro* studies should be discussed in terms of the role of transporters in the absorption, distribution and elimination of the drug, and in terms of possible interactions both with the drug as a victim and a perpetrator. Exposure information in different sub-populations, *e.g.* in those with renal impairment, will also be required. The assessor responsible for reviewing the *in vitro* inhibition data varies among the member states; although most countries now have pharmaco-kinetic assessors who perform this role, it may be the responsibility of the non-clinical or clinical assessor, depending on the national agency.

Similar transporter information is required for both US and Japanese NDA submissions (Table 11.1). The relevant sections where the information is recorded is also detailed in Table 11.1. In the USA, both preclinical and clinical outcomes are reviewed by the Office of Clinical Pharmacology. In Japan, the review is performed by the PMDA. The review considers the need for dose adjustments for subpopulations or patients taking specific types of co-medications, or the need for additional transporter studies.

11.2.5   Scientific Advice on New Drug Applications

In addition to meetings held as part of the submission (usually pre-submission and clarification meetings during the procedure), scientific advice can be received from the agencies during the development of a drug or prior to submission.

In the EU, advice can be sought from the CHMP, where the advice represents a consolidated view across the national competent authorities in Europe. Scientific advice can also be requested from the national competent authorities individually, *e.g.* the Medicines and Healthcare Products Regulatory Agency (MHRA) or Medical Products Agency (MPA). In the case of advice from a national competent authority, this is specific to the authority: for the MHRA it takes the form of written advice following a face-to-face meeting, for the MPA it usually takes the form of a face-to-face meeting only. For advice from the CHMP, the company will receive written advice; if issues are raised with the proposed development programme, a face-to-face clarification meeting then takes place in order to finalise the written advice.

Table 11.1 Transporter information in regulatory submissions.

Regulatory authority	NDAs	Sections in assessment report containing transporter information	Guidelines containing transporter information	Scientific advice
EMA	Marketing Authorisation Centralised through CHMP scientific committee Decentralised or mutual recognition through national competent authorities	Non-clinical section: 3.2 Absorption 3.3 Distribution 3.4 Metabolism 3.5 Excretion 3.6 Pharmacokinetic Drug Interactions Clinical pharmacology: 2.1.3 Absorption 2.1.4 Distribution 2.1.5 Elimination 2.1.10 Interactions	Investigation of Drug Interactions: CPMP/EWP/560/95/Rev. 1.2012 Guideline on the Use of Pharmacogenetic Methodologies in the Pharmacokinetic Evaluation of Medicinal Products: EMA/CHMP/37646/2009 Concept Paper on Qualification and Reporting of PBPK Modelling and Analyses: EMA/CHMP/211243/2014	EMA scientific advice or national scientific advice
FDA	NDA Submitted to and reviewed by FDA	7 Drug Interactions: Effect of Other Drugs Effect on Other Drugs 12 Clinical Pharmacology: Pharmacokinetics – Absorption – Distribution – Elimination – Drug Interactions	Guidance for Industry: Drug Interaction Studies—Study Design, Data Analysis, Implications for Dosing, and Labeling Recommendations 2012 Clinical Pharmacogenomics: Premarket Evaluation in Early-Phase Clinical Studies and Recommendations for Labeling 2013	Published documents, meetings upon request or public meetings

PMDA/ MHLW	NDAs	Non-clinical section	Clinical Pharmacokinetic Studies of Pharmaceuticals:	PMDA scientific advice (face to face or written consultation on non-clinical safety test/studies and clinical trials
	PMDA – Scientific review – Post marketing safety measures MHLW – Confirm approval of applications – Advisory Committee – (Pharmaceutical Affairs and Food Sanitation Council)	3.(ii) Pharmacokinetic studies 3.(ii).A Submitted data 3.(ii).A (1) Absorption (2) Distribution (3) Metabolism (4) Excretion (5) Pharmacokinetic drug interactions (DIs) (6) Other pharmacokinetic studies 3.(ii).B Outline of the PMDA review Clinical Pharmacology 4.(ii) Clinical pharmacology (CP) studies 4.(ii).A Submitted data 4.(ii).A (1) Studies using human biomaterials (5) Assessment of DI 4.(ii).B Outline of the PMDA review	MHLW/Pharmaceuticals Medical Safety Bureau/ Notification No.796/2001 Guideline on Non-clinical Pharmacokinetic Studies: MHLW/Pharmaceuticals Medical Safety Bureau/ Notification No. 496/1998 Methods of Drug Interaction Studies: MHLW/Pharmaceuticals Medical Safety Bureau/ Notification No. 813/2001 Drug Interaction Guideline for Drug Development and Labelling Recommendations (final draft): MHLW/Pharmaceuticals and Food Safety Bureau/2014	

The FDA also gives scientific advice in various ways, including face-to-face meetings, planned telephone conferences, public meetings, or domestic and international published documents.[12]

Similarly, the PMDA offers scientific advice in the form of face-to-face or written consultation on non-clinical safety tests/studies and clinical trials.

11.3   Regulatory Guidelines

11.3.1   History of Transporters in Regulatory Guidelines

Regulatory guidance on transporters has increased substantially in the last decade. The main driver has been the exponential increase in knowledge in the field of DDIs and an increasing number of interactions in NDAs and in the literature that have been attributed to effects on transporters.

The CHMP Drug Interaction guideline published in 1997 (CPMP/EWP/560/95), in terms of transporters, focused primarily on P-gp in the intestine, with some general information on transporters involved in elimination, *e.g.* renal transporters. However, much has changed since then. A revised CHMP guideline on the Investigation of Drug Interactions was approved for use in Europe in 2012 and came into effect on January 1, 2013.[1] This guidance outlines recommended information on a number of transporters now known to be involved in the absorption, distribution, and hepatic and renal elimination of drugs.

The FDA draft guidance on drug interaction studies published in 2006 included discussion on transporter-based DDIs, but the focus was mainly on P-gp. The FDA published updated draft guidance on drug interactions for comment in February 2012,[2] and this is currently being revised. The 2012 draft guidance includes an expanded list of clinically-relevant transporters, and methodologies to evaluate interaction potentials between new drugs and these transporters, as recommended by the International Transporter Consortium (ITC).[13,14]

Regulatory documents on drug interaction studies, and other pharmacokinetic studies including toxicokinetics and repeated-dose tissue distribution studies, are published regionally in Japan. The regulatory document entitled "Methods of Drug Interaction Studies" issued by the MHLW in 2001 focussed not only on metabolising enzymes but also transporters as an important mechanism of drug interactions to be considered and examined during drug development.[15] This resulted in transporter-mediated drug interactions being studied *in vitro* between 1997 and 2008, mostly for P-gp.

More recently in Japan, transporter-mediated non-clinical (*in vitro*) drug interaction studies became almost routinely examined for the development of new oral drug candidates in the 2009–2013 fiscal year.[16,17] Transporters other than P-gp have recently been recognised and investigated during drug development and *in vitro* study data regarding BCRP, OATP1B1, OATP1B3, OAT1, OAT3, OCT2 and MATE drug interactions have also been filed in the NDAs of some drugs. Clinical studies of transporter-mediated drug interactions have been conducted and filed for new drug submissions consistently: between 10 and 20% of NDAs for all NMEs, and about 30% of NDAs for NMEs as oral dosage forms in the Japanese 2009–2013 fiscal year.[16,17] P-gp is the most commonly targeted transporter during the clinical development phase.

The initial document, "Methods of Drug Interaction Studies" has been revised and the final draft of the new regulatory document, entitled "Drug Interaction Guideline for Drug Development and Labelling Recommendations" was published in July 2014.[3] This draft is currently being revised and updated, based on advances in science and technology, and the aim of international harmonisation.

In all of the guidelines outlined above, the importance of transporters in drug interactions is recognised and guidance is given on the information that is required to support their role in describing the pharmacokinetics of the NME in all individuals and to support product labelling. It is important to note that even though many aspects of the guideline recommendations from each agency overlap, there are some aspects where the guidelines differ in both recommendations and requirements.

The other field in which the importance of drug transporters has been recognised is in the consideration of exposure in special populations, and particularly populations with polymorphisms in transporter proteins. This is detailed in the CHMP guideline on the use of pharmacogenetic methodologies in the pharmacokinetic evaluation of medicinal products (EMA/CHMP/37646/2009)[18] and the FDA guidance, Clinical Pharmacogenomics: Premarket Evaluation in Early-Phase Clinical Studies and Recommendations for Labeling.[19] In evaluating the benefit–risk assessment in sub-populations at risk of altered exposure, knowledge of pharmacogenetic differences is important. Inter-individual variability in genes that may influence the outcome of drug treatment (*e.g.* genes encoding drug transporters) is studied in relation to the efficacy of drug treatment and adverse drug reactions.

11.3.2   European (EMA) Guidance

Information on transporters is contained in the CHMP Drug Interaction guideline.[1] However, it is recognised that the *in vitro–in vivo* extrapolation of drug transporter interactions is currently less mature than for metabolising enzymes and requires additional experience and continued scientific developments. Thus, the approach defined for drug–transporter interactions is likely to continue to evolve.

Interactions involving active transport are discussed in the guideline under the different physiological processes.

11.3.2.1   *The Drug as a Potential Transporter Substrate*

11.3.2.1.1   Absorption. The involvement of transport proteins is evaluated to enable predictions of interactions where the absorption of the drug is altered due to inhibition or induction of these proteins. It is recommended that the involvement of transporters in drug absorption is evaluated *in vitro* in Caco-2 cells, taking into account the physicochemistry and permeability of the investigational drug. If the *in vitro* transport and permeability data indicate that active intestinal transport may affect the bioavailability of the new drug, attempts should be made to identify the transporter involved *in vitro* (Figure 11.2). Detailed recommendations on how to study intestinal transporter involvement and to determine the apparent permeability constant *in vitro* are given in Appendices II and III of the guideline.[1] When a candidate transporter has been identified, and interactions through inhibition are likely to be clinically relevant, an *in vivo* study with a strong inhibitor is recommended if known inhibitors are registered as medicinal products in the EU. If the candidate transporter is subject to genetic polymorphism, *e.g.* OATP1B1,[20] studies in subjects

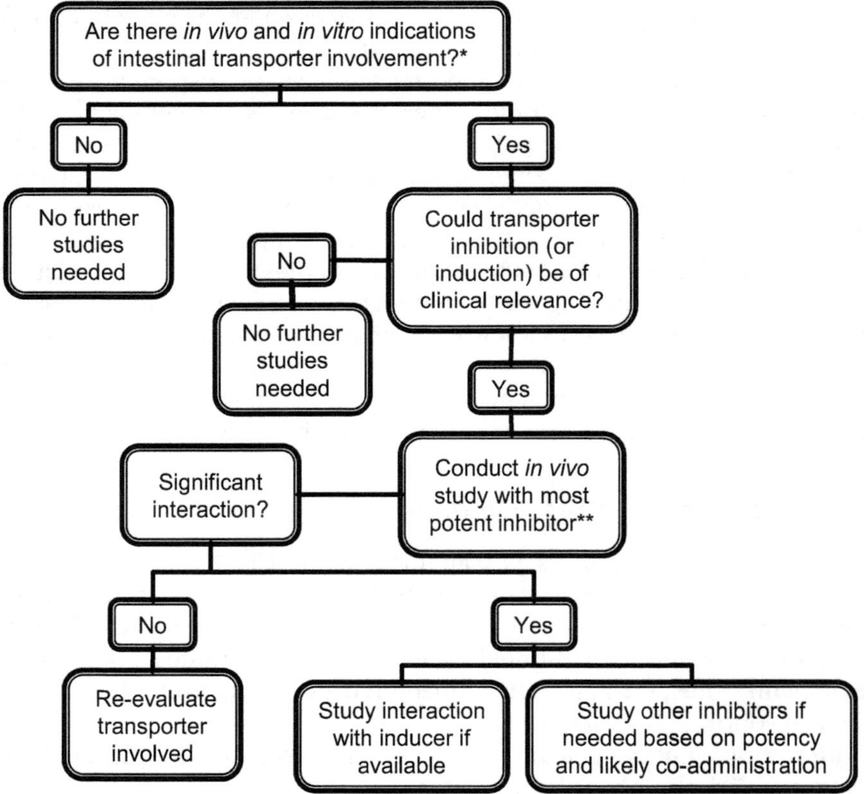

Figure 11.2   *In vivo* investigations of intestinal transporter involvement from the EMA guideline.

of certain genotypes giving rise to markedly altered expression or activity of the transporter may be useful for investigating the involvement of the transporter *in vivo* and the estimation of the potential for pharmacokinetic interactions *via* inhibition (or induction) of the transporter.

11.3.2.1.2 Distribution.   Distribution interactions include effects through modulation of active uptake or efflux transport of the drug, as well as interactions due to displacement of a substrate of a transporter. At a transport protein level, these interactions are expected to give rise to altered distribution/transport of drug to organs where these transporters are expressed, although an alteration in drug transport may not be fully reflected by observed changes in plasma concentrations. Therefore, the inclusion of pharmacodynamic markers to reflect altered distribution to the organs expressing the transporter should be considered whenever possible.

Knowledge regarding distribution interactions due to transporter inhibition is increasing.[21] If the investigational drug is a substrate for transport proteins, the potential for clinically relevant distribution interactions should be discussed in light of all of the available data. This should include data on: the tissue specific expression, *in vivo* importance of the transporter in the particular organ, drug permeability characteristics and distribution data in preclinical species (taking potential species differences into account). Available clinical safety data in patients with reduced transport caused by a genetic polymorphism or interactions, as well as the expected clinical consequences of an altered distribution, should also be considered. If indicated, and feasible, *in vivo* studies investigating the effect of transporter inhibition on the pharmacokinetics as well as pharmacodynamics (including pharmacodynamic markers for the potential effect on the transporter-expressing organ) are recommended. Distribution imaging techniques could be useful to study altered distribution.[22]

Both target organs for the clinical effect and potential target organs for safety should be considered. As an example, inhibition of transporter-mediated efflux of a hepatotoxic drug from the liver could in theory give rise to increased hepatocyte drug exposure and, therefore, increase the frequency of concentration-dependent hepatotoxicity. If the transporter potentially controlling target tissue exposure is subject to marked genetic polymorphisms, investigations of the effect of a genotype giving rise to reduced transporter activity on the target organ safety (or efficacy, if relevant) in Phase III trials could indicate the consequences of transporter inhibition by a concomitant drug.

11.3.2.1.3 Elimination.   Information on transporters involved in major elimination processes, whether uptake or efflux, should be gained as early as possible during drug development. The need for data at different phases is driven by the predicted magnitude of the systemic exposure increase or decrease, if the transporter is inhibited or induced, and the clinical consequences.

In vitro data may be sufficient before Phase III, provided that the use of potentially interacting drugs can be restricted in the study protocol. Inhibition of OATPs has been reported to result in marked increases in the systemic exposure of drugs subject to hepatic uptake transport, and involvement of these transporters may occur without any indications from the plasma pharmacokinetic information.[23] Therefore, the possible involvement of OATP1B1 and 1B3 uptake transport should be investigated *in vitro* for drugs estimated to have ≥25% hepatic elimination (clearance by hepatic metabolism and biliary secretion together contributing to ≥25%). As scientific knowledge evolves, other hepatic uptake transporters may need to be screened if their inhibition generally has been observed to lead to large effects on drug elimination.

Investigations of transporters involved in drug elimination are indicated if available *in vivo* data show that active renal, biliary or gut wall secretion of unchanged drug is involved in a main part of the drug elimination and thus modulation of the transporter involved may be of clinical relevance. Decision trees from the guidelines are shown in Figure 11.3. In line with the

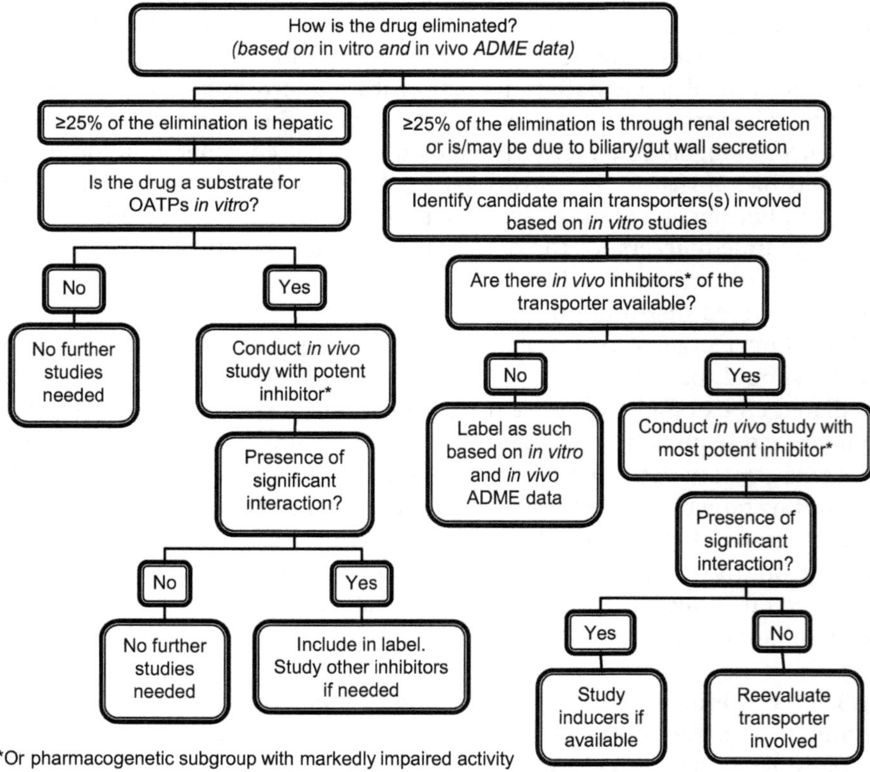

Figure 11.3 Investigation of transporter involvement in drug elimination from the EMA Guideline. ADME: absorption, distribution, metabolism and excretion.

requirements for enzyme identification, if either renal secretion or biliary/gut wall secretion, separately, is estimated to account for more than 25% of drug elimination, attempts should be made to identify the transporter(s) involved in the active secretion. The importance of renal secretion is estimated by comparing total renal clearance to the renal filtration clearance. Expectations are that if a drug is determined to be actively cleared then it will be investigated as a substrate of possible transporters identified by the applicant, *e.g.* OAT1, OAT3 and OCT2. If biliary excretion of the drug is indicated then, again, appropriate transporters should be identified, *e.g.* Pgp, BCRP and BSEP. Other transporters may need to be considered depending on the drug structure, *e.g.* for peptide or nucleoside analogues.

Depending on the information at hand, it may be difficult to estimate the quantitative importance of biliary and gut wall secretion for total elimination. This should be based on the mass balance data, and supported by available interaction data, potential pharmacogenetic information, data in patients with hepatic impairment and permeability data. An intravenous mass balance study can provide important information for quantifying the importance of biliary/gut wall secretion of orally administered drugs. Data on absolute bioavailability may also inform the estimation of the extent of elimination through these elimination routes. If a large fraction of an oral dose is recovered as unchanged drug in faeces, biliary excretion cannot be excluded unless an intravenous mass balance study or an absolute bioavailability study is used to estimate the amount of drug that is absorbed.

11.3.2.2   *The Drug as a Potential Transporter Inhibitor*

The effect of the drug on transporters also needs to be considered in terms of the drug as a perpetrator. The CHMP regulatory guidance for studying inhibition of transport proteins is described in Section 5.3.4.1 of the guideline.[1] In addition, a decision tree on the investigation of the transporter inhibitory potential of a drug is provided in Figure 11.4. *In vitro* data on transporter inhibition should preferably be available before initiating Phase III, to allow for more adequate exclusion criteria in the Phase III study protocols. As a result, the risk for drug interactions leading to adverse events or lack of efficacy of concomitantly administered products may be mitigated. Currently, *in vitro* inhibition data of transporters are generally only required for the parent drug and not metabolites.

As the field of drug transporters is still relatively young and rapidly evolving, the clinical relevance of the already identified transporters may be partly unknown and there are probably still clinically-relevant drug transporters to discover. Therefore, the list of transporters to study for drug inhibition given in the guideline may be updated as science evolves. The CHMP guideline includes transporters that have been shown to be relevant *in vivo* and deliberately presents very limited information on suitable specific *in vitro* and *in vivo* model substrates and inhibitors for transporters to

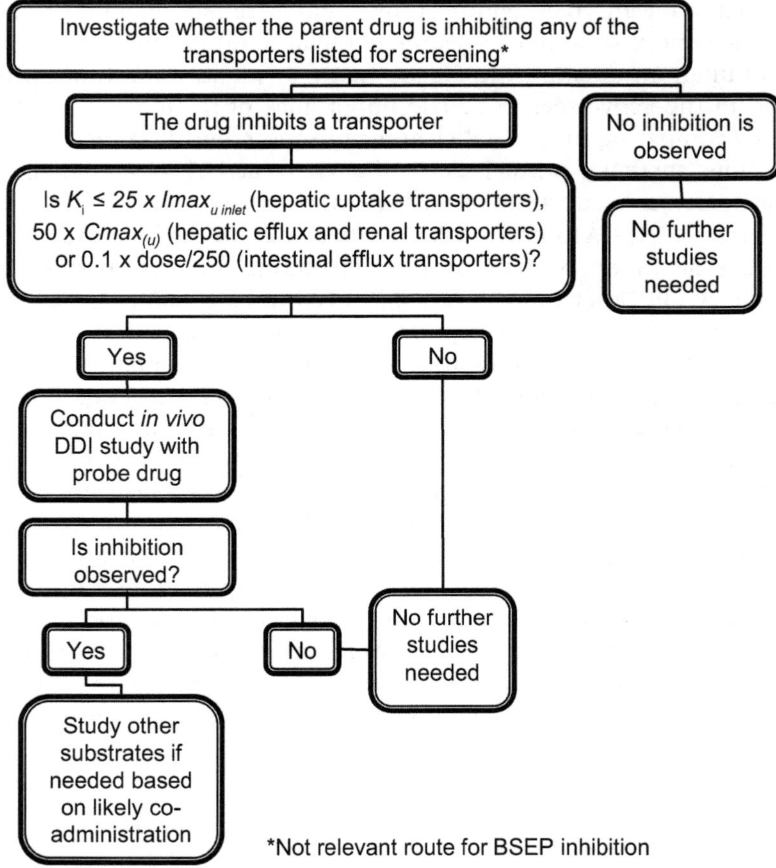

Figure 11.4   Investigation of the transporter inhibitory potential of a drug from the EMA guideline. $Imax_{u\ inlet}$: maximum unbound inhibitor concentration at the inlet to the liver.

avoid becoming rapidly outdated. Instead, it was intended that suitable specific controls should be chosen based on current scientific evidence and best practice. This is in contrast to some other regulatory guidelines, where extensive lists of model substrates and inhibitors for both *in vitro* and *in vivo* use are presented.[2] In the future, the potential for a shared list of substrates and inhibitors is being considered, which would involve a collaboration between the EMA, FDA and PMDA.

Compared with clinical inhibition of drug metabolising enzymes, which indiscriminately leads to increased exposure of the victim drug, inhibition of transporters in the clinical setting may lead to increased or decreased plasma exposure of the victim drug. Even though the most common case is an increase in plasma exposure due to a transporter interaction, there may be cases where the interaction leads to a decrease in victim drug plasma exposure (*e.g.* inhibition of an intestinal uptake transporter or a transporter responsible for reuptake in the kidney). In addition, the drug concentration in a specific tissue may be altered due to transporter inhibition without a corresponding change in drug plasma concentrations, *e.g.* at the blood–brain barrier. Therefore, inhibition of transporters may lead to alteration of plasma and/or tissue concentrations *in vivo* that subsequently may result in either safety concerns or lack of efficacy.

The focus of the CHMP guideline with regard to the investigational drug as a transporter inhibitor is based on clinical knowledge. The transporters that need to be investigated for transporter inhibition are those where interactions are known to alter the pharmacokinetics of transporter sub-strates *in vivo*.[24,25] The most evident example to date is the *in vivo* increase in statin exposure and, consequently, an increased risk for myopathy and rhabdomyolysis due to inhibition of OATP1B1 and/or 1B3.[26,27]

Presently, the transporters that need to be investigated for transporter inhibition *in vitro* include P-gp (*ABCB1*), OATP1B1 (*SLCO1B1*), OATP1B3 (*SLCO1B3*), OCT2 (*SLC22A2*), OAT1 (*SLC22A6*), OAT3 (*SLC22A8*) and BCRP (*ABCG2*). The respective location of the specific transporters is shown in Figure 1.3 in Chapter 1. Investigations of the inhibitory effect on OCT1 (*SLC22A1*), MATE1 (*SLC47A1*) and MATE2 (*SLC47A2*) could also be con-sidered. As BSEP (*ABCB11*) plays an important role in bile acid transport from the hepatocytes into bile, clinically relevant inhibition of BSEP may lead to increased levels of bile acids in the hepatocytes and the systemic circulation. Thus, BSEP *in vitro* inhibition should also preferably be investigated during drug development. If clinically relevant BSEP inhibition is indicated *in vitro*, adequate biochemical monitoring of serum bile salts is recommended during further drug development. In addition, if there are indications that other transporters are involved in an observed, unexpected *in vivo* interaction, *in vitro* inhibition data for this transporter may also be needed.

Specifically for P-gp, a high inter-laboratory variability in *in vitro* inhibition parameter estimation has been shown by Bentz *et al.* 2013[28] and thus it is currently recommended to use two separate *in vitro* model systems for in-vestigation of P-gp inhibition *in vitro*.

When studying transporter inhibition, the effect of multiple concen-trations of the investigational drug on a substrate for the specific transporter should be investigated. Single-point concentration measurements are not considered satisfactory due to their uncertainty and poor accuracy. The concentration range should clearly cover the concentration cut-off relevant for the site of interaction for the respective transporter (see below) and allow for generation of an inhibitory constant (K_i) below that concentration cut-off. When the K_i is not possible to obtain, the concentration of inhibitor required to inhibit transport by 50% (IC_{50}) may be used if linear conditions and a lack of time dependency of the inhibition can be demonstrated. If the concen-trations used in the assay are below the cut-off, the results are considered inconclusive.

Inhibition of intestinal transporters, *e.g.* P-gp and BCRP, should be investigated in relation to the expected maximum concentration on the apical side of the enterocytes that can be estimated by eqn (11.1).[2] However, if the drug has a low solubility profile, the maximum possible soluble concentration at the pH of the intestine, or preferably the solubility in fasted state simulated intestinal fluid (FaSSIF) or fed state simulated intestinal fluid (FeSSIF), could be used instead. The decision to use either FaSSIF or FeSSIF solubility data will depend on how the drug in question will be dosed in relation to food intake.

Hepatic uptake transporters, *e.g.* OATP1B1 and 1B3, will be subjected to higher blood/plasma drug concentrations during the absorption phase of an orally administered drug compared with the systemic exposure based on pharmacokinetic sampling after first passage of the drug. These higher concentrations can be estimated by applying eqn (11.2).[29] The 25-fold safety factor is applied to account for the variability in the K_i estimation.

Renal transporters (OAT1, OAT2 and OCT2), hepatic canalicular efflux transporters (P-gp and BCRP) and hepatic uptake transporters after intravenous administration (OATP1B1 and 1B3) have their main drug input *via* the systemic circulation. Consequently, for these transporters, the cut-off value is based on the steady state mean unbound maximum plasma concentration ($C_{\max(u)}$) obtained during treatment with the highest dose—see eqn (11.3). The applied 50-fold safety factor is chosen to account for the variability in K_i estimation, intracellular drug accumulation in the liver and kidney, higher liver concentrations during absorption, *etc.*

$$\text{Intestinal cut-off} = 0.1 \times \frac{\text{Dose}}{250\,\text{ml}} \tag{11.1}$$

$$\text{Hepatic inlet cut-off} = 25 \times f_{u(b)} \times \left(C_{\max(b)} + \left(\frac{f_a \times f_g \times k_a \times \text{Dose}}{Q_H} \right) \right) \tag{11.2}$$

$$\text{Systematic cut-off} = 50 \times C_{\max(u)} \tag{11.3}$$

$F_{u(b)}$ is the unbound fraction in the blood, $C_{\max(b)}$ the maximal total inhibitor concentration in the blood at steady state, f_a the fraction of dose absorbed, f_g the fraction of absorbed dose escaping gut wall extraction, k_a the absorption rate constant, Q_H the total hepatic blood flow.

No reliable *in vitro–in vivo* correlation is available for transporter data to date, therefore the safety factors applied for the equations above are deliberately chosen to be conservative and are based on experience gained in the enzyme inhibition area. However, with increasing knowledge, *e.g.* actual tissue abundance data of different transporter proteins, more reliable *in vitro–in vivo* correlations may become available. Estimations of the fraction unbound (f_u) in plasma are uncertain below a free fraction of 1%; therefore a free fraction of 1% should be used in the equations above if the actual estimated free fraction is below 1%. If possible, k_a should be determined but otherwise be set, as a worst case estimate, to 0.1 min^{-1}.

In vitro transporter data are used to inform decision making on the need for further assessment of transporter interactions, including *in vivo* drug interaction studies. The risk for clinically relevant *in vivo* transporter inhibition at a certain site can be excluded if the observed K_i value is above or equal to the cut-off concentrations given above. In contrast, if the *in vitro* results are below the cut-off, this indicates that a clinically relevant interaction at a specific transporter cannot be excluded. Then, the indication, target, patient population and drugs commonly used with the investigational drug, as well as the therapeutic window of the affected drug should be taken into consideration to inform the best way forward [*e.g.* *in vivo* interaction studies, summary of product characteristics (SmPC) recommendations].

Transporters can be induced similarly to metabolising enzymes (see *Drug Transporters: Volume 2: Recent Advances and Emerging Technologies*, Chapter 2 for more information on transporter regulation pathways). If an investigational drug has been observed to be an inducer of enzymes *via* nuclear receptors such as pregnane X receptor (PXR) and constitutive androstane receptor (CAR), it is likely that transporters regulated through these receptors will be induced. If PXR and/or CAR-mediated induction is observed *in vivo*, a study investigating the *in vivo* induction of P-gp mediated transport is recommended. The need for *in vivo* studies of the potential inducing effect on other transporters regulated through the same pathways should also be considered. If the investigational drug is likely to be combined with a drug whose pharmacokinetics are significantly influenced by a PXR or CAR regulated transporter, an interaction study with that drug is recommended to enable specific treatment recommendations for the combination.

The translation of *in vitro* data to *in vivo* data for transporters is not well established.[30] Of note, in the 2013 review of NDAs by the FDA,[4] 85% showed a positive *in vitro* interaction with at least one transporter, either as a substrate or inhibitor; however, few interactions translated through to *in vivo* data. As more *in vivo* data become available, this translation should improve.

11.3.2.3   In vitro *Studies: General Considerations*

In vitro data are important to understand the role of transporters in drug interactions. Some drug transporters are known to display inter-species differences, *e.g.* due to differences in amino acid sequence, substrate/inhibitor recognition, expression levels and tissue distribution.[31] Therefore, it is important that *in vitro* data are generated utilising an assay in which human transporters are studied. *In vitro* data on transporters from other species are considered, at best, supportive. Of note, many of the immortalised cell lines used for *in vitro* transporter studies are of animal origin, *e.g.* MDCK (canine) and CHO (chinese hamster). This is adequate as long as the inserted overexpressed transporter being studied is of human origin.

There are a number of different assay systems used to study transporters *in vitro*, as described in Chapter 7 of this book. These include transfected cell lines (*e.g.* HEK293 and MDCK), human cell lines with endogenous transporter expression (*e.g.* Caco-2), hepatocytes, membrane vesicles overexpressing transporters and oocytes injected with transporter cDNA. Also, specific inhibitors, gene knock-out technology and silencing mRNA techniques may be utilised in combination with assay systems as tools for modulating transporter function *in vitro*. There are many factors that influence the choice of assay for a specific study, these include, but are not limited to: which transporter to study, what to study (*e.g.* substrate and inhibition potential), drug properties such as passive permeability and lipophilicity, in-house knowledge, and availability. It is important to ensure that there is no compensatory upregulation of other transporters if one is silenced or downregulated. The demands for validation of and controls for these different assay types differ and are further discussed below.

As a general rule, *in vitro* transporter assays should include cells or vesicles expressing the transporter of interest, and control substrates and inhibitors specific for the investigated transporter. When studying a specific transporter, assay setups that include both overexpressed and empty vector (mock) transfected cells/vesicles are in general preferred over systems without overexpressed transporters. However, a cell or vesicle assay with endogenously-expressed transporters is acceptable but requires higher demands and more specific controls. Investigation of recovery and/or non-specific binding of the investigated drug should be included in the *in vitro* assay setup, especially for lipophilic drugs. The choice of assay systems and controls, *e.g.* specific substrates and inhibitors, should be based on appropriate scientific references or in-house data.

In line with requirements for metabolising enzyme involvement, if active secretion is the major elimination pathway of a metabolite with significant target activity (estimated contribution to the *in vivo* pharmacological effect ≥50% of the total effect), attempts should be made to identify the transporter(s) involved. The need to investigate transporter involvement in renal or biliary/gut wall excretion of metabolites should also be considered when available preclinical and clinical information indicate that the metabolite has a major contribution to off-target (adverse) effects.

Physiologically-based pharmacokinetic modelling (PBPK) approaches, described in more detail in Chapter 9, are recognised as being useful for understanding interactions due to metabolising enzymes and can replace an *in vivo* study. However, due to a lack of knowledge within the transporter field, including key information on the abundance of the proteins, and the lack of adequate qualification of the models, there is currently low confidence in regulatory groups for PBPK modelling of transporters. This conclusion was endorsed recently by the IQ Consortium group.[32]

11.3.2.4 In vivo *Studies: General Considerations*

When a candidate transporter has been identified using *in vitro* studies, an *in vivo* study with a strong inhibitor of the transporter at the site of interest is recommended, providing that the interactions through inhibition are likely to be clinically relevant and that known inhibitors are marketed within the EU. *In vivo* studies in subjects of genotypes that can give rise to markedly reduced expression or activity of a transporter may be useful to verify and quantify the involvement of a transporter known to have clinically relevant polymorphisms, *e.g.* OATP1B1.[20] However, quantitative extrapolation of such data to drug interactions with inhibitors should be justified based on the published literature. As transporter inhibition may alter drug distribution, inclusion of pharmacodynamic markers is encouraged in the *in vivo* studies if relevant and possible.

Interactions with unstudied *in vivo* inhibitors should be predicted based on the acquired *in vivo* information and the scientific literature. If there are commonly used drug combinations where an interaction is expected, it is recommended to investigate the interaction *in vivo*. Similarly, if there are inducers of the transporter marketed within the EU, an interaction study with such an inducer is recommended. The possible effect of transporter inhibition and induction on the availability of the investigational drug for metabolism (transporter–enzyme interplay), such as the interplay observed between P-gp and CYP3A,[33] should be discussed and, if needed, an *in vivo* study should be considered. For P-gp, renal inhibition can be determined using the renal clearance of digoxin. Inhibition of intestinal P-gp may be assessed in an *in vivo* DDI study with a sufficiently specific P-gp substrate with low oral bioavailability, *e.g.* dabigatran etexilate,[34] which seems to be more sensitive to intestinal P-gp inhibition than oral digoxin.

The design of an *in vivo* study falls under the general recommendations for *in vivo* interactions outlined in the Drug Interaction guideline.[1] An *in vivo* interaction study would usually be of cross-over or sequential design. A parallel group design is generally not recommended due to the confounding inter-individual variability, and comparisons with historical controls are generally not acceptable. An open study is satisfactory, but blinding should be considered if pharmacodynamic markers are included in the study. *In vivo* studies performed to investigate whether the investigational drug inhibits or induces a drug transporter *in vivo* should be performed with a well-validated probe drug for the transporter. The probe drug should ideally be exclusively or almost exclusively eliminated by or inhibited through one specific transporter *in vivo*. The use of subpopulations with polymorphic transporters may be an alternative if there is no well-validated probe drug for the transporter.

11.3.2.5   Labelling

The CHMP guideline on SmPC (September 2009—Eudralex vol. 2C)[35] gives advice on how to present information about interactions in the label. Information about drug interactions should be presented in Sections 4.5 and 5.2 of the SmPC, and cross-referenced to Sections 4.2, 4.3 or 4.4, if relevant. Section 4.5 should contain all detailed information on drug interactions and only the recommendation should be given in the cross-referenced sections. In Section 4.5, interactions affecting the investigational drug should be given first, followed by interactions resulting in effects on other drugs. In these sub-sections, the order of presentation should be contraindicated combinations, those where concomitant use is not recommended, followed by others. The text should be as mechanistically clear as possible to enable mechanism-based predictions of interactions with drugs not mentioned in the text.

Brief information about the transporters with a major impact on absorption, distribution or elimination of the drug, as well as the effect of the investigational drug on transporters should be summarised in Section 4.5 as a mechanistic basis for the interaction information. *In vivo* transporter inhibition or induction information should be presented, including relevant details of the study, *e.g.* dose and duration of treatment. All data for probe drugs, even if the interaction is not clinically relevant, should be included. The drug should be classified by giving an estimation of how much the systemic exposure of sensitive substrates of the transporters could be affected by the investigational product. Results of interaction studies should be extrapolated to other drugs in the form of a list of drugs likely to be affected to a clinically relevant extent. For transporters where the list will be long (*e.g.* P-gp), drugs should be selected for inclusion based on clinical relevance and the severity of the clinical consequences of the interaction. If *in vitro* data indicate that a medicinal product affects a transporter, but the available scientific knowledge does not allow predictions of interactions *in vivo*, it is recommended to include the *in vitro* information in Section 5.2 of the SmPC for future use.

Clear treatment recommendations should be given to the prescriber. Wording such as "caution is advised" should be avoided in favour of a recommendation on proposed actions. If a drug combination is stated to be "not recommended", practical recommendations should be presented in case concomitant treatment cannot be avoided. The need for time-specific information and recommendations should be considered. Such situations include interactions where the time interval between perpetrator and victim drug administration is of particular importance and a certain time interval between administrations has been studied, time-dependent interactions such as induction, and drugs with long half-lives, *etc.* The estimated course of onset of the interaction and the time course after ending concomitant treatment should be given, as well as, when relevant, time-specific recommendations. If it is likely that the interaction effect would be different with another dose or at another time point than the one studied, this should be reflected in the recommendations.

In special circumstances, where there are very limited therapeutic alternatives due to marked interactions with most drugs of the same class, examples of drugs with smaller interactions could be given to assist the prescriber. When relevant, the interaction potential in specific populations, such as children or patients with impaired renal function, should be addressed.

Occasionally, at the time of the application, the required information on the role of transporters in the disposition and elimination of the drug may be lacking. Additionally, information on the effect of the drug on transporters, *e.g.* inhibition, may also be lacking. This would be highlighted by the assessors and the information requested. It would normally be expected that this information would be produced prior to the granting of approval for the application. In some circumstances (*e.g.* an accelerated assessment), the data may not be available at the time of granting approval for the application. In this case, the applicant will be expected to commit to performing these studies as a post-authorisation commitment and this will usually be included in the risk management plan. Interim wording for the SmPC will be agreed upon to highlight the missing information and to recommend actions, *e.g.* caution or avoidance of particular substrates or inhibitors.

11.3.3   Draft FDA Guidance

The draft FDA guidance (FDA Guidance for Industry: *Drug Interaction Studies—Study Design, Data Analysis, Implications for Dosing, and Labeling Recommendations*, 2012)[2] recommends a similar approach to the EMA guideline and has been discussed in a number of publications.[36–38] The list of transporters that were considered desirable for *in vitro* drug interaction studies are as follows, based on the recommendations from the ITC[13]: P-gp, BCRP, OATP1B1, OATP1B3, OCT2, OAT1 and OAT3.

11.3.3.1   The Drug as a Potential Transporter Substrate

For transporter substrates, the need to conduct *in vitro* evaluations is based on the contribution of each pathway to the overall elimination of an investigational drug. As the absorption process is an important determinant of pharmacokinetics for orally administered drugs, it is recommended that all of these investigational drugs should be evaluated *in vitro* to determine substrate specificities of P-gp and BCRP, while *in vivo* evaluations may be obviated for highly permeable and soluble drugs (Biopharmaceutics Classification System (BCS) class I, refer to the FDA's Guidance for Industry on *Waiver of In vivo Bioavailability and Bioequivalence Studies for Immediate-Release Solid Oral Dosage Forms Based on a Biopharmaceutics Classification System*[39]). Preferable test systems include Caco-2 cells or transporter-overexpressed cell lines. For investigational drugs that undergo hepatic metabolism or biliary secretion for ≥25% of their systemic clearance, and that have active hepatocyte uptake or physiological properties supporting the importance of active uptake (*e.g.* high hepatic concentration, charged at physiological pH, *etc.*), substrate specificities of OATP1B1 and 1B3 should be tested using transporter-overexpressed cell lines. For investigational drugs that undergo active tubular secretion for ≥25% of their systemic clearance, their substrate specificities should be tested for OCT2, OAT1 and OAT3, using transporter-expressing cell lines (Figure 11.5). The guidance suggests

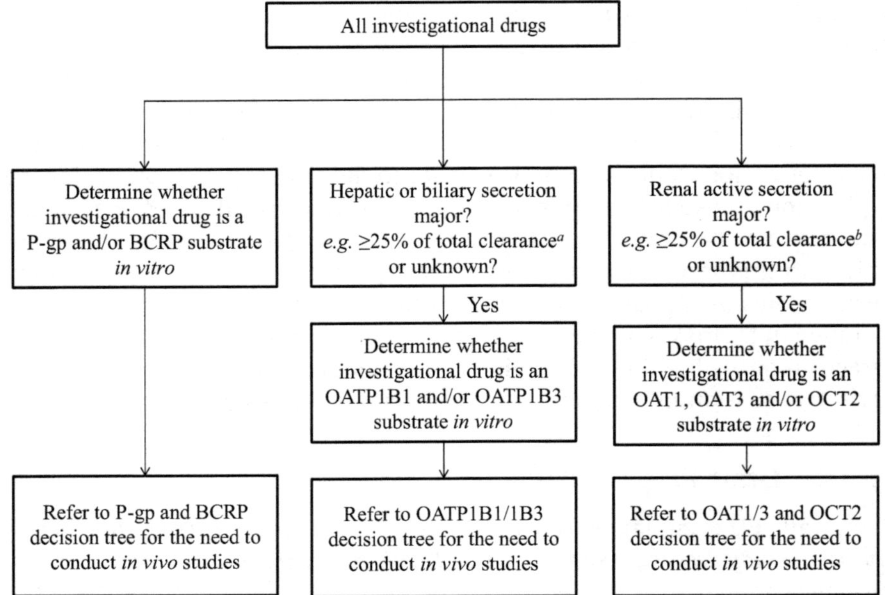

Figure 11.5	Evaluation of investigational drugs as substrates for P-gp, BCRP, OATP1B1, OATP1B3, OAT1, OAT3 and OCT2 transporters from the FDA guidance. [a]Investigational drugs should be evaluated *in vitro* to determine whether they are a substrate of hepatic uptake transporters OATP1B1/OATP1B3 when their hepatic pathway is significant. [b]Investigational drugs should be evaluated *in vitro* to determine whether they are a substrate of OAT1/3 and OCT2 when their renal active secretion is important.

that if the net flux ratio is ≥2 and efflux is inhibited by specific inhibitors (*e.g.* for P-gp/BCRP), or uptake in transporter-overexpressed cell lines is ≥2-fold that of empty vector-transfected cell lines and can be inhibited by specific inhibitors of hepatic or renal transporters, an investigational drug is considered a substrate, and the *in vivo* significance of this should be tested in clinical DDI studies with probe inhibitors. The guidance also states that other cut-off values may be used if supported by prior experience with the cell system used.

11.3.3.2   *The Drug as a Potential Transporter Inhibitor*

In general, it is recommended that all investigational drugs should be tested *in vitro* for their potential as inhibitors of the clinically important transporters shown above regardless of their elimination routes,[2] since reports have shown clinically-relevant transporter-mediated DDIs with the transporter substrates[13,30] P-gp (*e.g.* digoxin), BCRP (*e.g.* rosuvastatin), OATP1B1/1B3 (*e.g.* statins), OAT1/3 (*e.g.* methotrexate, tenofovir) and OCT2 (*e.g.* metformin). Test systems include transporter-overexpressed cell lines for all of the transporters, in addition to Caco-2 cells for P-gp and BCRP. Based on the *in vitro* kinetic parameters, the potential of drugs as *in vivo* inhibitors are evaluated with different numerical methods for the respective transporters based on calculated maximum plasma, hepatic inlet or gut concentrations, as per the European guideline, but sometimes with different safety margins.

11.3.3.3   In vivo *Studies*

When a clinical transporter-mediated DDI study is deemed appropriate, the draft FDA guidance gives similar recommendations to the EMA guideline on the design of clinical studies, in which cross-over and sequential designs are suggested. A parallel group design is also suggested in the FDA guidance, while it is generally not recommended in the CHMP guideline. In addition to study design considerations, the FDA guidance and website have tables of common substrates, inhibitors and inducers, and known interactions. The substrates include digoxin, dabigatran etexilate and fexofenadine for P-gp, rosuvastatin and sulfasalazine for BCRP, pitavastatin, pravastatin and rosuvastatin for OATP1B1, telmisartan for OATP1B3, adefovir, methotrexate and zidovudine for OAT1, acyclovir, methotrexate, zidovudine and cipro-floxacin for OAT3, and metformin for OCT2. The inhibitors include ver-apamil, quinidine, cyclosporine, itraconazole and amiodarone for P-gp, cyclosporine for BCRP, cyclosporine and rifampicin (single dose) for OATP1B1/1B3, probenecid for OAT1/OAT3, and cimetidine for OCT2.[2] In choosing the probe substrates and inhibitors in clinical DDI studies, one might consider the possibility of co-administration with an investigational drug. It is also important to note that many compounds are substrates or inhibitors of multiple transporters in the same or different organs. As seen in the above list of substrates and inhibitors, rosuvastatin is a substrate of both BCRP and OATP1B1, and cyclosporine is an inhibitor of P-gp, BCRP and OATP1B1/1B3. Therefore, selection of probe compounds should be based on the specific questions being addressed (*e.g.* the contribution of or effect on which transporter should be examined).

11.3.3.4   *Labelling*

The draft FDA guidance states that, on product labels, drug interaction in-formation is usually included in two sections; Drug Interactions and Clinical Pharmacology. There is also individual guidance on the information re-quired on drug labels.[40]

In general, the Drug Interactions section gives practical instructions for preventing or managing the effect of co-medications, such as avoidance of co-administrations or dosing adjustments. This section gives a com-prehensive view of DDIs involving the drug of interest, either as a victim or a perpetrator. Some information regarding the mechanisms of DDIs is also mentioned in this section. The Clinical Pharmacology section usually gives more detailed, mechanism-based information of each clinical DDI study with information on drug elimination pathways (metabolic enzymes and/or transporters). Information on drug interactions may appear in other labelling sections. If specific combinations of drugs should be avoided, this information appears in the Contraindications section. Specific instructions for dosing adjustments are summarised in the Dosage and Administration section. Information on serious or otherwise clinically significant outcomes is emphasised in the Warnings and Precautions section.

11.3.4   Draft MHLW Guideline

In Japan, the final draft of the latest drug interaction guideline "Drug Interaction Guideline for Drug Development and Labeling Recommendations" was issued in July 2014.[3] In this draft guideline, transporter species to be examined during drug development, and evaluation methods of drug interactions regarding these transporters, are described in detail with reference to recent research findings.

A summary of the transporter sections in the final 2014 draft of the Japanese drug interaction guideline are described in the following sections.

Drug interactions between a drug under development (an investigational drug) and already approved drugs that may be used with the investigational drug should be studied from two aspects, *i.e.* the effect of the concomitant drug on the investigational drug (as a victim) and the effect of the investigational drug on the concomitant drug (as a perpetrator). The basic factors contributing to drug interactions are investigated in non-clinical (*in vitro*) studies, prior to the implementation of clinical drug interaction studies. For evaluation of transport using *in vitro* study systems, it is recommended that assessments should also be carried out using typical substrates and typical inhibitors, and it is necessary to evaluate the investigational drug with a study system that has been shown to adequately manifest the function of the transporter under investigation. In the case of a study to examine the possibility of an investigational drug as a substrate of a particular transporter, drug concentrations should be sufficiently low, relative to the estimated Michaelis constant (K_m) value, to avoid saturation of the transporter. In the case of a study to examine the possibility of an investigational drug as an inhibitor of a particular transporter, the K_i value should be calculated. When a drug with a known K_m value is used as the substrate, $IC_{50} = K_i$ can be assumed under the condition where substrate concentrations are sufficiently lower than the K_m value. It is feasible to use the apparent IC_{50} value based on the concentration in the assay medium for studies examining efflux transporters in a cellular study system.

Figure 11.6 summarises the evaluation of a drug as a substrate of transporters from the Japanese guideline.

11.3.4.1   *P-gp and BCRP*

It has been demonstrated that both P-gp and BCRP, expressed at the apical membrane of epithelial cells in the gastrointestinal tract, function as efflux

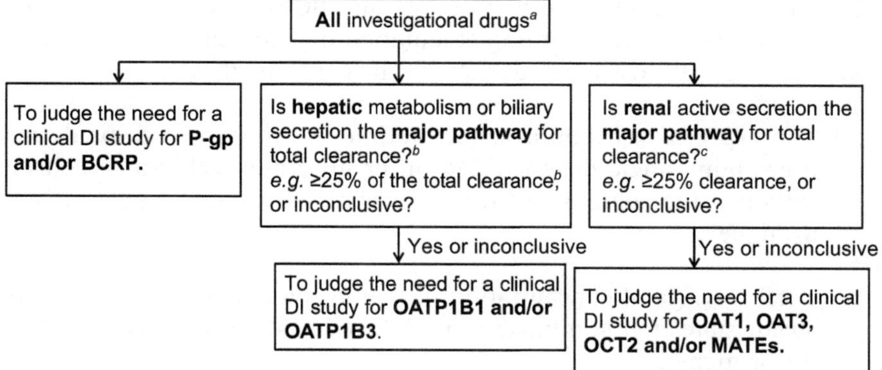

Figure 11.6   The evaluation of a drug as a substrate of transporters, from the Japanese guideline. ADME: absorption, distribution, metabolism and excretion; DI: drug interaction. [a]Interaction between transporters and metabolites is considered to be studied based on the metabolite investigation. [b]When an investigational drug for which the hepatic pathway is important (example: clearance *via* hepatic or biliary secretion accounts for 25% or more of the total clearance), it should be evaluated whether the drug is a substrate of hepatic uptake transporters OATP1B1 and OATP1B3. Biliary secretion can be estimated from the preclinical data (*in vitro* hepatocyte uptake data or *in vivo* ADME data obtained from radiolabelled compounds) and from *in vivo* non-renal clearance data. [c]When an investigational drug for which renal tubular secretion is important (secretion clearance accounts for 25% or more of the total clearance), it should be evaluated whether the drug is a substrate of OAT1, OAT3, OCT2 and MATEs. The fraction of secretion clearance (%) should be estimated from the formula $(CL_r - f_u \times GFR)/CL_{total}$ (where CL_r is renal clearance, f_u is the unbound fraction of drug in plasma, GFR is the glomerular filtration rate and CL_{total} is body clearance).

transporters that decrease the absorption of their substrate drugs. Therefore, the guidance recommends that all investigational drugs should be evaluated in *in vitro* studies to examine the possibility of them being substrates for P-gp or BCRP. Because these transporters are also expressed in the liver, kidney and brain, and can affect the elimination and distribution of drugs to the brain, investigation of drugs intended for administration routes other than oral administration may also be needed.

For P-gp substrates, the need for a clinical drug interaction study should be judged taking into consideration F_a, F_g, the presence/absence of renal tubular secretion and the risk of neuronal toxicity, because P-gp is known to be involved in intestinal absorption, renal tubular secretion and brain distribution. In the case of BCRP substrates, only an indication that the drug is a BCRP substrate should be provided for a NDA and a clinical drug interaction study is not required, because it is currently difficult to design a clinical drug interaction study using typical *in vivo* BCRP inhibitors.

For the evaluation of inhibition of P-gp and BCRP by the investigational drug, it is recommended that implementation of a clinical drug interaction study with P-gp (BCRP) substrates should be considered, when:

(1) The expected maximum concentration in the gastrointestinal tract (maximum single dose per 250 ml or maximum achievable concentration if the solubility of the investigational drug is low)/IC_{50} is 10 or higher

or

(2) Total C_{max} (total concentration of bound plus unbound drug) in the steady state after administration of the clinical maximum dose/IC_{50} is 0.1 or higher.

11.3.4.2   OATP1B1 and 1B3 in the Liver

Blood concentrations of substrate drugs are known to increase if OATP1B1 and 1B3, which are transporters expressed on the blood side of hepatocytes and take up drugs from the blood into the hepatocytes, are inhibited.

Investigational drugs that are mainly eliminated *via* hepatic metabolism or biliary excretion (*i.e.* clearance *via* either route accounting for ≥25% of the total drug clearance) should be assessed for the possibility of them being substrates for the hepatic uptake transporters OATP1B1 and 1B3 (Figure 11.6). There may be instances where the need for *in vitro* studies can be judged based on examination of the distribution to the liver.

For studies to assess the inhibition of OATP1B1 and 1B3 by an investigational drug, it should be ensured that the *in vitro* method used covers a concentration range that allows judgement of whether or not the K_i value of the investigational drug is greater than four-fold the maximum blood concentration of the unbound drug at the inlet to the liver ($[I]_{u,inlet,max}$) at the estimated clinical dose.

11.3.4.3   OAT1, OAT3, OCT2 and MATEs in the Kidney

OAT1, OAT3 and OCT2 are transporters expressed on the blood side of proximal tubular epithelial cells, and transport drugs from the blood into the proximal tubular epithelial cells, whereas P-gp, MATE1, MATE-2K and BCRP are transporters expressed on the apical side of proximal tubular epithelial cells, mediating the excretion of drugs from the proximal tubular epithelial cells into the urine. Blood concentrations of the substrates of these transporters may be elevated if the transporters are inhibited. In addition, if P-gp, MATEs and BCRP are inhibited, the concentrations of drugs in the proximal tubular epithelial cells may be increased without any changes occurring in the blood concentrations.

For investigational drugs whose major route of elimination is active renal secretion (renal secretion clearance accounting for ≥25% of the total clearance), it is recommended that the possibility of the drug being a substrate for OAT1, OAT3, OCT2, MATE1 or MATE2-K should be explored *in vitro* (Figure 11.6).

For studies to examine the inhibition of OAT1, OAT3, OCT2, MATE1 or MATE2-K by an investigational drug, it should be ensured that the concentration range covered allows judgement of whether or not the K_i (IC_{50}) value of the investigational drug is greater than four-fold the C_{max} of the unbound drug at the estimated clinical dose.

11.3.4.4   Other Transporters

There are other well-known drug and endogenous substance transporters such as MRP2, MRP4, OCT1 and BSEP. Evaluations of drug interactions mediated by these transporters are not always recommended during the drug development process, but findings reported for drugs that are similar in chemical structure to the investigational drug should be considered.

In the kidney, MRP2 and MRP4 are expressed on the apical side of the proximal tubular epithelial cells, and mediate excretion of drugs from the proximal tubular epithelial cells into the urine.

In the liver, OCT1 is expressed on the blood side of the hepatocytes and transports drugs from the blood into the hepatocytes, and MRP2 is expressed on the apical side of the hepatocytes and mediates excretion of drugs from the hepatocytes into the bile.

Furthermore, in the case of MRP2 and BSEP, which are involved in the biliary excretion of endogenous substances such as bile acids and bilirubin, it is possible that their inhibition by drugs would increase the blood and tissue concentrations of the aforementioned endogenous substances.

11.3.4.5   Labelling

Information on drug interactions is generally provided in the Interactions section of the PI. The names of drugs and drug interaction information such as clinical symptoms, treatments, mechanisms and risk factors are mentioned in an easily understandable manner using tables. If concomitant use with other drugs causes enhancement or attenuation of the pharmacological actions, then investigations of adverse drug reactions of the investigational drug or of the concomitant drugs, or aggravation of the primary disease, together with a consideration of whether clinical precautions are necessary, should be implemented in these cases. Contraindications and precautions for concomitant use are determined based on the seriousness of the risk.

Drug interaction information is also provided in other sections. Contraindicated drugs should also be mentioned briefly in the Contraindications section. If a dose or regimen adjustment is necessary, the practical instruction should be described comprehensively in the Precautions for Dosage and Administration section, based on the quantitative information obtained from clinical drug interaction studies, *etc*. If the precaution is critical from the viewpoint of risk management, it should be described in the Warning or Important Precautions section.

The Pharmacokinetics section gives detailed information on the mechanisms of the interactions and supporting data from drug interaction studies. The result of clinical drug interaction studies should be illustrated with a simple outline of the quantitative changes in the most important pharmacokinetic and/or exposure–response parameters, using narrative text, tables and/or figures.

11.4   Conclusion

NDAs to the European, US and Japanese agencies require comprehensive *in vitro* and *in vivo* data to describe the role of transporters in the absorption, distribution and elimination of the drug. The effect of the drug on other drugs also needs to be understood in terms of possible inhibition of transporters involved in the disposition and clearance of concomitantly-administered drugs.

Guidance on transporter interactions is included in a number of regulatory guidelines from the EMA, FDA and MHLW. The guidelines on Drug Interactions from the respective agencies contain extensive information on expectations around the *in vitro* and *in vivo* data that are required. While these guidelines are broadly similar in their requirements, there are some differences in the importance given to some of the emerging, less well understood transporters, and in the utilisation of *in vitro* transporter inhibition data to inform the need for *in vivo* studies; with some agencies taking a more conservative approach. However, these differences are caused by a scarcity of data to inform decisions and it is recognised that the science is rapidly developing. There is presently an initiative to harmonise the requirements in the respective guidelines. As data sets become available these will be reviewed and requirements will likely be updated when needed, *e.g.* it is currently recognised that there is a substantial amount of data on OATP1B1/1B3 and these are being reviewed. It is expected that regulatory guidelines will be regularly updated, with future requirements of the different agencies likely to reach more of a consensus.

Disclaimer

The views expressed in this chapter do not necessarily represent the view or policies of the FDA or its staff. No official support or endorsement by the FDA is intended or should be inferred.

The views expressed in this chapter are not an official policy statement or requirement of the MHLW and PMDA. No official support or endorsement by MHLW and PMDA is intended or should be inferred.

Acknowledgements

The authors thank Dr Terry Shepard and Dr David Wright of the MHRA and Dr Eva Gil Berglund of the MPA for their review of the chapter.

Dr Kenta Yoshida was supported in part by an appointment to the ORISE Research Participation Program at the Center for Drug Evaluation and Research (CDER) administered by the Oak Ridge Institute for Science and Education through an agreement between the US Department of Energy and CDER.

References

1. EMA Guidance: *Investigation of Drug Interactions*, CPMP/EWP/560/95/ Rev. 1, 2012.
2. FDA Guidance for Industry: *Drug Interaction Studies—Study Design, Data Analysis, Implications for Dosing, and Labeling Recommendations*, 2012.
3. Ministry of Health, Labour and Welfare, Pharmaceuticals Medical Safety Bureau: *Drug Interaction Guideline for Drug Development and Labeling Recommendations (Final Draft)* 2014. http://www.pmda.go.jp/files/ 000206158.pdf. In Japanese.
4. J. Yu, T. K. Ritchie, A. Mulgaonkar and I. Ragueneau-Majlessi, Drug disposition and drug-drug interaction data in 2013 FDA new drug applications: a systematic review, *Drug Metab. Dispos.*, 2014, **42**(12), 1991–2001.
5. S. Agarwal *et al.*, Review of P-gp inhibition data in recently approved new drug applications: utility of the proposed [I(1)]/IC(50) and [I(2)]/IC(50) criteria in the P-gp decision tree, *J. Clin. Pharmacol.*, 2013, **53**(2), 228–233.
6. Jinjjing *et al.*, Key Findings from Preclinical and Clinical Drug Interaction Studies Presented In New Drug and Biological License Applications Approved by the FDA in 2014, *Drug Metab. Dispos.*, 2015, **44**, 83–101.
7. N. Nagai, Drug interaction studies on new drug applications: Current situations and regulatory view in Japan, *Drug Metab. Pharmacokinet.*, 2010, **25**(1), 3–15.
8. http://www.ema.europa.eu/ema/index.jsp?curl=pages/about_us/general/ general_content_000109.jsp&mid=WC0b01ac0580028a47.
9. http://www.fda.gov/Drugs/DevelopmentApprovalProcess/ HowDrugsareDevelopedandApproved/ApprovalApplications/ NewDrugApplicationNDA/.
10. http://www.fda.gov/forpatients/approvals/fast/ucm405405.htm.
11. http://www.pmda.go.jp/ (in Japanese), http://www.pmda.go.jp/english/ index.html (in English).
12. Sahajwalla *et al.*, The role of the FDA in guiding drug *development*, in *Principles of Clinical Pharmacology*, Academic Press, 3rd edn, 2012.
13. S. M. Huang and J. Woodcock, Transporters in drug development: advancing on the Critical Path, *Nat. Rev. Drug. Discovery*, 2010, **9**(3), 175.
14. K. M. Giacomini *et al.*, Membrane transporters in drug development, *Nat. Rev. Drug Discovery*, 2010, **9**(3), 215.

15. Ministry of Health, Labour and Welfare, Pharmaceuticals Medical Safety Bureau: *Method of Drug Interaction Studies*, PMSB/ELD Notification No. 813, June, 2001.

16. M. Sato, The 10th International Society for the Study of Xenobiotics (ISSX) Meeting. P166 *"Drug interaction studies during drug development: Current status and regulatory perspectives in Japan."* Toronto Canada, October, 2013.

17. D. Iwata, The 19th North American ISSX and 29th JSSX Meeting. P179" *Drug interaction studies during drug development and new drug applications (2): Current status and PMDA perspectives on drug interactions involving transporters."* San Francisco, USA, October, 2014.

18. *EMA Guidance, The use of pharmacogenetic methodologies in the pharmacokinetic evaluation of medicinal products* (EMA/CHMP/37646/2009).

19. FDA Guidance, *Clinical Pharmacogenomics: Premarket Evaluation in Early-Phase Clinical Studies and Recommendations for Labeling* 2013.

20. M. Niemi, Transporter pharmacogenetics and Statin toxicity, *Nature*, 2010, **87**(1), 130.

21. A. Grover and L. Bennet, Effects of drug transporters on volume of distribution, *AAPS J.*, 2008, **11**(2), 250.

22. H. Kusuhara, Imaging in the Study of Membrane Transporters, *Clin. Pharmacol. Ther.*, 2013, **94**(1), 33–36.

23. Y. Shitara *et al.*, Clinical significance of organic anion transporting polypeptides (OATPs) in drug disposition: their roles in hepatic clearance and intestinal absorption, *Biopharm. Drug Dispos.*, 2013, **34**(1), 45–78.

24. S. Härtter *et al.*, Decrease in the oral bioavailability of dabigatran etexilate after co-medication with rifampicin, *Br. J. Clin. Pharmacol.*, 2012, **74**(3), 490.

25. S. Johansson *et al.*, *Pharmacokinetic* Evaluations of the Co-Administrations of Vandetanib and Metformin, Digoxin, Midazolam, Omeprazole or Ranitidine, *Clin. Pharmacokinet.*, 2014, **53**, 837.

26. S. G. Simonson *et al.*, Rosuvastatin pharmacokinetics in heart transplant recipients administered an antirejection regimen including cyclosporine, *Clin. Pharmacol. Ther.*, 2004, **76**(2), 167.

27. Kiser *et al.*, Drug/Drug interaction between lopinavir/ritonavir and rosuvastatin in healthy volunteers, *J. Acquired Immune Defic. Syndr.*, 2008, **47**(5), 570.

28. J. Bentz *et al.*, Variability in P-glycoprotein inhibitory potency (IC_{50}) using various in vitro experimental systems: implications for universal digoxin drug-drug interaction risk assessment decision criteria, *Drug Metab. Dispos.*, 2013, **41**(7), 1347.

29. K. Ito *et al.*, Which concentration of the inhibitor should be used to predict in vivo drug interactions from in vitro data?, *AAPS PharmSci.*, 2002, **4**(4), E25.

30. Menochet *et al.*, Use of Mechanistic Modeling to Assess Interindividual Variability and Interspecies Differences in Active Uptake in Human and Rat Hepatocytes, *Drug Metab. Dispos.*, 2012, **40**(9), 1744.

31. C. Hilgendorf *et al.*, Expression of Thirty-six Drug Transporter Genes in Human Intestine, Liver, Kidney, and Organotypic, *Drug Metab. Dispos.*, 2007, **35**(8), 1333.

32. H. M. Jones *et al.*, Physiologically based pharmacokinetic modeling in drug discovery and development: A pharmaceutical industry perspective, *Clin. Pharmacol. Ther.*, 2015, **97**(3), 247–262.

33. R. van Waterschoot and A. Schinkel, A Critical Analysis of the Interplay between Cytochrome P450 3A and P-Glycoprotein: Recent Insights from Knockout and Transgenic Mice, *Pharmacol. Rev.*, 2011, **63**(2), 390–410.

34. G. Hankley and J. Eikelboom, Dabigatran Etexilate: A New Oral Thrombin Inhibitor, *Circulation*, 2011, **123**, 14366.

35. A guideline on Summary of product Characteristics, Rev 2, European Commission, 2009.

36. Tweedie *et al.*, Transporter studies in Drug Development: Experience to data and follow-up on decision trees from the international transporter consortium, *Clin. Pharmacol. Ther.*, 2013, **94**(1), 113.

37. L. Zhang, Y. Zhang, J. M. Strong, K. S. Reynolds and S.-M. Huang, A regulatory viewpoint on transporter-based drug interactions, *Xenobiotica*, 2008, **38**(7–8), 709–724.

38. K. M. Hillgren *et al.*, Emerging transporters of clinical importance: an update from the International Transporter Consortium, *Clin. Pharmacol. Ther.*, 2013, **94**(1), 52.

39. FDA Guidance for Industry: Waiver of In Vivo Bioavailability and Bioequivalence Studies for Immediate-Release Solid Oral Dosage Forms Based on a Biopharmaceutics Classification System, 2000.

40. FDA Guidance, Labelling of pharmaceutical drugs for human use, 2015.

Subject Index